Lecture Notes in Computer Science 5827

Commenced Publication in 1973
Founding and Former Series Editors:
Gerhard Goos, Juris Hartmanis, and Jan van Leeuwen

Rudolf Berghammer   Ali Mohamed Jaoua
Bernhard Möller (Eds.)

Relations
and Kleene Algebra
in Computer Science

11th International Conference on Relational Methods
in Computer Science, RelMiCS 2009
and 6th International Conference on Applications
of Kleene Algebra, AKA 2009
Doha, Qatar, November 1-5, 2009
Proceedings

 Springer

Volume Editors

Rudolf Berghammer
Christian-Albrechts-Universität zu Kiel
Institut für Informatik
Olshausenstraße 40
24098 Kiel, Germany
E-mail: rub@informatik.uni-kiel.de

Ali Mohamed Jaoua
Qatar University
College of Engineering
CSE 209, P.O. Box 2713
Doha, Qatar
E-mail: jaoua@qu.edu.qa

Bernhard Möller
Universität Augsburg
Institut für Informatik
Universitätsstr. 6a
86135 Augsburg, Germany
E-mail: bernhard.moeller@informatik.uni-augsburg.de

Library of Congress Control Number: 2009934859

CR Subject Classification (1998): I.1.1, I.1.3, I.2.4, F.1.1, G.2.1

LNCS Sublibrary: SL 1 – Theoretical Computer Science and General Issues

ISSN 0302-9743

ISBN 978-3-642-04638-4 Springer Berlin Heidelberg New York

springer.com

© Springer-Verlag Berlin Heidelberg 2009

Typesetting: Camera-ready by author, data conversion by Scientific Publishing Services, Chennai, India
Printed on acid-free paper   SPIN: 12763399   06/3180   5 4 3 2 1 0

Preface

This volume contains the proceedings of the 11th International Conference on Relational Methods in Computer Science (RelMiCS 11) and the 6th International Conference on Applications of Kleene Algebra (AKA 6). The joint conference took place in Doha, Quatar, November 1–5, 2009. Its purpose was to bring together researchers from various subdisciplines of computer science, mathematics and related fields who use the calculus of relations and/or Kleene algebra as methodological and conceptual tools in their work.

This conference is the joint continuation of two different strands of meetings. The seminars of the RelMiCS series were held in Schloss Dagstuhl (Germany) in January 1994, Parati (Brazil) in July 1995, Hammamet (Tunisia) in January 1997, Warsaw (Poland) in September 1998, Québec (Canada) in January 2000, and Oisterwijk (The Netherlands) in October 2001. The conference on Applications of Kleene Algebra started as a workshop, also held in Schloss Dagstuhl, in February 2001. To join these two themes in one conference was mainly motivated by the substantial common interests and overlap of the two communities. Over the years this has led to fruitful interactions and openened new and interesting research directions. Joint meetings have been held in Malente (Germany) in May 2003, in St Catherines (Canada) in February 2005, in Manchester (UK) in August/September 2006 and in Frauenwörth (Germany) in April 2008.

This volume contains 24 contributions by researchers from all over the world. In addition to the 22 regular papers there were the invited talks *Computational Social Choice Using Relation Algebra and* RELVIEW by Harrie de Swart (Tilburg University, Netherlands) and *Knowledge and Structure in Social Algorithms* by Rohit Parikh (Brooklyn College and CUNY Grad Center, USA). The papers show that relational and Kleene algebra methods have wide-ranging diversity and applicability in theory and practice.

In addition, for the third time, a PhD programme was offered. It included three invited tutorials by Marcelo Frías (University of Buenos Aires, Argentina), Ali Jaoua (University of Qatar at Doha) and Gunther Schmidt (University of the Armed Forces at Munich, Germany).

We are very grateful to the members of the Programme Committee and the external referees for their care and diligence in reviewing the submitted papers. We also want to thank Qatar University for having accepted to host the conference and Aws Al-Taie, Fatma Al-Baker, Zeina Hazem Al-Azmeh, Fatma Al-Bloushi and Mohamad Hussein for their assistance; they made organizing this meeting a pleasant experience. We also gratefully appreciate the excellent facilities offered by the EasyChair conference administration system. Finally, we cordially thank our sponsors Supreme Education Council (SEC), Qatar National Research Fund (QNRF) and Qatar University (QU) for their generous support.

November 2009

Rudolf Berghammer
Ali Jaoua
Bernhard Möller

Organization

Programme Committee

J. Al'Jaam	Doha, Qatar
R. Berghammer	Kiel, Germany
H. de Swart	Tilburg, Netherlands
J. Desharnais	Laval, Canada
R. Duwairi	Doha, Qatar
M. Frías	Buenos Aires, Argentina
H. Furusawa	Kagoshima, Japan
P. Höfner	Augsburg, Germany
A. Jaoua	Doha, Qatar
P. Jipsen	Chapman, USA
W. Kahl	McMaster, Canada
Y. Kawahara	Kyushu, Japan
L. Meinicke	Sydney, Australia
A. Mili	Tunis, Tunisia; New York, USA
B. Möller	Augsburg, Germany
C. Morgan	Sydney, Australia
E. Orłowska	Warsaw, Poland
S. Saminger-Platz	Linz, Austria
G. Schmidt	Munich, Germany
R. Schmidt	Manchester, UK
G. Struth	Sheffield, UK
M. Winter	Brock, Canada

External Referees

Bernd Braßel	Koki Nishizawa
Jean-Lou De Carufel	Viorel Preoteasa
Jan Christiansen	Ingrid Rewitzky
Han-Hing Dang	Pawel Sobocinski
Ernst-Erich Doberkat	Kim Solin
Roland Glück	Sam Staton

Table of Contents

Knowledge and Structure in Social Algorithms

Rohit Parikh

Brooklyn College and CUNY Graduate Center
City University of New York
365 Fifth Avenue
New York, NY 10016
http://sci.brooklyn.cuny.edu/cis/parikh/

Abstract. While contemporary Game Theory has concentrated much on strategy, there is somewhat less attention paid to the role of knowledge and information transfer. There are exceptions to this rule of course, especially starting with the work of Aumann [2], and with contributions made by ourselves with coauthors Cogan, Krasucki and Pacuit [17,13]. But we have still only scratched the surface and there is still a lot more that can be done. In this paper we point to the important role which knowledge plays in social procedures (colorfully called *Social Software* [15]).

Keywords: Knowledge, Society, Algorithms.

The peculiar character of the problem of a rational economic order is determined precisely by the fact that the knowledge of the circumstances of which we must make use never exists in concentrated or integrated form, but solely as the dispersed bits of incomplete and frequently contradictory knowledge which all the separate individuals possess. The economic problem of society is thus not merely a problem of how to allocate "given" resources – if "given" is taken to mean given to a single mind which deliberately solves the problem set by these "data." It is rather a problem of how to secure the best use of resources known to any of the members of society, for ends whose relative importance only these individuals know.

F. Hayek
Individualism and Economic Order

1 Introduction

The first third of the XXth century saw two important developments. One of these was Ramsey's tracing of personal probabilities to an agent's choices [22]. This was a precursor to the work of de Finetti, von Neumann and Morgenstern,

R. Berghammer et al. (Eds.): RelMiCS/AKA 2009, LNCS 5827, pp. 1–12, 2009.

and Savage [6,10,23]. The other one was Turing's invention of the Turing machine [26] and the formulation of the Church-Turing thesis according to which all computable functions on natural numbers were recursive or Turing computable.[1]

Game theory has depended heavily on the first of these developments, since of course von Neumann and Morgenstern can be regarded as the fathers of Game theory. But the other development has received less attention. That development led to the development and design of computers and also to fields like Logic of Programs, Complexity Theory and Analysis of Algorithms. It also resulted in much deeper understanding of algorithms, but only of computer algorithms. Social algorithms have remained largely unanalyzed mathematically except in special subfields like Social Choice Theory [1] or Fair Division [3]. These fields, however, do not tend to analyze *complex* social algorithms (even algorithms of modest complexity like the two thousand year old Euclid's algorithm) as is done in computer science.[2] The typical game theoretic example tends to be either a one shot game, or else such a game repeated.

A later development, going back to the work of Hintikka, Lewis and a little later Aumann [8,9,2], brought in the issue of *knowledge*. The notion of common knowledge is of course very important for Aumann as *common knowledge of rationality* can be seen as a justification for backward induction arguments.

But knowledge too has received less attention than it might. We all know that the Valerie Plame affair [21] had something to do with someone knowing something which they should not have, and someone revealing something which they should not have. But why should they not? Clearly because of certain possible consequences. Knowledge and knowledge transfer are ubiquitous in how social algorithms work. Note that the fact that the FBI bugged Burris's phone conversations with Blagojevich's brother played an important role, and the fact that we do not want the FBI to have unlimited right to listen in on conversations are extremely important knowledge considerations.

We will try in this paper to bring attention to the importance of the two issues of knowledge and logical structure of algorithms, and show the way to a broader arena in which game theorists might want to play. Hopefully, in fact almost certainly, there is a rich general theory to be developed.

---

[1] The research reported here was supported in part by a research grant from the PSC-CUNY program. Previous versions of this paper were presented at a workshop on knowledge at the University of Quebec in Montreal (2007), and at the World Game Theory meeting at Northwestern University (2008).

[2] But society itself is replete with extremely complex algorithms. Just consider the complexity involved in Obama's election to the presidency, the consequent vacating of his senate seat, Blagojevich's acquiring the right to name Obama's successor, Blagojevich naming Burris to Obama's vacant seat, Blagojevich's impeachment and removal from office, demands, so far unsuccessful, for Burris to step down, and, no doubt, quiet satisfaction on the part of the Republicans. And even Obama's election to the presidency is hardly a simple event since it involved factors like Hillary's association with her husband, a former president, an initital feeling on the part of African-Americans that Obama, having no ancestry in the institution of slavery was not "one of us," etc. etc.

The notion of algorithm is implicit in so many things which happen in everyday life. We humans are tool-making creatures (as are chimps to a somewhat smaller extent) and both individual and social life is over-run with routines, from cooking recipes (Savage's celebrated eggs to omelette example [23] for instance) to elections – a subject of much discussion going back to Condorcet.

Over the last ten years or so, a field called *Social Software* [15] has come into existence which carries out a systematic study of such questions, and the purpose of this paper is to give an introduction to the knowledge-theoretic issues. We will proceed by means of examples.

2   Structure

Normally, a piece of social software or social algorithm has a logical structure. As was argued in [15], this structure must address three important aspects, namely incentives, knowledge, and logical structure. For normally, an algorithm has logical structure, "*A* happens before *B*, which is succeeded by action *C* if condition *X* holds and by action *D* if *X* does not hold."

But quite often, the logical structure of the algorithm is parasitic on logical (or algorithmic) properties of existing physical or social structures. Clearly a prison needs a certain physical structure in order to be *able* to confine people, and a classroom needs a blackboard or a lectern in order for it to be usable as the venue of a lecture. Thus the teacher can now perform actions like write "No class tomorrow" on the blackboard and the students can read what she wrote or copy it in their notebooks. The physical properties of the blackboard enable certain actions with their own algorithmic properties. The fact that there is no class the next day now becomes common knowledge and the students can make plans to use the time that has been freed up.

2.1   Queues

A *social structure* with certain logical properties is a queue.

The queue is a very popular institution which occurs both in daily life and in computer programs. In a computer program, a queue is a FIFO structure, where FIFO means, "First in, first out." There are two operations, one by which an element is deleted from the front of the queue, and a second one where an element is added to the back of the queue. In real life, the queue could consist of people waiting at a bank to be served. The person *deleted* is the one who was at the front of the queue but is no longer in the queue, and who receives service from a teller. An element which is *added* is a new customer who has just arrived and who goes to the back of the queue.

Clearly the queue implements our notions of fairness, (which can be proved rigorously as a theorem) that someone who came earlier gets service earlier, and in a bank this typically does happen. If someone in a bank tries to rush to the head of the line, people will stop him. Thus *'violations easily detectable'* is a crucial knowledge property.

We also have queues at bus stops and quite often the queue breaks down; there is a rush for seats at the last moment. Presumably the difference arises because things happen much faster in a bus queue than they do in a bank. At a bus stop, when the bus arrives, everything happens very fast and people are more interested in getting *on* the bus than in enforcing the rules.

Consider now, by comparison, the problem of *parking*, which is a similar problem. A scarce resource needs to be allocated on the basis of some sort of priority, which, now, is more difficult to determine. When people are looking for parking in a busy area, they tend to cruise around until they find a space. There is no queue as such, but in general we still want that someone who arrives first should find a parking space and someone who arrives later may not. This is much more likely in a university or company parking lot, which is compact, and may even have a guard, rather than on the street, where parking is distributed, and priority does play *some* role but it is only probabilistic. Clearly the lack of information about where the parking space is, and who came first, plays an important role.

This fact has unfortunate consequences as Shoup [24] points out.

> *When my students and I studied cruising for parking in a 15-block business district in Los Angeles, we found the average cruising time was 3.3 minutes, and the average cruising distance half a mile (about 2.5 times around the block). This may not sound like much, but with 470 parking meters in the district, and a turnover rate for curb parking of 17 cars per space per day, 8,000 cars park at the curb each weekday. Even a small amount of cruising time for each car adds up to a lot of traffic.*
>
> *Over the course of a year, the search for curb parking in this 15-block district created about 950,000 excess vehicle miles of travel – equivalent to 38 trips around the earth, or four trips to the moon. And here's another inconvenient truth about underpriced curb parking: cruising those 950,000 miles wastes 47,000 gallons of gas and produces 730 tons of the greenhouse gas carbon dioxide. If all this happens in one small business district, imagine the cumulative effect of all cruising in the United States.*

Shoup regards this problem as one of incentive and suggests that parking fees be raised so that occupancy of street parking spaces is only 85%. But clearly this will penalize the less affluent drivers. The new fees will likely be still less than the cost of garage parking, affluent drivers will abandon garage parking for street parking, and the less affluent drivers will be priced out. Note by contrast that we do not usually charge people for standing in a queue. We could, and surely queues would also be shorter if people had to pay to stand in them. But this has not occurred to anyone as a solution to the 'standing in line problem.'

An algorithmic solution to the problem of parking might well be possible using something like a GPS system. If information about empty parking spaces was available to a central computer which could also accept requests from cars for parking spaces, and allocate spaces to arriving cars, then a solution could in fact be implemented. The information transfer and the allocation system would in effect convert the physically distributed parking spaces into the algorithmic equivalent of a queue. There would be little wasteful consumption of gasoline, and the drivers would save a great deal of time and frustration.

And here indeed is an implementation of the alternate solution

Find a Place to Park on Your GPS – Spark Parking Makes it Possible

Navigation Developers Can Access Spark Parking Points of Interest Through New Tele Atlas ContentLink Program

San Francisco, CA, March 21, 2007

Running late for a meeting and worried about finding a place to park? Unhappy about paying outrageous valet parking fees at your favorite restaurant? These headaches will soon be a thing of the past. Spark Parking's detailed parking location information data is now available through the newly released Tele Atlas ContentLinkSM portal for application developers to incorporate into a range of GPS devices and location-based services and applications.

Spark Parking's detailed parking information provides the locations of every paid parking facility in each covered city – from the enormous multi-level garages to the tiny surface lots hidden in alleys. In addition, Spark Parking includes facility size, operating hours, parking rates, available validations, and many more details not previously available from any source. As a result, drivers will easily be able to find parking that meets their needs and budgets.

http://www.pr.com/press-release/33381

SAN FRANCISCO
Where's the bus? NextMuni can tell you.
System uses GPS to let riders know when streetcar will arrive
Rachel Gordon, Chronicle Staff Writer

Thursday, March 29, 2007

San Francisco's Municipal Railway may have a hard time running on time, but at least the transit agency is doing more to let riders know when their next bus or streetcar is due to arrive.

The "NextMuni" system, which tracks the location of vehicles via satellite, is now up and running on all the city's electrified trolley bus lines. It had been available only on the Metro streetcar lines and the 22-Fillmore, a trolley bus line that served as an early test.

The whereabouts of the Global Positioning System-equipped vehicles are fed into a centralized computer system that translates the data into user-friendly updates available on the Internet and on cell phones and personal digital assistants.

http://www.sfgate.com/

Ultimately, the difference between queues and searching for parking is structural. In one case there is an easy algorithmic solution which respects priority (more or less) and in the other case such solutions are harder to find – except when we are dealing with parking lots or use sophisticated new technology.

2.2 Keys

Here is another example. When you rent an apartment, you receive a key from the landlord. The key serves two purposes. Its possession is *proof of a right*, the right to enter the apartment. But its possession is also a *physical enabler*. The two are not the same of course, since if you lose your key, you still have the *right*, for it is still your apartment. But you are not *enabled*, as you cannot get in. If some stranger finds the key, then he is enabled, but does not have the right. Thus the two properties of a key do not coincide perfectly. But normally the two do coincide.

There are other analogs of a key which perform similar functions to a key. A password to a computer account is like a key, but does not need to be carried in your pocket. An ID card establishes your right to enter, but you typically need a guard to be present, to see your card and to let you into the building. If the building is locked and the guard is not present, you are out of luck.

In any case, these various generalized keys differ algorithmically in some crucial ways. Stealing someone's identity was at one time very difficult. You had to look like that person, know some personal facts, and you had to stay away from that person's dog who knew perfectly well that you had the wrong smell. You needed a different 'ID' for the dog than you needed for people.

But nowadays identity theft is extremely easy. Lots of Social Security numbers, and mothers' maiden names are out there for the taking, and people who do not look like you at all can make use of them. Personal appearance or brass keys which originally provided proof of "right to entry," have been replaced by electronic items which are very easy to steal.[3]

Let x be an individual, and let $R(x)$ mean that x has the *right* to use the resources controlled by the key, and $E(x)$ mean that x is *enabled* by the key. Then we have two important conditions.

– **Safety:** $E(x) \to R(x)$.   Whoever is enabled has the right
– **Liveness:** $R(x) \to E(x)$.   Whoever has the right is enabled.

Of course safety could be thought of in terms of the contrapositive,
$$\sim R(x) \; \to \; \sim E(X)$$
namely, whoever does not have the right is not enabled. Usually, safety is more important than liveness. If you lose your key and someone finds it, you are in trouble. But liveness also matters. A good notion of *key* must provide for both properties.

At one time, university libraries tended to be open. People not connected to the university, even if they did not have the right, were still *able* to enter the library. There was open access corresponding to the fact that liveness was

---

[3] According to the Javelin Research and Strategy Center, identity theft affected some 10 million victims in 2008, a 22% increase from 2007.

thought of as more important than safety. But the trend in the last few decades has been in the opposite direction and entry to libraries is strictly controlled, at least in the US.

In any case the structural problem (of safety) can be addressed at the incentive level, for instance by instituting heavy penalties for stealing identities. But we could also look for a structural solution without seeking to penalize anyone.

Toddlers are apt to run away and get into trouble, but we do not solve the problem by punishing them – we solve it by creating barriers to such escape, e.g., safety gates. A magnetic card which you can swipe also serves as a purely structural solution to the safety problem.

Another interesting example is a fence. A farmer may have a fence to keep his sheep in, and the fence prevents certain kinds of movement – namely sheep running away. Here the fence is a physical barrier and implements the safety condition in a purely physical way. But sometimes, on a university campus, we will see a very low fence around a grassy area. Almost anyone can walk over the fence, so the fence is no longer a physical obstacle. Rather the value of the fence is now informational. It says, *Thou shalt not cross!* With the yellow tape which the police sometimes put up, perhaps around a crime scene, or perhaps simply to block off some intersection, the *Thou shalt not cross* acquires quite a bit of punch.

3 Crime and Punishment

We offer a simple model to explain certain common situations where knowledge plays a role and can be used for reward or punishment.

3.1 Prisoner's Dilemma

In this game, two men are arrested and invited to testify against each other. If neither testifies, then there is a small penalty since there is no real evidence. But if one *defects* (testifies) and the other does not, then the defector goes free and the other gets a large sentence. If both defect they both get medium sentences. Jointly they are better off (The payoffs are 3 each) if neither defects, but for both of them, defecting is the dominant strategy. It yields better payoffs regardless of how the other acts. But if they both defect, then they end up with (1,1) which is worse. If one defects and the other remains honest then the honest one suffers for his honesty. In the table below, the highest payoff of 4 corresponds to going free and the lowest payoff of 0 corresponds to the longest sentence.

	Coop	Def
Coop	3, 3	0, 4
Def	4, 0	1, 1

There is a unique, rather bad Nash equilibrium at SE with (1,1), while the (3,3) solution on NW, though better for *both*, is not a Nash equilibrium.

Let us change this now into a three person game, where the third agent S (Society) has a payoff equal to the sum of the payoffs of the two original agents.

Consider now the expanded game G. In G, after the first two players make their moves, the third player moves and can choose among p_r (punish Row), p_c (punish Column), p_b (punish both), and n (no action). p_r, as we might expect, results in a negative payoff for Row of say 5. *If* Row has defected, S can play p_r which results in a negative payoff of 5 for the Row player. Similarly for Column and p_c. G is a full information game in that after Row and Column have made their moves, S knows what moves they made. Since S also suffers when Row or Column betrays his partner, S has an incentive to punish the erring player and the threat of S's punishment will keep the two players honest. We now get (from the point of view of Row and Column)

	Coop	Def
Coop	3, 3	0, −1
Def	−1, 0	−4, −4

and the NE solution with payoffs of (3,3) becomes the unique Nash equilibrium.

We now introduce a slightly different game G'. G' is just like the game G, except that S *lacks information* as to who made which move. If the societal payoff is only 4 or less, S knows that one of Row and Column cheated but it does not know which one. Thus it has no way to punish, and cheating can take place with impunity.

Clearly socially responsible behavior is more likely in G than in G' and the difference arises from the fact that in G, S has some information which it does not have in G'.

This, of course is why the FBI taps the phones of suspected criminals. A social agency has the incentive to punish anti-social behavior, and in order to do this, it needs to get information and change a G'-like situation into a G-like situation.[4]

Naturally, the agency S might not be benign. S could easily be a Mafia boss who needs to know when some member of the mob "sings", i.e., betrays the oath of silence. The singer could then be punished if and when he comes out of prison.

The FBI could itself have non-benign reasons for tapping phones. For instance we know that Martin Luther King's phone was tapped in order for the FBI to have power over him. This situation can be represented game theoretically, by turning G into a *four payer game* where the FBI-like agent (call is S_1) which has the power to punish, is not society at large but an *agent* of society. And then society, while wishing to control anti-social behavior on the part of Row and Column, also needs to control its own agent (say the FBI) whose job it is to keep Row and Column in check, but who may have its own payoff function distinct from social welfare.

We shall not go more into this in this paper.

---

[4] Of course all this is rather obvious, but it is important to point to the game theoretic reason not only behind punishment, but behind the acquisition of information relevant to it.

4   Cooperative Knowledge

Distributed Algorithms are much studied by computer scientists. A lot of commercial activity which goes on on the web has the property of being a distributed algorithm with many players. And of course the market is itself a very old distributed algorithm.

In such algorithms, it is crucial to make sure that when agents have to act, they have the requisite knowledge. And models for calculating such knowledge have existed for some time; we ourselves have participated in constructing such models [20,19]. See also [4].

The notion of *common knowledge* as the route to consensus was introduced by Aumann in [2]. There is subsequent work by various people, including Geanakoplos and Polemarchakis [7] and ourselves [17]. Aumann simply assumed common knowledge, and showed that two agents would agree on the value of a random variable if they had common knowledge of their beliefs about it. [7] showed that even if the agents did not have common knowledge to start with, if they exchanged values, they would arrive at consensus, and common knowledge of that fact. Parikh and Krasucki [17] carried this one step further and considered many agents exchanging values in *pairwise interactions*. No common knowledge could now arise, as most agents would remain unaware of individual transactions they were not a party to. Nonetheless there would be consensus. Thus this exchange of values could be seen as a distributed algorithm which achieved a result.

Issues about how knowledge enters into social algorithms are discussed in [11,13,20].

[20] actually discusses how a framework for defining knowledge can be developed. A finite number of agents have some private information to start with, and they exchange messages. Each exchange of messages reveals something about the situation, or, in technical terms, it reduces the size of the relevant Kripke structure or Aumann structure. An agent who has seen some events but not others can make guesses as to what other events *could* have taken place and it knows some fact ϕ iff ϕ would be true *regardless* of how the unseen events went. This framework is used in both [13,11].

[11] discusses agents who are connected along some graph, and knowledge can move only along the edges of a graph. Thus if agent i is not connected to agent j, then i cannot directly obtain information from j, but might get such information via a third agent k, as in fact Novak got some information from Judith Miller. Such edges may be approved or disapproved, and if information transfer took place along a disapproved edge, then that could be cause for legal sanctions, not because harm had occurred, but because harm *could* occur and the algorithm was no longer secure.

It is shown in [11] that the graph completely determines the logical properties of possible states of knowledge, and vice versa. Indeed, an early version of that paper already discussed the Plame case before it hit the headlines.

In [13] we consider how obligations arise from knowledge. We consider the following examples:

Example 1: Uma is a physician whose neighbour is ill. Uma does not know and has not been informed. Uma has no obligation (as yet) to treat the neighbour.

Example 2: Uma is a physician whose neighbour Sam is ill. The neighbour's daughter Ann comes to Uma's house and tells her. Now Uma does have an obligation to treat Sam, or perhaps call in an ambulance or a specialist.

Example 3: Mary is a patient in St. Gibson's hospital. Mary is having a heart attack. The caveat which applied in case a) does not apply here. The hospital has an obligation to *be aware* of Mary's condition at all times and to provide emergency treatment as appropriate.

In such cases, when an agent cannot herself take a requisite action, it is incumbent upon her to provide such information to the agent who *can* take such action. Or, as in the case of the hospital, the agent has an obligation not only to act, but also to gather knowledge so as to be *able to act* when the occasion arises. A milder example of such situations consists of requiring homeowners to install fire alarms. Homeowners are not only required (advised) to take action when there is a fire, they are also required to set up a system such that *if there is a fire, they will know about it.*

Again the semantics from [20] is used. Various possible sequences of events are possible, depending on the actions taken by the agents. Some of these sequences are better than others, and some, indeed, are disastrous, as when a patient is not treated for lack of information. It is shown in [13] how *having information* creates obligations on the agents, and also how the need to *convey information* arises, when one knows that an agent who could carry out some required action lacks the requisite information.

5   Summary

We have given examples of situations where knowledge transfer and algorithmic structure can affect or even determine the sorts of social algorithms which are possible. As we have said earlier, understanding the role of knowledge in the working of society is a big project. The importance of knowledge has always been recognized, even in the Indian school of *Navya-Nyaya*, by Plato's Socrates (especially the dialogues *Meno*, and *Theaetetus*), and by Confucius. But its importance in the actual running of society has been only recently begun to be appreciated by those who do formal work. The work we described above indicates how rich the domain of interest is here.

6   Further Research

A topic we have not addressed is that of cheap talk [5,25] where an agent says something which might not be true, or might be deceptive. The listener then has to take the agent's motives (her payoff function) into account in order to properly interpret her words. In [25], Stalnaker analyzes the statement of the secretary of treasury, John Snow, to the effect that a cheap dollar helps exports. This possibly innocent remark by Snow led to a fall in the value of the dollar. Did Snow *intend* this fall because such a fall would help exports, or did he make a mistake? Experts differed in their interpretations, just as experts differ in their evaluation of Obama's recent remarks in the Gates-Crowley affair.

This is a fascinating topic but clearly for another paper.

Another example which would also fit, but must be postponed to a future paper is the knowledge property of wills (last will and testament). For a will to be legal, it must be witnessed by two people, which is rather like being a member of a recursively enumerable (r.e.) set. Indeed we often talk about *witnesses* in that context too. However, for a will to be *valid,* there must not be a subsequent will, which is like *not* having a counter witness and being a member of a co-r.e. set. These differences, which are knowledge-theoretic, have practical effects and hence also legal ones.

References

1. Arrow, K.: Social Choice and Individual Values. Wiley, Chichester (1963)
2. Aumann, R.: Agreeing to disagree. Annals of Statistics 4, 1236–1239 (1976)
3. Brams, S., Taylor, A.: The Win-Win Solution: guaranteeing fair shares to everybody. Norton (1999)
4. Fagin, R., Halpern, J., Moses, Y., Vardi, M.: Reasoning about Knowledge. MIT Press, Cambridge (1995)
5. Farrell, J., Rabin, M.: Cheap talk. Journal of Economic Perspectives 10, 103–118 (1996)
6. de Finetti, B.: Foresight: its Logical Laws, Its Subjective Sources (translation of his 1937 article in French). In: Kyburg, H.E., Smokler, H.E. (eds.) Studies in Subjective Probability. Wiley, New York (1964)
7. Geanakoplos, J., Polemarchakis, H.: We can't disagree forever. J. Economic Theory
8. Hintikka, J.: Knowledge Belief: an introduction to the logic of the two notions. Cornell University Press (1962)
9. Lewis, D.: Convention, a Philosophical Study. Harvard U. Press, Cambridge (1969)
10. von Neumann, J., Morgenstern, O.: Theory of Games and Economic Behavior. Princeton University Press, Princeton (1944); 2nd ed. (1947)
11. Pacuit, E., Parikh, R.: Reasoning about Communication Graphs. In: van Benthem, J., Gabbay, D., Löwe, B. (eds.) Interactive Logic: Selected Papers from the 7th Augustus de Morgan Workshop, Texts in Logic and Games, vol. 1, pp. 135–157 (2007)
12. Pacuit, E., Parikh, R.: Social Interaction, Knowledge, and Social Software. In: Goldin, D., Smolka, S., Wegner, P. (eds.) Interactive Computation: The New Paradigm. Springer, Heidelberg (2006)
13. Pacuit, E., Parikh, R., Cogan, E.: The Logic of Knowledge Based Obligation. Knowledge, Rationality and Action, a subjournal of Synthese 149(2), 311–341 (2006)
14. Parikh, R.: The Logic of Games and its Applications. Annals of Discrete Math. 24, 111–140 (1985)
15. Parikh, R.: Social Software. Synthese 132, 187–211 (2002)

16. Parikh, R.: Levels of knowledge, games, and group action. Research in Economics 57, 267–281 (2003)
17. Parikh, R., Krasucki, P.: Communication, Consensus and Knowledge. J. Economic Theory 52, 178–189 (1990)
18. Parikh, R., Parida, L., Pratt, V.: Sock Sorting, appeared in a volume dedicated to Johan van Benthem, University of Amsterdam, August 1999; reprinted in Logic J. of IGPL 9 (2001)
19. Parikh, R., Ramanujam, R.: Distributed Processing and the Logic of Knowledge. In: Parikh, R. (ed.) Logic of Programs 1985. LNCS, vol. 193, pp. 256–268. Springer, Heidelberg (1985)
20. Parikh, R., Ramanujam, R.: A knowledge based semantics of messages. J. Logic, Language, and Information 12, 453–467 (2003)
21. The Plame Affair, Wikipedia, http://en.wikipedia.org/wiki/Plame_affair
22. Ramsey, F.P.: Truth and probability. In: The Foundations of Mathematics and other Logical Essays (1926)
23. Savage, L.: The Foundations of Statistics. Wiley, Chichester (1954)
24. Shoup, D.: Gone Parkin', Op-Ed page, The New York Times, March 29 (2007)
25. Stalnaker, R.: Saying and Meaning, Cheap Talk and Credibility. In: Benz, A., Jager, G., van Rooij, R. (eds.) Game Theory and Pragmatics, pp. 83–100. Palgrave Macmillan, Basingstoke (2005)
26. Turing, A.M.: On Computable Numbers, with an Application to the Entscheidungsproblem. Proceedings of the London Mathematical Society, Series 2 42, 230–265 (1937)

Computational Social Choice Using Relation Algebra and RELVIEW*

Harrie de Swart[1], Rudolf Berghammer[2], and Agnieszka Rusinowska[3]

[1] Department of Philosophy, Tilburg University
P.O. Box 90153, 5000 LE Tilburg, The Netherlands
H.C.M.deSwart@uvt.nl
[2] Institut für Informatik, Christian-Albrechts-Universität Kiel
Olshausenstraße 40, 24098 Kiel, Germany
rub@informatik.uni-kiel.de
[3] GATE, Université Lumière Lyon 2 - CNRS
93 Chemin des Mouilles - B.P. 167, 69131 Ecully Cedex, France
rusinowska@gate.cnrs.fr

Abstract. We present an overview of the potential of relation algebra and the software tool RELVIEW, based on it, to compute solutions for problems from social choice. Using one leading example throughout the text, we subsequently show how the RELVIEW tool may be used to compute and visualize minimal winning coalitions, swingers of a given coalition, vulnerable winning coalitions, central players, dominant players, Banzhaf power indices of the different players, Hoede Bakker indices of the different players in a network, and finally stable coalitions/governments. Although problems from social choice and games are mostly exponential, due to the BDD implementation of RELVIEW, computations are feasible for the examples which appear in practice.

1 Introduction

While both Computer Science and Social Choice Theory are senior scientific disciplines, only recently links between them have been observed, which resulted in developing *Computational Social Choice*. The aim of this paper is to present applications of Relation Algebra and RELVIEW to Social Choice Theory, that consist of computing solutions for problems from social choice.

After elections in list systems of proportional representation a coalition has to be formed which is preferably stable, meaning that none of the coalition partners has an incentive to leave that coalition. In Section 5 we will give precise definitions of a feasible coalition and of a feasible stable government, the central notions in [16]. We will show how the BDD based RELVIEW tool uses relation algebra to compute and visualize all feasible stable coalitions, which is, in general, a complex task.

---

* Co-operation for this paper is supported by European Science Foundation EURO-CORES Programme - LogICCC.

R. Berghammer et al. (Eds.): RelMiCS/AKA 2009, LNCS 5827, pp. 13–28, 2009.

Related to stability of a government is the existence of some specific players in a political game. Given a parliament or a city council we are interested in computing which players are the central players, and which are the dominant players, if any. In Section 3 we will relation-algebraically specify these key players and apply the RELVIEW program to determine them. In order to measure the power of a player, the so called power indices have been defined. We will present relation-algebraic specifications of the Banzhaf indices, the well-known power indices related to the concept of a swinger, and will execute these specifications with the help of the RELVIEW tool.

The decision-making process in most (if not all) decision bodies, in particular in politics, can be represented by a social network with mutual influences among the players. When a collective decision is made, each player appears to be successful or not, respectively powerful or not, depending on whether the collective decision coincides with the player's wishes, and depending on the player's ability to affect the collective outcome, respectively. In Section 4 we will apply relation algebra and RELVIEW to compute the success of a player.

Before we present the applications of Relation Algebra to Social Choice Theory, first in Section 2 we recapitulate basic relation-algebraic notions that will be used in the paper.

2   Relation-Algebraic Preliminaries

We denote the set (in this context also called type) of all relations with domain M and range N by $[M \leftrightarrow N]$ and write $R : M \leftrightarrow N$ instead of $R \in [M \leftrightarrow N]$. As basic operations of relation algebra we use R^{T} (*transposition*), $\overline{R}$ (*complement*), $R \cup S$ (*union*), $R \cap S$ (*intersection*) and RS (*composition*), as special relations we use O (*empty relation*), L (*universal relation*), and I (*identity relation*), and as predicates we use $R \subseteq S$ (*inclusion*) and $R = S$ (*equality*). A Boolean matrix interpretation of relations is well suited for many purposes and also used as one of the graphical representations of relations within RELVIEW. Also we use Boolean matrix terminology and notation, i.e., speak of rows, columns and 1- and 0-entries of relations and write $R_{j,k}$ instead of $\langle j, k \rangle \in R$.

The symmetric quotient $\mathrm{syq}(R, S) = \overline{R^{\mathsf{T}} \overline{S}} \cap \overline{\overline{R}^{\mathsf{T}} S} : N_1 \leftrightarrow N_2$ is used to compare the columns of two relations R and S with the same domain M and possible different ranges N_1 and N_2, since $\mathrm{syq}(R, S)_{j,k}$ holds iff for all $x \in M$ we have $R_{x,j}$ iff $S_{x,k}$, i.e., the j-column of R equals the k-column of S.

A *vector* is a relation v with $v = v\mathsf{L}$. As for a vector the range is irrelevant, we consider only vectors $v : N \leftrightarrow \mathbf{1}$ with a specific singleton set $\mathbf{1}$ as range and omit the second subscript, i.e., write v_k. Such a vector can be interpreted as a Boolean column vector and *describes* the subset $\{k \in N \mid v_k\}$ of N. As another way to deal with sets we apply *membership relations* $\mathsf{E} : N \leftrightarrow 2^N$ between N and its powerset 2^N. These specific relations are defined by $\mathsf{E}_{k,S}$ iff $k \in S$. Finally, we use injective mappings generated by vectors for modeling sets. If $v : N \leftrightarrow \mathbf{1}$ describes the subset S of N in the sense above, then the injective mapping $\mathrm{inj}(v) : S \leftrightarrow N$ is obtained from the identity relation $\mathsf{I} : N \leftrightarrow N$ by removing all rows which correspond to a 0-entry in v. Hence, we have $\text{inj}(v)_{j,k}$ iff $j = k$. A combination of injective mappings generated by vectors with membership relations allows a *column-wise enumeration* of sets of subsets. More specifically, if $v : 2^N \leftrightarrow \mathbb{1}$ describes a subset $\mathcal{N}$ of 2^N in the sense above, i.e., $\mathcal{N}$ equals the set $\{S \in 2^N \mid v_S\}$, then for all $k \in N$ and $S \in \mathcal{N}$ we have $(\mathsf{E}\,\text{inj}(v)^\mathsf{T})_{k,S}$ iff $k \in S$. In Boolean matrix terminology this means, that the elements of $\mathcal{N}$ are described precisely by the columns/vectors of $\mathsf{E}\,\text{inj}(v)^\mathsf{T} : N \leftrightarrow \mathcal{N}$.

Given a Cartesian product $M \times N$ of two sets M and N, there are the two canonical projection functions which decompose a pair $u = \langle u_1, u_2 \rangle$ into its first component u_1 and its second component u_2. For a relation-algebraic approach it is useful to consider instead of these functions the corresponding *projection relations* $\pi : M \times N \leftrightarrow M$ and $\rho : M \times N \leftrightarrow N$ such that $\pi_{u,j}$ iff $u_1 = j$ and $\rho_{u,k}$ iff $u_2 = k$. Projection relations enable us to specify the well-known fork operation relation-algebraically. For $R : X \leftrightarrow M$ and $S : X \leftrightarrow N$ their *fork* is $[R, S] = R\pi^\mathsf{T} \cap S\rho^\mathsf{T} : X \leftrightarrow M \times N$. Then we have $[R, S]_{x,u}$ iff R_{x,u_1} and S_{x,u_2}. The above projection relations π and ρ also allow to define a Boolean lattice isomorphism between $[M \times N \leftrightarrow \mathbb{1}]$ and $[M \leftrightarrow N]$ via $v \mapsto \text{rel}(v)$, where $\text{rel}(v) = \pi^\mathsf{T}(\rho \cap v\mathsf{L}^\mathsf{T})$ with $\mathsf{L} : N \leftrightarrow \mathbb{1}$. We call $\text{rel}(v) : M \leftrightarrow N$ the (proper) relation corresponding to the vector $v : M \times N \leftrightarrow \mathbb{1}$ since v_u iff $\text{rel}(v)_{u_1,u_2}$.

3   Simple Games

Simple games are mainly used for modelling decision-making processes, for instance, in the field of political science. Some recent textbooks on game theory are [11] and [12]. In the following, we show how relation algebra and RELVIEW can be used for solving problems on simple games. For it, we use a definition via the set of winning coalitions.

Definition 3.1. *A simple game is a pair (N, W), where $N = \{1, 2, \ldots, n\}$ is a set of players (agents, parties) and W is a set of winning coalitions, i.e., subsets of N. The game (N, W) is* monotone *if for all $S, T \in 2^N$, if $S \in W$ and $S \subseteq T$, then $T \in W$.*

In the following, we consider a small example for a simple game that stems from real political live.

Example 3.1. In the period 2006 - 2010 the city council of the municipality of Tilburg (NL) consists of the 10 parties

$$\text{PvdA, CDA, SP, LST, VVD, GL, D66, TVP, AB, VSP}$$

with respectively 11, 7, 5, 5, 4, 3, 1, 1, 1, 1 seats. The total number of seats is 39. Decisions are taken by simple majority, i.e., in order that a proposal is accepted, 20 persons have to vote in favor. Parties typically vote en bloc. So, in this example N is the set of the 10 parties just mentioned and a coalition is winning if the total number of seats of the parties in the coalition is at least 20. A simple game like this is called a *weighted majority game* and usually denoted by $[20; 11, 7, 5, 5, 4, 3, 1, 1, 1, 1]$.

There are two obvious ways to model simple games $(N, \mathcal{W})$ relation-algebraically:

1. By a vector $v : 2^N \leftrightarrow \mathbf{1}$ that describes the set $\mathcal{W}$ as subset of 2^N, so that v_S iff $S \in \mathcal{W}$. We call v the *vector model* of the game.
2. By a relation $M : N \leftrightarrow \mathcal{W}$, with $M_{k,S}$ iff $k \in S$ and $S \in \mathcal{W}$. We call M the *membership model* of the game.

Given a simple game, in [5] it is shown that if $v : 2^N \leftrightarrow \mathbf{1}$ is the vector model, then $\mathsf{E}\,\mathrm{inj}(v)^\mathsf{T} : N \leftrightarrow \mathcal{W}$ is the membership model, and conversely, if $M : N \leftrightarrow \mathcal{W}$ is the membership model, then $\mathrm{syq}(\mathsf{E}, M)\mathsf{L} : 2^N \leftrightarrow \mathbf{1}$ (with $\mathsf{L} : \mathcal{W} \leftrightarrow \mathbf{1}$) is the vector model. In view of RELVIEW, the membership model is more suited for input and output, whereas the vector model is more suited for the development of relation-algebraic specifications that can be used for problem solving.

Definition 3.2. *Let $(N, \mathcal{W})$ be a monotone simple game.*

- *Coalition S is* minimal winning *if $S \in \mathcal{W}$, but $T \notin \mathcal{W}$ for all $T \subset S$.*
- *Player k is a* swinger *of coalition S if $S \in \mathcal{W}$, $k \in S$, but $S \setminus \{k\} \notin \mathcal{W}$.*
- *Coalition S is* vulnerable winning *if $S \in \mathcal{W}$ and it contains a swinger.*

To explain these notions, we consider again the example of the city council of the municipality of Tilburg.

Example 3.2. The 3-parties coalition

$$S = \{\text{PvdA, CDA, GL}\}$$

with 21 seats is a minimal winning coalition: every proper subset of it has less than 20 seats. In addition, every player in this coalition S is a swinger of S. As another example, the winning 4-parties coalition

$$S' = \{\text{PvdA, SP, VVD, GL}\}$$

with 23 seats is not minimal, because without GL (3 seats) it is still winning, but it is vulnerable because it contains the swingers PvdA (11 seats), SP (5 seats) and VVD (4 seats).

It is well-known that for the membership relation $\mathsf{E} : N \leftrightarrow 2^N$ by the relation $\mathsf{S} := \overline{\mathsf{E}^\mathsf{T}\, \overline{\mathsf{E}}} : 2^N \leftrightarrow 2^N$ set inclusion is specified, i.e., it holds that $\mathsf{S}_{S,T}$ iff $S \subseteq T$. In [5] it is shown that $\mathsf{R} := \mathrm{syq}([\overline{\mathsf{I}}, \mathsf{E}], \mathsf{E}) : N \times 2^N \leftrightarrow 2^N$ specifies the removal of an element from a set. The latter means that for all $\langle k, S \rangle \in N \times 2^N$ and $T \in 2^N$ it holds $\mathsf{R}_{\langle k,S \rangle, T}$ iff $S \setminus \{k\} = T$. Based on these facts, again in [5] the following result is proved.

Theorem 3.1. *Let $v : 2^N \leftrightarrow \mathbf{1}$ be the vector model of the simple game $(N, \mathcal{W})$. Then the vector*

$$\mathrm{minwin}(v) := v \cap \overline{(\mathsf{S} \cap \overline{\mathsf{I}})^\mathsf{T} v} : 2^N \leftrightarrow \mathbf{1}$$

describes the set $\mathcal{W}_{\min}$ of all minimal winning coalitions. Assuming additionally that the simple game $(N, \mathcal{W})$ is monotone and $\mathsf{L} : N \leftrightarrow \mathbf{1}$, for the relation

$$\mathrm{Swingers}(v) := \mathsf{E} \cap \mathsf{L}v^\mathsf{T} \cap \mathrm{rel}(\mathsf{R}\,\overline{v}) : N \leftrightarrow 2^N$$

it holds Swingers$(v)_{k,S}$ *iff k is a swinger of S and the vector*

$$\text{vulwin}(v) := \text{Swingers}(v)^{\mathsf{T}}\mathsf{L} : 2^N \leftrightarrow \mathbf{1}$$

describes the set of all vulnerable winning coalitions.

To give an impression how easy it is to transform the above relation-algebraic specifications into RELVIEW code, we present the RELVIEW program for the column-wise enumeration of all minimal winning coalitions:

```
Minwin(E,v)
    DECL S, I,
    BEG  S = -(E^ * -E);
         I = I(S);
         m = v & -((S & -I)^ * v)
         RETURN E * inj(m)^
    END.
```

This program expects the membership relation $\mathsf{E} : N \leftrightarrow 2^N$ and the membership model $v : 2^N \leftrightarrow \mathbf{1}$ as input. Then its first statement computes the set inclusion relation $\mathsf{S} : 2^N \leftrightarrow 2^N$, its second one the identity relation $\mathsf{I} : 2^N \leftrightarrow 2^N$ and its third one the vector minwin$(v) : 2^N \leftrightarrow \mathbf{1}$. Following the technique of Section 2, finally, by the expression of the RETURN-clause the set described by minwin(v) is column-wisely enumerated and this relation is delivered as result.

Example 3.3. In case of our running example of the city council of Tilburg, the above program yields as output the following picture with all 49 minimal winning coalitions described by the 49 columns of the 10×49 matrix.

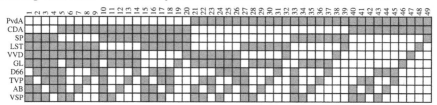

In such a matrix a black square means a 1-entry and a white square means a 0-entry. So, e.g., the first column describes the coalition consisting of the parties SP, LST, VVD, GL, TVP, AB and VSP. The following 10×7 RELVIEW matrix shows that the minimum number of parties needed to form a winning coalition is 3 and there are exactly 7 possibilities to form such 3-parties winning coalitions. From this result we also know that there exist winning coalitions with 20 seats, i.e., the minimum number of seats necessary to become winning.

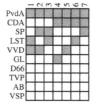

RELVIEW also computes 417 vulnerable winning coalitions, where the largest number of critical players in such a coalition is 7 (and 6 such cases exist).

An important player, which can be seen as 'policy oriented' or 'policy seeking', is considered next. In order to define it, all players must be ordered on a relevant policy dimension, normally from left to right. The particular position of the player we introduce now makes him very powerful.

Definition 3.3. *Given a simple game $(N, \mathcal{W})$ and a policy order of the players in the form of a linear strict order relation $P : N \leftrightarrow N$, a player $k \in N$ is central if the connected coalition $\{j \in N \mid P_{j,k}\}$ "to the left of k" as well as the connected coalition $\{j \in N \mid P_{k,j}\}$ "to the right of k" are not winning, but both can be turned into winning coalitions when k joins them.*

The following relation-algebraic expression enables RELVIEW immediately to compute the vector describing the set of central players (which contains at most one element); see again [5] for its formal development.

Theorem 3.2. *Let a simple game $(N, \mathcal{W})$ with a policy order $P : N \leftrightarrow N$ be given and assume that $v : 2^N \leftrightarrow 1$ is the game's vector model. Using $Q := P \cup \mathsf{I}$ as reflexive closure of P, the vector*

$$\mathrm{central}(v, P) := \overline{\mathrm{syq}(P, \mathsf{E})v} \cap \overline{\mathrm{syq}(P^\mathsf{T}, \mathsf{E})v} \cap \mathrm{syq}(Q, \mathsf{E})v \cap \mathrm{syq}(Q^\mathsf{T}, \mathsf{E})v$$

of type $[N \leftrightarrow \mathbf{1}]$ describes the set of all central players.

To demonstrate the use of RELVIEW for computing a central player, we consider again our running example of the city council of Tilburg.

Example 3.4. Assume that the left-to-right strict order relation $<$ of the parties of the city council of Tilburg is as follows:

GL < SP < PvdA < D66 < CDA < AB < VSP < VVD < LST < TVP

Then D66 with 1 seat is the central player, because the parties to the left of it have only 19 seats and also the parties to the right of it have only 19 seats. The left one of the two following pictures shows the strict order relation P, given by the placements of the parties in the above list, as depicted by REL-VIEW. When applying the RELVIEW program resulting from the above relation-algebraic specification to the membership model of our running example and P, the tool yields the right one of the pictures below as result.

Hence, also RELVIEW computes that D66 is the central player of our game.

The next definition introduces the last important class of players we consider in this paper. Such players may be seen as 'policy blind' or 'office seeking'.

Definition 3.4. *Let $(N, \mathcal{W})$ be a simple game.*

- *The player $k \in N$ dominates the coalition $S \in 2^N$, written as $k \gg S$, if $k \in S$ and for all $U \in 2^N$ with $U \cap S = \emptyset$, if $U \cup (S \setminus \{k\}) \in \mathcal{W}$, then $U \cup \{k\} \in \mathcal{W}$, while there exists $U \in 2^N$ such that $U \cap S = \emptyset$, $U \cup \{k\} \in \mathcal{W}$, but $U \cup (S \setminus \{k\}) \notin \mathcal{W}$.*
- *The player $k \in N$ is dominant if there exists a coalition $S \in \mathcal{W}$ such that $k \gg S$.*

If k dominates S, then k can form a winning coalition with players outside of S while $S \setminus \{k\}$ is not able to do this. The dominant players are the most powerful players of the game. Such players neither must exist nor must be unique. However, Peleg proved in [10] that in weak simple games and weighted majority games at most one dominant player may occur.

Example 3.5. In case of our running example of the city council of Tilburg, the party PvdA with 11 seats dominates the coalition

$$\{\text{PvdA, CDA, GL}\}$$

and hence is a dominant player: for any coalition U not containing PvdA, CDA, GL, if $U \cup \{\text{CDA, GL}\}$ is winning, then also $U \cup \{\text{PvdA}\}$ is winning, since PvdA has one more seat than CDA (7 seats) and GL (3 seats) together. But conversely, there is a coalition, viz.

$$U = \{\text{LST, VVD}\},$$

having 9 seats which together with PvdA is winning, but not together with CDA and GL. In [5] a relation-algebraic specification of type $[N \leftrightarrow 2^N]$ for the dominance relation $\gg$ is developed. Again, the usefulness of the RELVIEW tool becomes clear: for our running example, e.g., it computes that

- PvdA dominates 140 coalitions, 50 of them winning;
- CDA dominates 48 coalitions, none of them winning;
- SP and LST each dominate 22 coalitions, none of them winning;
- VVD dominates 16 coalitions, none of them winning;
- GL dominates 11 coalitions, none of them is winning;
- D66, TVP, AB and VSP each dominate 1 coalition, none of them winning.

Hence, PvdA, the party with the maximum number of seats, is the (only) dominant player. The RelView-picture for the vector of dominant players is below:

One of the most important elements of simple games is to measure the power of players. To this end, during the last decades some so-called power indices have been proposed. We present here the normalized Banzhaf index $B(k)$ of player k, introduced in [1]. The absolute (non-normalized) Banzhaf index $\overline{B}(k)$ of player k is the probability that player k is decisive for the outcome, that is the number of times that k is a swinger in a winning coalition, divided by the number (2^{n-1} if there are n players) of winning coalitions he belongs to, assuming that all coalitions are equally likely and that each player votes yes or no with probability $\frac{1}{2}$.

Definition 3.5. *Let $(N, \mathcal{W})$ be a monotone simple game and $k \in N$. Then the absolute Banzhaf index $\overline{B}(k)$ and the normalized Banzhaf index $B(k)$ of k are defined as follows, where n is the number of players:*

$$\overline{B}(k) := \frac{|\{S \in \mathcal{W} \mid k \ swinger \ of \ S\}|}{2^{n-1}} \qquad B(k) := \frac{\overline{B}(k)}{\sum_{j \in N} \overline{B}(j)}$$

In [5] the following descriptions are derived for the absolute and normalized Banzhaf indices. In it we denote for X and Y being finite, for $R : X \leftrightarrow Y$ and $x \in X$, the number of 1-entries of R with $|R|$ and the number of 1-entries of the x-row of R with $|R|_x$. Hence, $|R|$ equals the cardinality of R (as set of pairs) and $|R|_x$ equals the cardinality of the subset Y' of Y that is described by the transpose of the x-row.

Theorem 3.3. *Assume a monotone simple game $(N, \mathcal{W})$ with n players and its vector model $v : 2^N \leftrightarrow 1$. Then we have for all players $k \in N$:*

$$\overline{B}(k) = \frac{|\mathrm{Swingers}(v)|_k}{2^{n-1}} \qquad B(k) = \frac{|\mathrm{Swingers}(v)|_k}{|\mathrm{Swingers}(v)|}$$

If RELVIEW depicts a relation R as a Boolean matrix in its relation-window, then additionally in the window's status bar the number $|R|$ is shown. Furthermore, it is possible to mark rows and columns. So far, we only have shown the possibility to use strings or consecutive numbers as labels. But also the numbers $|R|_x$ automatically can be attached as labels. In combination with the above theorem this immediately allows to compute Banzhaf indices via the tool.

Example 3.6. Using the features just described, RELVIEW computes in case of our running example the following normalized Banzhaf indices:

PvdA: $\frac{332}{988}$ CDA: $\frac{160}{988}$ SP: $\frac{116}{988}$ LST: $\frac{116}{988}$ VVD: $\frac{96}{988}$

GL: $\frac{88}{988}$ D66: $\frac{20}{988}$ TVP: $\frac{20}{988}$ AB: $\frac{20}{988}$ VSP: $\frac{20}{988}$

For obtaining the absolute Banzhaf indices we only have to replace the denominator 988 (which equals the number of all swingers) by the number 512 (i.e., by 2^{10-1}). Notice that although the number of seats of PvdA is about 1.5 times that of CDA, the power of PvdA expressed by the Banzhaf index is more than twice the power of CDA.

4   Social Networks

Players (agents, decision makers, parties) are frequently operating in a social network, where there are mutual influences, such that the final decision of a player may be different from his original inclination. The generalized Hoede-Bakker (GHB) power index [9, 15] takes these aspects into account.

Consider a social network with a set N of n players (agents). Given a proposal, each player is supposed to have an inclination 'yes' (expressed by the truth value 1) or 'no' (expressed by the truth value 0). An *inclination vector* is a list of 0's or 1's of length n and $I = \{0,1\}^n$ is the set of all inclination vectors. We assume that, due to influences among the players, the final decision of a player may differ from his original inclination. Let $Bi \in \{0,1\}^n$ denote the *decision vector* that results from the inclination vector i. Then B is called the *influence function*. For instance, B may be defined by following only unanimous trend-setters or by following a majority of trend-setters. We also assume a *group decision* function gd, that assigns to each decision vector Bi the decision 'yes' (1) or 'no' (0) of the group. For instance, gd may be defined by $gd(Bi) = 1$ iff the majority of the players decides to vote 'yes'.

In order to define the generalized Hoede Bakker index of player k in the given social network, we first need the following sets.

Definition 4.1. *Based on the influence function B and the group decision function gd, we define for each $k \in N$ the following four sets:*

$$I_k^{++}(B,gd) := \{i \in I \mid i_k = 1 \ \wedge \ gd(Bi) = 1\}$$
$$I_k^{+-}(B,gd) := \{i \in I \mid i_k = 1 \ \wedge \ gd(Bi) = 0\}$$
$$I_k^{-+}(B,gd) := \{i \in I \mid i_k = 0 \ \wedge \ gd(Bi) = 1\}$$
$$I_k^{--}(B,gd) := \{i \in I \mid i_k = 0 \ \wedge \ gd(Bi) = 0\}$$

So, for instance, the set $I_k^{++}(B,gd)$ is the set of all inclination vectors i, where the player k has inclination 'yes' and the group decision, given the decision vector Bi, is also 'yes'.

Definition 4.2. *Given B and gd as above, the generalized Hoede-Bakker index* $\mathrm{GHB}_k(B,gd)$ *of player $k \in N$ is defined as follows:*

$$\mathrm{GHB}_k(B,gd) := \frac{|I_k^{++}| - |I_k^{+-}| + |I_k^{--}| - |I_k^{-+}|}{2^n}$$

The value of $\mathrm{GHB}_k(B,gd)$ measures a kind of 'net' Success, i.e., Success − Failure, where by a successful player, given i, B and gd, we mean a player k whose inclination i_k coincides with the group decision $gd(Bi)$.

With 10 players or parties, as in the case of our running Tilburg city council example from the last section, the set I of all inclination vectors contains 2^{10} = 1024 elements. In RELVIEW this set can be represented by a 10×2^{10} matrix, where the 10 rows are labeled with the 10 parties and the 1024 columns, considered as single vectors, represent the possible 1024 inclination vectors. Generalizing this idea to arbitrary social networks, the set I of all inclination vectors of the network with set N of players column-wisely is enumerated by the membership relation $\mathsf{E} : N \leftrightarrow 2^N$.

In most cases appearing in practice, the mutual influences among the players can be represented by a *dependency relation* $D : N \leftrightarrow N$ on the players; for an example see below. Given D and assuming the influence function B to be fixed via D, e.g., by 'following only unanimous trend-setters' as rule[1], in [4] a relation-algebraic specification $Dvec(D) : N \leftrightarrow 2^N$ that column-wisely enumerates all decision vectors is developed. The latter means that, if the X-column (with $X \in 2^N$) of the membership relation $\mathsf{E} : N \leftrightarrow 2^N$ represents the inclination vector i (that is, $i_k = 1$ iff $k \in X$), then the X-column of the relation $Dvec(D)$ represents the decision vector Bi. RELVIEW can easily deal with matrices of the size appearing in our running example of the city council of Tilburg.

Since group decisions are truth values, all group decisions can be modeled by a group decision vector $gdv(D) : 2^N \leftrightarrow 1$ such that, e.g., in the case of simple majority as group decision rule, the X-component of the vector $gdv(D)$ is 1 iff the X-column of $Dvec(D)$ contains strictly more 1-entries than 0-entries. But also other group decision rules are possible. We will consider such another rule below, viz. weighted simple majority. Here to each player $k \in N$ a natural number w_k is assigned as its weight and a further natural number q is given as quota. Then the X-component of $gdv(D)$ is 1 iff $\sum_{k \in X} w_k \geq q$.

Example 4.1. We assume in case of our running example of the city council of Tilburg a dependency relation D on the 10 parties that, drawn as directed graph, looks as follows:

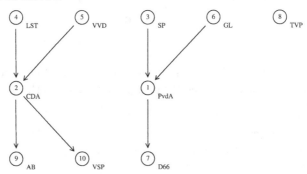

As this picture shows, we assume, for instance, $D_{\text{LST,CDA}}$ and $D_{\text{VVD,CDA}}$ meaning that CDA is dependent on the two trend-setters LST and VVD. If, again for instance, the inclination vector i is given by

$$i_{\text{PvdA}} = 1 \qquad i_{\text{CDA}} = 0 \qquad i_{\text{SP}} = 0 \qquad i_{\text{LST}} = 1 \qquad i_{\text{VVD}} = 0$$
$$i_{\text{GL}} = 0 \qquad i_{\text{D66}} = 1 \qquad i_{\text{TVP}} = 0 \qquad i_{\text{AB}} = 0 \qquad i_{\text{VSP}} = 1$$

---

[1] Under this rule, the vote of player k is equal to the inclination of his trend-setters (predecessors in the graph given by D) if they all have the same inclination. Otherwise, player k votes according to his own inclination.

then under 'following only unanimous trend-setters' as influence rule it is transformed into the decision vector

$$Bi_{\text{PvdA}} = 0 \qquad Bi_{\text{CDA}} = 0 \qquad Bi_{\text{SP}} = 0 \qquad Bi_{\text{LST}} = 1 \qquad Bi_{\text{VVD}} = 0$$
$$Bi_{\text{GL}} = 0 \qquad Bi_{\text{D66}} = 1 \qquad Bi_{\text{TVP}} = 0 \qquad Bi_{\text{AB}} = 0 \qquad Bi_{\text{VSP}} = 0$$

since PvdA follows its predecessors SP and GL (with $i_{\text{SP}} = i_{\text{GL}} = 0$) and VSP votes as its predecessor CDA (with $i_{\text{CDA}} = 0$). Hence, since for the weights (numbers of seats) of the approving parties LST and D66 and the quota we have $w_{\text{LST}} + w_{\text{D66}} = 5 + 1 < 20 = q$, the decision of the city council of Tilburg is 'no'. Note that also the original inclination would lead to a negative decision due to $w_{\text{PvdA}} + w_{\text{VSP}} + w_{\text{LST}} + w_{\text{D66}} = 18$.

In [4] one finds also a solution for computing the group decision vector as relational vector $gdv(D) : 2^N \leftrightarrow \mathbf{1}$ from the relation $Dvec(D) : N \leftrightarrow 2^N$ and also relation-algebraic specifications $ipp(p, g)$, $ipm(p, g)$, $imp(p, g)$, $imm(p, g)$, each of type $[2^N \leftrightarrow \mathbf{1}]$, for describing the sets $I_k^{++}(B, gd)$, $I_k^{+-}(B, gd)$, $I_k^{-+}(B, gd)$ and $I_k^{--}(B, gd)$ respectively, where the player k is described by a column point $p : N \leftrightarrow \mathbf{1}$ (i.e., a vector with exactly one 1-entry) and $g = gdv(D)$ is the group decision vector. Finally, from the numbers of 1-entries of these vectors (i.e., the sizes of the sets of inclination vectors they represent) automatically provided by RELVIEW when computing them, the generalized Hoede-Bakker (GHB) indices easily can be obtained for each player $k \in N$ via the tool.

Below we present the relation algebraic expressions just mentioned, for the influence rule 'following only unanimous trend-setters' and weighted group decision. In case of our running example this means that the city council votes 'yes' iff the parties that vote in favor have at least $q = 20$ seats. Relation-algebraically the quota q is given by the length of a vector w, i.e., the size of its domain. The weight of a party corresponds to the number of its seats. Relation-algebraically, the latter can be modeled as follows. We assume S to be the set of all seats and the distribution of the seats over the parties to be given by a relation $W : S \leftrightarrow N$ such that $W_{s,k}$ iff seat $s \in S$ is owned by party $k \in N$. The relation W is a mapping in the relational sense and for each $k \in N$ the k-column of W consists of exactly w_k 1-entries, with w_k being the weight (number of seats) of party k.

Theorem 4.1. *Let $D : N \leftrightarrow N$ be the dependency relation and $d = D^{\mathsf{T}}\mathsf{L} : N \leftrightarrow \mathbf{1}$ describe the set of dependent players, i.e., players which have predecessors with respect to D. Then, we obtain $Dvec(D) : N \leftrightarrow 2^N$ as*

$$Dvec(D) = (\mathsf{E} \cap (\overline{d\mathsf{L}} \cup (d\mathsf{L} \cap D^{\mathsf{T}}\mathsf{E} \cap D^{\mathsf{T}}\overline{\mathsf{E}}))) \cup (d\mathsf{L} \cap \overline{D^{\mathsf{T}}\overline{\mathsf{E}}}).$$

Furthermore, the group decision vector $gdv(D) : 2^N \leftrightarrow \mathbf{1}$ is given by

$$gdv(D) = \text{syq}(W\, Dvec(D), \mathsf{E})\, m,$$

where $\mathsf{E} : S \leftrightarrow 2^S$ is the membership relation on the seats and m is defined by $m = \overline{\text{cardfilter}(\mathsf{L}, w)}$. Finally, with g as abbreviation for $gdv(D)$ and $\mathsf{E} : N \leftrightarrow 2^N$ as the membership relation on the parties, we have:

$$ipp(p,g) = (p^\mathsf{T}\mathsf{E} \cap g)^\mathsf{T} \qquad ipm(p,g) = (p^\mathsf{T}\mathsf{E} \cap \bar{g})^\mathsf{T}$$
$$imp(p,g) = (p^\mathsf{T}\bar{\mathsf{E}} \cap g)^\mathsf{T} \qquad imm(p,g) = (p^\mathsf{T}\bar{\mathsf{E}} \cap \bar{g})^\mathsf{T}$$

The vector $m : 2^S \leftrightarrow \mathbf{1}$ used in this theorem fulfills for all $X \in 2^S$ that m_X iff $|X| \geq q$. Such a vector can be easily obtained with the help of the operation *cardfilter*. If $v : 2^S \leftrightarrow \mathbf{1}$ describes the subset $\mathcal{S}$ of 2^S and the size of the domain of $w : W \leftrightarrow \mathbf{1}$ is at most $|M| + 1$, then for all $X \in 2^S$ we have *cardfilter*$(v, w)_X$ iff $X \in \mathcal{S}$ and $|X| < |W|$. Hence, since $\mathsf{L} : 2^S \leftrightarrow \mathbf{1}$ describes entire powerset 2^S, in our theorem, the complement of *cardfilter*(L, w) describes the subset of 2^S whose elements have at least size q.

Example 4.2. Given the influence rule and group decision function mentioned in the last example, we obtain for our running example the following sizes of the sets underlying the definition of the generalized Hoede-Bakker index. Here are the values for $I^{++}(B, gd)$, $I^{+-}(B, gd)$, $I^{-+}(B, gd)$, $I^{--}(B, gd)$ and $\mathrm{GHB}_k(B, gd)$ for the ten parties.

party	$I^{++}(B,gd)$	$I^{+-}(B,gd)$	$I^{-+}(B,gd)$	$I^{--}(B,gd)$	GHB
PvdA	328	184	184	328	288/1024
CDA	280	232	232	280	96/1204
SP	392	120	120	392	544/1024
LST	328	184	184	328	288/1024
VVD	312	200	200	312	224/1024
GL	360	152	152	360	416/1024
D66	256	256	256	256	0/1024
TVP	280	232	232	280	96/1024
AB	256	256	256	256	0/1024
VSP	256	256	256	256	0/1024

Notice the high values of the generalized Hoede-Bakker indices for SP and GL. This may be explained by the assumption made that these parties are influencing PvdA and this latter party has most (viz. 11) seats.

5   Stable Coalitions

After elections with list systems of proportional representation a coalition has to be formed, which usually needs to have at least half of the seats. In general, there are many of such coalitions – in our example 3.3 there are already 49 minimal winning coalitions and 417 vulnerable winning coalitions. In practice, however, many parties do not see the possibility to cooperate with each other, and consequently many winning coalitions are not feasible. We call a coalition *feasible* if it is accepted by every party in the coalition. A feasible coalition S alone is not enough: it has to come with a policy p, where the latter may consist of several sub-policies. We call a pair $g = (S, p)$ with $S \in 2^N$ and a policy p a *government*. Of course, the policy p of a government (S, p) should be feasible for S, i.e., at least acceptable to all members of the coalition. So, the set G of all feasible governments is the set of all pairs (S, p) with a feasible coalition S and a policy p that is feasible for S. It seems reasonable to assume that every party i that belongs to at least one feasible government has a weak preference ordering R^i over the feasible governments.

To determine in practice which coalitions are feasible and which policies are feasible for these coalitions, one may apply another software tool, called Mac-Beth. How this tool may be used for this purpose is explained in [14].

Definition 5.1. Let $h = (S, p)$ and g be feasible governments. h dominates g, denoted by $h \succ g$, iff for every party $i \in S$, $hR^i g$ (or $R^i_{h,g}$) and for at least one party $i \in S$, not $gR^i h$ (i.e., not $R^i_{g,h}$). A feasible government g is called stable if there is no feasible government h that dominates g.

Also in this section we consider the city council of Tilburg as example.

Example 5.1. Let us assume that for the city council of Tilburg there are 3 feasible coalitions: $S_1 = \{\text{PvdA, SP, VVD, GL}\}$, $S_2 = \{\text{PvdA, CDA, GL}\}$ and $S_3 = \{\text{PvdA, CDA, LST, VVD}\}$. Further there are 3 policy sub-issues on which the coalition partners have to agree: building a city-ring, buying a theater and building a shopping mall. Suppose for coalition S_1 only policy $p_1 = $ (no, yes, yes) is feasible, for coalition S_2 only policy $p_2 = $ (yes, no, no) is acceptable, and for coalition S_3 only policy $p_3 = $ (no, no, no) is feasible. So, there are three potential feasible governments (S_1, p_1), (S_2, p_2), (S_3, p_3), and the parties involved in at least one such government are PvdA, CDA, SP, LST, VVD and GL. Assume now the following preference relations of these parties over the three feasible governments.

$$\text{PvdA: } g_2 > g_1 > g_3; \quad \text{CDA: } g_3 > g_2 > g_1; \quad \text{SP: } g_1 > g_2 > g_3;$$
$$\text{LST: } g_3 > g_2 > g_1; \quad \text{VVD: } g_3 > g_1 > g_2; \quad \text{GL: } g_2 > g_1 > g_3.$$

One can easily check that g_2 dominates g_1: every party in S_2 strongly prefers g_2 to g_1. Further, g_2 does not dominate g_3: party CDA in S_2 strongly prefers g_3 to g_2. Conversely, g_3 does not dominate g_2: party PvdA in S_3 strongly prefers g_2 to g_3. Some further thinking shows that g_2 and g_3 are the stable governments: they are not dominated by any other feasible government.

In the example above it was easy to compute the stable governments by hand. But if there are, for instance, 17 feasible governments, as in [2], each party involved in one of these 17 governments is supposed to have a weak preference ordering over these 17 governments. In such a case, in order to compute all stable governments, it is hardly - or not at all - feasible to do all computations by hand. For that reason we have delivered in [2] relation algebraic expressions for the notion of feasible government, for the dominance relation and for the stability predicate.

In order to develop a relation-algebraic specification of feasible governments, we need two 'acceptability' relations A and B to be given. Let P be the set of policies. We assume $A : N \leftrightarrow P$ such that

$$A_{i,p} \iff \text{party } i \text{ accepts policy } p$$

for all $i \in N$ and $p \in P$, and $B : N \leftrightarrow 2^N$ such that

$$B_{i,S} \iff \text{party } i \text{ accepts coalition } S$$

for all $i \in N$ and $S \in 2^N$.

Theorem 5.1. *Assume* $\mathsf{E} : N \leftrightarrow 2^N$ *to be the membership-relation on the set* N *of parties. Then for*

$$FeaC(B) = \overline{(\mathsf{E} \cap \overline{B})^{\mathsf{T}} \mathsf{L}} : 2^N \leftrightarrow 1$$

we have that $feaC(B)_S$ *iff for all* $i \in S$ *it holds* $B_{i,S}$, *for*

$$IsFea(A) = \overline{\mathsf{E}^{\mathsf{T}} \overline{A}} : 2^N \leftrightarrow P$$

we have that $isFea(A)_{S,p}$ *iff for all* $i \in S$ *it holds* $A_{i,p}$, *and for*

$$FeaG(A, B) = \overline{\mathsf{E}^{\mathsf{T}} \overline{A}} \cap \overline{(\mathsf{E} \cap \overline{B})^{\mathsf{T}} \mathsf{L}} \mathsf{L} : 2^N \leftrightarrow P$$

we have that $feaG(A, B)_{S,p}$ *iff* $feaC(B)_S$ *and* $isFea(A)_{S,p}$.

For the proof we refer to [2]. The next thing to do is to give a relation algebraic formulation of the dominance relation. To this end we should specify which parties are a member of which government. So, we suppose a membership relation $M : N \leftrightarrow G$ to be given, such that $M_{i,g}$ iff party i is a member of government g. We also suppose for each party i a weak preference ordering R^i over the governments to be given. It turns out to be convenient to work with a global utility relation $C : N \leftrightarrow G \times G$ such that $C_{i,<h,g>}$ iff $R^i_{h,g}$. This can be achieved by defining $C = \text{vec}(R^{i_1})^{\mathsf{T}} + \ldots + \text{vec}(R^{i_k})^{\mathsf{T}}$, where $i_1, \ldots, i_k$ are the parties involved in at least one feasible government, vec is the inverse of the function rel, and $+$ forms the sum of two relations. In our case the latter means that all row vectors $\text{vec}(R^{i_j})^{\mathsf{T}}$ are joined together to form the rows of C. Finally, in [2] the following results are shown.

Theorem 5.2. *Let* $\pi : G \times G \leftrightarrow G$ *and* $\rho : G \times G \leftrightarrow G$ *be the projection relations and* $E : G \times G \leftrightarrow G \times G$ *the exchange relation. If we define*

$$Dominance(M, C) = \overline{(\pi M^{\mathsf{T}} \cap \overline{C}^{\mathsf{T}}) \mathsf{L}} \cap (\pi M^{\mathsf{T}} \cap E \overline{C}^{\mathsf{T}}) \mathsf{L},$$

then we have for all $u = \langle h, g \rangle \in G \times G$ *that* $Dominance(M, C)_u$ *if and only if* $h \succ g$. *And if we define* $stable(M, C) : G \leftrightarrow 1$ *by*

$$Stable(M, C) = \overline{\rho^{\mathsf{T}} Dominance(M, C)},$$

then we have $stable(M, C)_g$ *iff there is no* $h \in G$ *such that* $h \succ g$ *(i.e., iff* g *is stable).*

Example 5.2. We will apply the RELVIEW software to the city council of Tilburg, as described above in example 5.1. As input for the program we have to specify the membership relation M with $M_{i,g}$ iff party i is a member of government g and the global comparison relation C with $C_{i,<h,g>}$ iff party i weakly prefers h to g. Below we specify M and C by the following RELVIEW matrices.

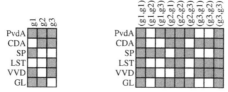

Applying the RELVIEW programs for the dominance and the stability relations, we get the following outputs in the form of a graph and a matrix respectively, as one wishes. Notice that from the dominance graph it already becomes clear that g_2 and g_3 are the stable, i.e., not dominated, governments.

6   Conclusions

We have shown that many problems from social choice and game theory can be formulated in relation algebraic terms, which subsequently can be used in extremely short programs for the RELVIEW tool. Due to its very efficient implementation of relations based on Binary Decision Diagrams (BDDs), this tool can deal with numbers of players that appear in practical applications (e.g., in real political life), although many of the problems are of exponential size.

In this overview paper, we have focused on several applications of Relation Algebra and RELVIEW to game theoretical concepts, but we did not present all of our results on this issue. In [5], for instance, we have delivered relation-algebraic specifications of key players in a simple game, i.e., of dummies, dictators, vetoers, and null players, and we have also calculated power indices different from the Banzhaf indices. In [4], based e.g. on the sets $I_k^{++}(B, gd)$, $I_k^{+-}(B, gd)$, $I_k^{-+}(B, gd)$ and $I_k^{--}(B, gd)$ introduced in Section 4, we have calculated modifications of the generalized Hoede-Bakker index GHB_k that coincide with some standard power indices, e.g., the Coleman indices [6, 7] and the Rae index [13]. In [4] we have also applied the relation-algebraic approach to compute some concepts defined in [8], like influence indices, sets of followers of a coalition, and the kernel of an influence function. In [3] we have used techniques from graph theory to let

RELVIEW compute coalitions / governments 'as stable as possible', in case there turn out to be no stable ones.

There are still many topics from Game Theory and Social Choice Theory to which an application of the relation algebraic approach can be very useful. For the near future we plan to apply relation algebra and RELVIEW for computing all kinds of solution concepts of games, such as the core, the uncovered set, etc.

References

[1] Banzhaf, J.F.: Weighted Voting Doesn't Work: A Mathematical Analysis. Rutgers Law Review 19, 317–343 (1965)

[2] Berghammer, R., Rusinowska, A., de Swart, H.: Applying Relational Algebra and RELVIEW to Coalition Formation. European Journal of Operational Research 178, 530–542 (2007)

[3] Berghammer, R., Rusinowska, A., de Swart, H.: An Interdisciplinary Approach to Coalition Formation. European Journal of Operational Research 195, 487–496 (2009)

[4] Berghammer, R., Rusinowska, A., de Swart, H.: Applying Relation Algebra and RELVIEW to Measures in a Social Network. European Journal of Operational Research (to appear)

[5] Berghammer, R., Rusinowska, A., de Swart, H.: A Relation-algebraic Approach to Simple Games (submitted for publication)

[6] Coleman, J.S.: Control of Collectivities and the Power of a Collectivity to Act. In: Lieberman, Gordon, Breach, Social Choice, London (1971)

[7] Coleman, J.S.: Individual Interests and Collective Action: Selected Essays. Cambridge University Press, Cambridge (1986)

[8] Grabisch, M., Rusinowska, A.: A Model of Influence in a Social Network. Theory and Decision (to appear)

[9] Hoede, C., Bakker, R.: A Theory of Decisional Power. Journal of Mathematical Sociology 8, 309–322 (1982)

[10] Peleg, B.: Coalition Formation in Simple Games with Dominant Players. International Journal of Game Theory 10, 11–33 (1981)

[11] Peleg, B., Sudhölter, P.: Introduction to the Theory of Cooperative Games. Springer, New York (2003)

[12] Peters, H.: Game Theory: A Multi-leveled Approach. Springer, Berlin (2008)

[13] Rae, D.: Decision Rules and Individual Values in Constitutional Choice. American Political Science Review 63, 40–56 (1969)

[14] Roubens, M., Rusinowska, A., Swart, H.: Using MACBETH to Determine Utilities of Governments to Parties in Coalition Formation. European Journal of Operational Research 172, 588–603 (2006)

[15] Rusinowska, A., de Swart, H.: Generalizing and Modifying the Hoede-Bakker Index. In: de Swart, H., Orłowska, E., Schmidt, G., Roubens, M. (eds.) TARSKI 2006. LNCS (LNAI), vol. 4342, pp. 60–88. Springer, Heidelberg (2006)

[16] Rusinowska, A., de Swart, H., van der Rijt, J.W.: A New Model of Coalition Formation. Social Choice and Welfare 24, 129–154 (2005)

A Model of Internet Routing Using Semi-modules

John N. Billings and Timothy G. Griffin

Computer Laboratory, University of Cambridge
{John.Billings,Timothy.Griffin}@cl.cam.ac.uk

Abstract. Current Internet routing protocols exhibit several types of anomalies that can reduce network reliability. In order to design more robust protocols we need better formal models to capture the complexities of Internet routing. In this paper we develop an algebraic model that clarifies the distinction between *routing tables* and *forwarding tables*. We hope that this suggests new approaches to the design of routing protocols.

1 Introduction

Internet data traffic traverses a sequence of links and routers as it travels from source to destination. Routers employ *forwarding tables* to control traffic at each step. Typically, forwarding tables are constructed automatically from *routing tables*, which in turn are generated by routing protocols that dynamically discover network paths.

We attempt to clarify the distinction between forwarding and routing from a high-level perspective, ignoring implementation details. To model routing, we use an algebraic approach based on idempotent semirings (see for example [1]). For this paper, a (network-wide) routing table is simply a matrix $\mathbf{R}$ that satisfies an equation

$$\mathbf{R} = (\mathbf{A} \otimes \mathbf{R}) \oplus \mathbf{I},$$

where $\mathbf{A}$ is an adjacency matrix associated with a graph weighted over a (well-behaved) semiring S. Various algorithms, distributed or not, can be used to compute a routing table $\mathbf{R} = \mathbf{A}^*$ from the adjacency matrix. Each entry $\mathbf{R}(i, j)$ is (implicitly) associated with a set of optimal paths from node i to node j.

In Section 2 we model a forwarding table as a matrix $\mathbf{F}$ where each entry $\mathbf{F}(i, d)$ is (implicitly) associated with a set of paths from node i to destination d. Here destinations are assumed to be in a namespace disjoint from nodes. We then treat the construction of a forwarding table $\mathbf{F}$ as the process of solving an equation

$$\mathbf{F} = (\mathbf{A} \rhd \mathbf{F}) \,\square\, \mathbf{M},$$

where $\mathbf{F}$ and $\mathbf{M}$ contain entries in a semi-module $(\square, \rhd)$ over the semiring S. Entries in the *mapping table* $\mathbf{M}(i, d)$ contain metrics associated with the attachment of external destination d to infrastructure node i.

The solution $\mathbf{F} = \mathbf{R} \rhd \mathbf{M}$ tells us how to combine routing and mapping to produce forwarding. We present several semi-module constructions that model common Internet forwarding idioms such as *hot-* and *cold-potato* forwarding. Section 3 shows how

R. Berghammer et al. (Eds.): RelMiCS/AKA 2009, LNCS 5827, pp. 29–43, 2009.

mapping tables can themselves be generated from forwarding tables. This provides a model of one simple type of *route redistribution* between distinct routing protocols.

In Section 4 we discuss how this model is related to current rethinking of the Internet's addressing architecture (see for example John Day's book [2]), and to existing problems with route redistribution [3,4,5]. For completeness, Appendix A supplies definitions of semirings and semi-modules.

2   Routing versus Forwarding

With Internet technologies we can make a distiction between *routing* and *forwarding*. We will consider routing to be a function that establishes and maintains available paths within a specified routing domain. How such paths are actually used to carry traffic is for us a question of forwarding.

Of course, routing and forwarding are intimately related, and in practice the two terms are often used as if they were synonyms. Indeed, in the simplest case the distinction may seem pointless: when the forwarding function causes traffic to flow on exactly the paths provided by the routing function. However, even in this simple case the distinction must be made because of the possibility of multiple *equal cost* paths within a network. There are many possible choices for forwarding with equal cost paths, such as randomly choosing a path or dynamically balancing load between paths.

In this section we model a network's *infrastructure* as a directed graph $G = (V, E)$. Given a pair of nodes $i, j \in V$, routing computes a set of paths in G that can be used to transit data from i to j. We model routing with a $V \times V$ *routing matrix* $\mathbf{R}$. Entry $\mathbf{R}(i, j)$ in fact corresponds to the minimal-cost path *weight* from i to j, although under certain assumptions (see later) it is straightforward to recover the associated paths.

In addition, we suppose that there is a set of *external destinations* D that are independent of the network. Destinations can be *directly attached* to any number of nodes in the network. We model the attachment information using a $V \times D$ *mapping matrix* $\mathbf{M}$. Forwarding then consists of finding minimal-cost paths from nodes $i \in V$ to destinations $d \in D$. We model forwarding using a $V \times D$ *forwarding matrix* $\mathbf{F}$. We shall see that there are several different ways to combine routing and mapping matrices to produce a forwarding matrix, with each such method potentially leading to a different set of forwarding paths; the examples in this section all share the same routing matrix, yet have distinct forwarding paths.

2.1   Algebraic Routing

In this section we provide a basic overview of algebraic routing with semirings. Let $S = (S, \oplus, \otimes)$ be an idempotent semiring (Appendix A.1). Associate arcs with elements of the semiring using a weight function $w \in E \to S$. Let $\mathbf{A}$ be the $V \times V$ adjacency matrix induced by w. Denote the set of all paths in G from node i to node j by $P(G, i, j)$. Given a path $\mu = \langle u_0, u_1, \cdots, u_n \rangle \in P(G, i, j)$ with $u_0 = i$ and $u_n = j$, define the weight of μ as $w(\mu) = w(u_0, u_1) \otimes w(u_1, u_2) \otimes \cdots w(u_{n-1}, u_n)$. For paths $\mu, \nu \in P(G, i, j)$, summarise their weights as $w(\mu) \oplus w(\nu)$. Define the *shortest-path weight* from i to j as

$$\delta(i, j) = \sum_{p \in P(G, i, j)}^{\oplus} w(p).$$

We seek to find a $V \times V$-matrix $\mathbf{R}$ satisfying $\mathbf{R}(i, j) = \delta(i, j)$. We term this matrix a *routing solution* because it models the shortest-path weights across a network's infrastructure. How might we compute the value of this matrix? It is straightforward to show that such a matrix $\mathbf{R}$ also satisfies the *routing equation*

$$\mathbf{R} = (\mathbf{A} \otimes \mathbf{R}) \oplus \mathbf{I}. \tag{1}$$

Now, define the *closure* of $\mathbf{A}$ as $\mathbf{A}^* = \mathbf{I} \oplus \mathbf{A} \oplus \mathbf{A}^2 \oplus \cdots$. It is well-known that if this closure exits, then it satisfies equation 1 (see the classic reference [6], or the recent survey of the area [1]). Various sufficient conditions can be assumed to hold on the semiring S which imply that $\mathbf{A}^*$ will always exist (and so the set of adjacency matrices becomes a Kleene algebra), but we will not explore these conditions here.

Note that we use the standard notation of identifying the operators from the semiring S with those from the same semiring lifted to operate over (square) matrices with elements from S. The additive operator for the lifted semiring is defined as

$$(\mathbf{X} \oplus \mathbf{Y})(i, j) = \mathbf{X}(i, j) \oplus \mathbf{Y}(i, j),$$

whilst the multiplicative operator is defined as

$$(\mathbf{X} \otimes \mathbf{Y})(i, j) = \sum_{k \in V}^{\oplus} \mathbf{X}(i, k) \otimes \mathbf{Y}(k, j).$$

Hence we see that the latter operator in fact uses both the underlying $\oplus$ and $\otimes$ operators from S. This distinction will become more significant when we consider lifting certain semi-modules to operate over matrices.

The example in Figure 1 illustrates the algebraic approach. Figure 1(a) presents a simple five node graph with integer labels and Figure 1(b) shows the associated adjacency matrix. We assume that arc weights are symmetric. We wish to compute *shortest-distances* between each pair of nodes and therefore we compute the closure using the semiring MinPlus $= (\mathbb{N}^\infty, \min, +)$, where $\mathbb{N}^\infty = \mathbb{N} \cup \{\infty\}$. The resulting matrix is given in Figure 1(c). It is straight-forward to recover the corresponding paths because MinPlus is *selective*. That is, for all x, $y \in S$ we have $x \oplus y \in \{x, y\}$. Hence the computed weights actually correspond to the weights of individual paths. The bold arrows in Figure 1(a) denote the shortest-paths tree rooted at node 1; the corresponding path weights are given in the first row of the matrix in Figure 1(c).

2.2   Importing External Destinations

As before, suppose that our network is represented by the graph $G = (V, E)$, labelled with elements from the semiring S. Let the external nodes be chosen from some set D, satisfying $V \cap D = \emptyset$. Attach external nodes to G using the *attachment edges* $E' \subseteq V \times D$. In the simplest case, the edges in E' have weights from S, although in

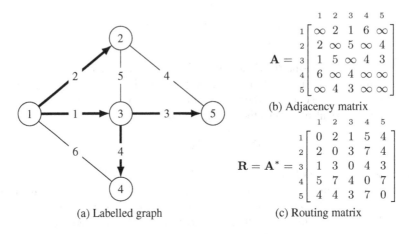

(a) Labelled graph

$$\mathbf{A} = \begin{array}{c} \\ 1 \\ 2 \\ 3 \\ 4 \\ 5 \end{array} \begin{array}{ccccc} 1 & 2 & 3 & 4 & 5 \\ \left[\begin{array}{ccccc} \infty & 2 & 1 & 6 & \infty \\ 2 & \infty & 5 & \infty & 4 \\ 1 & 5 & \infty & 4 & 3 \\ 6 & \infty & 4 & \infty & \infty \\ \infty & 4 & 3 & \infty & \infty \end{array}\right] \end{array}$$

(b) Adjacency matrix

$$\mathbf{R} = \mathbf{A}^* = \begin{array}{c} \\ 1 \\ 2 \\ 3 \\ 4 \\ 5 \end{array} \begin{array}{ccccc} 1 & 2 & 3 & 4 & 5 \\ \left[\begin{array}{ccccc} 0 & 2 & 1 & 5 & 4 \\ 2 & 0 & 3 & 7 & 4 \\ 1 & 3 & 0 & 4 & 3 \\ 5 & 7 & 4 & 0 & 7 \\ 4 & 4 & 3 & 7 & 0 \end{array}\right] \end{array}$$

(c) Routing matrix

Fig. 1. Algebraic routing example using the MinPlus semiring

the next section we show how to relax this assumption. Let the $V \times D$ *mapping matrix* $\mathbf{M}$ represent the attachment edges.

We now wish to compute the $V \times D$ matrix $\mathbf{F}$ of shortest-path weights from nodes in V to nodes in D. We term $\mathbf{F}$ a *forwarding solution* because it comprises the information required to reach destinations, instead of other infrastructure nodes. We compute $\mathbf{F}$ by post-multiplying the routing solution $\mathbf{R}$ by the mapping matrix $\mathbf{M}$. That is, for $i \in V$ and $d \in D$, we have

$$\mathbf{F}(i, d) = (\mathbf{R} \otimes \mathbf{M})(i, d) = \sum_{k \in V}^{\oplus} \mathbf{R}(i, k) \otimes \mathbf{M}(k, d) = \sum_{k \in V}^{\oplus} \delta(i, k) \otimes \mathbf{M}(k, d).$$

Hence we see that $\mathbf{F}(i, d)$ corresponds to the shortest total path length from i to d. In other words, $\mathbf{F}$ solves the *forwarding equation*

$$\mathbf{F} = (\mathbf{A} \otimes \mathbf{F}) \oplus \mathbf{M}. \tag{2}$$

Note that we are able to change the value of $\mathbf{M}$ and recompute $\mathbf{F}$ without recomputing $\mathbf{R}$. From an Internet routing perspective this is an important property; if the external information is dynamically computed (by another routing protocol, for example) then it may frequently change, and in such instances it is desirable to avoid recomputing routing solutions.

We illustrate this model of forwarding in Figure 2. The labelled graph of Figure 2(a) is based upon that in Figure 1(a), with the addition of two external nodes: d_1 and d_2. The adjacency matrix $\mathbf{A}$ remains as before, whilst the mapping matrix $\mathbf{M}$, given in Figure 2(b), contains the attachment information for d_1 and d_2. The forwarding solution $\mathbf{F}$ that results from multiplying $\mathbf{R}$ by $\mathbf{M}$ is given in Figure 2(c). Again, it is easy to verify that the elements of $\mathbf{F}$ do indeed correspond to the weights of the shortest paths from nodes in V to nodes in D.

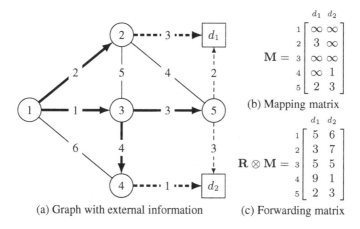

(a) Graph with external information

(b) Mapping matrix

(c) Forwarding matrix

Fig. 2. Example of combining routing and mapping to create *forwarding*

2.3 A General Import Model Using Semi-modules

Within Internet routing, it is common for the entries in routing and forwarding tables to have distinct types, and for these types to be associated with distinct order relations. We therefore generalise the import model of the previous section to allow this possibility. In particular, we show how to solve this problem using algebraic structures known as *semi-modules* (Appendix A.2).

Assume that we are using the semiring $S = (S, \oplus, \otimes)$ and suppose that we wish to construct forwarding matrices with elements from the idempotent, commutative semi-group $N = (N, \square)$. Furthermore, suppose that the mapping matrix $\mathbf{M}$ contains entries over N. In order to compute forwarding entries, it is necessary to combine routing entries with mapping entries, as before. However, we can no longer use the multiplicative operator from S because the mapping entries are of a different type. Therefore we introduce an operator $\triangleright \in (S \times N) \to N$ for this purpose. We can now construct forwarding entries as

$$\mathbf{F}(i, d) = (\mathbf{R} \triangleright \mathbf{M})(i, d) = \sum_{k \in V}^{\square} \mathbf{R}(i, k) \triangleright \mathbf{M}(k, d). \tag{3}$$

It is also possible to equationally characterize the resulting forwarding entries, as before. Assume that $\mathbf{R}$ is a routing solution i.e. it satisfies Equation 1. Then, providing that the algebraic structure $N = (N, \square, \triangleright)$ is a semi-module, $\mathbf{F} = \mathbf{R} \triangleright \mathbf{M}$ is a solution to the forwarding equation

$$\mathbf{F} = (\mathbf{A} \triangleright \mathbf{F}) \,\square\, \mathbf{M}.$$

In other words, we can solve for $\mathbf{F}$ with $\mathbf{F} = \mathbf{A}^* \triangleright \mathbf{M}$. Significantly, we are able to use semi-modules to model the mapping information whilst still retaining a semiring model of routing.

We now develop two important semi-module constructions that model the most common manner in which routing and mapping are combined: the *hot-potato* and *cold-potato* semi-modules. First define an *egress* node for a destination d as a node k within the routing domain that is directly attached to d. Hot-potato forwarding to d first selects paths to the closest egress nodes for d and then breaks ties using the mapping information. In contrast, cold-potato forwarding first selects paths to the egress nodes for d with the most preferred mapping values, and then breaks ties using the routing distances.

We now formally define the hot-potato semi-module. Let $S = (S, \oplus_S, \otimes_S)$ be an idempotent semiring with $(S, \oplus_S)$ selective and let $T = (T, \oplus_T)$ be a monoid. The hot-potato semi-module over S is defined as

$$\text{Hot}(S, T) = ((S \times T) \cup \{\infty\}, \vec{\oplus}, \rhd_{\text{fst}}),$$

where $s_1 \rhd_{\text{fst}} (s, t) = (s_1 \otimes_S s, t)$ and $s_1 \rhd_{\text{fst}} \infty = \infty$. The *left* lexicographic product semigroup $((S \times T) \cup \{\infty\}, \vec{\oplus})$ is defined in Appendix A.3. In common with semirings, we can lift semi-modules to operate over (non-square) matrices. When lifting the hot-potato semi-module, we rename the multiplicative operator from $\rhd_{\text{fst}}$ to $\rhd_{\text{hp}}$. This is because the lifted multiplicative operator no longer simply applies its left argument to the first component of its right argument. In fact, we shall shortly see that the cold-potato semi-module uses the same underlying multiplicative operator but with a different additive operator, and therefore has a different lifted multiplicative operator.

The behaviour of the hot-potato semi-module can be algebraically characterised as follows. Suppose that for all $j \in V$ and $d \in D$, $\mathbf{M}(j, d) \in \{(1_S, t), \infty_T\}$ where 1_S is the multiplicative identity for S and t is some element of T. Then from Equation 3 it is easy to check that

$$(\mathbf{R} \rhd_{\text{hp}} \mathbf{M})(i, d) = \sum_{\substack{j \in V \\ M(j, d) = (1_S, t)}}^{\vec{\oplus}} (\mathbf{R}(i, j), t).$$

That is, as desired, the mapping metric is simply used to tie-break over otherwise minimal-weight paths to the edge of the routing domain.

We illustrate the hot-potato model of forwarding in Figure 3. This example uses the semi-module $\text{Hot}(\text{MinPlus}, \text{Min})$, where $\text{Min} = (\mathbb{N}^\infty, \min)$. The graph of Figure 3(a) is identical to that of Figure 2(a), but now the attachment arcs of d_1 and d_2 are weighted with elements of the hot-potato semi-module. The associated mapping matrix is given in Figure 3(b), whilst the resulting forwarding table is shown in Figure 3(c). Note that 0 is the multiplicative identity of the $(\min, +)$ semiring. Comparing this example to Figure 2, we see that node 1 reaches d_2 via egress node 5 instead of node 4. This is because the mapping information is only used for tie-breaking, instead of being directly combined with the routing distance. Also, in this particular example, it is never the case that there are multiple paths of minimum cost to egress nodes, and therefore no tie-breaking is performed by the mapping information.

Turning to cold-potato forwarding, the associated semi-module again combines routing and attachment information using the lexicographic product, but with priority now given to the attachment component. As before, let $S = (S, \oplus_S, \otimes_S)$ be a semiring and $T = (T, \oplus_T)$ be a monoid, but now with T idempotent and selective. The cold-potato semi-module over S is defined as

$$\text{Cold}(S, T) = ((S \times T) \cup \{\infty\}, \overleftarrow{\oplus}, \rhd_{\text{fst}}).$$

Note that the *right* lexicographic product semigroup $(S \times T, \overleftarrow{\oplus})$ is defined in Appendix A.3. Again, when lifting the cold-potato semi-module to operate over matrices we rename the multiplicative operator from $\triangleright_{\text{fst}}$ to $\triangleright_{\text{cp}}$.

Figure 4 illustrates the cold-potato model of forwarding. This example uses the cold-potato semi-module Cold(MinPlus, Min), but is otherwise identical to Figure 3. It is easy to verify that priority is now given to the mapping information when selecting egress nodes.

Within Internet routing, hot-potato forwarding corresponds to choosing the closest egress point from a given routing domain. This is the default behaviour for the Border Gateway Protocol (BGP) routing protocol [7] because it tends to minimise resource usage for outbound traffic within the domain. In contrast, cold-potato forwarding allows the mapping facility to select egress nodes, and hence can lead to longer paths being chosen within the domain. As a result, cold-potato forwarding is less commonly observed

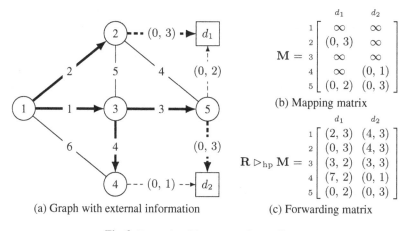

(a) Graph with external information   (c) Forwarding matrix

Fig. 3. Example of *hot-potato* forwarding

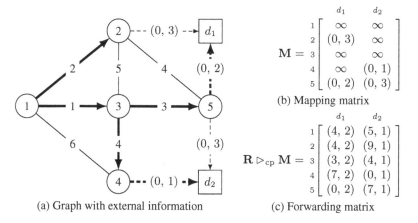

(a) Graph with external information   (c) Forwarding matrix

Fig. 4. Example of *cold-potato* forwarding

in general on the Internet. However, one specific use is within client-provider peering relations in order to minimise the use of the client's network resources for inbound traffic (at the possible expense of increased resource usage on the provider's network).

2.4   Idealized OSPF: An Example of Combined Mappings

We now present a highly-idealized account that attempts to tease out an algebraic description of the construction of forwarding tables in the OSPF routing protocol [8]. The specification of this protocol [9] runs for 244 pages and is primarily focused on implementation details. For simplicity we ignore OSPF areas.

Destinations are attached in three different ways in OSPF. Type 0 (our terminology) destinations are directly attached to a node, while Type 1 and Type 2 destinations (terminology of [9]) represent two ways of attaching external destinations. These may be statically-configured or learned via other routing protocols. The OSPF specification defines the relative preference for destination types to be used in constructing a forwarding table: Type 0 are preferred to Type 1, and Type 1 are preferred to Type 2.

In addition, Type 2 destinations are associated with a metric that is to be inspected before the internal routing metric In other words, cold-potato forwarding is used for Type 2 destinations. We generalize OSPF and assume that Type 2 metrics come from a commutative, idempotent monoid. $U = (U, \oplus_U)$. We use the network in Figure 5(a) as an ongoing example; here we have $U = (\mathbb{N}^\infty, \max)$, and we think of this as a *bandwidth* metric (note that here $\infty_U = 0$).

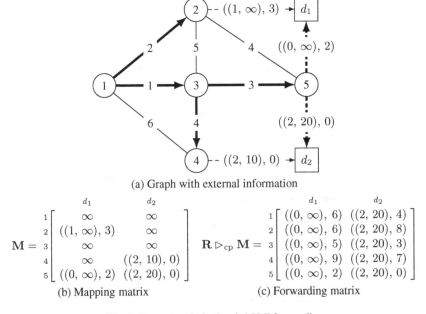

(a) Graph with external information

$$\mathbf{M} = \begin{array}{c} \\ 1 \\ 2 \\ 3 \\ 4 \\ 5 \end{array} \begin{array}{c} d_1 \\ \left[\begin{array}{c} \infty \\ ((1, \infty), 3) \\ \infty \\ \infty \\ ((0, \infty), 2) \end{array} \right. \end{array} \begin{array}{c} d_2 \\ \left. \begin{array}{c} \infty \\ \infty \\ \infty \\ ((2, 10), 0) \\ ((2, 20), 0) \end{array} \right] \end{array}$$

(b) Mapping matrix

$$\mathbf{R} \rhd_{cp} \mathbf{M} = \begin{array}{c} \\ 1 \\ 2 \\ 3 \\ 4 \\ 5 \end{array} \begin{array}{c} d_1 \\ \left[\begin{array}{c} ((0, \infty), 6) \\ ((0, \infty), 6) \\ ((0, \infty), 5) \\ ((0, \infty), 9) \\ ((0, \infty), 2) \end{array} \right. \end{array} \begin{array}{c} d_2 \\ \left. \begin{array}{c} ((2, 20), 4) \\ ((2, 20), 8) \\ ((2, 20), 3) \\ ((2, 20), 7) \\ ((2, 20), 0) \end{array} \right] \end{array}$$

(c) Forwarding matrix

Fig. 5. Example of *idealized-OSPF* forwarding

We use the following set into which we embed each destination type,

$$W = ((\{0, 1, 2\} \times U) \times \mathbb{N}^\infty) \cup \{\infty\}.$$

Each destination type is embedded into W as follows:

Type	Metric	Embedding
0	$m \neq \infty$	$((0, \infty_U), m)$
1	$m \neq \infty$	$((1, \infty_U), m)$
2	$u \neq \infty_U$	$((2, u), 0)$

The elements of W are then ordered using the (right) lexicographic product (see Appendix A.3),

$$\vec{\oplus} = (\{1, 2, 3\}, \min) \; \vec{\times} \; (U, \oplus_U) \; \vec{\times} \; (\mathbb{N}^\infty, \min).$$

Hence the order amongst metrics of the same destination type remains unchanged, whilst the order between different destination types respects the ordering defined within the OSPF specification.

Assume we start with one mapping matrix for each type of destination (how these matrices might actually be produced is ignored),

	$\mathbf{M_0}$			$\mathbf{M_1}$			$\mathbf{M_2}$	
	d_1	d_2		d_1	d_2		d_1	d_2
1	∞	∞	1	∞	∞	1	∞	∞
2	∞	∞	2	$((1, \infty), 3)$	∞	2	$((2, 40), 0)$	∞
3	∞	∞	3	∞	∞	3	∞	∞
4	∞	∞	4	∞	∞	4	∞	$((2, 10), 0)$
5	$((0, \infty), 2)$	∞	5	$((1, \infty), 17)$	∞	5	$((2, 10), 0)$	$((2, 20), 0)$

We then construct a combined mapping matrix $\mathbf{M}$ by summing the individual matrices as $\mathbf{M} = \mathbf{M_0} \oplus \mathbf{M_1} \oplus \mathbf{M_2}$. The resulting mapping matrix is shown in Figure 5(b).

We define the OSPF semi-module as

$$\text{OSPF}(U) = (W, \vec{\oplus}, \rhd_{\text{snd}})$$

where

$$m \rhd_{\text{snd}} ((l, u), m') = ((l, u), m + m')$$
$$m \rhd_{\text{snd}} \infty = \infty.$$

The OSPF semi-module is a variant of the cold-potato semi-module; here, instead of combining routing data with the first component of the mapping information and using the right lexicographic order, we instead combine it with the second component and use the left lexicographic order. Hence we refer to the lifted multiplicative operator as $\rhd_{\text{cp}}$.

Figure 5(c) illustrates the resulting forwarding matrix, $\mathbf{F} = \mathbf{R} \rhd_{\text{cp}} \mathbf{M}$. For d_1, we see that the Type 0 route is given priority over the Type 1 route. In contrast, there are two Type 2 routes for d_2, and hence the bandwidth component is used as a tie-breaker.

3 Simple Route Redistribution

In this section we show how to allow forwarding between multiple domains by generalizing the import model from Section 2 (here, we limit ourselves to modelling the case where there are two routing domains, although it is possible to generalize to a greater number). In particular, we show how the forwarding matrix from one domain can be used within the mapping matrix of another. This models *redistribution*, where a routing solution from one routing protocol is used within another. Additionally, we demonstrate that it is possible for each routing/forwarding domain to use a different semiring/semi-module pair.

Begin by assuming that there are two routing domains, $G_1 = (V_1, E_1)$ and $G_2 = (V_2, E_2)$. Also, assume that there is a set of destinations, D, with V_1, V_2 and D pairwise disjoint. Let G_1 be connected to G_2 with the attachment arcs $E_{1,2} \subseteq V_1 \times V_2$, represented as the $V_1 \times V_2$ *bridging matrix* $\mathbf{B}_{1,2}$. Similarly, let G_2 be connected to D with the attachment arcs $E_{2,d} \subseteq V_2 \times D$, represented as the $V_2 \times D$ attachment matrix $\mathbf{M}_2$. Let $\mathbf{F}_2$ be the forwarding matrix for G_2. We demonstrate how to construct a forwarding matrix from V_1 to D.

We shall use Figure 6 as a running example. Figure 6(a) illustrates two graphs, G_1 and G_2. The second graph, G_2, is directly connected to destinations d_1 and d_2 and therefore we are able to compute the forwarding matrix for G_2 using the method from Section 2. We model the routing in G_2 using the bandwidth semiring $\mathrm{MaxMin} = (\mathbb{N}^\infty, \max, \min)$ and the forwarding using the cold-potato semi-module $\mathrm{Cold}(\mathrm{MaxMin}, \mathrm{Min})$, where $\mathrm{Min} = (\mathbb{N}^\infty, \min)$. The mapping matrix is given in Figure 6(b), whilst the routing and forwarding matrices are given in Figure 6(c) and Figure 6(d) respectively.

In order to compute a forwarding matrix from V_1 to D, we must first construct a mapping matrix $\mathbf{M}_1$ from V_1 to D by combining the forwarding matrix $\mathbf{F}_2$ from G_2 with the bridging matrix $\mathbf{B}_{1,2}$. Let the forwarding in G_2 be modelled using the semi-module $N_2 = (N_2, \square_2, \rhd_2)$, and that the bridging matrix is modelled using the semigroup $(N_1, \square_1)$. We construct a *right* semi-module $(N_1, \square_1, \lhd_1)$ over N_2 i.e. with $\lhd_1 \in (N_2 \times N_1) \to N_1$. Then we compute the mapping matrix from G_1 as $\mathbf{M}_1 = \mathbf{B}_{1,2} \lhd_1 \mathbf{F}_2$.

Returning to Figure 6, the bridging matrix $\mathbf{B}_{1,2}$ is illustrated in Figure 6(e). We combine the forwarding matrix $\mathbf{F}_2$ with $\mathbf{B}_{1,2}$ using the right version of the semi-module

$$\mathrm{Hot}(\mathrm{MinPlus}, \mathrm{Cold}(\mathrm{MaxMin}, \mathrm{Min})).$$

The resulting mapping matrix $\mathbf{M}_1$ is illustrated in Figure 6(f).

Finally, we must combine the mapping matrix $\mathbf{M}_1$ with the routing solution $\mathbf{R}_1$. Suppose that $\mathbf{R}_1$ has been computed using the semiring S. Then we construct a *left* semi-module $(N_1, \square_1, \rhd_1)$ over S. Compute the forwarding matrix for G_1 as

$$\mathbf{F}_1 = \mathbf{R}_1 \rhd_1 \mathbf{M}_1 = \mathbf{R}_1 \rhd_1 (\mathbf{B}_{1,2} \lhd_1 \mathbf{F}_2)$$

Hence we see that we have in fact used a pair of semi-modules with identical additive components: a left semi-module $(N_1, \square_1, \lhd_1)$ over N_2 and a right semi-module $(N_1, \square_1, \rhd_1)$ over S.

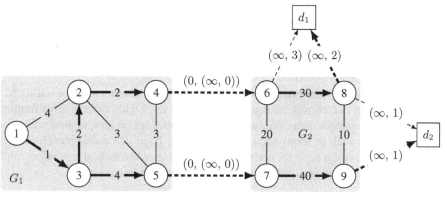

(a) Multiple graphs with external information

$$\mathbf{M}_2 = \begin{array}{c} \\ 6 \\ 7 \\ 8 \\ 9 \end{array} \begin{array}{c} d_1 \quad\quad d_2 \\ \begin{bmatrix} (\infty, 3) & \infty \\ \infty & \infty \\ (\infty, 2) & (\infty, 1) \\ \infty & (\infty, 1) \end{bmatrix} \end{array}$$

(b) G_2 mapping matrix

$$\mathbf{R}_2 = \begin{array}{c} \\ 6 \\ 7 \\ 8 \\ 9 \end{array} \begin{array}{c} 6 \quad 7 \quad 8 \quad 9 \\ \begin{bmatrix} \infty & 20 & 30 & 20 \\ 20 & \infty & 20 & 40 \\ 30 & 20 & \infty & 20 \\ 20 & 40 & 20 & \infty \end{bmatrix} \end{array}$$

(c) G_2 routing matrix

$$\mathbf{F}_2 = \mathbf{R}_2 \triangleright_{cp} \mathbf{M}_2$$
$$= \begin{array}{c} \\ 6 \\ 7 \\ 8 \\ 9 \end{array} \begin{array}{c} d_1 \quad\quad d_2 \\ \begin{bmatrix} (30, 2) & (30, 1) \\ (20, 2) & (40, 1) \\ (\infty, 2) & (\infty, 1) \\ (20, 2) & (\infty, 1) \end{bmatrix} \end{array}$$

(d) G_2 forwarding matrix

$$\mathbf{M}_1 = \mathbf{B}_{1,2} \triangleleft_{hp} \mathbf{F}_2$$

$$\mathbf{B}_{1,2} = \begin{array}{c} \\ 1 \\ 2 \\ 3 \\ 4 \\ 5 \end{array} \begin{array}{c} 6 \quad\quad 7 \quad\quad 8 \quad 9 \\ \begin{bmatrix} \infty & \infty & \infty & \infty \\ \infty & \infty & \infty & \infty \\ \infty & \infty & \infty & \infty \\ (0, (\infty, 0)) & \infty & \infty & \infty \\ \infty & (0, (\infty, 0)) & \infty & \infty \end{bmatrix} \end{array}$$

(e) G_1 to G_2 bridging matrix

$$= \begin{array}{c} \\ 1 \\ 2 \\ 3 \\ 4 \\ 5 \end{array} \begin{array}{c} d_1 \quad\quad\quad d_2 \\ \begin{bmatrix} \infty & \infty \\ \infty & \infty \\ \infty & \infty \\ (0, (30, 2)) & (0, (30, 1)) \\ (0, (20, 2)) & (0, (40, 1)) \end{bmatrix} \end{array}$$

(f) G_1 mapping matrix

$$\mathbf{F}_1 = \mathbf{R}_1 \triangleright_{hp} \mathbf{M}_1$$

$$\mathbf{R}_1 = \begin{array}{c} \\ 1 \\ 2 \\ 3 \\ 4 \\ 5 \end{array} \begin{array}{c} 1 \quad 2 \quad 3 \quad 4 \quad 5 \\ \begin{bmatrix} 0 & 3 & 1 & 5 & 5 \\ 3 & 0 & 2 & 2 & 3 \\ 1 & 2 & 0 & 4 & 4 \\ 5 & 2 & 4 & 0 & 3 \\ 5 & 3 & 4 & 3 & 0 \end{bmatrix} \end{array}$$

(g) G_1 routing matrix

$$= \begin{array}{c} \\ 1 \\ 2 \\ 3 \\ 4 \\ 5 \end{array} \begin{array}{c} d_1 \quad\quad\quad d_2 \\ \begin{bmatrix} (5, (30, 2)) & (5, (40, 1)) \\ (2, (30, 2)) & (2, (30, 1)) \\ (4, (30, 2)) & (4, (40, 1)) \\ (0, (30, 2)) & (0, (30, 1)) \\ (0, (20, 2)) & (0, (40, 1)) \end{bmatrix} \end{array}$$

(h) G_1 forwarding matrix

Fig. 6. Example of *simple route redistribution*

.

Completing the example of Figure 6, the routing matrix for G_1 is computed using the semiring MinPlus. The resulting matrix $\mathbf{R}_1$ is shown in Figure 6(g). We combine $\mathbf{R}_1$ with the mapping matrix $\mathbf{M}_1$ using the left version of the semi-module

$$\text{Hot}(\text{MinPlus}, \text{Cold}(\text{MaxMin}, \text{Min})).$$

The resulting forwarding table $\mathbf{F}_1$ is given in Figure 6(h). The bold arrows in Figure 6(a) denote the forwarding paths from node 1 to destinations d_1 and d_2. Note that the two egress nodes (4 and 5) from G_1 are at identical distances from 1, and therefore the bandwidth components from G_2 are used as tie-breakers. This results in a different egress point for each destination.

This model is significant because it is the first algebraic account of route redistribution – an area that is currently treated as a 'black art' even within the Internet routing community (for example, there are only informal guidelines on how to avoid redistribution anomalies such as loops and oscillations). We hope that our approach can be generalized to provide a basis for understanding existing redistribution techniques, and also for developing new approaches to protocol inter-operations.

4   Related Work and Open Problems

4.1   Locators and Identifiers

Our term *mapping table* has been borrowed (slightly loosely) from recent work attempting to differentiate between infrastructure addresses (called locators) and end-user addresses (called identifiers), as with the Locator/ID Separation Protocol (LISP) [10]. This effort has been motivated by a perceived need to reduce the size of routing and forwarding tables in the Internet's backbone (the world of inter-domain routing [11]).

Restated in our abstract setting, the essential problem that LISP is attempting to solve is that a mapping table $\mathbf{M}$ may be many orders of magnitude larger than the routing table $\mathbf{R}$, leading to a very large forwarding table $\mathbf{F} = \mathbf{R} \rhd \mathbf{M}$. Since there is no separation between mapping and routing today, such table growth is in fact becoming a real operational problem. (Note that we are using the terms *routing table* and *forwarding table* in a rather unconventional, network-wide, sense. In a distributed setting, the entries $\mathbf{F}(i, \_)$ make up the forwarding table at node i.)

LISP proposes that forwarding tables $\mathbf{F}$ be only partially constructed using an on-demand approach – an entry $\mathbf{F}(i, d)$ is not constructed until router i receives traffic destined for d. This in turn requires some type of distributed mapping service, for which there are several proposals currently under consideration.

In this paper we have used the separation of locators and identifiers to provide an algebraic view of the distinction between routing and forwarding tables. This model suggests that the Locator/ID split might be usefully applied to intra-domain routing.

4.2   Route Redistribution

Examples of somewhat *ad hoc* mechanisms and techniques added to routing are *route redistribution* for distributing routes between distinct routing protocols (as already discussed), and *administrative distance* (discussed below). Recent research has documented their widespread use and illustrated routing anomalies that can arise as a result [3,4,5]. From our point of view, that work represents a *bottom-up* approach that starts with the complex implementation details of current legacy software. We hope that we have initiated a complementary, *top-down*, approach. The algebraic model has the advantage of making clear *what problem* is being solved, as distinct from *what algorithm* is being implemented to solve a problem. We assert that the Internet routing literature is severely hobbled by the way that these distinct issues are often hopelessly tangled together.

Our model suggests that new protocols should be designed with a clear distinction between routing, mapping, and forwarding. Furthermore, mechanisms for constructing mapping tables and forwarding tables should be elevated from proprietary implementations to first-class status and standardized.

4.3 Loss of Distributivity

We may attempt to solve the equation $\mathbf{F} = (\mathbf{A} \rhd \mathbf{F}) \square \mathbf{M}$ using an iterative method,

$$\mathbf{F}^{[0]} = \mathbf{M},$$
$$\mathbf{F}^{[k+1]} = (\mathbf{A} \rhd \mathbf{F}^{[k]}) \square \mathbf{M}.$$

When $\mathbf{A}^*$ exists, and $(N, \square, \rhd)$ is a semi-module over S, then it is not too hard to see that $\lim_{k \to \infty} \mathbf{F}^{[k]} = \mathbf{A}^* \rhd \mathbf{M}$. However, when modeling current Internet routing protocols several problems may be encountered.

The first is that S may not in fact be a semiring due to violations of the distributivity laws. The situation may not be as hopeless at it might seem. Recent research [12] has pointed out that distributivity that is so essential to semiring theory may have to be abandoned to model some types of Internet routing. Even without distributivity, it may be possible to use an iterative method to arrive at a (locally optimal) routing solution [13].

On the other hand, it may be that S is a well-behaved semiring, but that the structure $(N, \square, \rhd)$ is not a semi-module over S. Again, this seems to arise with violations of semi-module distributivity. We conjecture that it would be straightforward to extend the results of [13] to iterations of $\mathbf{F}^{[k]}$. That is, that if (1) the natural order is a total order, (2) $m < \infty_N \implies m < a \rhd m$ and (3) only simple paths are considered, then the iterative method will converge to a (locally optimal) solution to the equation $\mathbf{F} = (\mathbf{A} \rhd \mathbf{F}) \square \mathbf{M}$.

The lack of a global notion of optimality may in fact be perfectly reasonable in the wide Internet where competing Internet Service Providers need to share routes but their local commercial relationships prevent agreement as to what represents a *best* route. However, we suspect that a less obvious source of this type of routing may have evolved with administrative distance.

4.4 Administrative Distance

Administrative distance [3,4,5] is used for determining which entries are placed in a forwarding table when distinct protocols running on the same router have routes to common destinations. Algebraic modeling of administrative distance remains open for a careful formal treatment.

One fundamental problem seems to be that capturing the way routers currently implement administrative distance leads to algebraic structures that are not distributive. In fact, we conjecture that the technique is inherently non-distributive in some sense.

If this is the case, then we can model current routing as implementing an iterative method attempting to find a fixed-point over a non-distributive structure. Perhaps the way forward is again to elevate this procedure from proprietary code to the level of a first class protocol and design constraints sufficient to guarantee convergence.

Acknowledgements

We are grateful for the support of the Engineering and Physical Sciences Research Council (EPSRC, grant EP/F002718/1) and a grant from Cisco Systems. We also thank M. Abdul Alim, Arthur Amorim, Alexander Gurney, Vilius Naudžiūnas, Philip Taylor, and the conference reviewers for their helpful feedback.

References

1. Gondran, M., Minoux, M.: Graphs, Dioids, and Semirings: New Models and Algorithms. Springer, Heidelberg (2008)
2. Day, J.: Patterns in Network Architectures: A return to fundamentals. Prentice Hall, Englewood Cliffs (2008)
3. Le, F., Xie, G., Zhang, H.: Understanding route redistribution. In: Proc. Inter. Conf. on Network Protocols (2007)
4. Le, F., Xie, G., Pei, D., Wang, J., Zhang, H.: Shedding light on the glue logic of the internet routing architecture. In: Proc. ACM SIGCOMM (2008)
5. Le, F., Xie, G., Zhang, H.: Instability free routing: Beyond one protocol instance. In: Proc. ACM CoNext (December 2008)
6. Carré, B.: Graphs and Networks. Oxford University Press, Oxford (1979)
7. Rekhter, Y., Li, T.: A Border Gateway Protocol. RFC 1771 (BGP version 4) (March 1995)
8. Moy, J.: OSPF: Anatomy of an Internet Routing Protocol. Addison-Wesley, Reading (1998)
9. Moy, J.: OSPF version 2. RFC 2328 (1998)
10. Farinacci, D., Fuller, V., Meyer, D., Lewis, D.: Locator/ID separation protocol (LISP). draft-ietf-lisp-02.txt. Work In Progress (2009)
11. Halabi, S., McPherson, D.: Internet Routing Architectures, 2nd edn. Cisco Press (2001)
12. Sobrinho, J.L.: An algebraic theory of dynamic network routing. IEEE/ACM Transactions on Networking 13(5), 1160–1173 (2005)
13. Griffin, T.G., Gurney, A.J.T.: Increasing bisemigroups and algebraic routing. In: 10th International Conference on Relational Methods in Computer Science (RelMiCS10) (April 2008)
14. Gurney, A.J.T., Griffin, T.G.: Lexicographic products in metarouting. In: Proc. Inter. Conf. on Network Protocols (October 2007)

A   Basic Definitions

Our definitions of semirings and semi-modules are fairly standard, taken from [1]. The definitions of lexicographic operations are from [14].

A.1   Semirings

A semiring is a structure $S = (S, \oplus, \otimes)$ where $(S, \oplus)$ is a commutative semigroup, $(S, \otimes)$ is semigroup, and the following conditions hold: (1) $\otimes$ distributes over $\oplus$,

$$x \otimes (y \oplus z) = (x \otimes y) \oplus (x \otimes z), \quad (y \oplus z) \otimes x = (y \otimes x) \oplus (z \otimes x),$$

(2) there exists an identity for $\oplus$, $0_S \in S$, and an identity for $\otimes$, $1_S \in S$, and (3) it is assumed that 0_S is an annihilator for $\otimes$ i.e. $0_S \otimes x = x \otimes 0_S = 0_S$.

For routing, we are normally working with idempotent semirings where $x \oplus x = x$. In this case $(S, \oplus)$ is a semi-lattice, and we use one of the natural orders – $x \leq_\oplus^L y \equiv x = x \oplus y$ and $x \leq_\oplus^R y \equiv y = x \oplus y$. For routing, we will stick with the order $\leq_\oplus^L$ since it corresponds well with the notion of least cost paths. For this reason we use ∞_S rather than 0_S, since $x \leq_\oplus^L \infty_S$ for all $x \in S$.

A.2   Semi-modules

Let $S = (S, \oplus, \otimes)$ be a semiring. A (left) semi-module over S is a structure $N = (N, \square, \triangleright)$, where $(N, \square)$ is a commutative semigroup and $\triangleright$ is a function $\triangleright \in (S \times N) \to N$ that satisfies the following distributivity laws,

$$x \triangleright (m \square n) = (x \triangleright m) \square (x \triangleright n), \quad (x \oplus y) \triangleright m = (x \triangleright m) \square (y \triangleright m).$$

In addition, we assume that the identity for $\square$ exists, 0_N, and that $0_S \triangleright m = 0_N$, $x \triangleright 0_N = 0_N$, and $1_S \triangleright m = m$. Again, in our applications $\square$ is often idempotent, and we use the notation ∞_N rather than 0_N.

A.3   Lexicographic Product

Suppose that we have two semigroups $S = (S, \oplus_S)$ and $T = (T, \oplus_T)$, with S *selective* (i.e. for all s_1, $s_2 \in S$ we have $s_1 \oplus s_2 \in \{s_1, s_2\}$). Then the left lexicographic product of S and T is defined to be the semigroup $S \vec{\times} T = ((S \times T) \cup \{\infty\}, \vec{\oplus})$, where for $(s_1, t_1), (s_2, t_2) \in S \times T$ we have

$$(s_1, t_1) \vec{\oplus} (s_2, t_2) = \begin{cases} (s_1, t_1 \oplus_T t_2) & s_1 = s_1 \oplus_S s_2 = s_2 \\ (s_1, t_1) & s_1 = s_1 \oplus_S s_2 \neq s_2 \\ (s_2, t_2) & s_1 \neq s_1 \oplus_S s_2 = s_2. \end{cases}$$

The right lexicographic product $\overleftarrow{\times}$ is similar, except that the semigroup T is assumed to be selective and the order of comparison is reversed.

Visibly Pushdown Kleene Algebra and Its Use in Interprocedural Analysis of (Mutually) Recursive Programs[*]

Claude Bolduc and Béchir Ktari

Département d'informatique et de génie logiciel
Université Laval, Québec, QC, G1K 7P4, Canada
claude.bolduc.1@ulaval.ca, ktari@ift.ulaval.ca

Abstract. Kleene algebra is a great formalism for doing intraprocedural analysis and verification of programs, but it seems difficult to deal with interprocedural analysis where the power of context-free languages is often needed to represent both the program and the property. In the model checking framework, Alur and Madhusudan defined visibly pushdown automata, which accept a subclass of context-free languages called visibly pushdown languages, to do some interprocedural analyses of programs while remaining decidable. We present visibly pushdown Kleene algebra, an extension of Kleene algebra that axiomatises exactly the equational theory of visibly pushdown languages. The algebra is simply Kleene algebra along with a family of implicit least fixed point operators. Some interprocedural analyses of (mutually) recursive programs are possible in this formalism and it can deal with some non-regular properties.

1 Introduction

Kleene algebra is the algebraic theory of finite automata and regular expressions. It has been used successfully to do intraprocedural analysis of programs [3,9]. However, how do we deal with interprocedural analysis when we have procedures and local scopes? In these analyses, the representation of the control flow may be a context-free language if there are (mutually) recursive procedures. So, Kleene algebra alone is not well suited to do it.

Some work has been done to extend Kleene algebra to handle subclasses of context-free languages [2,10,11,13]. However, these extensions do not seem to be satisfactory for interprocedural analysis: the complexity of the equational theory of these extensions is unknown or undecidable and frameworks for interprocedural analysis and verification, based on these extensions, are only able to deal with regular properties. This second argument is disturbing: non-regular properties are interesting for interprocedural analysis. In particular, the ability to use the nesting structure of procedure calls and returns (procedural context) when defining properties is useful. For example, here are some non-regular properties[1]:

---

[*] This research is supported by NSERC and FQRNT.
[1] This list is mostly inspired by [6].

R. Berghammer et al. (Eds.): RelMiCS/AKA 2009, LNCS 5827, pp. 44–58, 2009.
© Springer-Verlag Berlin Heidelberg 2009

Secure file manipulation policies like "whenever a secret file is opened in a secure procedural context, it must be closed before control exits the current context";

Stack-sensitive security properties like "a program must not execute a sensitive operation at any point when an untrusted procedure is currently on the stack or has ever been on the stack";

Logging policies like "whenever a procedure returns an error value, the error must be logged via a log procedure before control leaves the current procedural context".

In the model checking framework, Alur and Madhusudan defined visibly pushdown automata that accept a subclass of context-free languages they called visibly pushdown languages [1]. The idea behind visibly pushdown automata is to drive the stack manipulations of the automaton according to the current "type" of input symbol it reads. The class of visibly pushdown languages is surprisingly robust and the language equivalence problem is EXPTIME-complete [1]. So, it is possible to use visibly pushdown automata in model checking to represent both the (mutually) recursive program and the property to be checked. In fact, some non-regular properties like those above can be expressed with visibly pushdown automata.

This paper presents visibly pushdown Kleene algebra, which is an extension of Kleene algebra that axiomatises exactly the equational theory of visibly pushdown languages. So, the complexity of the equational theory of visibly pushdown Kleene algebra is EXPTIME-complete. A family of implicit least fixed point operators is simply added to the standard operators of Kleene algebra. Some interprocedural analyses of (mutually) recursive programs are possible in this formalism and it can deal with some non-regular properties.

2   Visibly Pushdown Automata

Visibly pushdown automata were introduced by Alur and Madhusudan [1]. Visibly pushdown automata are a particular case of pushdown automata in which the stack manipulations are driven (made "visible") by the input word (which is thought of as a string representing an execution of an interprocedural program). To allow this, the input alphabet Σ of a visibly pushdown automaton is divided in three disjoint sets Σ_i, Σ_c and Σ_r which represent, respectively, the set of internal actions, the set of calls and the set of returns of a program. The idea behind this is: when a visibly pushdown automaton reads (i) an internal action, it cannot modify the stack; (ii) a call action, it must push a symbol on the stack; (iii) a return action, it must read the symbol on the top of the stack and pop it (unless it is the bottom-of-stack symbol). This idea is quite simple but useful in program analysis: since a word is an execution of a program and the program's source code is usually available, it is easy to infer the type of an action in a word.

Visibly pushdown automata define a class of languages, called *visibly pushdown languages*, which is a strict subclass of deterministic context-free languages and a strict superclass of regular languages and balanced languages. For example, if $a \in \Sigma_i$, $c \in \Sigma_c$ and $r \in \Sigma_r$, then $\{c^n r^n \mid n \in \mathbb{N}\}$ is a visibly pushdown language, but not $\{a^n r^n \mid n \in \mathbb{N}\}$. Visibly pushdown languages are closed under union, concatenation, Kleene star, intersection, complementation and prefix-closure. Also, nondeterministic visibly pushdown languages are as expressive as deterministic ones. Moreover, the language equivalence problem is EXPTIME-complete [1]. Recall that the language equivalence problem is undecidable for context-free grammars.

3   Visibly Pushdown Regular Expressions

To our knowledge, no one seems to have defined a concept of "visibly pushdown regular expression" to denote exactly the visibly pushdown languages. This is a problem for the definition of an algebra, so we fill the gap here. Some set operations on visibly pushdown languages are first introduced.

Let Σ_i, Σ_c and Σ_r be three disjoint finite sets. The set of finite words on the alphabet $\Sigma_i \cup \Sigma_c \cup \Sigma_r$ is denoted by $(\Sigma_i \cup \Sigma_c \cup \Sigma_r)^*$. Let $S, T \subseteq (\Sigma_i \cup \Sigma_c \cup \Sigma_r)^*$. The concatenation operation on sets of words is defined as usual: $S \bullet T \overset{\text{def}}{=} \{st \mid s \in S \wedge t \in T\}$. The power S^n with respect to $\bullet$ is defined inductively by $S^0 \overset{\text{def}}{=} \{\varepsilon\}$ (where ε is the empty word) and $S^{n+1} \overset{\text{def}}{=} S \bullet S^n$. This allows to define the Kleene star operator by $S^* \overset{\text{def}}{=} (\cup\, n \mid n \in \mathbb{N} : S^n)$.

The above standard operators of Kleene algebra are not sufficient to generate any visibly pushdown language from $\emptyset$, $\{\varepsilon\}$, and the singletons $\{a\}$ for $a \in \Sigma_i \cup \Sigma_c \cup \Sigma_r$. So, other operators have to be defined in order to generate any visibly pushdown language. To do this, first note that a finite word $w \overset{\text{def}}{=} \sigma_1 \sigma_2 \sigma_3 \ldots \sigma_n$ in $(\Sigma_i \cup \Sigma_c \cup \Sigma_r)^*$, where each σ_i is a letter from $\Sigma_i \cup \Sigma_c \cup \Sigma_r$, may have *pending calls* and *pending returns*. Intuitively, a pending call is a call action $\sigma_i \in \Sigma_c$ that is not matched with a return action $\sigma_j \in \Sigma_r$ where $i < j$ and a pending return is a return action $\sigma_i \in \Sigma_r$ that is not matched with a call action $\sigma_j \in \Sigma_c$ where $j < i$. For example, in the word arbbcdaaasb for which $\Sigma_i \overset{\text{def}}{=} \{a, b\}$, $\Sigma_c \overset{\text{def}}{=} \{c, d\}$ and $\Sigma_r \overset{\text{def}}{=} \{r, s\}$, the first action r is a pending return and the first action c is a pending call. Obviously, d and s are well matched since d occurs before s in the word and there is no other pending call or pending return between d and s.

A visibly pushdown language can contain words that have pending calls and pending returns. Pending calls and pending returns are necessary to have a class of languages closed under the prefix-closure operator. It is also an interesting tool to model non-halting programs. The number of pending calls and pending returns in a word can be calculated [4]. A word $w \in (\Sigma_i \cup \Sigma_c \cup \Sigma_r)^*$ is said to be *well matched* if and only if it does not have pending calls or pending returns.

It is not difficult to see that visibly pushdown languages differ from regular languages mostly for their well-matched words. More precisely, if we are able to generate any visibly pushdown language that contains only well-matched words, we can use the standard operators of Kleene algebra on these languages and the singletons $\{a\}$ for $a \in \Sigma_c \cup \Sigma_r$ to generate any visibly pushdown language. So, a way is needed to generate any visibly pushdown language that contains only well-matched words. An infinite family of operators for doing that is defined.

The family works on *finite* lists of "blocks". There are two kinds of blocks:

Unary blocks of the form $[_x M]^y$, where $M \in \{\{a\} \mid a \in \Sigma_i\} \cup \{\{\varepsilon\}, \emptyset\}$. The labels x and y are respectively called the *starting label* and the *ending label*. The set M is called the *operand* of the unary block;

Binary blocks of the form $[_x \{c\} \downarrow_z \uparrow^w \{r\}]^y$, where $c \in \Sigma_c$ and $r \in \Sigma_r$. The labels x, y, z and w are respectively called the *starting label*, the *ending label*, the *call label* and the *return label*. The sets $\{c\}$ and $\{r\}$ are respectively called the *left operand* and the *right operand* of the binary block.

Let $\mathcal{B}$ be a *finite* list of unary and binary blocks that use a *finite* set of labels E, and let $x, x' \in E$. An operator of the family has the form $(_x \mathcal{B})^{x'}$. The operator's arity is the number of unary blocks in $\mathcal{B}$ plus twice the number of binary blocks in $\mathcal{B}$. Also, note that the labels in an operator are not variables but just a way to identify which operator of the family is used. To see this more clearly, let $\Sigma_i \overset{\mathrm{def}}{=} \{a, b\}$, $\Sigma_c \overset{\mathrm{def}}{=} \{c, d\}$, $\Sigma_r \overset{\mathrm{def}}{=} \{r, s\}$ and $E \overset{\mathrm{def}}{=} \{v, w, x, y, z\}$, and rewrite the expression $(_x [_y \{c\} \downarrow_w \uparrow^v \{r\}]^z, [_y \{b\}]^z, [_w \{d\} \downarrow_x \uparrow^z \{s\}]^v, [_x \{a\}]^y)^z$ by the expression $f_{(x,(y,w,v,z),(y,z),(w,x,z,v),(x,y),z)}(\{c\}, \{r\}, \{b\}, \{d\}, \{s\}, \{a\})$ in which $f_{(x,(y,w,v,z),(y,z),(w,x,z,v),(x,y),z)}$ is a 6-ary operator. Moreover, understand that each operator of the family is partial. An operand cannot be any set of words; just those allowed by the definition of blocks.

The idea behind an expression $(_x \mathcal{B})^{x'}$ is to generate any well-matched word that can be produced by a correct "travelling" of the list of blocks, starting the travel in any block that has x as starting label and ending it in any block that has x' as ending label. A correct travelling starting with y, ending with y' and producing a set of well-matched words S is a finite sequence $b_1 b_2 \ldots b_n$ of blocks of $\mathcal{B}$ (where $n > 0$), such that $n = 1$ and b_n is a unary block of the form $[_y M]^{y'}$ and $S = M$, or $n > 1$ and (there are three possible cases):

- b_1 is a unary block of the form $[_y M]^{v'}$ for $v' \in E$ (including y') and $b_2 \ldots b_n$ is a correct travelling starting with v', ending with y' and producing T, and $S = M \bullet T$;

- b_1 is a binary block of the form $[_y \{c\} \downarrow_z \uparrow^w \{r\}]^{y'}$ and $b_2 \ldots b_n$ is a correct travelling starting with z, ending with w and producing T, and $S = \{c\} \bullet T \bullet \{r\}$;

- b_1 is a binary block of the form $[_y \{c\} \downarrow_z \uparrow^w \{r\}]^{v'}$ for $v' \in E$ (including y') and there exists an $i \in \mathbb{N}$ such that $1 < i < n$ and $b_2 \ldots b_i$ is a correct travelling starting with z, ending with w and producing T and $b_{i+1} \ldots b_n$ is a correct travelling starting with v', ending with y' and producing U, and $S = \{c\} \bullet T \bullet \{r\} \bullet U$.

Here are some examples. Let $\Sigma_i \overset{\mathrm{def}}{=} \{a, b\}$, $\Sigma_c \overset{\mathrm{def}}{=} \{c, d\}$, $\Sigma_r \overset{\mathrm{def}}{=} \{r, s\}$ and $E \overset{\mathrm{def}}{=} \{v, w, x, y, z\}$. The expression

- $\left(_x \left[_x \{a\}\right]^x \right)^x$ denotes the visibly pushdown language $\{a^n \mid n > 0\}$;
- $\left(_x \left[_x \{a\}\right]^y, \left[_x \{a\}\right]^y \right)^x$ denotes the visibly pushdown language $\emptyset$;
- $\left(_x \left[_x \{c\} \downarrow_x \uparrow^y \{r\}\right]^y, \left[_x \{\varepsilon\}\right]^y \right)^y$ denotes the visibly pushdown language $\{c^n r^n \mid n \in \mathbb{N}\}$;
- $\left(_x \left[_y \{c\} \downarrow_w \uparrow^v \{r\}\right]^z, \left[_y \{b\}\right]^z, \left[_w \{d\} \downarrow_x \uparrow^z \{s\}\right]^v, \left[_x \{a\}\right]^y \right)^z$ denotes the visibly pushdown language $\{a(cda)^n b(sr)^n \mid n \in \mathbb{N}\}$.

Note from the examples that the position of a block in the list is not important, neither if it appears more than once in the list. This states that the list of blocks works in fact like a set.

Let $\mathcal{B}$ be a finite list of unary and binary blocks using a finite set of labels E. Define $\mathcal{B}^1$ as the set of unary blocks of $\mathcal{B}$ and $\mathcal{B}^2$ as the set of binary blocks of $\mathcal{B}$. For $n \in \mathbb{N}$, define the *power-recursion operator* $\left(_x \mathcal{B} \right)^y_n$, where $x, y \in E$, by induction on n: $\left(_x \mathcal{B} \right)^y_0 \stackrel{\text{def}}{=} (\cup M \mid \left[_x M\right]^y \in \mathcal{B}^1 : M)$, and

$$
\begin{aligned}
\left(_x \mathcal{B} \right)^y_{n+1} \stackrel{\text{def}}{=} \ & (\cup M, v \mid \left[_x M\right]^v \in \mathcal{B}^1 : M \bullet \left(_v \mathcal{B} \right)^y_n) \\
& \cup \ (\cup c, z, r, w \mid \left[_x \{c\} \downarrow_z \uparrow^w \{r\}\right]^y \in \mathcal{B}^2 : \{c\} \bullet \left(_z \mathcal{B} \right)^w_n \bullet \{r\}) \\
& \cup \ (\cup c, z, r, w, v, n_1, n_2 \mid \left[_x \{c\} \downarrow_z \uparrow^w \{r\}\right]^v \in \mathcal{B}^2 \wedge n_1, n_2 \in \mathbb{N} \\
& \qquad \wedge \ n_1 + n_2 = n - 1 : \{c\} \bullet \left(_z \mathcal{B} \right)^w_{n_1} \bullet \{r\} \bullet \left(_v \mathcal{B} \right)^y_{n_2}) \ .
\end{aligned}
$$

Intuitively, $\left(_x \mathcal{B} \right)^y_n$ denotes the set of all well-matched words that can be generated by any correct travelling of $\mathcal{B}$ of length $n + 1$ starting with x and ending with y. With this definition, it is easy to define an operator $\left(_x \mathcal{B} \right)^y$ by $\left(_x \mathcal{B} \right)^y \stackrel{\text{def}}{=} (\cup n \mid n \in \mathbb{N} : \left(_x \mathcal{B} \right)^y_n)$.

What is the relationship of the family of operators with fixed points? Let $\mathcal{B}_{\text{ex}} \stackrel{\text{def}}{=} \left[_y \{c\} \downarrow_w \uparrow^v \{r\}\right]^z, \left[_y \{b\}\right]^z, \left[_w \{d\} \downarrow_x \uparrow^z \{s\}\right]^v, \left[_x \{a\}\right]^y$. Take the expression $\left(_x \mathcal{B}_{\text{ex}} \right)^z$. One can view (although it is not entirely true) the list of blocks like a way to encode a special case of context-free grammars in a linear way. The preceding blocks encode the grammar:

$$
\begin{array}{lll}
X_{(y,z)} \to c X_{(w,v)} r, & X_{(y,t)} \to c X_{(w,v)} r X_{(z,t)} & \text{for } t \in \{v, w, x, y, z\}, \\
X_{(y,z)} \to b, & X_{(y,t)} \to b X_{(z,t)} & \text{for } t \in \{v, w, x, y, z\}, \\
X_{(w,v)} \to d X_{(x,z)} s, & X_{(w,t)} \to d X_{(x,z)} s X_{(v,t)} & \text{for } t \in \{v, w, x, y, z\}, \\
X_{(x,y)} \to a, & X_{(x,t)} \to a X_{(y,t)} & \text{for } t \in \{v, w, x, y, z\}.
\end{array}
$$

Nonterminals of the form $X_{(t,t')}$ are used because it is important to remember not only the current label t that is expanded but also the label t' that must be reached in the end. With this grammar-based view, the value of $\left(_x \mathcal{B}_{\text{ex}} \right)^z$ is simply the language generated by the grammar when starting with the nonterminal $X_{(x,z)}$ (a least fixed point). Clearly, it is the visibly pushdown language $\{a(cda)^n b(sr)^n \mid n \in \mathbb{N}\}$. Note that these special cases of context-free grammars are not strong enough to be the basis of our algebra as we will discuss in Sect. 6.

We are now ready to define visibly pushdown regular expressions. A *visibly pushdown regular expression* is defined over three disjoint sets Σ_i, Σ_c and Σ_r by the following propositions:

- 0 and 1 are visibly pushdown regular expressions;
- if $a \in \Sigma_i \cup \Sigma_c \cup \Sigma_r$, then a is a visibly pushdown regular expression;
- if $\mathcal{B}$ is a finite list of unary blocks each containing one element of $\Sigma_i \cup \{0, 1\}$, and binary blocks each containing one element of Σ_c as left operand and one element of Σ_r as right operand, where all blocks use a finite set of labels E, and $x, y \in E$, then $(\!|_x \mathcal{B} |\!)^y$ is a visibly pushdown regular expression;
- if p and q are visibly pushdown regular expressions, then $p \cdot q$, $p + q$ and p^* are visibly pushdown regular expressions.

The family of operators $(\!|_x \mathcal{E} |\!)^y$, where $\mathcal{E}$ denotes a finite list of unary blocks without operand and binary blocks without left and right operands using any finite set of labels E, and where $x, y \in E$, is denoted by $\mathcal{F}_{(\!|)}$. The language denoted by a visibly pushdown regular expression p is noted by $\mathcal{L}(p)$ and is defined by

$$\mathcal{L}(0) \stackrel{\text{def}}{=} \emptyset, \qquad \mathcal{L}(1) \stackrel{\text{def}}{=} \{\varepsilon\}, \qquad \mathcal{L}(a) \stackrel{\text{def}}{=} \{a\} \text{ for any } a \in \Sigma_i \cup \Sigma_c \cup \Sigma_r,$$

and extends over the structure of visibly pushdown regular expressions where $\cdot$ becomes $\bullet$, + becomes $\cup$, and $*$ and $(\!|_x \mathcal{E} |\!)^y \in \mathcal{F}_{(\!|)}$ become the set operators $*$ and $(\!|_x \mathcal{E} |\!)^y$.

The class of visibly pushdown regular expressions is rich enough to denote exactly the visibly pushdown languages as shown by the following theorem.

Theorem 1 (Theorem à la Kleene for visibly pushdown regular expressions [4]). *Let Σ_i, Σ_c and Σ_r be three disjoint finite sets. Let $L \subseteq (\Sigma_i \cup \Sigma_c \cup \Sigma_r)^*$ be a language. The following propositions are equivalent:*

(i) L is accepted by a visibly pushdown automaton;
(ii) L is denoted by a visibly pushdown regular expression.

4   Visibly Pushdown Kleene Algebra (**VPKA**)

We now want to define an algebra that characterizes exactly the equality of the languages denoted by two visibly pushdown regular expressions. Before showing the axioms, we recall Kozen's definition of Kleene algebra [7] and give some intuition about the axioms.

Definition 1 (Kleene algebra). *A Kleene algebra is an algebraic structure $(K, +, \cdot, *, 0, 1)$ satisfying the following axioms[2].*

$$
\begin{array}{lll}
p + (q + r) = (p + q) + r & p(qr) = (pq)r & p + 0 = p \\
p(q + r) = pq + pr & p + q = q + p & p0 = 0 = 0p \\
(p + q)r = pr + qr & p + p = p & p1 = p = 1p \\
qp + r \leqslant p \rightarrow q^*r \leqslant p & 1 + p^*p \leqslant p^* & p \leqslant q \leftrightarrow p + q = q \\
pq + r \leqslant p \rightarrow rq^* \leqslant p & 1 + pp^* \leqslant p^* &
\end{array}
$$

---

[2] In the sequel, we write pq instead of $p \cdot q$. The increasing precedence of the operators is $+$, $\cdot$ and $*$.

The axiomatisation of VPKA proposed below adds seven axioms to Kleene algebra. The first two axioms are unfolding axioms for unary and binary blocks. The third axiom is also an unfolding axiom but it is called the maximality of $(\!|\,|\!)$-travel axiom, since it represents the fact that if a $(\!|\,|\!)$-expression is forced to travel through an intermediate label y' for any correct travelling, then it is more restricted than if this $(\!|\,|\!)$-expression is not forced to travel through it.

Let $\mathcal{B}$ be a finite list of unary and binary blocks using labels from a finite set E and let $x, y \in E$. Axioms (4) and (5) are induction axioms (equational implication axioms) that define an expression $(\!|_x \mathcal{B}|\!)^y$ as the least solution for a component $X_{(x,y)}$ of an inequational system like the one generated by the grammar on page 48. These axioms are similar to the two Kleene star induction axioms, but (4) and (5) have to deal with an inequational system instead of a simple formula. So, axioms (4) and (5) use a set of solutions of the form $s_{(u,u')}$, for $u, u' \in E$, for the equational system instead of a single solution. This raises an interesting point. Remember from the example on page 48 that not every nonterminal $X_{(t,t')}$ is needed for the calculation of the result. For example, the nonterminal $X_{(\mathsf{x},\mathsf{w})}$ is never used in the calculation when starting with $X_{(\mathsf{x},\mathsf{z})}$. To approximate the needed nonterminals, two functions are defined. First, it is easy to calculate such an approximation by travelling "forward" in the list of blocks. A function $\mathsf{F}^1_{\mathcal{B}} : 2^{E \times E} \to 2^{E \times E}$ is defined for any $V \subseteq E \times E$ by:

$$\mathsf{F}^1_{\mathcal{B}}(V) \stackrel{\text{def}}{=} V \cup \{(y,y') \mid (\exists\, z,m \mid (z,y') \in V : [_z m]^y \in \mathcal{B}^1)\}$$
$$\cup \{(y,y'),(w,w') \mid (\exists\, z,c,r \mid (z,y') \in V : [_z c \downarrow_w \uparrow^{w'} r]^y \in \mathcal{B}^2)\} \ .$$

Of course, the list of blocks can also be travelled "backward". So, a function $\mathsf{B}^1_{\mathcal{B}} : 2^{E \times E} \to 2^{E \times E}$ is also defined for any $V \subseteq E \times E$ by:

$$\mathsf{B}^1_{\mathcal{B}}(V) \stackrel{\text{def}}{=} V \cup \{(y,y') \mid (\exists\, z,m \mid (y,z) \in V : [_{y'} m]^z \in \mathcal{B}^1)\}$$
$$\cup \{(y,y'),(w,w') \mid (\exists\, z,c,r \mid (y,z) \in V : [_{y'} c \downarrow_w \uparrow^{w'} r]^z \in \mathcal{B}^2)\} \ .$$

It is easy to see that these two functions are monotone. So, their least fixed point exist and we respectively call them $\mathsf{F}^*_{\mathcal{B}}$ and $\mathsf{B}^*_{\mathcal{B}}$. Note that $\mathsf{F}^*_{\mathcal{B}}$ and $\mathsf{B}^*_{\mathcal{B}}$ do not coincide. For example, $(\mathsf{y},\mathsf{z}) \in \mathsf{F}^*_{\mathcal{B}_{\mathrm{ex}}}(\{(\mathsf{x},\mathsf{z})\})$ and $(\mathsf{x},\mathsf{y}) \notin \mathsf{F}^*_{\mathcal{B}_{\mathrm{ex}}}(\{(\mathsf{x},\mathsf{z})\})$, but $(\mathsf{y},\mathsf{z}) \notin \mathsf{B}^*_{\mathcal{B}_{\mathrm{ex}}}(\{(\mathsf{x},\mathsf{z})\})$ and $(\mathsf{x},\mathsf{y}) \in \mathsf{B}^*_{\mathcal{B}_{\mathrm{ex}}}(\{(\mathsf{x},\mathsf{z})\})$. The functions $\mathsf{F}^*_{\mathcal{B}}$ and $\mathsf{B}^*_{\mathcal{B}}$ are used in axioms (4) and (5) to restrict the inequational system to solve.

Axioms (6) and (7) are equational implications called $(\!|\,|\!)$-simulation axioms. They are inspired both by the definition of a simulation relation between automata and by the bisimulation rule of Kleene algebra: $pq = rp \to pq^* = r^*p$. The goal of (6) is to verify if a finite list $\mathcal{B}$ of unary and binary blocks using only labels from a finite set E can be simulated (when travelling "forward") by another finite list $\mathcal{C}$ of unary and binary blocks using only labels from a finite set E'. The simulation relation is encoded by a set of expressions like $b_{z_2}^{(z',z)}$, where $z, z' \in E$ and $z_2 \in E'$, which are usually (but not necessarily) the constant 0 or 1, stating if the label z is simulated by z_2 when the first label processed just after the last unmatched call is z' (such a z' is useful to show some results like the determinization of visibly pushdown automata). The simulation relation is correct if it commutes with internal actions, calls and returns for any block of the list. However, note that any label simulated just before and after a call must be remembered until its matching return is reached. Also, axiom (7) is similar to (6), but the simulation is done by travelling "backward".

Definition 2 (Visibly pushdown Kleene algebra). *Let Σ_i, Σ_c and Σ_r be three disjoint finite sets of "atomic elements" such that at least one of the sets contains at least one element. A visibly pushdown Kleene algebra is a structure $(P, +, \cdot, {}^{*}, \mathcal{F}_{(|\,|)}, 0, 1)$ generated by Σ_i, Σ_c and Σ_r under the axioms of Kleene algebra (in other words, the structure $(P, +, \cdot, {}^{*}, 0, 1)$ is a Kleene algebra) and the following additional laws are satisfied by all expressions $(|_x \mathcal{B} |)^y$ and $(|_{x_2} \mathcal{C} |)^{y_2}$ where $\mathcal{B}$ and $\mathcal{C}$ are finite lists of unary blocks each containing one element of $\Sigma_i \cup \{0, 1\}$ as operand and binary blocks each containing one element of Σ_c as left operand and one element of Σ_r as right operand, where all blocks of $\mathcal{B}$ use a finite set of labels E and all blocks of $\mathcal{C}$ use a finite set of labels E', $x, y, y' \in E$, $x_2, y_2 \in E'$, $s_{(u,u')}, b^{(u,u')}_{x'_2} \in P$ for all $u, u' \in E$, $x'_2 \in E'$:*

$$m \leqslant (|_x \mathcal{B} |)^y, \qquad \text{for } [\,{}_x m\,]^y \in \mathcal{B}^1 \,, \tag{1}$$

$$c \cdot (|_z \mathcal{B} |)^w \cdot r \leqslant (|_x \mathcal{B} |)^y, \qquad \text{for } [\,c \!\downarrow_x \!\uparrow_z^{\,w}\, r\,]^y \in \mathcal{B}^2 \,, \tag{2}$$

$$(|_x \mathcal{B} |)^{y'} \cdot (|_{y'} \mathcal{B} |)^y \leqslant (|_x \mathcal{B} |)^y \,, \tag{3}$$

$$\begin{array}{l}
\Big(\wedge u, u' \mid (u, u') \in \mathsf{F}_{\mathcal{B}}^{*}(\{(x, y)\}) : \\[4pt]
\quad (\wedge m \mid [\,{}_u m\,]^{u'} \in \mathcal{B}^1 : m \leqslant s_{(u,u')}) \\[4pt]
\quad \wedge \; (\wedge m, v \mid [\,{}_u m\,]^{v} \in \mathcal{B}^1 : m \cdot s_{(v,u')} \leqslant s_{(u,u')}) \\[4pt]
\quad \wedge \; (\wedge c, z, r, w \mid [\,c \!\downarrow_u \!\uparrow_z^{\,w}\, r\,]^{u'} \in \mathcal{B}^2 : c \cdot s_{(z,w)} \cdot r \leqslant s_{(u,u')}) \\[4pt]
\quad \wedge \; (\wedge c, z, r, w, v \mid [\,c \!\downarrow_u \!\uparrow_z^{\,w}\, r\,]^{v} \in \mathcal{B}^2 : c \cdot s_{(z,w)} \cdot r \cdot s_{(v,u')} \leqslant s_{(u,u')}) \Big) \\[4pt]
\rightarrow \; (|_x \mathcal{B} |)^y \leqslant s_{(x,y)} \,,
\end{array} \tag{4}$$

$$\begin{array}{l}
\Big(\wedge u, u' \mid (u, u') \in \mathsf{B}_{\mathcal{B}}^{*}(\{(x, y)\}) : \\[4pt]
\quad (\wedge m \mid [\,{}_u m\,]^{u'} \in \mathcal{B}^1 : m \leqslant s_{(u,u')}) \\[4pt]
\quad \wedge \; (\wedge m, v \mid [\,{}_v m\,]^{u'} \in \mathcal{B}^1 : s_{(u,v)} \cdot m \leqslant s_{(u,u')}) \\[4pt]
\quad \wedge \; (\wedge c, z, r, w \mid [\,c \!\downarrow_u \!\uparrow_z^{\,w}\, r\,]^{u'} \in \mathcal{B}^2 : c \cdot s_{(z,w)} \cdot r \leqslant s_{(u,u')}) \\[4pt]
\quad \wedge \; (\wedge c, z, r, w, v \mid [\,c \!\downarrow_v \!\uparrow_z^{\,w}\, r\,]^{u'} \in \mathcal{B}^2 : s_{(u,v)} \cdot c \cdot s_{(z,w)} \cdot r \leqslant s_{(u,u')}) \Big) \\[4pt]
\rightarrow \; (|_x \mathcal{B} |)^y \leqslant s_{(x,y)} \,,
\end{array} \tag{5}$$

$$\left(\wedge \ u, u', u'', x_2', m \mid u'' \in E \wedge x_2' \in E' \wedge [\, m\, \overset{u'}{\underset{u}{]}} \in \mathcal{B}^1 : \right.$$

$$\left. b_{x_2'}^{(u'',u)} \cdot m \leqslant (\sum\ y_2'' \mid [\, m\, \overset{y_2''}{\underset{x_2'}{]}} \in \mathcal{C}^1 : m \cdot b_{y_2''}^{(u'',u')}) \right)$$

$$\wedge \left(\wedge \ u, u', u'', x_2', c, z, r, w \mid u'' \in E \wedge x_2' \in E' \wedge [\, c \overset{w}{\underset{u}{\downarrow}} \overset{u'}{\underset{z}{\uparrow}} r\,] \in \mathcal{B}^2 : \right.$$

$$\left. b_{x_2'}^{(u'',u)} \cdot c \leqslant (\sum\ z', w', y_2'' \mid [\, c \overset{w'}{\underset{x_2'}{\downarrow}} \overset{y_2''}{\underset{z'}{\uparrow}} r\,] \in \mathcal{C}^2 : b_{x_2'}^{(u'',u)} \cdot c \cdot b_{z'}^{(z,z)} \cdot b_{z'}^{(z,z)}) \right) \tag{6}$$

$$\wedge \left(\wedge \ u, u', u'', x_2', c, z, r, w, z', w'' \mid u'' \in E \wedge x_2', z', w'' \in E' \wedge [\, c \overset{w}{\underset{u}{\downarrow}} \overset{u'}{\underset{z}{\uparrow}} r\,] \in \mathcal{B}^2 : \right.$$

$$(\sum\ w', y_2'' \mid [\, c \overset{w'}{\underset{x_2'}{\downarrow}} \overset{y_2''}{\underset{z'}{\uparrow}} r\,] \in \mathcal{C}^2 : b_{x_2'}^{(u'',u)} \cdot c \cdot b_{z'}^{(z,z)} \cdot (\!(\, c\, \overset{w''}{\underset{z'}{)}}\!) \cdot b_{w''}^{(z,w)} \cdot r)$$

$$\leqslant (\sum\ y_2'' \mid [\, c \overset{w''}{\underset{x_2'}{\downarrow}} \overset{y_2''}{\underset{z'}{\uparrow}} r\,] \in \mathcal{C}^2 : c \cdot (\!(\, c\, \overset{w''}{\underset{z'}{)}}\!) \cdot r \cdot b_{y_2''}^{(u'',u')}) \right)$$

$$\to b_{x_2}^{(y',x)} \cdot (\!(\, \mathcal{B}\, \overset{y}{\underset{x}{)}}\!) \leqslant (\sum\ y_2' \mid y_2' \in E' : (\!(\, \mathcal{C}\, \overset{y_2'}{\underset{x_2}{)}}\!) \cdot b_{y_2'}^{(y',y)}) \ ,$$

$$\left(\wedge \ u, u', x_2', y_2', m \mid u, u' \in E \wedge [\, m\, \overset{y_2'}{\underset{x_2'}{]}} \in \mathcal{C}^1 : \right.$$

$$\left. m \cdot b_{y_2'}^{(u,u')} \leqslant (\sum\ u'' \mid [\, m\, \overset{u'}{\underset{u''}{]}} \in \mathcal{B}^1 : b_{x_2'}^{(u,u'')} \cdot m) \right)$$

$$\wedge \left(\wedge \ u, u', x_2', y_2', c, z, r, w \mid u, u' \in E \wedge [\, c \overset{w}{\underset{x_2'}{\downarrow}} \overset{y_2'}{\underset{z}{\uparrow}} r\,] \in \mathcal{C}^2 : \right.$$

$$\left. r \cdot b_{y_2'}^{(u,u')} \leqslant (\sum\ u'', z', w' \mid [\, c \overset{w'}{\underset{u''}{\downarrow}} \overset{u'}{\underset{z'}{\uparrow}} r\,] \in \mathcal{B}^2 : b_w^{(z',w')} \cdot b_{x_2'}^{(u,u'')} \cdot r) \right) \tag{7}$$

$$\wedge \left(\wedge \ u, u', x_2', y_2', c, z, r, w, u'', z', w' \mid u \in E \wedge [\, c \overset{w}{\underset{x_2'}{\downarrow}} \overset{y_2'}{\underset{z}{\uparrow}} r\,] \in \mathcal{C}^2 \right.$$

$$\wedge [\, c \overset{w'}{\underset{u''}{\downarrow}} \overset{u'}{\underset{z'}{\uparrow}} r\,] \in \mathcal{B}^2 : (\sum\ z'' \mid z'' \in E : c \cdot b_z^{(z',z'')} \cdot (\!(\, \mathcal{B}\, \overset{w'}{\underset{z''}{)}}\!) \cdot b_{x_2'}^{(u,u'')} \cdot r)$$

$$\leqslant b_{x_2'}^{(u,u'')} \cdot c \cdot (\!(\, \mathcal{B}\, \overset{w'}{\underset{z'}{)}}\!) \cdot r \right)$$

$$\to (\!(\, \mathcal{C}\, \overset{y_2}{\underset{x_2}{)}}\!) \cdot b_{y_2}^{(y',y)} \leqslant (\sum\ x' \mid x' \in E : b_{x_2}^{(y',x')} \cdot (\!(\, \mathcal{B}\, \overset{y}{\underset{x'}{)}}\!)) \ .$$

To get a better grip on the axioms, we prove that $(\!(_x\, [_x\, c \downarrow_x \uparrow^y r\,]^y, [_x\, a\,]^y\,)\!)^y \leqslant (\!(_x\, [_x\, c \downarrow_x \uparrow^y r\,]^y, [_x\, a\,]^x, [_x\, a\,]^y\,)\!)^y$ for $\Sigma_i \overset{\text{def}}{=} \{a\}$, $\Sigma_c \overset{\text{def}}{=} \{c\}$, $\Sigma_r \overset{\text{def}}{=} \{r\}$ and $E \overset{\text{def}}{=} \{x, y\}$. Let $\mathcal{C}_{ex} \overset{\text{def}}{=} [_x\, c \downarrow_x \uparrow^y r\,]^y, [_x\, a\,]^x, [_x\, a\,]^y$. By $F^*_{[_x\, c \downarrow_x \uparrow^y r\,]^y, [_x\, a\,]^y}(\{(x, y)\}) = \{(x, y), (y, y)\}$ and axiom (4), it suffices to prove, for $s_{(x,y)} \overset{\text{def}}{=} (\!(_x\, \mathcal{C}_{ex}\,)\!)^y$ and $s_{(y,y)} \overset{\text{def}}{=} 0$, that $a + a \cdot 0 + c \cdot (\!(_x\, \mathcal{C}_{ex}\,)\!)^y \cdot r + c \cdot (\!(_x\, \mathcal{C}_{ex}\,)\!)^y \cdot r \cdot 0 \leqslant (\!(_x\, \mathcal{C}_{ex}\,)\!)^y$ and $0 + 0 + 0 + 0 \leqslant 0$. The second inequality is trivial by Kleene algebra. For the first inequality, using simple Kleene algebraic reasoning and axioms (1) and (2):

$$a + a \cdot 0 + c \cdot (\!(_x\, \mathcal{C}_{ex}\,)\!)^y \cdot r + c \cdot (\!(_x\, \mathcal{C}_{ex}\,)\!)^y \cdot r \cdot 0 = a + c \cdot (\!(_x\, \mathcal{C}_{ex}\,)\!)^y \cdot r \leqslant (\!(_x\, \mathcal{C}_{ex}\,)\!)^y \ .$$

Let $\mathcal{B}$ be a finite list of unary and binary blocks, where all blocks of $\mathcal{B}$ use a finite set of labels E, and let $x, y \in E$. A theorem of VPKA, inspired by the definition of $([_x \mathcal{B}])^y$ for visibly pushdown languages, can be proved using Kleene algebra and axioms (1) to (4) (see [4] for the proof):

$$
\begin{aligned}
([_x \mathcal{B}])^y = &(\textstyle\sum m \mid [_x m]^y \in \mathcal{B}^1 : m) \\
&+ (\textstyle\sum m, v \mid [_x m]^v \in \mathcal{B}^1 : m \cdot ([_v \mathcal{B}])^y) \\
&+ (\textstyle\sum c, z, r, w \mid [_x c \downarrow_z \uparrow^w r]^y \in \mathcal{B}^2 : c \cdot ([_z \mathcal{B}])^w \cdot r) \\
&+ (\textstyle\sum c, z, r, w, v \mid [_x c \downarrow_z \uparrow^w r]^v \in \mathcal{B}^2 : c \cdot ([_z \mathcal{B}])^w \cdot r \cdot ([_v \mathcal{B}])^y) \; .
\end{aligned}
\tag{8}
$$

Note that any VPKA is a partial algebra because each operator of $\mathcal{F}_{([])}$ is defined only on atomic elements. However, the axiomatic system is still powerful. In fact, the axiomatic system is sound and complete for valid equations between languages denoted by visibly pushdown regular expressions and the equational theory of VPKA is EXPTIME-complete [4]. Note also that axioms (6) and (7) are not used often in proofs. In fact, most of the proofs of [4] do not use them.

4.1 Visibly Pushdown Kleene Algebra with Tests

Tests are an essential ingredient to analyze imperative programs. We add them in a way similar to Kleene algebra with tests [8]: a Boolean algebra $(B, +, \cdot, \overline{}, 0, 1)$ generated by atomic tests $\mathbf{B}$ is added to VPKA, where $B \subseteq P$. However, we would like to use tests in operators of $\mathcal{F}_{([])}$. How can we interpret a test? It seems natural to think of tests as a subset of internal actions. So, we extend the definition of unary blocks to allow tests. As any Boolean expression can be put in negation normal form, it is sufficient (see Sect. 4.2) to allow additional unary blocks where their operand is an element of $\{b, \overline{b} \mid b \in \mathbf{B}\}$. Also, axioms (1) to (7) are adapted in a natural way: consider each element of $\{b, \overline{b} \mid b \in \mathbf{B}\}$ like an internal action.

4.2 Metablocks

The definition of a list of blocks for $\mathcal{F}_{([])}$ is simple, but it can be tedious to write such a list for a large expression. To simplify this process, we define *metablocks* which are abbreviations (similar to regular expressions) of a list of blocks.

Let Σ_i, Σ_c, Σ_r and $\mathbf{B}$ be finite sets. Let E be a finite set of labels. A metablock is an expression $[_x e]^y$ where $x, y \in E$ and e is an expression of the set MBexp that is defined as the smallest set containing 0, 1, a for each $a \in \Sigma_i \cup \{b, \overline{b} \mid b \in \mathbf{B}\}$, $(c \downarrow_z \uparrow^w r)$ for each $c \in \Sigma_c$, $r \in \Sigma_r$ and $z, w \in E$, and closed under "operators" $\cdot$, $+$ and *. We allow to write pq instead of $p \cdot q$. A metablock is reduced to a list of unary and binary blocks by the function mb defined inductively by:

- $\mathrm{mb}([_x a]^y) \overset{\text{def}}{=} [_x a]^y$ for $a \in \{0, 1\} \cup \Sigma_i \cup \{b, \overline{b} \mid b \in \mathbf{B}\}$ and $x, y \in E$;
- $\mathrm{mb}([_x (c \downarrow_z \uparrow^w r)]^y) \overset{\text{def}}{=} [_x c \downarrow_z \uparrow^w r]^y$ for $c \in \Sigma_c$, $r \in \Sigma_r$ and $x, y, z, w \in E$;
- $\mathrm{mb}([_x p \cdot q]^y) \overset{\text{def}}{=} \mathrm{mb}([_x p]^z), \mathrm{mb}([_z q]^y)$ for $x, y \in E$ and a fresh label z;
- $\mathrm{mb}([_x p + q]^y) \overset{\text{def}}{=} \mathrm{mb}([_x p]^y), \mathrm{mb}([_x q]^y)$ for $x, y \in E$;
- $\mathrm{mb}([_x p^*]^y) \overset{\text{def}}{=} [_x 1]^y, \mathrm{mb}([_x p]^y), \mathrm{mb}([_x p]^z), \mathrm{mb}([_z p]^z), \mathrm{mb}([_z p]^y)$ for $x, y \in E$ and a fresh label z.

5 Interprocedural Analysis of (Mutually) Recursive Programs

We are building a framework for static analysis like the framework defined in [3] or in [9]. Currently, the framework is not fixed, but this is a work in progress.

A (mutually) recursive program allows the program constructs:

$$p; q, \qquad \text{if } b \text{ then } p \text{ else } q, \qquad \text{while } b \text{ do } p,$$

and the procedure program construct:

$$\text{procedure } f \ (\text{ var } l_1, \text{var } l_2, \ldots, \text{var } l_n \ ) \text{ begin } p \text{ end}$$

for programs p, q, test b, function name f and local variables $l_1, l_2, \ldots, l_n$. Intuitively, the procedure program construct defines a procedure named f that has body p defined using local variables $l_1, l_2, \ldots, l_n$. A procedure that will act as the first procedure called by the program when it starts must be defined. Note that a program construct can contain call of procedures.

Static analysis of (mutually) recursive programs follows the process in Fig. 1. First, encode the program in VPKA. This is done in three steps:

1. Define the desired abstraction for atomic program instructions and variables through the sets Σ_i, Σ_c, Σ_r and $\mathbf{B}$;
2. Encode the program's control flow by an expression p;
3. Encode the desired semantics of atomic program instructions, variables and variable passing mechanism by a set of equational hypotheses $\mathcal{H}$.

These steps are semiautomatic. In particular, the encoding in step 2 of the program constructs gives an expression of MBexp:

$$p; q \overset{\text{def}}{=} pq, \qquad \text{if } b \text{ then } p \text{ else } q \overset{\text{def}}{=} bp + \bar{b}q, \qquad \text{while } b \text{ do } p \overset{\text{def}}{=} (bp)^* \bar{b},$$

and the body of the procedure program construct gives a metablock:

$$\text{procedure } f \ (\text{ var } h_1, \text{var } h_2, \ldots, \text{var } h_n \ ) \text{ begin } p \text{ end} \overset{\text{def}}{=} [_f p]^\tau,$$

where τ is the end of any procedure. For more flexibility, the encoding of atomic instructions, call instructions and variables is left at the user's discretion.

Currently, the framework deals only with halting programs, since non-halting programs need specific mechanisms (see for example [9]). Halting programs are obtained by restricting recursive procedures and loops to simple cases.

Step 3 is a powerful step, but it is also the most tedious one. The hypotheses usually represent an abstraction of the data flow of the program. Some classes of hypotheses are already studied in Kleene algebra with tests (see for example [8]). However, it is also useful to have some hypotheses that represent the variable passing mechanism of the procedure. We present some of these in page 55.

Currently, the exact encoding of the property in VPKA is not fixed. Since the framework is restricted to halting programs, the full power of visibly pushdown

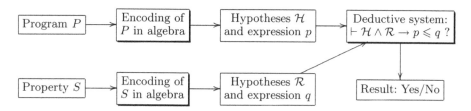

Fig. 1. Framework of static analysis in visibly pushdown Kleene algebra

automata can be used to define the property (this is not the case with non-halting programs). However, defining a property by such automata does not seem always easy. In our experiments, we found it clearer to define the property directly in VPKA. This has the drawback of not having a clear encoding like in [3] or [9]. We used an encoding of the visibly pushdown language, representing the set of executions of any program on Σ_i, Σ_c, Σ_r and **B** that satisfies the property, that is composed of an expression q of VPKA and a set $\mathcal{R}$ of "refinement hypotheses" that help to sharpen the program's abstraction according to the desired property. Some refinement hypotheses are presented in page 56.

The static analysis ends by verifying if $\mathcal{H} \wedge \mathcal{R} \rightarrow p \leqslant q$ is a theorem of VPKA. The program is said to satisfy the property if and only if the formula is a theorem of VPKA.

Example. Take the abstract program of Fig. 2(a). In it, the action $o(i)$ represents the opening of the file named i with a writing to i and $c(i)$ represents the writing to the file named i with the closing of i. For the example, suppose that $o(i)$ and $c(i)$ are internal actions. We want to show the non-regular property ϕ: any file opened in a procedure must be closed exactly in this procedure. We prove it for the cases where $i \in \{0, 1\}$ (it is a drastic abstraction of the possible values of i).

For the encoding of the program, use an atomic test b to represent $i = 0$ and $\bar{b}$ to represent $i = 1$. So, $\mathbf{B} \stackrel{\text{def}}{=} \{b\}$. Also, use $\Sigma_i \stackrel{\text{def}}{=} \{o, c\}$, $\Sigma_c \stackrel{\text{def}}{=} \{^{\langle}f\}$, $\Sigma_r \stackrel{\text{def}}{=} \{f^{\rangle}\}$ and $E \stackrel{\text{def}}{=} \{f, f', \tau\}$ where f' is used as the main procedure. Note that actions $o(i)$ and $c(i)$ are abstracted by actions o and c. The information that they depend on the value of i is lost. This will be taken care of in the refinement hypotheses. The encoding of the program's control flow p is given in Fig. 2(b).

The encoding of the desired semantics is given only for hypotheses used in the proof. So, $\mathcal{H}$ is bo = ob $\wedge$ $\bar{b} \cdot {}^{\langle}f \leqslant {}^{\langle}f \cdot b$ $\wedge$ $\bar{b} \cdot {}^{\langle}f \cdot (\![_f \mathcal{B}]\!)^{\tau} \cdot f^{\rangle} = {}^{\langle}f \cdot (\![_f \mathcal{B}]\!)^{\tau} \cdot f^{\rangle} \cdot \bar{b}$ where $\mathcal{B} \stackrel{\text{def}}{=} [_{f'} {}^{\langle}f \downarrow_f \uparrow^{\tau} f^{\rangle}]^{\tau}$, $[_f \text{boc} + \text{bo}({}^{\langle}f \downarrow_f \uparrow^{\tau} f^{\rangle})c]^{\tau}$. The first hypothesis states that action o does not modify the test b. The second and the third hypotheses abstractly encode the passing of the variable i by value: the second is a Hoare triple that states that if $i = 1$ just before calling f, then the value of i at the beginning of the newly called f is now 0, and the third states that the value of i just before calling f is remembered just after returning from f.

As we saw earlier, the property is relative to the exact file used. But the encoding of the program's control flow does not give enough information about

procedure f (int i) begin
if $(i \leqslant 0)$ then
$o(i)$;
$c(i)$
else
$o(i)$;
call $f(i-1)$;
$c(i)$
end if
end procedure

(a) Abstract program

$$\big(\!\!\big|_{f'} [_{f'}\, {}^{\langle}f \downarrow_f \uparrow^\tau f^{\rangle}\,]^\tau, [_f \mathsf{boc} + \bar{\mathsf{bo}}({}^{\langle}f\downarrow_f\uparrow^\tau f^{\rangle})\mathsf{c}]^\tau\, \big|\!\!\big)^\tau$$

(b) Program's control flow in algebra

$$\big(\!\!\big|_x [_x\Big(C(\mathsf{x}) + \mathsf{o}_1\big(C(\mathsf{y}_1) + \mathsf{o}_2(C(\mathsf{y}_{1,2}))^*\mathsf{c}_2\big)^*\mathsf{c}_1$$
$$+\ \mathsf{o}_2\big(C(\mathsf{y}_2) + \mathsf{o}_1(C(\mathsf{y}_{1,2}))^*\mathsf{c}_1\big)^*\mathsf{c}_2\Big)]^\tau,$$
$$[_{\mathsf{y}_1}\big(C(\mathsf{y}_1) + \mathsf{o}_2(C(\mathsf{y}_{1,2}))^*\mathsf{c}_2\big)^*]^\tau,$$
$$[_{\mathsf{y}_2}\big(C(\mathsf{y}_2) + \mathsf{o}_1(C(\mathsf{y}_{1,2}))^*\mathsf{c}_1\big)^*]^\tau,$$
$$[_{\mathsf{y}_{1,2}}(C(\mathsf{y}_{1,2}))^*]^\tau\, \big|\!\!\big)^\tau$$

where $C(z) \overset{\text{def}}{=} ({}^{\langle}f\downarrow_z\uparrow^\tau f^{\rangle})$ for $z \in \{\mathsf{x}, \mathsf{y}_1, \mathsf{y}_2, \mathsf{y}_{1,2}\}$

(c) The property in algebra

Fig. 2. Elements for the example of interprocedural analysis

which file is really used. To allow this, add four internal actions, namely o_1, o_2, c_1 and c_2, to Σ_i and define $\mathcal{R}$ to be $\mathsf{bo} \leqslant \mathsf{o}_1 \wedge \mathsf{bo} \leqslant \mathsf{o}_2 \wedge \mathsf{bc} \leqslant \mathsf{c}_1 \wedge \bar{\mathsf{bc}} \leqslant \mathsf{c}_2$. These hypotheses state that actions o and c depend on b and $\bar{\mathsf{b}}$.

The encoding of the property ϕ by an expression q is given in Fig. 2(c). Intuitively, this expression states that, at the beginning of the context x, no file is opened; at the beginning of a context y_j for $j \in \{1,2\}$, only the file $j-1$ is opened; at the beginning of the context $\mathsf{y}_{1,2}$, the two files are opened.

The program satisfies ϕ if $\mathcal{H} \wedge \mathcal{R} \to p \leqslant q$ is a theorem of VPKA. We prove it. Let $\mathcal{B} \overset{\text{def}}{=} [_{f'}\, {}^{\langle}f \downarrow_f \uparrow^\tau f^{\rangle}\,]^\tau, [_f \mathsf{boc} + \bar{\mathsf{bo}}({}^{\langle}f\downarrow_f\uparrow^\tau f^{\rangle})\mathsf{c}]^\tau$. By (8), $\big(\!\!\big|_\tau \mathcal{B}\big|\!\!\big)^\tau = 0$. Now,

$\big(\!\!\big|_{f'} \mathcal{B}\big|\!\!\big)^\tau$

$=$ 　$\{\!\!\{$ Equation (8) & Result: $\big(\!\!\big|_\tau \mathcal{B}\big|\!\!\big)^\tau = 0$ & Zero of $\cdot$ & Identity of $+$ $\}\!\!\}$

　　$^{\langle}f \cdot \big(\!\!\big|_f \mathcal{B}\big|\!\!\big)^\tau \cdot f^{\rangle}$

$=$ 　$\{\!\!\{$ Metablocks & Equation (8) & Result: $\big(\!\!\big|_\tau \mathcal{B}\big|\!\!\big)^\tau = 0$ & Zero of $\cdot$ &
　　　Identity of $+$ $\}\!\!\}$

　　$^{\langle}f \cdot (\mathsf{boc} + \bar{\mathsf{bo}} \cdot {}^{\langle}f \cdot \big(\!\!\big|_f \mathcal{B}\big|\!\!\big)^\tau \cdot f^{\rangle} \cdot \mathsf{c}) \cdot f^{\rangle}$

$\leqslant$ 　$\{\!\!\{$ Idempotency of tests & Hypotheses in $\mathcal{H}$ & Kleene algebra with
　　　tests: $\mathsf{bo} = \mathsf{ob} \leftrightarrow \bar{\mathsf{bo}} = \mathsf{o}\bar{\mathsf{b}}$ & Monotonicity of $\cdot$ and $+$ $\}\!\!\}$

　　$^{\langle}f \cdot (\mathsf{bobc} + \bar{\mathsf{bo}} \cdot {}^{\langle}f \cdot \mathsf{b} \cdot \big(\!\!\big|_f \mathcal{B}\big|\!\!\big)^\tau \cdot f^{\rangle} \cdot \bar{\mathsf{bc}}) \cdot f^{\rangle}$

$=$ 　$\{\!\!\{$ Metablocks & Equation (8) & Result: $\big(\!\!\big|_\tau \mathcal{B}\big|\!\!\big)^\tau = 0$ & Zero of $\cdot$ &
　　　Identity of $+$ $\}\!\!\}$

　　$^{\langle}f \cdot (\mathsf{bobc} + \bar{\mathsf{bo}} \cdot {}^{\langle}f \cdot \mathsf{b} \cdot (\mathsf{boc} + \bar{\mathsf{bo}} \cdot {}^{\langle}f \cdot \big(\!\!\big|_f \mathcal{B}\big|\!\!\big)^\tau \cdot f^{\rangle} \cdot \mathsf{c}) \cdot f^{\rangle} \cdot \bar{\mathsf{bc}}) \cdot f^{\rangle}$

$=$ 　$\{\!\!\{$ Distributivity of $\cdot$ over $+$ & Contradiction of tests & Zero of $\cdot$ &
　　　Identity of $+$ & Hypothesis: $\mathsf{bo} = \mathsf{ob}$ $\}\!\!\}$

　　$^{\langle}f \cdot (\mathsf{bobc} + \bar{\mathsf{bo}} \cdot {}^{\langle}f \cdot \mathsf{bobc} \cdot f^{\rangle} \cdot \bar{\mathsf{bc}}) \cdot f^{\rangle}$

$\leqslant$ 　$\{\!\!\{$ Hypotheses in $\mathcal{R}$ & Monotonicity of $\cdot$ and $+$ $\}\!\!\}$

　　$^{\langle}f \cdot (\mathsf{o}_1\mathsf{c}_1 + \mathsf{o}_2 \cdot {}^{\langle}f \cdot \mathsf{o}_1\mathsf{c}_1 \cdot f^{\rangle} \cdot \mathsf{c}_2) \cdot f^{\rangle}$

$\leqslant$ 　$\{\!\!\{$ Kleene algebra: $s_1 \leqslant s_1 + s_2$ & Equation (8) & Metablocks $\}\!\!\}$

　　q .

6   Related Work and Discussion

The axiomatisation of subclasses of context-free languages is not new (see for example [2,10,11]). Leiß proposed Kleene algebra with recursion which is essentially an idempotent semiring with an explicit general least fixed point operator μ. Bloom and Ésik did something similar to Leiß when they defined iteration algebras, but they also developed a robust theory for fixed point operators. In contrast, we define a family $\mathcal{F}_{(\!\!|\,\,|\!\!)}$ of partial operators that are implicit least fixed points and deal only with a restricted set of fixed point formulae.

There are some reasons why we use the family of operators instead of context-free grammars. Any expression $(\!\!|_x \mathcal{B} |\!\!)^y$ is a compact way to express a starting (x) and ending goal (y), and the definition of the operator allows to reach them by mixing forward and backward travelling. It is like defining several grammars at the same time and using the one needed on-the-fly with the ability to switch between these grammars at will. This ability is important when analysing programs and in the algebra itself. Note that (4) to (7) are committed to the direction of the travelling. Also, an expression $(\!\!|_x \mathcal{B} |\!\!)^y$ allows to make sure that the inequational system represented by $\mathcal{B}$ is "confined" to the expression, since $(\!\!|_x \mathcal{B} |\!\!)^y$ is the least solution for a *single component* of this system and not the entire system itself!

There already exist variants of Kleene algebra for statically analysing programs with or without procedures [3,9,13]. However, all these frameworks allow only to verify regular properties whereas we are able to deal with some non-regular properties. Of these frameworks, the only one that can deal with (mutually) recursive programs is the work of Mathieu and Desharnais [13]. It uses pushdown systems as programs and uses an extension of omega algebra with domain (that adds laws to represent explicit stack manipulations) along with matrices on this algebra to represent programs.

The model checking community works on the verification of (mutually) recursive programs [1,5,14]. However, most of the tools developed so far can just deal with regular properties. Of course, tools using visibly pushdown automata are able to deal with some non-regular properties. Note that our work is different because (i) the verification process looks for the existence of a proof in the algebraic system, (ii) we can encode proofs *à la* Proof-Carrying Code, and (iii) we are not limited to decidable problems when using the proof method.

7   Conclusion

We presented visibly pushdown Kleene algebra, an extension of Kleene algebra that axiomatises exactly the equational theory of visibly pushdown languages. The algebra adds a family of operators to Kleene algebra. These operators are implicit least fixed points that generate exactly any set of well-matched words. The algebraic system is sound and complete for the equational theory of visibly pushdown languages and its complexity is EXPTIME-complete.

We showed that VPKA extended with tests is suitable to do interprocedural analysis of (mutually) recursive programs. We sketched a framework to do these analyses. The framework is similar to [3] or [9], but procedures are added to the definition of programs and the verification of non-regular properties is possible.

For future work, one idea is to add to VPKA the omega operator along with an "infinite" version (infinite expansion, but a finite number of possibly pending calls at any time) of the family of operators to represent visibly pushdown ω-languages as defined in [1]. Also, it seems interesting to add a domain operator since it can give an algebra related to the extension of PDL defined in [12]. Moreover, other applications for VPKA need to be investigated, like the verification of the correct transformation of a recursive algorithm in an iterative algorithm.

Acknowledgements. We are grateful to Jules Desharnais and the anonymous referees for detailed comments.

References

1. Alur, R., Madhusudan, P.: Visibly pushdown languages. In: STOC 2004: Proc. of the 36th ACM symp. on Theory of computing, pp. 202–211. ACM, New York (2004)
2. Bloom, S.L., Ésik, Z.: Iteration theories: the equational logic of iterative processes. Springer, New York (1993)
3. Bolduc, C., Desharnais, J.: Static analysis of programs using omega algebra with tests. In: MacCaull, W., Winter, M., Düntsch, I. (eds.) RelMiCS 2005. LNCS, vol. 3929, pp. 60–72. Springer, Heidelberg (2006)
4. Bolduc, C., Ktari, B.: Visibly pushdown Kleene algebra. Technical Report DIUL-RR-0901, Laval university, QC, Canada (2009)
5. Bouajjani, A., Esparza, J., Maler, O.: Reachability analysis of pushdown automata: Application to model-checking. In: Mazurkiewicz, A., Winkowski, J. (eds.) CONCUR 1997. LNCS, vol. 1243, pp. 135–150. Springer, Heidelberg (1997)
6. Chaudhuri, S., Alur, R.: Instrumenting C programs with nested word monitors. In: Bošnački, D., Edelkamp, S. (eds.) SPIN 2007. LNCS, vol. 4595, pp. 279–283. Springer, Heidelberg (2007)
7. Kozen, D.: A completeness theorem for Kleene algebras and the algebra of regular events. Information and Computation 110(2), 366–390 (1994)
8. Kozen, D.: Kleene algebra with tests. Transactions on Programming Languages and Systems 19(3), 427–443 (1997)
9. Kozen, D.: Kleene algebras with tests and the static analysis of programs. Technical Report TR2003-1915, Cornell University, NY, USA (November 2003)
10. Leiß, H.: Kleene modules and linear languages. Journal of Logic and Algebraic Programming 66(2), 185–194 (2006)
11. Leiß, H.: Towards Kleene algebra with recursion. In: Kleine Büning, H., Jäger, G., Börger, E., Richter, M.M. (eds.) CSL 1991. LNCS, vol. 626, pp. 242–256. Springer, Heidelberg (1992)
12. Löding, C., Lutz, C., Serre, O.: Propositional dynamic logic with recursive programs. Journal of Logic and Algebraic Programming 73, 51–69 (2007)
13. Mathieu, V., Desharnais, J.: Verification of pushdown systems using omega algebra with domain. In: MacCaull, W., Winter, M., Düntsch, I. (eds.) RelMiCS 2005. LNCS, vol. 3929, pp. 188–199. Springer, Heidelberg (2006)
14. Schwoon, S.: Model-Checking Pushdown Systems. PhD thesis, Technische Universität München (2002)

Towards Algebraic Separation Logic

Han-Hing Dang, Peter Höfner, and Bernhard Möller

Institut für Informatik, Universität Augsburg, D-86135 Augsburg, Germany
{h.dang,hoefner,moeller}@informatik.uni-augsburg.de

Abstract. We present an algebraic approach to separation logic. In particular, we give algebraic characterisations for all constructs of separation logic. The algebraic view does not only yield new insights on separation logic but also shortens proofs and enables the use of automated theorem provers for verifying properties at a more abstract level.

1 Introduction

Two prominent formal methods for reasoning about the correctness of programs are Hoare logic [9] and the wp-calculus of Dijkstra [7]. These approaches, although foundational, lack expressiveness for shared mutable data structures, i.e., structures where updatable fields can be referenced from more than one point (e.g. [19]). To overcome this deficiency Reynolds, O'Hearn and others have developed *separation logic* for reasoning about such data structures [17]. Their approach extends Hoare logic by assertions to express separation within memory, both in the store and the heap. Furthermore the command language is enriched by some constructs that allow altering these separate ranges. The introduced mechanisms have been extended to concurrent programs that work on shared mutable data structures [16].

This paper presents an algebraic approach to separation logic. As a result many proofs become simpler while still being fully precise. Moreover, this places the topic into a more general context and therefore allows re-use of a large body of existing theory.

In Section 2 we recapitulate syntax and semantics of expressions in separation logic and give a formal definition of an update-operator for relations. Section 3 gives the semantics of assertions. After providing the algebraic background in Section 4, we shift from the validity semantics of separation logic to one based on the set of states satisfying an assertion. Abstracting from the set view yields an algebraic interpretation of assertions in the setting of semirings and quantales. In Section 6 we discuss special classes of assertions: pure assertions do not depend on the heap at all; intuitionistic assertions do not specify the heap exactly. We conclude with a short outlook.

2 Basic Definitions

Separation logic, as an extension of Hoare logic, does not only allow reasoning about explicitly named program variables, but also about anonymous variables

R. Berghammer et al. (Eds.): RelMiCS/AKA 2009, LNCS 5827, pp. 59–72, 2009.
© Springer-Verlag Berlin Heidelberg 2009

in dynamically allocated storage. Therefore a program state in separation logic consists of a *store* and a *heap*. In the remainder we consistently write s for stores and h for heaps.

To simplify the formal treatment, one defines values and addresses as integers, stores and heaps as partial functions from variables or addresses to values and states as pairs of stores and heaps:

$$Values = \mathbb{Z} \; ,$$
$$\{\text{nil}\} \uplus Addresses \subseteq Values \; ,$$
$$Stores = V \rightsquigarrow Values \; ,$$
$$Heaps = Addresses \rightsquigarrow Values \; ,$$
$$States = Stores \times Heaps \; ,$$

where V is the set of all variables, $\uplus$ denotes the disjoint union on sets and $M \rightsquigarrow N$ denotes the set of partial functions between M and N. With this definition, we slightly deviate from [19] where stores are defined as functions from variables to values of $\mathbb{Z}$ and heaps as functions from addresses into values of $\mathbb{Z}$, while addresses are also values of $\mathbb{Z}$.

The constant nil is a value for pointers that denotes an improper reference like null in programming languages like JAVA; by the above definitions, nil is not an address and hence heaps do not assign values to nil.

As usual we denote the domain of a relation (partial function) R by $\text{dom}(R)$:

$$\text{dom}(R) =_{df} \{x : \exists \, y.(x,y) \in R\} \; .$$

In particular, the domain of a store denotes all currently used program variables and $\text{dom}(h)$ is the set of all currently allocated addresses on a heap h.

As in [14] and for later definitions we also need an *update* operator. It is used to model changes in stores and heaps. We will first give a definition and then explain its meaning.

Let R and S be partial functions. Then we define

$$R \mid S =_{df} R \cup \{(x,y) \mid (x,y) \in S \wedge x \notin \text{dom}(R)\} \; . \tag{1}$$

The relation R updates the relation S with all possible pairs of R in such a way that $R \mid S$ is again a partial function. The domain of the right hand side of $\cup$ above is disjoint from that of R. In particular, $R \mid S$ can be seen as an extension of R to $\text{dom}(R) \cup \text{dom}(S)$. In later definitions we abbreviate an update $\{(x,v)\} \mid S$ on a single variable or address by omitting the set-braces and simply writing $(x,v) \mid S$ instead.

Expressions are used to denote values or Boolean conditions on stores and are independent of the heap, i.e., they only need the store component of a given state for their evaluation. Informally, *exp*-expressions are simple arithmetical expressions over variables and values, while *bexp*-expressions are Boolean expressions over simple comparisons and true, false. Their syntax is given by

$$var ::= x \mid y \mid z \mid \ldots$$
$$exp ::= 0 \mid 1 \mid 2 \mid \ldots \mid var \mid exp \pm exp \mid \ldots$$
$$bexp ::= \text{true} \mid \text{false} \mid exp = exp \mid exp < exp \mid \ldots$$

The semantics e^S of an expression e w.r.t. a store s is straightforward (assuming that all variables occurring in e are contained in $\text{dom}(s)$). For example,

$$c^S = c \quad \forall c \in \mathbb{Z} , \quad \text{true}^S = \text{true} \quad \text{and} \quad \text{false}^S = \text{false} .$$

3   Assertions

Assertions play an important rôle in separation logic. They are used as predicates to describe the contents of heaps and stores and as pre- or postconditions in programs, like in Hoare logic:

$$assert ::= bexp \mid \neg\, assert \mid assert \vee assert \mid \forall\, var.\, assert \mid$$
$$\text{emp} \mid exp \mapsto exp \mid assert * assert \mid assert \mathbin{-\!*} assert .$$

In the remainder we consistently write p, q and r for assertions. Assertions are split into two parts: the "classical" ones from predicate logic and four new ones that express properties of the heap. The former are supplemented by the logical connectives $\wedge$, $\rightarrow$ and $\exists$ that are defined, as usual, by $p \wedge q =_{df} \neg(\neg p \vee \neg q)$, $p \rightarrow q =_{df} \neg p \vee q$ and $\exists v.\, p =_{df} \neg \forall v.\, \neg p$.

The semantics of assertions is given by the relation $s, h \models p$ of *satisfaction*. Informally, $s, h \models p$ holds if the state (s, h) satisfies the assertion p; an assertion p is called *valid* iff p holds in every state and, finally, p is *satisfiable* if there exists a state (s, h) which satisfies p. The semantics is defined inductively as follows (e.g. [19]).

$$
\begin{aligned}
s, h &\models b & &\Leftrightarrow_{df} \ b^S = \text{true} \\
s, h &\models \neg p & &\Leftrightarrow_{df} \ s, h \not\models p \\
s, h &\models p \vee q & &\Leftrightarrow_{df} \ s, h \models p \ \text{ or } \ s, h \models q \\
s, h &\models \forall v.\, p & &\Leftrightarrow_{df} \ \forall x \in \mathbb{Z} : (v, x) \,|\, s, h \models p \\
s, h &\models \text{emp} & &\Leftrightarrow_{df} \ h = \emptyset \\
s, h &\models e_1 \mapsto e_2 & &\Leftrightarrow_{df} \ h = \{(\, e_1^S ,\, e_2^S \,)\} \\
s, h &\models p * q & &\Leftrightarrow_{df} \ \exists\, h_1, h_2 \in \mathit{Heaps} : \text{dom}(h_1) \cap \text{dom}(h_2) = \emptyset \text{ and} \\
& & & \qquad h = h_1 \cup h_2 \text{ and } s, h_1 \models p \text{ and } s, h_2 \models q \\
s, h &\models p \mathbin{-\!*} q & &\Leftrightarrow_{df} \ \forall\, h' \in \mathit{Heaps} : (\text{dom}(h') \cap \text{dom}(h) = \emptyset \text{ and } s, h' \models p) \\
& & & \qquad \text{implies } s, h' \cup h \models q .
\end{aligned}
$$

Here, b is a *bexp*-expression, p, q are assertions and e_1, e_2 are *exp*-expressions. The first four clauses do not make any assumptions about the heap and only carry it along without making any changes to it; they are well known from predicate logic or Hoare logic [9].

The remaining lines describe the new parts in separation logic: For an arbitrary state (s, h), emp ensures that the heap h is empty and contains no

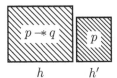 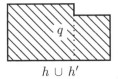

Fig. 1. Separating implication[1]

addressable cells. An assertion $e_1 \mapsto e_2$ characterises states with the singleton heap that has exactly one cell at the address e_1^s with the value e_2^s. To reason about more complex heaps, the *separating conjunction* $*$ is used. It allows expressing properties of heaps that result from merging smaller disjoint heaps, i.e., heaps with disjoint domains.

The *separating implication* $p \mathbin{-\!\!*} q$ guarantees, that if the current heap h is extended with a heap h' satisfying p, the merged heap $h \cup h'$ satisfies q (cf. Figure 1). If the heaps are not disjoint, the situation is interpreted as an error case and the assertion is not satisfied.

4   Quantales and Residuals

To present our algebraic semantics of separation logic in the next section we now prepare the algebraic background.

A *quantale* [20] is a structure $(S, \leq, 0, \cdot, 1)$ where $(S, \leq)$ is a complete lattice and $\cdot$ is completely disjunctive, i.e., $\cdot$ distributes over arbitrary suprema. Moreover 0 is the least element and 1 is the identity of the $\cdot$ operation. The infimum and supremum of two elements $a, b \in S$ are denoted by $a \sqcap b$ and $a + b$, resp. The greatest element of S is denoted by $\top$. The definition implies that $\cdot$ is strict, i.e., that $0 \cdot a = 0 = a \cdot 0$ for all $a \in S$. The notion of a quantale is equivalent to that of a *standard Kleene algebra* [3] and a special case of the notion of an idempotent semiring.

A quantale is called *Boolean* if its underlying lattice is distributive and complemented, whence a Boolean algebra. Equivalently, a quantale S is Boolean if it satisfies the Huntington axiom $a = \overline{\overline{a} + b} + \overline{\overline{a} + b}$ for all $a, b \in S$ [12,11]. The infimum is then defined by the de Morgan duality $a \sqcap b =_{df} \overline{\overline{a} + \overline{b}}$. An important Boolean quantale is REL, the algebra of binary relations over a set under set inclusion, relation composition and set complement.

A quantale is called *commutative* if $\cdot$ commutes, i.e., $a \cdot b = b \cdot a$ for all a, b.

In any quantale, the *right residual* $a \backslash b$ [1] exists and is characterised by the Galois connection

$$x \leq a \backslash b \Leftrightarrow_{df} a \cdot x \leq b \,.$$

Symmetrically, the *left residual* b/a can be defined. However, if the underlying quantale is commutative then both residuals coincide, i.e., $a \backslash b = b/a$. In REL,

---

[1] The right picture might suggest that the heaps are adjacent after the join. But the intention is only to bring out abstractly that the united heap satisfies q.

one has $R \backslash S = \overline{R^\smile \, ; \, \overline{S}}$ and $R/S = \overline{\overline{R} \, ; \, S^\smile}$, where $^\smile$ denotes relational converse and ; is relational composition.

In a Boolean quantale, the *right detachment* $a \lfloor b$ can be defined based on the left residual as

$$a \lfloor b =_{df} \overline{\overline{a}/b} \, .$$

In REL, $R \lfloor S = R \, ; \, S^\smile$. By de Morgan's laws, the Galois connection for $/$ transforms into the exchange law

$$a \lfloor b \leq x \Leftrightarrow \overline{x} \cdot b \leq \overline{a} \qquad \text{(exc)}$$

for $\lfloor$ that generalises the Schröder rule of relational calculus. An important consequence is the Dedekind rule [13]

$$a \sqcap (b \cdot c) \leq (a \lfloor c \sqcap b) \cdot c \, . \qquad \text{(Ded)}$$

5 An Algebraic Model of Assertions

We now give an algebraic interpretation for the semantics of separation logic. The main idea is to switch from the satisfaction-based semantics for single states to an equivalent set-based one where every assertion is associated with the set of all states satisfying it. This simplifies proofs considerably.

For an arbitrary assertion p we therefore define its set-based semantics as

$$[\![p]\!] =_{df} \{(s,h) : s,h \models p\} \, .$$

The sets $[\![p]\!]$ of states will be the elements of our algebra. By this we then have immediately the connection $s,h \models p \Leftrightarrow (s,h) \in [\![p]\!]$. This validity assertion can be lifted to set of states by setting, for $A \subseteq States$, $A \models p \Leftrightarrow A \subseteq [\![p]\!]$. The embedding of the standard Boolean connectives is given by

$$[\![\neg p]\!] = \{(s,h) : s,h \not\models p\} = \overline{[\![p]\!]} \, ,$$
$$[\![p \vee q]\!] = [\![p]\!] \cup [\![q]\!] \, ,$$
$$[\![\forall v.p\,]\!] = \{(s,h) : \forall x \in \mathbb{Z} \, . \, (v,x) \,|\, s,h \models p\} \, .$$

Using these definitions, it is straightforward to show that

$$[\![p \wedge q]\!] = [\![p]\!] \cap [\![q]\!] \, , \qquad [\![p \rightarrow q]\!] = \overline{[\![p]\!]} \cup [\![q]\!] \, , \qquad \text{and}$$
$$[\![\exists v.p\,]\!] = \overline{[\![\forall v. \neg p\,]\!]} = \{(s,h) : \exists x \in \mathbb{Z} \, . \, (v,x) \,|\, s,h \models p\} \, ,$$

where $|$ is the update operation defined in (1).

The emptiness assertion emp and the assertion operator $\mapsto$ are given by

$$[\![\mathsf{emp}]\!] =_{df} \{(s,h) : h = \emptyset\}$$
$$[\![e_1 \mapsto e_2]\!] =_{df} \{(s,h) : h = \{(e_1^s, e_2^s)\}\} \, .$$

Next, we model the separating conjunction $*$ algebraically by

$$[\![\, p * q \,]\!] =_{df} [\![\, p \,]\!] \uplus [\![\, q \,]\!], \text{ where}$$
$$P \uplus Q =_{df} \{(s, h \cup h') : (s, h) \in P \wedge (s, h') \in Q \wedge \mathsf{dom}(h) \cap \mathsf{dom}(h') = \emptyset\} \ .$$

In this way inconsistent states as well as "erroneous" merges of non-disjoint heaps are excluded.

These definitions yield an algebraic embedding of separation logic.

Theorem 5.1. *The structure* $\mathsf{AS} =_{df} (\mathcal{P}(States), \subseteq, \emptyset, \uplus, [\![\, \mathsf{emp} \,]\!])$ *is a commutative and Boolean quantale with* $P + Q = P \cup Q$.

The proof is by straightforward calculations; it can be found in [4]. It is easy to show that $[\![\mathsf{true}]\!]$ is the greatest element in the above quantale, i.e., $[\![\mathsf{true}]\!] = \top$, since every state satisfies the assertion true. This implies immediately that $[\![\mathsf{true}]\!]$ is the neutral element for $\sqcap$. However, in contrast to addition $\cup$, multiplication $\uplus$ is in general not idempotent.

Example 5.2. In AS,

$$[\![\, (x \mapsto 1) * (x \mapsto 1) \,]\!] = [\![\, x \mapsto 1 \,]\!] \uplus [\![\, x \mapsto 1 \,]\!] = \emptyset \ .$$

This can be shown by straightforward calculations using the above definitions.

$$\begin{aligned}
&\quad [\![\, (x \mapsto 1) * (x \mapsto 1) \,]\!] \\
&= [\![\, (x \mapsto 1) \,]\!] \uplus [\![\, (x \mapsto 1) \,]\!] \\
&= \{(s, h \cup h') : (s, h), (s, h') \in [\![\, x \mapsto 1 \,]\!] \wedge \mathsf{dom}(h) \cap \mathsf{dom}(h') = \emptyset\} \\
&= \emptyset \ .
\end{aligned}$$

$[\![\, x \mapsto 1 \,]\!]$ is the set of all states that have the single-cell heap $\{(s(x), 1)\}$. The states (s, h) and (s, h') have to share this particular heap. Hence the domains of the merged heaps would not be disjoint. Therefore the last step yields the empty result. $\Box$

As a check of the adequacy of our definitions we list a couple of properties.

Lemma 5.3. *In separation logic, for assertions* p, q, r, *we have*

$$\frac{}{(p \wedge q) * r \Rightarrow (p * r) \wedge (q * r)} \qquad \text{and} \qquad \frac{p \Rightarrow r \quad q \Rightarrow s}{p * q \Rightarrow r * s} \ .$$

The second property denotes isotony of separating conjunction. Both properties together are, by standard quantale theory, equivalent to isotony of separating conjunction.

More laws and examples can be found in [4].

For the separating implication the set-based semantics extracted from the definition in Section 3 is

$$[\![\, p \mathbin{-\!\!*} q \,]\!] =_{df} \{(s, h) : \forall\, h' \in Heaps : (\mathsf{dom}(h) \cap \mathsf{dom}(h') = \emptyset \wedge (s, h') \in [\![\, p \,]\!])$$
$$\Rightarrow (s, h \cup h') \in [\![\, q \,]\!]\} \ .$$

This implies that separating implication corresponds to a residual.

Lemma 5.4. *In* AS, $[\![\,p \twoheadrightarrow q\,]\!] = [\![\,p\,]\!]\backslash[\![\,q\,]\!] = [\![\,q\,]\!]/[\![\,p\,]\!]$.

Proof. We first show the claim for a single state. By definition above, set theory and definition of $\cup$, we have

$$(s, h) \in [\![\,p \twoheadrightarrow q\,]\!]$$
$$\Leftrightarrow \forall h' : ((s, h') \in [\![\,p\,]\!] \wedge \mathsf{dom}(h) \cap \mathsf{dom}(h') = \emptyset \Rightarrow (s, h \cup h') \in [\![\,q\,]\!])$$
$$\Leftrightarrow \{(s, h \cup h') : (s, h') \in [\![\,p\,]\!] \wedge \mathsf{dom}(h) \cap \mathsf{dom}(h') = \emptyset\} \subseteq [\![\,q\,]\!]$$
$$\Leftrightarrow \{(s, h)\} \cup [\![\,p\,]\!] \subseteq [\![\,q\,]\!] .$$

and therefore, for arbitrary set R of states,

$$R \subseteq [\![\,p \twoheadrightarrow q\,]\!]$$
$$\Leftrightarrow \forall (s, h) \in R : (s, h) \in [\![\,p \twoheadrightarrow q\,]\!]$$
$$\Leftrightarrow \forall (s, h) \in R : \{(s, h)\} \cup [\![\,p\,]\!] \subseteq [\![\,q\,]\!]$$
$$\Leftrightarrow R \cup [\![\,p\,]\!] \subseteq [\![\,q\,]\!] .$$

Hence, by definition of the residual, $[\![\,p \twoheadrightarrow q\,]\!] = [\![\,p\,]\!]\backslash[\![\,q\,]\!]$. The second equation follows immediately since multiplication $\cup$ in AS commutes (cf. Section 2). $\quad\Box$

Now all laws of [19] about $\twoheadrightarrow$ follow from the standard theory of residuals (e.g. [2]). Many of these laws are proved algebraically in [4]. For example, the two main properties of separating implication, namely the currying and decurrying rules, are nothing but the transcriptions of the defining Galois connection for right residuals.

Corollary 5.5. *In separation logic the following inference rules hold:*

$$\frac{p * q \Rightarrow r}{p \Rightarrow (q \twoheadrightarrow r)} , \qquad \text{(currying)} \qquad \frac{p \Rightarrow (q \twoheadrightarrow r)}{p * q \Rightarrow r} . \qquad \text{(decurrying)}$$

This means that $q \twoheadrightarrow r$ is the weakest assertion guaranteeing that a state in $[\![\,q \twoheadrightarrow r\,]\!]$ merged with a state in $[\![\,q\,]\!]$ yields a state in $[\![\,r\,]\!]$.

As far as we know, in his works Reynolds only states that these laws follow directly from the definition. We are not aware of any proof of the equalities given in Lemma 5.4, although many authors state this claim and refer to Reynolds.

As a further example we prove the algebraic counterpart of the inference rule

$$\overline{q * (q \twoheadrightarrow p) \Rightarrow p} .$$

Lemma 5.6. *Let S be a quantale. For $a, b \in S$ the inequality $q \cdot (q \backslash p) \leq p$ holds.*

Proof. By definition of residuals we immediately get
$$q \cdot (q \backslash p) \leq p \Leftrightarrow q \backslash p \leq q \backslash p \Leftrightarrow \text{true} .$$
$\quad\Box$

6   Special Classes of Assertions

Reynolds distinguishes different classes of assertions [19]. We will give algebraic characterisations for three main classes, namely *pure*, *intuitionistic* and *precise assertions.* Pure assertions are independent of the heap of a state and therefore only express conditions on store variables. Intuitionistic assertions do not describe the domain of a heap exactly. Hence, when using these assertions one does not know whether the heap contains additional anonymous cells. In contrast, precise assertions point out a unique subheap which is relevant to its predicate.

6.1   Pure Assertions

An assertion p is called *pure* iff it is independent of the heaps of the states involved, i.e.,

$$p \text{ is pure} \Leftrightarrow_{df} (\forall h, h' \in Heaps : s, h \models p \Leftrightarrow s, h' \models p) .$$

Theorem 6.1. *In* AS *an element* $[\![p]\!]$ *is pure iff it satisfies, for all* $[\![q]\!]$ *and* $[\![r]\!]$,

$$[\![p]\!] \cup [\![\text{true}]\!] \subseteq [\![p]\!] \quad \text{and} \quad [\![p]\!] \cap ([\![q]\!] \cup [\![r]\!]) \subseteq ([\![p]\!] \cap [\![q]\!]) \cup ([\![p]\!] \cap [\![r]\!]) .$$

Before we give the proof, we derive a number of auxiliary laws. The above theorem motivates the following definition.

Definition 6.2. In an arbitrary Boolean quantale S an element p is called *pure* iff it satisfies, for all $a, b \in S$,

$$p \cdot \top \leq p , \tag{2}$$
$$p \sqcap (a \cdot b) \leq (p \sqcap a) \cdot (p \sqcap b) . \tag{3}$$

The first equation models upwards closure of pure elements. It can be strengthened to an equation since its converse holds for arbitrary Boolean quantales. The second equation enables pure elements to distribute over meet and is equivalent to downward closure.

Lemma 6.3. *Property* (3) *is equivalent to* $p \lfloor \top \leq p$, *where* $p \lfloor \top$ *forms the downward closure of* p.

Proof. ($\Leftarrow$): Using Equation (Ded), isotony and the assumption, we get

$$p \sqcap a \cdot b \leq (p \lfloor b \sqcap a) \cdot b \leq (p \lfloor \top \sqcap a) \cdot b \leq (p \sqcap a) \cdot b$$

and the symmetric formula $p \sqcap a \cdot b \leq a \cdot (p \sqcap b)$. From this the claim follows by

$$p \sqcap (a \cdot b) = p \sqcap p \sqcap (a \cdot b) \leq p \sqcap ((p \sqcap a) \cdot b) \leq (p \sqcap a) \cdot (p \sqcap b) .$$

($\Rightarrow$): By Axiom (3) we obtain $p \sqcap (\overline{p} \cdot \top) \leq (p \sqcap \overline{p}) \cdot (p \sqcap \top) = 0$ and hence, by shunting and the exchange law (exc), $p \lfloor \top \leq p$. $\square$

Corollary 6.4. p *is pure iff* $p \cdot \top \leq p$ *and* $\overline{p} \cdot \top \leq \overline{p}$.

Corollary 6.5. *Pure elements form a Boolean lattice, i.e., they are closed under* $+, \sqcap$ *and* $\overline{}$.

Moreover we get a fixed point characterisation if the underlying quantale commutes.

Lemma 6.6. *In a Boolean quantale, an element p is pure iff $p = (p \sqcap 1) \cdot \top$ holds.*

Proof. We first show that $p = (p \sqcap 1) \cdot \top$ follows from Inequations (2) and (3). By neutrality of $\top$ for $\sqcap$, neutrality of 1 for $\cdot$, meet-distributivity (3) and isotony, we get

$$p = p \sqcap \top = p \sqcap (1 \cdot \top) \le (p \sqcap 1) \cdot (p \sqcap \top) \le (p \sqcap 1) \cdot \top .$$

The converse inequation follows by isotony and Inequation (2):

$$(p \sqcap 1) \cdot \top \le p \cdot \top \le p .$$

Next we show that $p = (p \sqcap 1) \cdot \top$ implies the two inequations $p \cdot \top \le p$ and $\overline{p} \cdot \top \le \overline{p}$ which, by Corollary 6.4, implies the claim. The first inequation is shown by the assumption, the general law $\top \cdot \top = \top$ and the assumption again:

$$p \cdot \top = (p \sqcap 1) \cdot \top \cdot \top = (p \sqcap 1) \cdot \top = p .$$

For the second inequation, we note that in a Boolean quantale the law $\overline{s \cdot \top} = (\overline{s} \sqcap 1) \cdot \top$ holds for all subidentities s ($s \le 1$) (e.g. [6]). From this we get

$$\overline{p} \cdot \top = \overline{(p \sqcap 1) \cdot \top} \cdot \top = (\overline{p} \sqcap 1) \cdot \top \cdot \top = (\overline{p} \sqcap 1) \cdot \top = \overline{(p \sqcap 1) \cdot \top} = \overline{p} . \quad \square$$

Corollary 6.7. *The set of pure elements forms a complete lattice.*

Proof. Lemma 6.6 characterises the pure elements as the fixed points of the isotone function $f(x) = (x \sqcap 1) \cdot \top$ on the quantale. By Tarski's fixed point theorem these form a complete lattice. $\quad \square$

Proof of Theorem 6.1. By Lemma 6.6 and definition of the elements of AS it is sufficient to show that the following formulas are equivalent in separation logic

$$\forall s \in Stores, \forall h, h' \in Heaps : (s, h \models p \Leftrightarrow s, h' \models p) , \tag{4}$$

$$\forall s \in Stores, \forall h \in Heaps : (s, h \models p \Leftrightarrow s, h \models (p \wedge \mathsf{emp}) * \mathsf{true}) . \tag{5}$$

Since both assertions are universally quantified over states we omit that quantification in the remainder and only keep the quantifiers on heaps. Before proving this equivalence we simplify $s, h \models (p \wedge \mathsf{emp}) * \mathsf{true}$. Using the definitions of Section 3, we get for all $h \in Heaps$

$$\begin{aligned}
& s, h \models (p \wedge \mathsf{emp}) * \mathsf{true} \\
\Leftrightarrow{} & \exists h_1, h_2 \in Heaps : \mathsf{dom}(h_1) \cap \mathsf{dom}(h_2) = \emptyset \text{ and } h = h_1 \cup h_2 \\
& \text{and } s, h_1 \models p \text{ and } s, h_1 \models \mathsf{emp} \text{ and } s, h_2 \models \mathsf{true} \\
\Leftrightarrow{} & \exists h_1, h_2 \in Heaps : \mathsf{dom}(h_1) \cap \mathsf{dom}(h_2) = \emptyset \text{ and } h = h_1 \cup h_2 \\
& \text{and } s, h_1 \models p \text{ and } h_1 = \emptyset \\
\Leftrightarrow{} & \exists h_2 \in Heaps : h = h_2 \text{ and } s, \emptyset \models p \\
\Leftrightarrow{} & s, \emptyset \models p .
\end{aligned}$$

The last line shows that a pure assertion is independent of the heap and hence, in particular, has to be satisfied for the empty heap. Next we show the implication (4) $\Rightarrow$ (5). Instantiating Equation (4) and using the above result immediately imply the claim:

$$\forall\, h, h' \in Heaps : (s, h \models p \Leftrightarrow s, h' \models p)$$
$$\Rightarrow \forall\, h \in Heaps : (s, h \models p \Leftrightarrow s, \emptyset \models p)$$
$$\Leftrightarrow \forall\, h \in Heaps : (s, h \models p \Leftrightarrow s, h \models (p \wedge \mathsf{emp}) * \mathsf{true}) .$$

For the converse direction, we take two instances of (5). Then, using again the above result, we get

$$\forall\, h \in Heaps : (s, h \models p \Leftrightarrow s, h \models (p \wedge \mathsf{emp}) * \mathsf{true})$$
$$\text{and } \forall\, h' \in Heaps : (s, h' \models p \Leftrightarrow s, h' \models (p \wedge \mathsf{emp}) * \mathsf{true})$$
$$\Rightarrow \forall\, h, h' \in Heaps : (s, h \models p \Leftrightarrow s, h \models (p \wedge \mathsf{emp}) * \mathsf{true}$$
$$\text{and } s, h' \models p \Leftrightarrow s, h' \models (p \wedge \mathsf{emp}) * \mathsf{true})$$
$$\Leftrightarrow \forall\, h, h' \in Heaps : (s, h \models p \Leftrightarrow s, \emptyset \models p \text{ and } s, h' \models p \Leftrightarrow s, \emptyset \models p)$$
$$\Rightarrow \forall\, h, h' \in Heaps : (s, h \models p \Leftrightarrow s, h' \models p) .$$

$\square$

The complexity of this proof in predicate-logic illustrates the advantage that is gained by passing to an algebraic treatment. Logic-based formulas (in particular in separation logic) can become long and complicated. Calculating at the abstract level of quantales often shorten the proofs. Moreover the abstraction paves the way to using first-order off-the-shelf theorem provers for verifying properties; whereas a first-order theorem prover for separation logic has yet to be developed and implemented (cf. Section 7).

To conclude the paragraph concerning pure elements we list a couple of properties which can be proved very easily by our algebraic approach.

Lemma 6.8. *Consider a Boolean quantale S, pure elements $p, q \in S$ and arbitrary elements $a, b \in S$ Then*

(a) $p \cdot a = p \sqcap a \cdot \top$;
(b) $(p \sqcap a) \cdot b = p \sqcap a \cdot b$;
(c) $p \cdot q = p \sqcap q$; *in particular $p \cdot p = p$ and $p \cdot \bar{p} = 0$.*

Their corresponding counterparts in separation logic and the proofs can again be found in [4].

The following lemma shows a.o. that in the complete lattice of pure elements meet and join coincide with composition and sum, respectively.

As far as we know these closure properties are new and were not shown in separation logic so far.

6.2   Intuitionistic Assertions

Let us now turn to intuitionistic assertions. Following [19], an assertion p is *intuitionistic* iff

$$\forall\, s \in Stores, \forall\, h, h' \in Heaps : (h \subseteq h' \text{ and } s, h \models p) \text{ implies } s, h' \models p . \quad (6)$$

This means for a heap that satisfies an intuitionistic assertion p that it can be extended by arbitrary cells and still satisfies p.

Similar calculations as in the proof of Theorem 6.1 yield the equivalence of Equation (6) and

$$\forall s \in Stores,\ \forall h \in Heaps : (s, h \models p * \textsf{true} \Rightarrow s, h \models p)\ . \tag{7}$$

Lifting this to an abstract level motivates the following definition.

Definition 6.9. In an arbitrary Boolean quantale S an element i is called *intuitionistic* iff it satisfies

$$i \cdot \top \leq i\ . \tag{8}$$

Elements of the form $i \cdot \top$ are also called vectors or ideals.

Corollary 6.10. *Every pure element of a Boolean quantale is intuitionistic.*

As before we just give a couple of properties. The proofs are again straightforward at the algebraic level.

Lemma 6.11. *Consider a Boolean quantale S, intuitionistic elements $i, j \in S$ and arbitrary elements $a, b \in S$ Then*

(a) $(i \sqcap 1) \cdot \top \leq i$;
(b) $i \cdot a \leq i \sqcap (a \cdot \top)$;
(c) $(i \sqcap a) \cdot b \leq i \sqcap (a \cdot b)$;
(d) $i \cdot j \leq i \sqcap j$.

Using the quantale AS, it is easy to see that none of these inequations can be strengthened to an equation. In particular, unlike as for pure assertions, multiplication and meet need not coincide.

Example 6.12. Consider $i =_{df} j =_{df} [\![x \mapsto 1 * \textsf{true}]\!] = [\![x \mapsto 1]\!] \cup [\![\textsf{true}]\!]$. By this definition it is obvious that i and j are intuitionistic. The definitions of Section 3 then immediately imply

$$i \cap j = [\![x \mapsto 1]\!] \cup [\![\textsf{true}]\!]$$
$$i \cup j = [\![x \mapsto 1]\!] \cup [\![\textsf{true}]\!] \cup [\![x \mapsto 1]\!] \cup [\![\textsf{true}]\!] = \emptyset\ .$$

The last step follows from Example 5.2. □

Other classes of assertions for separation logic are given in [19] and most of their algebraic counterparts in [4].

6.3   Precise Assertions

An assertion p is called *precise* if and only if for all states (s, h), there is at most one subheap h' of h for which $(s, h') \models p$. According to [18], this definition is equivalent to distributivity of $*$ over $\wedge$. Hence, using isotony of $*$ we can algebraically characterise precise assertions as follows.

Definition 6.13. In an arbitrary Boolean quantale S an element r is called *precise* iff for all p, q

$$(r * p) \sqcap (r * q) \leq r * (p \sqcap q) . \tag{9}$$

Next we give some closure properties for this assertion class.

Lemma 6.14. *If p and q are precise then also $p * q$ is precise.*

Proof. For arbitrary r_1 and r_2 we calculate

$$p * q * r_1 \sqcap p * q * r_2 \leq p * (q * r_1 \sqcap q * r_2) \leq (p * q) * (r_1 \sqcap r_2)$$

assuming p and q are precise.   □

Lemma 6.15. *If p is precise and $q \leq p$ then q is precise, i.e., precise assertions are downward closed.*

A proof can be found in [6].

Corollary 6.16. *For an arbitrary assertion q and precise p, also $p \sqcap q$ is precise.*

7   Conclusion and Outlook

We have presented a treatment towards an algebra of separation logic. For assertions we have introduced a model based on sets of states. By this, separating implication coincides with a residual and most of the inference rules of [19] are simple consequences of standard residual laws. For pure, intuitionistic and precise assertions we have given algebraic axiomatisations.

The next step will be to embed the command language of separation logic into a relational algebraic structure. A first attempt is given in [5] where we have defined a relational semantics for the heap-dependent commands and lifted the set-based semantics of assertions to relations. There, we are able to characterise the frame rule

$$\frac{\{p\} \, c \, \{q\}}{\{p * r\} \, c \, \{q * r\}} \,,$$

where p, q and r are arbitrary assertions and c is a command. The rule assumes that no free variable of r is modified by c. However, a complete algebraic proof of the frame rule is still missing, since we do not yet know how to characterise the conditions controlling the modification of free variables.

To underpin our approach we have algebraically verified one of the standard examples — an in-place list reversal algorithm. The details can be found in [4]. The term *in-place* means that there is no copying of whole structures, i.e., the reversal is done by simple pointer modifications.

So far we have not analysed situations where data structures share parts of their cells (cf. Figure 2). First steps towards an algebraic handling of such situations are given in [15,8]. In future work, we will adapt these approaches for our algebra of separation logic.

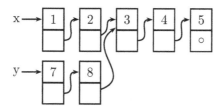

Fig. 2. Two lists with shared cells

Our algebraic approach to separation logic also paves the way to verifying properties with off-the-shelf theorem provers. Boolean semirings and quantales have proved to be reasonably well suitable for automated theorem provers [10]. Hence one of the next plans for future work is to analyse the power of such systems for reasoning with separation logic. A long-term perspective is to incorporate reasoning about concurrent programs with shared linked data structures along the lines of [16].

Acknowledgements. We are most grateful to the anonymous referees for their many valuable remarks.

References

1. Birkhoff, G.: Lattice Theory, 3rd edn., vol. XXV. American Mathematical Society, Colloquium Publications (1967)
2. Blyth, T., Janowitz, M.: Residuation Theory. Pergamon Press, Oxford (1972)
3. Conway, J.H.: Regular Algebra and Finite Machines. Chapman & Hall, Boca Raton (1971)
4. Dang, H.-H.: Algebraic aspects of separation logic. Technical Report 2009-01, Institut für Informatik (2009)
5. Dang, H.-H., Höfner, P., Möller, B.: Towards algebraic separation logic. Technical Report 2009-12, Institut für Informatik, Universität Augsburg (2009)
6. Desharnais, J., Möller, B.: Characterizing Determinacy in Kleene Algebras. Information Sciences 139, 253–273 (2001)
7. Dijkstra, E.: A discipline of programming. Prentice Hall, Englewood Cliffs (1976)
8. Ehm, T.: Pointer Kleene algebra. In: Berghammer, R., Möller, B., Struth, G. (eds.) RelMiCS 2003. LNCS, vol. 3051, pp. 99–111. Springer, Heidelberg (2004)
9. Hoare, C.A.R.: An axiomatic basis for computer programming. Communications of the ACM 12(10), 576–580 (1969)
10. Höfner, P., Struth, G.: Automated reasoning in Kleene algebra. In: Pfenning, F. (ed.) CADE 2007. LNCS (LNAI), vol. 4603, pp. 279–294. Springer, Heidelberg (2007)
11. Huntington, E.V.: Boolean algebra. A correction. Transaction of AMS 35, 557–558 (1933)
12. Huntington, E.V.: New sets of independent postulates for the algebra of logic. Transaction of AMS 35, 274–304 (1933)
13. Jónsson, B., Tarski, A.: Boolean algebras with operators, Part I. American Journal of Mathematics 73 (1951)

14. Möller, B.: Towards pointer algebra. Science of Computer Prog. 21(1), 57–90 (1993)
15. Möller, B.: Calculating with acyclic and cyclic lists. Information Sciences 119(3-4), 135–154 (1999)
16. O'Hearn, P.: Resources, concurrency, and local reasoning. Theoretical Computer Science 375, 271–307 (2007)
17. O'Hearn, P.W., Reynolds, J.C., Yang, H.: Local reasoning about programs that alter data structures. In: Fribourg, L. (ed.) CSL 2001 and EACSL 2001. LNCS, vol. 2142, pp. 1–19. Springer, Heidelberg (2001)
18. O'Hearn, P.W., Reynolds, J.C., Yang, H.: Separation and information hiding. ACM Trans. Program. Lang. Syst. 31(3), 1–50 (2009)
19. Reynolds, J.C.: An introduction to separation logic. Proceedings Marktoberdorf Summer School (2008) (forthcoming)
20. Rosenthal, K.: Quantales and their applications. Pitman Research Notes in Mathematics Series 234 (1990)

Domain and Antidomain Semigroups

Jules Desharnais[1], Peter Jipsen[2], and Georg Struth[3]

[1] Département d'informatique et de génie logiciel, Université Laval, Canada
`Jules.Desharnais@ift.ulaval.ca`
[2] Department of Mathematics and Computer Science, Chapman University, USA
`jipsen@chapman.edu`
[3] Department of Computer Science, University of Sheffield, UK
`G.Struth@dcs.shef.ac.uk`

Abstract. We axiomatise and study operations for relational domain and antidomain on semigroups and monoids. We relate this approach with previous axiomatisations for semirings, partial transformation semigroups and dynamic predicate logic.

1 Introduction

We axiomatise and study the(anti)domain and (anti)range operation on semigroups and monoids, generalising the concept of domain monoid in [JS08], and those of (anti)domain and (anti)range for semirings [DS08a] and a family of near-semirings [DS08b]. Our study of the antidomain operation is strongly based on Hollenberg's axioms [Hol97] which surely deserve more attention.

Our interest in these structures is threefold: First, they play a crucial role in the study of free algebras with (anti)domain operations, for representability results with respect to functions and relations, and for algebraising multimodal logics. Second, they form a basis for comparing and consolidating axiomatisations for categories, semigroups and Kleene algebras. Third, they provide a simple flexible basis for automated theorem proving in program and system verification.

Various expansions of semigroups with unary operations have been studied in semigroup theory (cf. [Sch70, JaS01, JaS04]), mostly motivated by the semigroups of partial transformations. Our primary model of interest is the algebra $Rel(X)$ of binary relations R on a set X with composition and unary (anti)domain and (anti)range operations given as subidentity relations. These are defined by

$$d(R) = \{(u, u) \in X^2 : (u, v) \in R \text{ for some } v \in X\},$$
$$a(R) = \{(u, u) \in X^2 : (u, v) \notin R \text{ for all } v \in X\},$$
$$r(R) = \{(v, v) \in X^2 : (u, v) \in R \text{ for some } u \in X\},$$
$$r'(R) = \{(v, v) \in X^2 : (u, v) \notin R \text{ for all } u \in X\}.$$

The algebra $Rel(X)$ is a standard semantic model for the input-output relation of nondeterministic programs and specifications, and the domain/range operations

R. Berghammer et al. (Eds.): RelMiCS/AKA 2009, LNCS 5827, pp. 73–87, 2009.

can be used to define pre- and postconditions and modal (program) operators on a state space. The (anti)domain and (anti)range operations induce a suitable *test algebra*—a state space—on the set of subidentity relations.

In the calculus of relations, partial and total functions, injections and surjections arise as special relations. Previous work in semigroup theory and category theory has investigated domain and antidomain predominantly in the context of (partial) functions. In the same way, domain axiomatisations for functions are specialisations of domain axiomatisations for relations. Therefore, the following subalgebras of $Rel(X)$ are of interest:

$PT(X)$,   the algebra of partial transformations (i.e. partial functions) on X;
$PI(X)$,   the algebra of partial injections on X;
$T(X)$,   the algebra of transformations (i.e. total functions) on X;
$S(X)$,   the algebra of permutations on X.

The first one corresponds to models of deterministic programs. The second and fourth case also consider a unary operation R^{-1} of converse. The domain and range operations are then definable as $d(R) = R;R^{-1} \cap \mathrm{id}_X$ and $r(R) = R^{-1};R \cap \mathrm{id}_X$. For each of these algebras it is natural to study the class of all algebras that can be embedded in them. Depending on the choice of unary operations in the signature, one obtains the class of groups, semigroups, inverse semigroups, and twisted domain semigroups (axiomatisations can be found below).

Many results in this paper have been obtained by automated theorem proving and automated model generation, using the tools Prover9 and Mace4 [McC07]. Instead of presenting these proofs we add input templates for domain semigroups and antidomain monoids to the paper and encourage the reader to replay our arguments with these tools. They are easy to install and use.

2   Motivation and Overview

We are interested in algebras where elements represent actions or computations of some system and where operations model the control flow in the system. Multiplication, for instance, could represent the sequential or parallel composition of actions and addition could represent nondeterministic choice. Special actions like multiplicative units could model ineffective actions—sometimes called *skip*— and additive units could model abortive actions. Examples of such algebras are semigroups or monoids that model sequential composition, and semirings that model sequential composition and nondeterministic choice. Concrete models of such algebras are partial and total functions, binary relations, languages, sets of paths in graphs or sets of traces.

In this context, a domain operation yields enabledness conditions for actions, that is, the domain $d(x)$ of an action x abstractly models those states from which the action x can be executed. Analogously, the antidomain $a(x)$ models those states from which the action x cannot be executed.

The starting point of the current investigation is a previous axiomatisation of domain and antidomain operations for semirings $(S, +, \cdot, 0, 1)$, in which a domain operation is a map $d : S \to S$ that satisfies

$$x = d(x) \cdot x, \qquad d(x \cdot y) = d(x \cdot d(y)), \qquad d(x) + 1 = 1,$$
$$d(0) = 0, \qquad d(x + y) = d(x) + d(y).$$

It can be shown that the *domain algebra* induced by this operation—the set $d(S)$—is a distributive lattice and that each domain semiring is automatically idempotent, that is, $x + x = x$ holds for all $x \in S$. If the semiring elements represent actions of some system, then $d(S)$ represents the states from which actions are enabled. Distributive lattices are suitable statespaces, but Boolean algebras are perhaps even better. To induce a Boolean domain algebra it is convenient to axiomatise a notion of antidomain (the Boolean complement of domain) as a map $a : S \to S$ that satisfies

$$a(x) \cdot x = 0, \qquad a(x \cdot y) = a(x \cdot a(a(y))), \qquad a(a(x)) + a(x) = 1.$$

Also antidomain semirings are automatically idempotent. Domain operations can be obtained in antidomain semirings by defining $d(x) = a(a(x))$. These definitions can readily be adapted to some weaker cases of semirings—so-called near-semirings—in which some of the semiring axioms are dropped [DS08b].

A natural generalisation is to investigate how these axioms can be adapted when the operation of addition is dropped and the domain algebras induced are still meant to yield useful state spaces. We consider a whole family of domain and antidomain axiomatisations for semigroups and monoids which is presented in Table 1 as an overview. Precise definitions are given in subsequent sections. In the case of domain, the weakest axiomatisations are so-called left-closure semigroups and monoids (LC-semigroups/monoids) (cf. [JaS01]). The domain algebras of these structures are meet-semilattices, but some natural properties of domain do not hold in this class. Domain semigroups and monoids (d-semigroups/monoids) capture some of the properties of domain for binary relations, while twisted d-semigroups/monoids capture precisely the quasiequational properties of domain for partial functions. In the case of antidomain, closable semilattice pseudo-complemented semigroups (closable SP-semigroups) were introduced in [JaS04]. Their axioms induce domain algebras that are Boolean algebras, but again some natural properties of antidomain do not hold. Antidomain monoids (a-monoids) capture all the equational properties of antidomain for binary relations, while twisted a-monoids capture some of the properties of antidomain for partial functions. A more thorough investigation of the whole family is the subject of this paper.

Table 1. Family of domain semigroups

domain		antidomain
LC-semigroups	LC-monoids	closable SP-semigroups
⋃⊦	⋃⊦	⋃⊦
d-semigroups	d-monoids	a-monoids
⋃⊦	⋃⊦	⋃⊦
twisted d-semigroups	twisted d-monoids	twisted a-monoids

3   Domain Semigroups

Our aim is to axiomatise domain and antidomain operations on semigroups and monoids that capture the fact that some computations may or may not be enabled from some set of states. We model the state space implicitly or internally through the images induced by the domain operations.

We consider semigroups $(S, \cdot)$ with an associative multiplication, usually left implicit, and monoids with a left and right multiplicative unit 1.

A *domain semigroup*, or *d-semigroup*, is a semigroup $(S, \cdot)$ extended by a domain operation $d : S \to S$ that satisfies the following axioms.

(D1) $d(x)x = x$
(D2) $d(xy) = d(xd(y))$
(D3) $d(d(x)y) = d(x)d(y)$
(D4) $d(x)d(y) = d(y)d(x)$

A monoid that satisfies these axioms is called a *domain monoid* or *d-monoid*.

It is easy to check that the axioms (D1)-(D4) hold in $Rel(X)$ and, in fact, in all domain semirings. The axiom (D2) has been called *locality axiom* in the context of domain semirings. In semigroup theory, it has previously been called *left-congruence condition* [JaS01].

The axioms (D1)-(D4) are irredundant in the classes of d-semigroups and d-monoids: Mace4 found models that satisfy the semigroup or monoid axiom and three of the domain axioms, but not the fourth one, for each combination of domain axioms.

The class of right closure semigroups is defined in [JaS01]. The intended models are functions under composition. We present a dual set of axioms for relational composition.

A *left closure semigroup*, or *LC-semigroup*, is a semigroup that satisfies the following axioms.

(D1) $d(x)x = x$
(L2) $d(d(x)) = d(x)$
(L3) $d(x)d(xy) = d(xy)$
(D4) $d(x)d(y) = d(y)d(x)$

Analogously, an *LC-monoid* is an LC-semigroup that is also a monoid.

Again, it can be shown that the domain axioms of LC-semigroups and LC-monoids are irredundant.

Lemma 1. *The class of d-semigroups is strictly contained in the class of LC-semigroups.*

Prover9 has shown that the axioms (L2) and (L3) follow from the domain axioms (D1), (D3) and (D4). Mace4 presented a four-element LC-semigroup in which (D3) does not hold. It is easy to prove the same result for classes of monoids.

Lemma 2. *d-semigroups are LC-semigroups that satisfy the locality axiom.*

A *domain element* of an LC-semigroup, domain semigroup or the corresponding monoid S is an element of $d(S) = \{d(x) : x \in S\}$. The next lemma presents a very useful characterisation of domain elements.

Lemma 3. *The domain elements of an LC-semigroup are precisely the fixed points of the domain operation.*

Proof. If $x \in d(S)$, then $x = d(y)$ for some $y \in S$ and $d(x) = d(d(y)) = d(y) = x$ by (L2). If $x \in S$ satisfies $d(x) = x$, then $x \in d(S)$ by definition.   □

This fixed point characterisation of domain elements in LC-semigroups, which a fortiori holds in domain semirings, is a key to checking closure properties of domain elements and describing the algebra of domain elements. It allows us to express the fact that x is a domain element within the language as $d(x) = x$.

Lemma 4

(a) *For any LC-semigroup S, the set $d(S)$ is a meet-subsemilattice of S. If S has a right unit 1, then $d(1) = 1$ is the top element of $d(S)$.*
(b) *Every meet-semilattice is a domain semigroup if $d(x) = x$ is imposed, and similarly every meet-semilattice with a top element is a domain monoid.*

Proof. (a) To show that domain elements are closed under composition, we use the fixed point lemma to verify that $d(d(x)d(y)) = d(x)d(y)$. Hence, by (D4), $d(S)$ is a commutative subsemigroup. Moreover $d(x)d(x) = d(x)$ holds, which implies that $d(S)$ is a subsemilattice. If $x1 = x$ holds in S, then $d(1) = d(1)1 = 1$ by (D1). The semilattice-order is defined, as usual, by $d(x) \leq d(y) \Leftrightarrow d(x)d(y) = d(x)$. Thus $d(x) \leq 1$ immediately follows from the monoidal right unit axiom.

(b) This fact is well known for LC-semigroups [JaS01]. In the case of d-semigroups we must verify that (D1)-(D4), with $d(s) = s$ for each element s, hold in every semilattice, which is trivial.   □

Because of these algebraic properties we call $d(S)$ the *domain algebra* of S. It can be shown [JaS01] that the semilattice-order on the domain algebra can be extended to a partial order—called the *fundamental order*—on the whole LC-semigroup by

$$x \leq y \Leftrightarrow x = d(x)y.$$

On partial functions, the dual of the fundamental order is called the *refinement order*.

Lemma 5.   *In any LC-semigroup the fundamental order coincides with the semilattice-order on the domain algebra, and is preserved by multiplication on the right.*

Note that preservation by multiplication on the left need not hold even on d-semigroups. This reflects the situation in $Rel(X)$, whereas for partial functions the fundamental order coincides with inclusion and is preserved by multiplication on both sides.

It seems interesting to compare the fundamental order $\leq$ with the usual natural order on domain semi*rings*, which is defined as $x \sqsubseteq y \Leftrightarrow x + y = y$.

Lemma 6. *The ordering $\leq$ is contained in $\sqsubseteq$ of domain semirings, but not necessarily equal.*

Proof. Prover9 has shown that $x = d(x)y \Rightarrow x + y = y$ holds in all domain semirings; Mace4 has found a three-element counterexample for the converse implication. $\square$

We also note that the usual relational demonic refinement ordering can be defined in this framework:

$$x \text{ refines } y \quad \text{iff} \quad d(y) \leq d(x) \text{ and } d(y)x \leq y.$$

We now outline a calculus of domain semigroups and we study some properties of their domain algebras. To formulate statements as strongly as possible from a logical point of view, we state positive properties for LC-semigroups and negative ones for domain semigroups. Automated theorem proving easily verifies the following basic laws.

Lemma 7. *Let S be an LC-semigroup and let $x, y \in S$. Then*

(a) $d(xy) \leq d(x)$.
(b) $d(x)y \leq y$, but not necessarily $yd(x) \leq y$.
(c) $x \leq d(x) \Leftrightarrow x = d(x)$.
(d) $x \leq 1 \Leftrightarrow x = d(x)$ if 1 is a right unit.
(e) $x \leq y \Rightarrow d(x) \leq d(y)$.
(f) $x \leq px \Leftrightarrow d(x) \leq p$ and $x = px \Leftrightarrow d(x) \leq p$ hold for all $p \in d(S)$.

Case (d) implies that, in d-monoids, the set of all domain elements is precisely the set of all subidentities. This is in contrast to the situation in domain semirings, where the domain elements can form a strict subset. There is no contradiction, since the subidentities on domain semigroups are taken with respect to $\leq$ whereas the subidentities on domain semirings are taken with respect to $\sqsubseteq$, which may admit more subidentities than $\leq$.

Case (f) captures a natural property of domain, namely

$$d(x) = \inf\{p \in d(S) : x \leq px\}.$$

Hence $d(x) = \inf\{p \in d(S) : x = px\}$ since all domain elements are left subidentities by (b). Accordingly, $d(x)$ is the least element in $d(S)$ which left preserves x, and the least domain element satisfying (D1). The assumption in (f) that $p \in d(S)$ cannot be much relaxed. The property fails if p is just a subidentity or an idempotent subidentity.

Lemma 8

(a) Every monoid can be expanded to a d-monoid.
(b) Some semigroups cannot be expanded to d-semigroups.
(c) Domain algebras of d-monoids need not be unique.

Proof

(a) The map $d(x) = 1$ for all $x \in S$ satisfies (D1) to (D4).

(b) The semigroup of positive integers under addition has no idempotents. Hence there are no candidates for membership in the domain algebra, and it is impossible to define a domain operation.

(c) The two d-monoids defined by

$$
\begin{array}{c|cc}
\cdot & 0 & 1 \\
\hline
0 & 0 & 0 \\
1 & 0 & 1
\end{array}
$$

with domain operations $d_1(x) = x$ and $d_2(x) = 1$ prove the claim.   □

An expansion of idempotent semirings to d-semirings is not always possible. There is a three-element counterexample.

An interesting question is whether the axiomatisation of d-monoids captures all the properties of the domain operation of binary relations. A d-monoid is called *representable* if it can be embedded in $Rel(X)$ for some set X such that $\cdot$, d and 1 correspond to composition, relational domain and id_X. By Schein's fundamental theorem for relation algebras [Sch70] the class of representable d-monoids is a quasivariety.

Proposition 9. *The following quasiidentity fails in a 4-element d-monoid but holds in $Rel(X)$:*

$$xy = d(x) \text{ and } yx = x \text{ and } d(y) = 1 \quad \text{imply} \quad x = d(x).$$

Proof. Finding the counterexample for d-monoids is easy with Mace4. To prove that the result holds for binary relations, consider $x, y \in Rel(X)$ and $(a, b) \in x$. Then $d(y) = 1$ implies $(b, c) \in y$ for some c. It follows from $xy = d(x)$ that $c = a$, hence $(b, a) \in y$. Now $yx = x$ implies that $(b, b) \in x$. Finally $xy = d(x)$ yields $(b, a) \in d(x)$, whence $b = a$. Since (a, b) is arbitrary it follows that $x = d(x)$.   □

Corollary 10. *The quasivariety of representable d-monoids is not a variety.*

4   Twisted Domain Semigroups

Partial functions under composition satisfy another equational property called the *twisted law* in [JaS01]:

$$xd(y) = d(xy)x.$$

This identity fails in $Rel(X)$ if we take x to be any relation that is not deterministic. However it is satisfied if composition is the relational demonic composition (defined below in the section on antidomain). A d-semigroup/monoid or LC-semigroup/monoid is *twisted* if it satisfies the twisted law. The next lemma follows easily by automated theorem proving and counterexample search.

Lemma 11. *The classes of twisted LC-semigroups and twisted d-semigroups coincide, and they are strictly contained in the class of d-semigroups.*

The results of this section characterise part of the spectrum between LC-semigroups and twisted semigroups. LC-semigroups, on the one hand, yield a uniform basis for characterising domain operations for relations and functions, but they do not capture locality, which holds in relational models. Twisted semigroups, on the other hand, satisfy locality, but capture only deterministic relations, that is, partial functions. Domain semigroups are located between these two extremes and capture relations better than LC-semigroups.

The domain semigroup axioms, but not the twisted axiom (Mace4 presented a five element counterexample) hold in all domain semirings, hence domain semigroups are a natural generalisation of domain semirings. Partial functions, of course, are not closed under union hence do not form a semiring.

We have seen in Section 3 that the fundamental order $\leq$ is preserved by multiplication on the right. If the twisted identity $d(zx)z = zd(x)$ is imposed on an LC-semigroup then it is preserved by multiplication on the left as well since $x \leq y$ implies $x = d(x)y$, hence $d(zx)zy = zd(x)y = zx$, i.e. $zx \leq zy$.

Various representation theorems have been proved for families of semigroups with respect to partial functions. For example, every group is embedded in the symmetric group $S(X)$ of all permutations of a set X. Similarly, every semigroup is embedded in the transformation semigroup $T(X)$ of all functions on a set X. *Inverse semigroups* are semigroups with a unary operation $^{-1}$ that satisfies the identities $x^{-1-1} = x$, $xx^{-1}x = x$ and $xx^{-1}yy^{-1} = yy^{-1}xx^{-1}$. It is a standard result of semigroup theory (independently due to Vagner 1952 and Preston 1954) that every inverse semigroup is embedded in the symmetric inverse semigroup $PI(X)$ of all partial injections on X. We recall below a fourth instance of such an embedding due to Trokhimenko [Tro73] (cf. [JaS0a]). We present a concise variant of the proof for the domain setting because it uses a general construction that should be of interest for the RelMiCS/AKA community.

Theorem 12. *[Tro73, JaS01] Every twisted d-semigroup can be embedded in a partial transformation semigroup. If the semigroup has a unit, it is mapped to the identity function.*

Proof. Let S be a twisted d-semigroup and consider the partial transformation semigroup $PT(S)$. For $a \in S$ define

- $D_a = \{xd(a) : x \in S\} = \{y \in S : yd(a) = y\}$,
- $f_a : D_a \to S$ by $f_a(x) = xa$, and
- $h : S \to PT(S)$ by $h(a) = f_a$.

The map h is called the *Cayley embedding* and it remains to check that

(a) $d(f_a) = f_{d(a)}$,
(b) $f_a;f_b = f_{ab}$, and
(c) h is injective.

By definition, $d(f_a) = \{(xd(a), xd(a)) : x \in S\}$, whereas $f_{d(a)}$ is defined on $D_{d(a)} = \{xd(d(a)) : x \in S\}$ by $f_{d(a)}(x) = xd(a)$. Since $d(d(a)) = d(a)$ and $xd(a)d(a) = xd(a)$, it follows that (a) holds.

To see that (b) holds, note that $(f_a; f_b)(x) = f_b(f_a(x)) = xab = f_{ab}(x)$, so it suffices to show that both functions have the same domain. Note that under the assumption of the twisted law we have $x = xd(y) \Leftrightarrow d(x) = d(xy)$.

Now x is in the domain of $f_a; f_b$ if and only if $x \in D_a$ and $xa \in D_b$, which means $xd(a) = x$ and $xad(b) = xa$. This can be expressed by $d(x) = d(xa)$ and $d(xa) = d(xab)$. Hence $d(x) = d(xab)$, and by the above equivalence we obtain $xd(ab) = x$, which shows that x is in the domain of f_{ab}. Conversely, if $xd(ab) = x$ then $d(x) = d(xd(ab)) = d(xd(ab)d(a)) = d(xd(a)) = d(xa)$ by (L3), (D4) and (D2), and likewise $d(xa) = d(xd(ab)a) = d(xd(ab)d(a)) = d(xd(ab)) = d(xab)$.

So h is a d-semigroup homomorphism, and it is injective since if $f_a = f_b$ then $x = xd(a)$ is equivalent to $x = xd(b)$. It follows that $d(a) = d(a)d(b) = d(b)$, whence $a = d(a)a = f_a(d(a)) = f_b(d(a)) = d(a)b = d(b)b = b$.

Finally, if S has a unit it follows immediately from the definitions that $D_1 = S$ and therefore $h(1) = f_1 = \mathrm{id}_S$.   □

Corollary 13. *Every commutative d-semigroup is twisted, and can be embedded in a partial transformation semigroup.*

5   Domain-Range Semigroups

A range operation can be defined on arbitrary semigroups by exploiting semi-group duality (with respect to opposition).

A *domain-range semigroup*, or *dr-semigroup* for short, is a semigroup with two unary operations d and r that satisfy the following axioms.

(D1)	$d(x)x = x$	(R1)	$xr(x) = x$
(D2)	$d(xy) = d(xd(y))$	(R2)	$r(xy) = r(r(x)y)$
(D3)	$d(d(x)y) = d(x)d(y)$	(R3)	$r(xr(y)) = r(x)r(y)$
(D4)	$d(x)d(y) = d(y)d(x)$	(R4)	$r(x)r(y) = r(y)r(x)$
(D5)	$d(r(x)) = r(x)$	(R5)	$r(d(x)) = d(x)$

Mace4 can show that the axioms (D5) and (R5) are not implied by the other axioms. This means that without these axioms, the domain algebra and the range algebra can be different. By the fixed point lemma for domain and its dual, the axioms (D5) and (R5) enforce that the domain algebra and the range algebra coincide, and both these axioms are needed for this result. (D4) and (R4) can be merged into the equivalent identity $d(x)r(y) = r(y)d(x)$.

By duality, it is clear that the identity $x = yr(x)$ also induces an ordering on S, but Mace4 can show that the order induced by domain and that by range need not coincide.

Again, the main examples of dr-semigroups are $Rel(X)$ and $PT(X)$. Inverse semigroups are also examples if we define $d(x) = xx^{-1}$, $r(x) = x^{-1}x$. In fact the twisted law holds for d, and its dual holds for r. The above representation theorem of Trokhimenko reduces to the Vagner-Preston representation theorem for inverse semigroups. However the twisted law does not hold for r in arbitrary partial transformation semigroups (simply because not all functions are injective).

Schweizer and Sklar [SS67] have provided an axomatisation for abstract function systems, using the following domain and range axioms.

$$d(x)x = x \qquad d(xd(y)) = d(xy) \qquad d(r(x)) = r(x) \qquad d(x)r(y) = r(y)d(x)$$
$$xr(x) = x \qquad r(r(x)y) = r(xy) \qquad r(d(x)) = d(x) \qquad xd(y) = d(xy)x$$

Schein has shown that adding the quasiidentity

$$xy = xz \Rightarrow r(x)y = r(x)z$$

axiomatises precisely the quasivariety of dr-semigroups of partial transformations (cf. [Sch70]). Prover9 easily shows that the first set of axioms without Schein's quasi-identity implies the axioms (D3) and (R3).

An interesting question is whether every dr-semigroup can be embedded into $Rel(X)$ for some set X. We leave it open.

6   Antidomain

We have seen in Section 2 that domain semirings admit a very compact axiomatisation that induces a Boolean domain algebra. It is based on a notion of antidomain from which domain can be obtained. In this setting, antidomain is a more fundamental notion than domain.

This section shows how this approach can be generalised to the semigroup or monoid case. We use the abbreviation $x' = a(x)$ for the antidomain operation, and define an *antidomain monoid*, or *a-monoid*, $(S, \cdot, 1, ')$ as a monoid $(S, \cdot, 1)$ that satisfies

(A1) $x'x = 0$
(A2) $x0 = 0$
(A3) $x'y' = y'x'$
(A4) $x''x = x$
(A5) $x' = (xy)'(xy')'$
(A6) $(xy)'x = (xy)'xy'$.

This axiomatisation is essentially due to Hollenberg [Hol97]. The one presented here is slightly more compact, and axiom (A5) is new, though essentially dual to one of Huntington's axioms for Boolean algebras. The axioms (A5) and (A6) might deserve further explanation. Intuitively, an expression $(xy)'$ can be understood as a modal box operator $[x]y'$, and it describes the set of states from which each x-step must lead to a state from which y is not enabled. Under this interpretation, an intuitive reading of (A5) is $x' = ([x]y') \cdot ([x]y'')$. This is a special case of the multiplicativity law $[x](p \cdot q) = ([x]p) \cdot ([x]q)$ for boxes, since $x' = [x]0$.

(A6) can be rewritten as $([x]y')x = ([x]y')xy'$, which says that executing x from those states from which each x-step must lead into y', leads into y'.

We write S' for the set $\{x' : x \in S\}$ of all *antidomain elements* of S. The constants 0 and 1 can be omitted from the language if we replace (A3) by $x'x = y'y$, (A2) by $xx'x = x'x$, and the monoid unit laws by $x(x'x)' = x$ (the left-unit law can be deduced from these axioms). In this sense the terminology *antidomain semigroup* is appropriate. However we prefer to use the more readable notation that makes the constants explicit. We can also define

$$x + y = (x'y')'$$

as an abbreviation. Mace4 can show that the antidomain axioms are irredundant.

Lemma 14

(a) *A monoid is trivial if it can be extended by an antidomain operation that has a fixed point.*
(b) *The map $d(x) = x''$ is a domain operation.*
(c) *The antidomain elements of an a-monoid are the fixed points of domain.*

The fixed point lemma in (c) is again a powerful tool for analysing the structure of antidomain elements.

Proposition 15. *Let S be an a-monoid. Then $(S', +, \cdot, ', 0, 1)$ is a Boolean subalgebra.*

Proof. We automatically verified the following properties. First, antidomain elements are are closed under addition and multiplication: $(x' + y')'' = x' + y'$ and $(x'y')'' = x'y'$. Closure under antidomain is trivial. Second, Huntington's axioms for Boolean algebras hold: $x + y = y + x$, $(x + y) + z = x + (y + z)$, and $x' = (x + y)' + (x + y')'$. Finally, $x'x'' = 0 = x''x'$ and $x' + x'' = 1$.   □

In fact, Lemma 14 and Proposition 15 follow already from the antidomain axioms without (A6).

In Boolean domain semirings, the domain algebra is uniquely determined. It is the maximal Boolean subalgebra of the subalgebra of subidentities.

Lemma 16. *The antidomain algebra of an a-monoid need not be unique.*

Mace4 presented a five-element model with two different antidomain operations.

Another interesting observation is that using antidomain, demonic composition $\square$ can be defined (it is associative in the presence of the twisted law):

$$x \square y = (xy')'xy.$$

The following lemma collects some further properties of antidomain. Note that $\leq$ is the fundamental order which on the subidentities coincides with the lattice order.

Lemma 17. *Let S be an a-monoid. For all $x, y, z \in S$, the following laws hold.*

(a) $0x = 0$
(b) $(xy'')' = (xy)'$
(c) $(x'y)'' = x'y''$
(d) $x \leq 1$ *implies* $xy = 0 \Leftrightarrow x \leq y'$
(e) $x \leq 1 \Leftrightarrow x'' = x$.
(f) $x' \leq (xy)'$
(g) $xy = 0 \Leftrightarrow xy'' = 0$

As in the case of d-monoids, by Schein's fundamental theorem, the class of representable a-monoids forms a quasivariety. Hollenberg has shown the following two additional results for a-monoids.

Theorem 18. *[Hol97]*

(a) The variety of a-monoids and the variety generated by all representable a-monoids are the same.
(b) The quasivariety of representable a-monoids is not a variety.

Hollenberg's counterexample for (b) is a 5-element Heyting algebra which fails the quasiidentity $x''y = x'' \wedge x'y = x' \Rightarrow y = 1$ that holds in all *representable a*-monoids. Since each Heyting algebra is a commutative a-monoid, it is twisted by Corollary 13. Consequently, in contrast to the case of d-monoids, the antidomain operation need not be represented correctly by the Cayley map. This indicates why the construction from Theorem 12 cannot even be adapted to *twisted a*-monoids. The question whether the quasivariety of representable a-monoids is finitely axiomatisable is open.

A weaker axiomatisation of an antidomain operation for semigroups is obtained as a subvariety of semilattice pseudo-complemented semigroups defined in [JaS04]. Recall that a pseudo-complement on a meet-semilattice is a unary operation $'$ that satisfies

$$xy = 0 \Leftrightarrow y \leq x'.$$

In the variety of semilattices with a unary operation, this formula is equivalent to the identities $x'x = 0$, $x0' = x$ and $x(xy)' = xy'$. The following result is proved in [Fri62].

Theorem 19. *For any pseudocomplemented meet-semilattice S, the set $B(S) = \{x'' : x \in S\}$ is a Boolean algebra with operations x', xy and $x'' + y'' = (x'y')'$.*

A *semilattice pseudo-complemented semigroup* or *SP-semigroup* is a semigroup that satisfies the following identities.

(A1) $x'x = 0$
(S2) $x0' = x$
(A3) $x'y' = y'x'$
(S4) $x'(x'y)' = x'y'$

In any SP-semigroup S the set $B(S) = \{x'' : x \in S\}$ is a meet-subsemilattice that is pseudo-complemented by the antidomain operation. As in Theorem 19, the set $B(S)$ is a Boolean algebra with join given by $x'' + y'' = (x'''y''')'$.

An SP-semigroup is called *closable* in [JaS04] if (A4), that is, $x''x = x$, holds.

Lemma 20

(a) Every closable SP-semigroup is a d-semigroup with $d(x) = x''$.
(b) A closable SP-semigroup is an a-monoid if and only if (A2) and (A6) hold.

Therefore, (A5) could be replaced by (S4) in the a-monoid axioms.

The proper superclass of a-monoids defined by (A1)-(A5) is interesting in its own right. Note that (A6) holds in every antidomain semi*ring*, since

$$(xy)'x = (xy)'x(y' + y'') = (xy)'xy' + (xy'')'xy'' = (xy)'xy' + 0 = (xy)'xy'.$$

Modal box and diamond operators can be defined already in this weaker setting.

Let $\langle x \rangle p = (xp)''$ and let $[x]p = (xp')'$, where $p = p''$. Then the diamond operator is strict and additive and the box operator is costrict and multiplicative:

$$\langle x \rangle 0 = 0, \quad \langle x \rangle (p + q) = \langle x \rangle p + \langle x \rangle q, \quad [x]1 = 1, \quad [x](p \cdot q) = ([x]p) \cdot ([x]q).$$

Also $[x]p = (\langle x \rangle p')'$ and $\langle x \rangle p = ([x]p')'$. This definition of modal operators is not possible in the weaker setting of closable SP-semigroups. Hence SP-semigroups have Boolean domain algebras, but are too weak to obtain Boolean algebras with operators.

Modal algebras allow one to define a notion of determinism as $\langle x \rangle p \leq [x]p$. We therefore call an a-monoid *deterministic* if it satisfies

$$(xy'')'' \leq (xy')'.$$

Proposition 21. *An a-monoid is deterministic if and only if it is twisted.*

Note that the twisted law implies (A6), but not every a-monoid is twisted or deterministic, and determinism does not imply (A6).

Finally, a notion of antirange can be axiomatised dually to that of antidomain. Because the antidomain and the antirange algebra automatically coincide, they need no further linking. In this setting, forward box and diamond operators $|x]$ and $|x\rangle$ can be defined from antidomain, and backward operators $[x|$ and $\langle x|$ from antirange. We have the following laws.

demodalisation	$\|x\rangle p \leq q \Leftrightarrow q'xp = 0$ and $\langle x\|p \leq q \Leftrightarrow pxq' = 0$
conjugation	$(\|x\rangle p)q = 0 \Leftrightarrow p(\langle x\|q) = 0$
Galois connections	$\|x\rangle p \leq q \Leftrightarrow p \leq [x\|q$ and $\langle x\|p \leq q \Leftrightarrow p \leq \|x]q$

In this setting, (A6) and its dual for antirange become derivable. For instance, (A6) is just the cancellation law $\langle x\|x\rangle p \leq p$ of the Galois connection. Note that because of the Galois connection, diamond operators are even completely additive, and box operators are completely multiplicative. In conclusion, monoids with antidomain and antirange allow us to define and calculate with modal operators.

7 Templates for Prover9 and Mace4

```
op(400, infix, ";").
op(500, infix, "+").
op(300, postfix, "'").

formulas(assumptions).  % domain semigroups

x;(y;z)=(x;y);z.
d(x);x=x.
d(x;y)=d(x;d(y)).
d(d(x);y)=d(x);d(y).
d(x);d(y)=d(y);d(x).

x<=y <-> x=d(x);y.

end_of_list.

formulas(assumptions).  % antidomain monoid

x;(y;z)=(x;y);z.
x;1=x.
x';x=0.
x;0=0.
x';y'=y';x'.
x'';x=x.
x'=(x;y)';(x;y')'.
(x;y)';x=((x;y)';x);y'.

x<=y <-> x=x'';y.

end_of_list.

formulas(goals). % insert goal here

end_of_list.
```

8 Conclusion

We have axiomatised operations for relational domain and antidomain for semi-groups and monoids, studied the structure of the domain algebras, developed the basic calculi, and compared these algebras with previous axiomatisations. Our approach continues and also generalises previous work on axiomatisations of domain for semirings and Kleene algebras. It forms the basis for further investigations, for instance, representation theorems, free algebras and other domain algebras.

Partial and total functions and deterministic programs are central to computer science applications, while relations and nondeterminism are important for specifications and for modelling more general computing systems. But the algebraic background that has been developed in semigroup theory over the last fifty years does not seem to be widely known and, to our knowledge, no link between functional and relational domain axiomatisations has so far been provided.

Besides closing this gap, a benefit of the abstract algebraic approach is also that the analysis of functions and relations with (anti)domain can—to a large extent—be automated. This allowed us to condense the paper and focus on the conceptual development.

Acknowledgement

We would like to thank Robin Hirsch, Marcel Jackson and Szabolcs Mikulás for interesting discussions. Special thanks to Tadeusz Litak for the very interesting pointer to Marco Hollenberg's work.

References

[DS08a]   Desharnais, J., Struth, G.: Modal semirings revisited. In: Audebaud, P., Paulin-Mohring, C. (eds.) MPC 2008. LNCS, vol. 5133, pp. 360–387. Springer, Heidelberg (2008)

[DS08b]   Desharnais, J., Struth, G.: Domain Axioms for a Family of Near-Semirings. In: Meseguer, J., Roşu, G. (eds.) AMAST 2008. LNCS, vol. 5140, pp. 330–345. Springer, Heidelberg (2008)

[Fri62]   Frink, O.: Pseudo-complements in semilattices. Duke Mathematics Journal 29, 500–515 (1962)

[Hol97]   Hollenberg, M.: An equational axiomatization of dynamic negation and relational composition. Journal of Logic, Language and Information 6, 381–401 (1997)

[JaS01]   Jackson, M., Stokes, T.: An invitation to C-semigroups. Semigroup Forum 62, 279–310 (2001)

[JaS04]   Jackson, M., Stokes, T.: Semilattice pseudo-complements on semigroups. Comm. Algebra 32, 2895–2918 (2004)

[JaS0a]   Jackson, M., Stokes, T.: Partial maps with domain and range: extending Schein's representation. Comm. Algebra (to appear)

[JS08]   Jipsen, P., Struth, G.: The structure of the one-generated free domain semiring. In: Berghammer, R., Möller, B., Struth, G. (eds.) RelMiCS/AKA 2008. LNCS, vol. 4988, pp. 234–242. Springer, Heidelberg (2008)

[McC07]   McCune, W.: Prover9 / Mace4 (2007), `www.prover9.org`

[MS06]   Möller, B., Struth, G.: Algebras of modal operators and partial correctness. Theoretical Computer Science 351, 221–239 (2006)

[Sch70]   Schein, B.M.: Relation algebras and function semigroups. Semigroup Forum 1, 1–62 (1970)

[SS67]   Schweizer, B., Sklar, A.: Function systems. Math. Annalen 172, 1–16 (1967)

[Tro73]   Trokhimenko, V.S.: Menger's function systems. Izv. Vysš. Učebn. Zaved. Matematika 11(138), 71–78 (in Russian)

Composing Partially Ordered Monads

Patrik Eklund and Robert Helgesson

Umeå University, Department of Computing Science, SE-90187 Umeå, Sweden
{peklund,rah}@cs.umu.se

Abstract. Composition of the many-valued powerset partially ordered monad with the term monad provides extensions to non-classical relations and also new examples for Kleene algebras.

Keywords: Kleene algebra, partially ordered monad.

1 Introduction

Monads equipped with order structures extend to partially ordered monads. In this paper we show how partially ordered monads can be composed building upon the underlying monad compositions. In particular we focus on composing the partially ordered many-valued powerset monad with the term monad. The order structure for the composed partially ordered monad is inherited from the partially ordered many-valued powerset monad. Partially ordered monads with some additional conditions are further shown to establish Kleene algebras, thus providing a generalized notion of powerset Kleene algebras extending the examples of Kleene algebras beyond strings strings [22] and relations [37]. Kleene algebras are used e.g. in formal languages [36] and analysis of algorithms [1,25].

Previous work on monads based on many-valued set functors include compactifications of generalised convergence spaces based on double powerset [7,8]. Further work for extension structures using partially ordered monads are found in [15,16]. Composition of monads involving many-valued set functors was first developed in [10].

Monads and monad compositions have been used in functional programming for structuring of functional programs [31,32]. In particular for parsing and type checking monad compositions have been useful [38]. A folklore example in logic programming is most general unifiers identified as co-equalisers in Kleisli categories of term monads [35].

Partially ordered monads are due to [14,15] and evolve from studies around filter based convergence structures and Cauchy structures [23,21]. More general set functors for convergence were considered in [12]. Empowering general structures with monads was initiated in [13]. More examples involving the fuzzy filter monad were developed in [7]. Monad and partially ordered monad techniques for compactification were developed in [8,16] originally inspired by a compactification construction [34] for filter based limit spaces.

The present paper is organized as follows. Section 2 describes the partially ordered many-valued powerset monads, followed by partially ordered monad

R. Berghammer et al. (Eds.): RelMiCS/AKA 2009, LNCS 5827, pp. 88–102, 2009.

compositions in Section 3. Partially ordered monads in a Kleene algebra setting is then described in Section 4.

2 Partially Ordered Monads

Before introducing partially ordered monads we recall the regular, unordered, monad and give two important examples. These examples, the powerset monad and term monad, respectively will later be used in exemplifying partially ordered monads and their subconstructions. Further, a brief exposition will be given on the Kleisli categories.

A monad over a category C is a structure $\mathbf{F} = (F, \eta, \mu)$, where $F : C \longrightarrow C$ is a (covariant) functor, and $\eta : id \longrightarrow F$ and $\mu : F \circ F \longrightarrow F$ are natural transformations for which $\mu \circ F\mu = \mu \circ \mu F$ and $\mu \circ F\eta = \mu \circ \eta F = id_F$ hold.

The Kleisli category $C_{\mathbf{F}}$ for $\mathbf{F}$ over C is given by $Ob(C_{\mathbf{F}}) = Ob(C)$ and $Hom_{C_{\mathbf{F}}}(X, Y) = Hom_C(X, FY)$. Morphisms $f : X \longrightarrow Y$ in $C_{\mathbf{F}}$ are morphisms $f : X \longrightarrow FY$ in C, with $\eta_X^F : X \longrightarrow FX$ the identity morphism. Composition of morphisms in $C_{\mathbf{F}}$ is given by

$$(X \xrightarrow{f} Y) \circ (Y \xrightarrow{g} Z) = X \xrightarrow{\mu_Z^F \circ Fg \circ f} FZ.$$

2.1 The Many-Valued Powerset Monads

Let L be a completely distributive lattice. For $L = \{0, 1\}$ we write $L = 2$. The covariant powerset functor L is defined by LX being the set of mappings $A : X \longrightarrow L$, and for morphisms $f : X \longrightarrow Y$ in Set we define ([18])

$$Lf(A)(y) = \bigvee_{f(x)=y} A(x).$$

Further, $\eta_X : X \longrightarrow LX$ is given by

$$\eta_X(x)(x') = \begin{cases} 1 & \text{if } x = x' \\ 0 & \text{otherwise} \end{cases} \tag{1}$$

and $\mu : L \circ L \longrightarrow L$ by

$$\mu_X(\mathcal{M})(x) = \bigvee_{A \in LX} A(x) \wedge \mathcal{M}(A). \tag{2}$$

This makes $\mathbf{L} = (L, \eta, \mu)$ a monad [29].

We may write $\mathbf{2}$ for the usual covariant powerset monad $(2, \eta, \mu)$, where $2X$ is the powerset of X, $\eta_X(x) = \{x\}$ and $\mu_X(\mathcal{B}) = \bigcup \mathcal{B}$. Further, note that the transitivity condition, relationally viewed as $f \circ f \subseteq f$, translates to $\bigcup 2f(f(x)) \subseteq f(x)$ for all $x \in X$. The category of 'sets and relations', i.e. where objects are sets and morphisms $f : X \longrightarrow Y$ are ordinary relations $f \subseteq X \times Y$ with composition of morphisms being relational composition, is isomorphic to the Kleisli category Set$_2$.

2.2   The Term Monad

Monads equip functors with algebraic structure. Godement [17] and Huber [20] showed that adjoint pairs give rise to monads. Lawvere [27] introduced universal algebra into category theory thereby introducing the term monad. In 1966/67 monads and their applications were further developed during seminars at the Forschungsinstitut für Mathematik at ETH in Zürich, including Beck's developments of distributive laws for monad compositions [4].

We will use a purely functorial description of term over a signature describing the term monad in a more formal way. A conventional (almost verbal) inductive definition of terms is not formal enough to yield a precise functorial notation, and further, does not reveal any substructures.

To begin with, for a set A, i.e. an object in Set, the constant set functor A_{Set} is the covariant set functor that assigns sets X to A, and mappings f to the identity map id_A. The coproduct $\coprod_{i \in I} \mathsf{F}_i$ of covariant set functors F_i assigns to each set X the disjoint union $\bigcup_{i \in I}(\{i\} \times \mathsf{F}_i X)$, and to each morphism $X \xrightarrow{f} Y$ in Set the mapping $(i, m) \mapsto (i, \mathsf{F}_i f(m))$, where $(i, m) \in (\coprod_{i \in I} \mathsf{F}_i)X$.

We will restrict to one-sorted signatures, and we therefore let $(\Omega_n)_{n \leq k}$, k a cardinal number, be a family of sets representing the operator domain, Here Ω_n contains n-ary operators. We write $\Omega_n id^n$ instead of $(\Omega_n)_{\mathsf{Set}} \times id^n$. Note that $\coprod_{n \leq k} \Omega_n id^n X$ is the set of all triples $(n, \omega, (x_i)_{i \leq n})$ with $n \leq k$, $\omega \in \Omega_n$ and $(x_i)_{i \leq n} \in X^n$.

Given these notations, the term functor T_Ω is now defined by transfinite induction. Firstly,

$$\mathsf{T}_\Omega^0 = id$$

and then

$$\mathsf{T}_\Omega^\iota = (\coprod_{n \leq k} \Omega_n id^n) \circ \bigcup_{\kappa < \iota} \mathsf{T}_\Omega^\kappa$$

for each positive ordinal ι. Induction is then in

$$\mathsf{T}_\Omega = \bigcup_{\iota < \bar{k}} \mathsf{T}_\Omega^\iota$$

where $\bar{k}$ is the least cardinal greater than k and $\aleph_0$.

We have that $(\mathsf{T}_\Omega X, (\sigma_\omega)_{\omega \in \Omega})$ is an Ω-algebra, if $\sigma_\omega((m_i)_{i \leq n}) = (n, \omega, (m_i)_{i \leq n})$ for $\omega \in \Omega_n$ and $m_i \in \mathsf{T}_\Omega X$. Morphisms $X \xrightarrow{f} Y$ in Set are extended in the usual way to the corresponding Ω-homomorphisms

$$(\mathsf{T}_\Omega X, (\sigma_\omega)_{\omega \in \Omega}) \xrightarrow{\mathsf{T}_\Omega f} (\mathsf{T}_\Omega Y, (\tau_\omega)_{\omega \in \Omega}),$$

where $\mathsf{T}_\Omega f$ is given as the Ω-extension of $X \xrightarrow{f} Y \hookrightarrow \mathsf{T}_\Omega Y$ associated with $(\mathsf{T}_\Omega Y, (\tau_{n\omega})_{(n,\omega) \in \Omega})$.

To obtain the term monad [29], define $\eta_X^{\mathsf{T}_\Omega}(x) = x$, and let $\mu_X^{\mathsf{T}_\Omega} = id_{\mathsf{T}_\Omega X}^\star$ be the Ω-extension of $id_{\mathsf{T}_\Omega X}$ with respect to $(\mathsf{T}_\Omega X, (\sigma_{n\omega})_{(n,\omega) \in \Omega})$.

2.3   Basic Triples and Partially Ordered Monads

Let acSLAT be the category of almost complete semilattices, i.e. partially ordered sets $(X, \preceq)$ such that the suprema $\bigvee \mathcal{A}$ of all non-empty subsets $\mathcal{A}$ of X exists. Morphisms $f : (X, \preceq) \longrightarrow (Y, \preceq)$ satisfy $f(\bigvee \mathcal{A}) = \bigvee f[\mathcal{A}]$ for non-empty $\mathcal{A} \in 2X$.

A *basic triple* ([14]) is a triple $\Phi = (\mathsf{F}, \preceq, \eta)$, where $(\mathsf{F}, \preceq) : \mathsf{Set} \longrightarrow \mathsf{acSLAT}$, $X \mapsto (\mathsf{F}X, \preceq)$ is a covariant functor, with $\mathsf{F} : \mathsf{Set} \longrightarrow \mathsf{Set}$ as the underlying set functor, and $\eta : id \longrightarrow \mathsf{F}$ is a natural transformation. Note, it follows immediately from the definition of $(\mathsf{F}, \preceq)$ that

$$\mathsf{F}f(\bigvee \mathcal{B}) = \bigvee_{B \in \mathcal{B}} \mathsf{F}f(B) \tag{3}$$

for all Set-morphisms $f : X \longrightarrow Y$ and non-empty $\mathcal{B} \in 2\mathsf{F}X$. That is, morphisms under F preserve non-empty suprema.

If $(\mathsf{F}, \preceq, \eta^F)$ and $(\mathsf{G}, \preceq, \eta^G)$ are basic triples, then $(\mathsf{F} \circ \mathsf{G}, \preceq, \eta^F \mathsf{G} \circ \eta^G)$ is also a basic triple.

Consider L as a functor from Set to acSLAT with $A \leq B$, $A, B \in \mathsf{L}X$, meaning $A(x) \leq B(x)$ for all $x \in X$. Then $(\mathsf{L}, \leq, \eta)$ is a basic triple where $\eta_X : X \longrightarrow \mathsf{L}X$ is given by (1).

A *partially ordered monad* is a quadruple $\mathbf{F} = (\mathsf{F}, \preceq, \eta, \mu)$, such that

(i) $(\mathsf{F}, \preceq, \eta)$ is a basic triple.
(ii) $\mu : \mathsf{FF} \longrightarrow \mathsf{F}$ is a natural transformation such that (F, η, μ) is a monad.
(iii) For all mappings $f, g : Y \longrightarrow \mathsf{F}X$, $f \preceq g$ implies $\mu_X \circ \mathsf{F}f \preceq \mu_X \circ \mathsf{F}g$, where $\preceq$ is defined argumentwise with respect to the partial ordering of $\mathsf{F}X$.
(iv) $\mu_X : (\mathsf{FF}X, \preceq) \longrightarrow (\mathsf{F}X, \preceq)$ preserves non-empty suprema. That is,

$$\mu_X(\bigvee \mathcal{M}) = \bigvee_{M \in \mathcal{M}} \mu_X(M) \tag{4}$$

for non-empty $\mathcal{M} \in 2\mathsf{FF}X$.

We observe that condition (iii) together with the known existence of a suprema implies that

$$\mu_X \circ \mathsf{F}(\vee_i f_i) = \vee_i(\mu_X \circ \mathsf{F}f_i) \tag{5}$$

given any family of morphisms $\{f_i : Y \longrightarrow \mathsf{F}X\}_{i \in I}$ and where the morphism $\vee_i f_i : Y \longrightarrow \mathsf{F}X$ is defined by $(\vee_i f_i)(x) = \bigvee_{i \in I} f_i(x)$. This holds since $g \leq \vee_i f_i$ for all $g \in \{f_i : Y \longrightarrow \mathsf{F}X\}_{i \in I}$ and the resulting inequality $\mu_X \circ \mathsf{F}g \leq \mu_X \circ \mathsf{F}(\vee_i f_i)$ subsequently give the identity of (5).

The basic triple $(\mathsf{L}, \leq, \eta^L)$ can be extended to a partially ordered monad [16] using the multiplication μ as given in (2). In the following we illustrate in some detail that this monad have the properties given in (3), (4), and (5).

First, given $f : X \longrightarrow Y$ and non-empty $\mathcal{B} \in 2LX$ we show (3) by

$$Lf(\bigvee \mathcal{B})(y) = Lf(\bigvee_{B \in \mathcal{B}} B)(y)$$

$$= \bigvee_{f(x)=y} (\bigvee_{B \in \mathcal{B}} B)(x)$$

$$= \bigvee_{f(x)=y} \bigvee_{B \in \mathcal{B}} B(x)$$

$$= \bigvee_{B \in \mathcal{B}} \bigvee_{f(x)=y} B(x)$$

$$= \bigvee_{B \in \mathcal{B}} Lf(B)(y).$$

Second, given $\mathcal{M} \in 2LLX$ and $x \in X$ we have (4) by

$$\mu_X(\bigvee \mathcal{M})(x) = \bigvee_{B \in LX} B(x) \wedge \bigvee_{M \in \mathcal{M}} M(B)$$

$$= \bigvee_{B \in LX} \bigvee_{M \in \mathcal{M}} B(x) \wedge M(B)$$

$$= \bigvee_{M \in \mathcal{M}} \bigvee_{B \in LX} B(x) \wedge M(B)$$

$$= \bigvee_{M \in \mathcal{M}} \mu_X(M)(x).$$

Finally, for $B \in LX$ and $x \in X$, (5) is shown by

$$[\mu_X \circ L(\vee_i f_i)](B)(x) = \bigvee_{B' \in LX} B'(x) \wedge \bigvee_{(\vee_i f_i)(y)=B'} B(y)$$

$$= \bigvee_{B' \in LX} \bigvee_{(\vee_i f_i)(y)=B'} [B'(x) \wedge B(y)]$$

$$= \bigvee_{y \in Y} [\bigvee_{i \in I} f_i(y)(x) \wedge B(y)]$$

$$= \bigvee_{i \in I} \bigvee_{y \in Y} [f_i(y)(x) \wedge B(y)]$$

$$= \bigvee_{i \in I} \bigvee_{B' \in LX} \bigvee_{f_i(y)=B'} [f_i(y)(x) \wedge B(y)]$$

$$= \bigvee_{i \in I} (\bigvee_{B' \in LX} B'(x) \wedge Lf_i(B)(B'))$$

$$= (\vee_i(\mu_X \circ Lf_i))(B)(x).$$

3   Composing Partially Ordered Monads

We may go further and consider composed monads as given in [10]. This definition, reproduced in Definition 1, state the requirements necessary when constructing a new monad based on the composition of two monads' underlying functors.

Definition 1. *Given monads* $\mathbf{F} = (\mathsf{F}, \eta^F, \mu^F)$ *and* $\mathbf{G} = (\mathsf{G}, \eta^G, \mu^G)$, *a distributive law is given by a natural transformation* $\sigma : \mathsf{G} \circ \mathsf{F} \longrightarrow \mathsf{F} \circ \mathsf{G}$ *(the swapper) such that*

(i) $\sigma_{\mathsf{G}X} \circ \mathsf{G}\eta_X^{FG} = \eta_{\mathsf{GG}X}^F \circ \eta_{\mathsf{G}X}^G$
(ii) $\mathsf{F}\mu_X^G \circ \sigma_{\mathsf{G}X} \circ \mathsf{G}\mu_X^{FG} = \mu_X^{FG} \circ \mathsf{F}\mu_{\mathsf{FG}X}^G \circ \sigma_{\mathsf{GFG}X}$
(iii) $\sigma_X \circ \eta_{\mathsf{F}X}^G = \mathsf{F}\eta_X^G$

The conditions on the swapper in a distributive law are precisely what we need so that we can compose monads to get a monad: $\mathbf{F} \bullet \mathbf{G} = (\mathsf{F} \circ \mathsf{G}, \eta^{FG}, \mu^{FG})$ *where*

$-\ \eta_X^{FG} = \eta_{\mathsf{G}X}^F \circ \eta_X^G$ *and*

$-\ \mu_X^{FG} : \mathsf{FGFG}X \xrightarrow{\ \mathsf{F}\sigma_{\mathsf{G}X}\ } \mathsf{FFGG}X \xrightarrow{\ \mu_{\mathsf{GG}X}^F\ } \mathsf{FGG}X \xrightarrow{\ \mathsf{F}\mu_X^G\ } \mathsf{FG}X$

Now, consider monad composition where the participating monads are partially ordered. Will the composition then also be ordered? The following proposition shows that this is indeed the case, if the swapper uphold an ordering condition. In fact, as the proposition shows, the "inner" monad is not required to be a partially ordered monad.

Proposition 1. *Let* $\mathbf{F} = (\mathsf{F}, \preceq^F, \eta^F, \mu^F)$ *be a partially ordered monad and* $\mathbf{G} = (\mathsf{G}, \eta^G, \mu^G)$ *be any monad such that* $\mathbf{F} \bullet \mathbf{G}$ *exists and the swapper is such that if* $f, g : Y \longrightarrow \mathsf{F}X$, $f \preceq^F g$, *then* $\sigma_X \circ \mathsf{G}f \preceq^F \sigma_X \circ \mathsf{G}g$. *Then* $(\mathsf{F} \circ \mathsf{G}, \preceq^F, \eta^{FG}, \mu^{FG})$ *is a partially ordered monad.*

Proof. First, to show that $(\mathsf{F} \circ \mathsf{G}, \preceq^F, \eta^{FG})$ is a basic triple we endow $\mathbf{G}$ with some trivial partial order, e.g. the diagonal. This in turn equips $\mathbf{G}$ with an underlying basic triple. The result is then immediate from composition of basic triples.

It remains to show that morphisms $f, g : Y \longrightarrow \mathsf{FG}X$ such that $f \preceq^F g$ imply that $\mu_X^{FG} \circ \mathsf{FG}f \preceq^F \mu_X^{FG} \circ \mathsf{FG}g$ and that μ^{FG} preserves non-empty suprema. The latter is immediate since morphisms under F preserve non-empty suprema and μ^F does so by definition.

For the implication, we observe that expansion of μ^{FG} gives

$$\mathsf{F}\mu_X^G \circ \mu_{\mathsf{GG}X}^F \circ \mathsf{F}\sigma_{\mathsf{G}X} \circ \mathsf{FG}f \preceq^F \mathsf{F}\mu_X^G \circ \mu_{\mathsf{GG}X}^F \circ \mathsf{F}\sigma_{\mathsf{G}X} \circ \mathsf{FG}g.$$

Using the naturality of μ^G and μ^F we rewrite to

$$\mu_{\mathsf{G}X}^F \circ \mathsf{FF}\mu_X^G \circ \mathsf{F}\sigma_{\mathsf{G}X} \circ \mathsf{FG}f \preceq^F \mu_{\mathsf{G}X}^F \circ \mathsf{FF}\mu_X^G \circ \mathsf{F}\sigma_{\mathsf{G}X} \circ \mathsf{FG}g$$

that – since functors distribute over morphism composition – we may equivalently state in the form

$$\mu_{\mathsf{G}X}^F \circ \mathsf{F}(\mathsf{F}\mu_X^G \circ \sigma_{\mathsf{G}X} \circ \mathsf{G}f) \preceq^F \mu_{\mathsf{G}X}^F \circ \mathsf{F}(\mathsf{F}\mu_X^G \circ \sigma_{\mathsf{G}X} \circ \mathsf{G}g).$$

Since $\mathbf{F}$ is a partially ordered monad it therefore suffices to show that

$$F\mu_X^G \circ \sigma_{GX} \circ Gf \preceq^F F\mu_X^G \circ \sigma_{GX} \circ Gg.$$

The result now immediately follows by applying the proposition condition on σ and preservation of non-empty suprema of morphisms under F. □

Recalling the $\mathbf{L}$ and $\mathbf{T}_\Omega$ monads we may form the composition $\mathbf{L} \bullet \mathbf{T}_\Omega$ as given in ([10]). For brevity we say $\mathbf{T}$ instead of $\mathbf{T}_\Omega$. In the composition σ is such that $\sigma_{X|T^0LX} = id_{LX}$ for the base case and for a term $t = (n, \omega, (t_i)_{i \leq n}) \in T^\alpha LX$, $\alpha > 0$, $t_i \in T^{\beta_i}LX$, $\beta_i < \alpha$.

$$\sigma(t)((n', \omega', (t_i')_{i \leq n'})) = \begin{cases} \bigwedge_{i \leq n} \sigma_X(t_i)(t_i') & \text{if } n = n' \text{ and } \omega = \omega' \\ 0 & \text{otherwise.} \end{cases}$$

We may now extend $\mathbf{L}$ such that it becomes partially ordered by letting $A \leq B$ with $A, B \in LX$ if $A(x) \leq B(x)$ for all $x \in X$. We may show that σ uphold the condition of Proposition 1 by determining that if $f, g : Y \longrightarrow LX$, $f(y) \leq g(y)$ for all $y \in Y$, then $(\sigma_X \circ Tf)(t) \leq (\sigma_X \circ Tg)(t)$ for all $t \in TY$. The result is immediate if $t \in T^0Y$. If, however, $t = (n, \omega, (t_i)_{i \leq n}), t' = (n, \omega, (t_i')_{i \leq n}) \in T^\alpha Y$ with $\alpha > 0$ then assume $\sigma_X \circ Tf(t_i) \leq \sigma_X \circ Tg(t_i)$ for each t_i. By induction we then have

$$\sigma_X(Tf(t))(t') = \bigwedge_{i \leq n} \sigma_X(Tf(t_i))(t_i') \leq \bigwedge_{i \leq n} \sigma_X(Tg(t_i))(t_i') = \sigma_X(Tg(t))(t').$$

Thus, the composed monad $\mathbf{L} \bullet \mathbf{T}_\Omega$ is partially ordered and share its order relation with $\mathbf{L}$.

4 Kleene Monads

An *idempotent semiring* is defined by $(K, +, \cdot, ^*, 0, 1)$ satisfying the conditions

$$p + (q + r) = (p + q) + r \tag{6}$$
$$p + q = q + p \tag{7}$$
$$p + 0 = p \tag{8}$$
$$p + p = p \tag{9}$$
$$p \cdot (q \cdot r) = (p \cdot q) \cdot r \tag{10}$$
$$1 \cdot p = p \tag{11}$$
$$p \cdot 1 = p \tag{12}$$
$$p \cdot (q + r) = p \cdot q + p \cdot r \tag{13}$$
$$(p + q) \cdot r = p \cdot r + q \cdot r \tag{14}$$
$$0 \cdot p = 0 \tag{15}$$
$$p \cdot 0 = 0 \tag{16}$$

A *Kleene algebra* [25,26,36] is an idempotent semiring $(K, +, \cdot, ^*, 0, 1)$ satisfying the conditions

$$1 + p \cdot p^* = p^* \tag{17}$$
$$1 + p^* \cdot p = p^* \tag{18}$$
$$q + p \cdot r \leq r \Rightarrow p^* \cdot q \leq r \tag{19}$$
$$q + r \cdot p \leq r \Rightarrow q \cdot p^* \leq r \tag{20}$$

Instead of (19) and (20) we may use the equivalent conditions

$$p \cdot r \leq r \Rightarrow p^* \cdot r \leq r \tag{21}$$
$$r \cdot p \leq r \Rightarrow r \cdot p^* \leq r \tag{22}$$

Definition 2. *The partially ordered monad* $\mathbf{F} = (\mathsf{F}, \preceq, \eta, \mu)$ *over* Set *is said to be a* Kleene monad, *if there exists a natural transformation* $0 : id \longrightarrow \mathsf{F}$ *such that the conditions*

$$f \circ 0_X = 0_X \tag{23}$$
$$0_X \circ f = 0_X \tag{24}$$

are fulfilled for any morphism $f : id \longrightarrow \mathsf{F}$. *When ambiguity is a risk, we say* $0^{\mathbf{F}}$ *rather than* 0.

As will be seen later these conditions will precisely encode the requirements necessary to define a Kleene algebra over a monad. Before then we first demonstrate that the monad $\mathbf{L}$ fulfill these conditions.

Proposition 2. *The monad* $\mathbf{L}$ *is a Kleene monad with* $0_X(x) : X \longrightarrow LX$ *being the constant* 0 *function for all* X, $x \in X$.

Proof. We begin by showing that 0 is a natural transformation. We have for some function $f : X \longrightarrow Y$ and $y \in Y$

$$(Lf \circ 0_X)(y) = \bigvee_{f(x)=y} 0_X(x) = 0 = (0_X \circ f)(y).$$

We must also show that 0 uphold (23) and (24). This is done as follows.

(23) Given any function $f : X \longrightarrow LX$ and $x, y \in X$ we have

$$(f \diamond 0_X)(x)(y) = (\mu_X \circ L0_X \circ f)(x)(y)$$
$$= \bigvee_{A \in LX} A(y) \wedge \Big(\bigvee_{0_X(x')=A} f(x)(x') \Big)$$

which gives two cases. If A is the constant zero function then we have

$$\bigvee_{A \in LX} 0 \wedge \Big(\bigvee_{0_X(x')=A} f(x)(x') \Big) = 0$$

otherwise we have

$$\bigvee_{A \in LX} A(y) \wedge 0 = 0$$

since $0_X(x') = A$ is never satisfied.

(24) Given any function $f : X \longrightarrow LX$ and $x, y \in X$ we have

$$
\begin{aligned}
(0_X \diamond f)(x)(y) &= (\mu_X \circ Lf \circ 0_X)(x)(y) \\
&= \bigvee_{A \in LX} A(y) \wedge \left(\bigvee_{f(x')=A} 0_X(x)(x') \right) \\
&= \bigvee_{A \in LX} A(y) \wedge \left(\bigvee_{f(x')=A} 0 \right) \\
&= \bigvee_{A \in LX} A(y) \wedge 0 = 0.
\end{aligned}
$$

All requirements from Definition 2 have been shown and we may conclude that **L** indeed is a Kleene monad. □

We have previously seen that the composition of a partially ordered monad with any other monad is itself partially ordered. This makes the composition eligible for being a Kleene monad. The following proposition establishes that this is the case, provided that is the composition swapper respects the natural transformation 0.

Proposition 3. Let $\mathbf{F} = (F, \preceq^F, \eta^F, \mu^F)$ be a Kleene monad and $\mathbf{G} = (G, \eta^G, \mu^G)$ be any monad such that the composition $\mathbf{F} \bullet \mathbf{G}$ exists. Then $\mathbf{F} \bullet \mathbf{G}$ is a Kleene monad with $0 = 0^{\mathbf{FG}} = 0^{\mathbf{F}} \star \eta^G$, provided that $f \circ 0^{\mathbf{F}}_{GX} = 0^{\mathbf{FG}}_X$, for all morphisms $f : X \longrightarrow FGX$, and the swapper, σ, is such that $\sigma \circ G0^{\mathbf{F}} = 0^{\mathbf{F}}G$.

Proof. Naturality of 0 is immediate from the naturality of $0^{\mathbf{F}}$ and η^G. The remaining conditions are shown as follows.

(23) Given any function $f : X \longrightarrow FGX$ we have

$$
\begin{aligned}
(f \diamond 0_X) &= \mu^{FG}_X \circ FG0_X \circ f \\
&= F\mu^G_X \circ \mu^F_{GGX} \circ F\sigma_{GX} \circ FG0_X \circ f \\
&= \mu^F_{GX} \circ FF\mu^G_X \circ F\sigma_{GX} \circ FG0_X \circ f \\
&= \mu^F_{GX} \circ FF\mu^G_X \circ F\sigma_{GX} \circ FGF\eta^G_X \circ FG0^{\mathbf{F}}_X \circ f \\
&= \mu^F_{GX} \circ FF\mu^G_X \circ FFG\eta^G_X \circ F\sigma_X \circ FG0^{\mathbf{F}}_X \circ f \\
&= \mu^F_{GX} \circ F\sigma_X \circ FG0^{\mathbf{F}}_X \circ f \\
&= \mu^F_{GX} \circ F(\sigma_X \circ G0^{\mathbf{F}}_X) \circ f \\
&= \mu^F_{GX} \circ F0^{\mathbf{F}}_{GX} \circ f \\
&= f \diamond 0^{\mathbf{F}}_{GX} = 0^{\mathbf{FG}}_X
\end{aligned}
$$

(24) Given any function $f : X \longrightarrow FGX$ we have

$$0_X \circ f = \mu_X^{FG} \circ FGf \circ 0_X$$
$$= F\mu_X^G \circ \mu_{GGX}^F \circ F\sigma_{GX} \circ FGf \circ 0_X$$
$$= \mu_{GX}^F \circ FF\mu_X^G \circ F\sigma_{GX} \circ FGf \circ 0_X$$
$$= \mu_{GX}^F \circ F(F\mu_X^G \circ \sigma_{GX} \circ Gf) \circ 0_X$$
$$= \mu_{GX}^F \circ F(F\mu_X^G \circ \sigma_{GX} \circ Gf) \circ 0_{GX}^F \circ \eta_X^G$$
$$= (0_{GX}^F \circ (F\mu_X^G \circ \sigma_{GX} \circ Gf)) \circ \eta_X^G$$
$$= 0_{GX}^F \circ \eta_X^G = 0_X.$$

Since all required properties are upheld, we conclude that $\mathbf{F} \bullet \mathbf{G}$ is a Kleene monad. □

We apply Proposition 3 in showing that the composition $\mathbf{L} \bullet \mathbf{T}_\Omega$ is a Kleene monad.

Proposition 4. *The partially ordered monad* $\mathbf{L} \bullet \mathbf{T}_\Omega$ *is a Kleene monad.*

Proof. We must show that $f \circ 0_{TX}^L = 0_X^{LT}$, for all $f : X \longrightarrow LTX$, and that $\sigma_X \circ T0_X^L = 0_{TX}^L$. We begin by showing the former. Let $x \in X$, $y \in TX$, and $f : X \longrightarrow LTX$ be any function, we then have

$$(f \circ 0_{TX}^L)(x)(y) = (\mu_{TX}^L \circ L0_{TX}^L \circ f)(x)(y)$$
$$= \bigvee_{A \in LTX} A(y) \wedge (\bigvee_{0_{TX}^L(x') = A} f(x)(x'))$$
$$= 0 = (0_{TX}^L \circ \eta_X^T)(x)(y).$$

We now show that $\sigma_X \circ T0_X^L = 0_{TX}^L$. Let 0 denote the constant zero function. Consider now a term $t \in T^0$, we have

$$(\sigma_X \circ T0_X^L)(t) = (id_{LX} \circ 0_{TX}^L)(t) = 0_{TX}^L(t).$$

Thus, the result holds for the base case. Let

$$t = (n, \omega, (t_i)_{t \leq n}), t' = (n, \omega, (t'_i)_{i \leq n}) \in T^\alpha X$$

with $\alpha > 0$ and assume $(\sigma_X \circ T0_X^L)(t_i) = 0_{TX}^L(t_i)$ for each t_i. By induction we have

$$\sigma_X(0_X^L(t))(t') = \bigwedge_{i \leq n} \sigma_X(0_X^L(t_i))(t'_i) = \bigwedge_{i \leq n} 0_{TX}^L(t_i)(t'_i) = 0 = 0_{TX}^L(t)(t')$$

which establishes our desired result. □

Having established the notion of Kleene monads we may now define a Kleene algebra over these monads. Let $1 = \eta_X$, and further, for $f_1, f_2 \in Hom(X, FX)$, define

$$f_1 + f_2 = f_1 \vee f_2,$$

i.e. pointwise according to $(f_1 + f_2)(x) = f_1(x) \vee f_2(x)$, and

$$f_1 \cdot f_2 = f_1 \circ f_2$$

where $f_1 \circ f_2 = \mu_X \circ F f_2 \circ f_1$ is the composition of morphisms in the corresponding Kleisli category of $\mathbf{F}$.

A partial order $\preceq$ on $Hom(X, FX)$ is defined pointwise, i.e. for $f_1, f_2 \in Hom(X, FX)$ we say $f_1 \preceq f_2$ whenever $f_1(x) \preceq f_2(x)$ for all $x \in X$. Note that $f_1 \preceq f_2$ if and only if $f_1 + f_2 = f_2$.

Proposition 5. *Let $\mathbf{F} = (F, \preceq, \eta, \mu)$ be a Kleene monad. Then $(Hom(X, FX), +, \cdot, 0, 1)$ is an idempotent semiring.*

Proof. We show each condition in turn:

(6) Follows immediately from associativity of $\vee$.

(7) Follows immediately from commutativity of $\vee$.

(8) Follows immediately from 0 being the identity element of $\vee$.

(9) Follows immediately from idempotency of suprema.

(10) Follows immediately from associativity of morphisms in the Kleisli category.

(11) By naturality of η and the definition of monads we have

$$1 \cdot f_1 = \eta_X \circ f_1 = \mu_X \circ F f_1 \circ \eta_X = \mu_X \circ \eta_{FX} \circ f_1 = f_1.$$

(12) The result follows directly from the definition of monads, i.e., we have

$$f_1 \cdot 1 = f_1 \circ \eta_X = \mu_X \circ F \eta_X \circ f_1 = f_1.$$

(13) By naturality of μ together with (5) we obtain

$$\begin{aligned}
f_1 \cdot (f_2 + f_3) &= \mu_X \circ F(f_2 + f_3) \circ f_1 \\
&= ([\mu_X \circ F f_2] + [\mu_X \circ F f_3]) \circ f_1 \\
&= [\mu_X \circ F f_2 \circ f_1] + [\mu_x F f_3 \circ f_1] \\
&= f_1 \cdot f_2 + f_1 \cdot f_3.
\end{aligned}$$

(14) By naturality of μ together with (3) we obtain

$$\begin{aligned}
(f_1 + f_2) \cdot f_3 &= \mu_X \circ F f_3 \circ (f_1 + f_2) \\
&= \mu_X \circ ([F f_3 \circ f_1] + [F f_3 \circ f_2]) \\
&= [\mu_X \circ F f_3 \circ f_1] + [\mu_X \circ F f_3 \circ f_2] \\
&= f_1 \cdot f_3 + f_2 \cdot f_3.
\end{aligned}$$

(15) Follows immediately from (24) since

$$0 \cdot f_1 = 0_X \circ f_1 = 0_X.$$

(16) Follows immediately from (23) since

$$f_1 \cdot 0 = f_1 \circ 0_X = 0_X$$

And with each necessary condition shown to hold, we conclude that Kleene monads are idempotent semirings. □

The introduction of Kleene asterates is now obvious. For mappings $f : X \longrightarrow FX$, define

$$f^* = \bigvee_{k=0}^{\infty} f^k$$

where $f^0 = 1$ and $f^{k+1} = f \circ f^k = \mu_X \circ F f^k \circ f$. Suprema of mappings $g_i : X \longrightarrow Y$ is given by $(\bigvee g_i)(x) = \bigvee g_i(x)$.

Theorem 1. Let $\mathbf{F} = (F, \preceq, \eta, \mu)$ be a Kleene monad. Then $(Hom(X, FX), +, \cdot, ^*, 0, 1)$ is a Kleene algebra.

Proof. The remaining conditions are proved as follows.

(17) We have

$$1 + f \cdot f^* = 1 \vee (\mu_X \circ F f^* \circ f)$$

$$= 1 \vee (\mu_X \circ F \bigvee_{k=0}^{\infty} f^k \circ f)$$

$$\overset{(5)}{=} 1 \vee \bigvee_{k=0}^{\infty} f^k \cdot f$$

$$= f^0 \vee \bigvee_{k=0}^{\infty} f^{k+1}$$

$$= \bigvee_{k=0}^{\infty} f^k = f^*.$$

(18) Similarly we have

$$1 + f^* \cdot f = 1 \vee (\mu_X \circ F f \circ f^*)$$

$$= 1 \vee (\mu_X \circ F f \circ \bigvee_{k=0}^{\infty} f^k)$$

$$\overset{(3),(4)}{=} 1 \vee \bigvee_{k=0}^{\infty} f^k \cdot f$$

$$= f^0 \vee \bigvee_{k=0}^{\infty} f^{k+1}$$

$$= \bigvee_{k=0}^{\infty} f^k = f^*.$$

(21) Firstly, note that $f_1 \le f_2$ implies $\mu_X \circ F f_1 \le \mu_X \circ F f_2$, and therefore $\mu_X \circ F f_1 \circ g \le \mu_X \circ F f_2 \circ g$, i.e. $g \cdot f_1 \le g \cdot f_2$. Therefore $f \cdot g \le g$ implies $f \cdot f \cdot g \le f \cdot g \le g$, and then also $f^k \cdot g \le g$, for all k. Thus $f \cdot g \le g$ implies

$$f^* \cdot g = \mu_X \circ F g \circ f^*$$

$$= \mu_X \circ F g \circ \bigvee_{k=0}^{\infty} f^k$$

$$= \bigvee_{k=0}^{\infty} \mu_X \circ F g \circ f^k$$

$$= \bigvee_{k=0}^{\infty} f^k \cdot g \le g$$

(22) Similarly, note that $f_1 \le f_2$ implies $\mu_X \circ F g \circ f_1 \le \mu_X \circ F g \circ f_2$, i.e. $f_1 \cdot g \le f_2 \cdot g$, by which we will have that $g \cdot f \le g$ implies $g \cdot f^k \le g$, for all k. Thus $g \cdot f \le g$ implies

$$g \cdot f^* = \mu_X \circ F f^* \circ g$$

$$= \mu_X \circ F \bigvee_{k=0}^{\infty} f^k \circ g$$

$$= \bigvee_{k=0}^{\infty} \mu_X \circ F f^k \circ g$$

$$= \bigvee_{k=0}^{\infty} g \cdot f^k \le g \qquad \square$$

5   Conclusion

Extending monads to include partial order opens up a range of applications in topology and algebra, in particular as partially ordered monads can be composed basically in the same way as ordinary monads. The application towards semirings and Kleene algebras clearly invites to further investigations on language constructions and their semantics. Partially ordered monads also become natural extensions to the term monad as used in general logics. Both these directions should be further developed.

Acknowledgement

We wish to thank anonymous referees for valuable comments improving the content of this paper.

References

1. Aho, A.V., Hopcroft, J.E., Ullman, J.D.: The Design and Analysis of Computer Algorithms. Addison-Wesley, Reading (1975)
2. Adámek, J., Herrlich, H., Strecker, G.: Abstract and concrete categories. Wiley-Interscience, New York (1990)
3. Barr, M., Wells, C.: Toposes, Triples and Theories. Springer, Heidelberg (1985)
4. Beck, J.: Distributive laws, Seminars on Triples and Categorical Homology Theory, 1966/1967. Lecture Notes in Mathematics, vol. 80, pp. 119–140. Springer, Heidelberg (1969)
5. Eklund, P., Galán, M.A., Gähler, W.: Partially ordered monads for monadic topologies, Kleene algebras and rough sets. Electronic Notes in Theoretical Computer Science 225(5), 67–81 (2009)
6. Eklund, P., Gähler, W.: Generalized Cauchy spaces. Math. Nachr. 147, 219–233 (1990)
7. Eklund, P., Gähler, W.: Fuzzy Filter Functors and Convergence. In: Rodabaugh, S.E., Klement, E.P., Höhle, U. (eds.) Applications of category theory to fuzzy subsets. Theory and Decision Library B, pp. 109–136. Kluwer, Dordrecht (1992)
8. Eklund, P., Gähler, W.: Completions and Compactifications by Means of Monads. In: Lowen, R., Roubens, M. (eds.) Fuzzy Logic, State of the Art, pp. 39–56. Kluwer, Dordrecht (1993)
9. Eklund, P., Gähler, W.: Partially ordered monads and powerset Kleene algebras. In: Proc. 10th Information Processing and Management of Uncertainty in Knowledge Based Systems Conference, IPMU 2004 (2004)
10. Eklund, P., Galán, M.A., Ojeda-Aciego, M., Valverde, A.: Set functors and generalised terms. In: Proc. 8th Information Processing and Management of Uncertainty in Knowledge-Based Systems Conference (IPMU 2000), pp. 1595–1599 (2000)
11. Eklund, P., Galán, M.A., Medina, J., Ojeda Aciego, M., Valverde, A.: A categorical approach to unification of generalised terms. Electronic Notes in Theoretical Computer Science 66(5) (2002),
http://www.elsevier.nl/locate/entcs/volume66.html
12. Gähler, W.: A topological approach to structure theory. Math. Nachr. 100, 93–144 (1981)
13. Gähler, W.: Monads and convergence. In: Proc. Conference Generalized Functions, Convergences Structures, and Their Applications, Dubrovnik (Yugoslavia) 1987, pp. 29–46. Plenum Press, New York (1988)
14. Gähler, W.: General Topology – The monadic case, examples, applications. Acta Math. Hungar. 88, 279–290 (2000)
15. Gähler, W.: Extension structures and completions in topology and algebra. In: Seminarberichte aus dem Fachbereich Mathematik, Band 70, FernUniversität in Hagen (2001)
16. Gähler, W., Eklund, P.: Extension structures and compactifications. In: Categorical Methods in Algebra and Topology (CatMAT 2000), pp. 181–205 (2000)
17. Godement, R.: Topologie algébrique et théorie des faisceaux, appendix. Hermann, Paris (1958)
18. Goguen, J.A.: L-fuzzy sets. J. Math. Anal. Appl. 18, 145–174 (1967)
19. Helgesson, R.: A categorical approach to logics and logic homomorphisms, UMNAD 676/07, Umeå University, Department of Computing Science (2007)
20. Huber, P.J.: Homotopy theory in general categories. Math. Ann. 144, 361–385 (1961)

21. Keller, H.H.: Die Limesuniformisierbarkeit der Limesräume. Math. Ann. 176, 334–341 (1968)
22. Kleene, S.C.: Representation of events in nerve nets and finite automata. In: Shannon, C.E., McCarthy, J. (eds.) Automata Studies, pp. 3–41. Princeton University Press, Princeton (1956)
23. Kowalsky, H.-J.: Limesräume und Komplettierung. Math. Nachr. 12, 301–340 (1954)
24. Kozen, D.: On induction vs. *-continuity. In: Kozen, D. (ed.) Logic of Programs 1981. LNCS, vol. 131, pp. 167–176. Springer, Heidelberg (1982)
25. Kozen, D.: Kleene algebra with tests. ACM Transactions on Programming Languages and Systems 19, 427–443 (1999)
26. Kuich, W., Salomaa, A.: Semirings, Automata, and Languages. Springer, Berlin (1986)
27. Lawvere, F.W.: Functorial Semantics of Algebraic Theories, Dissertation. Columbia University (1963)
28. Loeckx, J., Ehrich, H.-D., Wolf, M.: Specification of Abstract Data Types. Wiley-Teubner, Chichester (1996)
29. Manes, E.G.: Algebraic Theories. Springer, Heidelberg (1976)
30. Meseguer, J.: General logics. In: Ebbinghaus, H.-D., et al. (eds.) Logic Colloquium 1987, pp. 275–329. Elsevier, North-Holland, Amsterdam (1989)
31. Moggi, E.: An Abstract View of Programming Languages, Edinburgh Univ., Dept. of Comp. Sci. (1989)
32. Moggi, E.: Notions of computation and monads. Information and Computation 93, 55–92 (1991)
33. Pratt, V.: Action Logic and Pure Induction. In: van Eijck, J. (ed.) JELIA 1990. LNCS, vol. 478, pp. 97–120. Springer, Heidelberg (1991)
34. Richardson, G.D.: A Stone-Čech compactification for limit spaces. Proc. Amer. Math. Soc. 25, 403–404 (1970)
35. Rydeheard, D.E., Burstall, R.M.: A categorical unification algorithm. In: Poigné, A., Pitt, D.H., Rydeheard, D.E., Abramsky, S. (eds.) Category Theory and Computer Programming. LNCS, vol. 240, pp. 493–505. Springer, Heidelberg (1986)
36. Salomaa, A.: Two complete axiom systems for the algebra of regular events. J. ACM 13, 158–169 (1966)
37. Tarski, A.: On the calculus of relations. J. Symbolic Logic 6, 65–106 (1941)
38. Wadler, P.: Comprehending monads. Mathematical Structures in Computer Science 2, 461–493 (1992)

A Relation-Algebraic Approach to Liveness of Place/Transition Nets

Alexander Fronk[1] and Rudolf Berghammer[2]

[1] MATERNA GmbH, 44141 Dortmund, Germany
Alexander.Fronk@materna.de
[2] Christian-Albrechts-Universität Kiel, 24098 Kiel, Germany
rub@informatik.uni-kiel.de

Abstract. We provide a relation-algebraic characterization of liveness in Petri nets based on a relation-algebraic definition of both the structure and the state space of Petri nets. Such an approach, compared to the common ones that apply predicate logic and set theory, shifts the formalization to a more abstract level. As a main benefit, Petri net properties can be proved in a rigorous mathematical style. Since the characterizations are executable relational specifications, they provide the possibility for tool support.

1 Introduction

Technical systems can be of different kinds. They may never stop (such as operating systems), they may willingly or unwillingly stop in parts due to a deadlock (an operating system being gracefully degraded or blocked by concurrent threads waiting on each other), they may stop after performing a certain task (an air bag), or they may not even work at all (such as a paper jammed or damaged printer). It is often difficult to verify such liveness properties in concurrent systems. As Petri nets allow one to model them, analyzing the liveness properties of a Petri net replaces the analysis of the underlying system in order to, for example, determine the reason why a deadlock occurs.

In Software Engineering, relations occur in many places. For instance, in [4] it is shown how the pipes-and-filters architecture can be seen as a system of relations. UML class diagrams, as another example, relate elements such as classes and interfaces via association, aggregation, or inheritance. This can be utilized to express the structure of design patterns and the semantics of three-dimensional design languages by means of relations; see [1]. Graph-like structures and graphs in essence are of particular interest for relational analysis and, since the static part of a Petri net is a bipartite graph, static properties can be analyzed with relation algebra as well; cf. [2,9].

However, literature on the relation-algebraic treatment of dynamic properties of Petri nets seems to be rare. To our knowledge, [2] is the first paper on this topic. But it only considers the restricted class of condition/event nets. For the first time it seems that general place/transition nets are discussed in [6], where

R. Berghammer et al. (Eds.): RelMiCS/AKA 2009, LNCS 5827, pp. 103–118, 2009.

especially the reachability relation is formalized with relation-algebraic means. This work is continued in [8] particularly with regard to tool support, and in [7] in view of the simultaneous evaluation of reachable markings in a single step. The present paper is based on [6]. The novelties are a relation-algebraic treatment of liveness properties and relation-algebraic specifications of deadlocks, traps, and the important deadlock/trap property.

Our view on general Petri nets, when compared to approaches using predicate logic and set theory, shifts their formalization to a more abstract level. It allows rigorously for the deduction of specifications of liveness, deadlocks and traps, the deadlock/trap property, and of further other qualities from net-theoretic formulae, which are thus correct by construction. Since the resulting relation-algebraic specifications also are executable, tool support is possible. We embedded the specifications discussed here into our Petri net tool PETRA [8] that is tailored for analyzing both static and dynamic qualities of Petri nets. The tool bases on the KURE-Java library, an efficient BDD-implementation of relation algebra that has been extracted from the RELVIEW tool [10,11]. The approach discussed in [12] introduces BDDs for the analysis of Petri nets and provides algorithms manipulating BDDs directly. The efficiency of this data structure is demonstrated by the Dining Philosophers example. In addition to this approach, we investigate a relation-algebraic characterization of Petri nets which allows us to manipulate BDDs by algorithms that are mathematically precise and correct by construction. That is, we encapsulate BDDs and provide a mathematical and thus abstract interface to state space analysis independent from the data structure representing relations.

2   Preliminaries

In this section, we introduce the basic notions of relations and Petri nets which are needed in the remainder of the paper. For more details, we have to refer to the literature, see e.g. [14] for relations and [13] for Petri nets.

2.1   Relations and Relation Algebra

Assuming a (heterogeneous) relation algebra as defined in [14], its elements are called *(abstract) relations*. If R is a *concrete (set-theoretic) relation* between sets X, Y, this is denoted by $R : X \leftrightarrow Y$. Instead of $2^{X \times Y}$, we write $[X \leftrightarrow Y]$ for the set of all relations between X and Y and $R_{x,y}$ instead of $\langle x, y \rangle \in R$ to express that $x \in X$ and $y \in Y$ are related via R. The latter notation is motivated by the fact that we frequently interpret relations as Boolean matrices.

If R and S are relations, $R \cup S$, $R \cap S$, $R; S$, and $R \subseteq S$ denote their *union*, *intersection*, *composition*, and *inclusion*, respectively. Furthermore, R^T denotes the *converse* of R, $\overline{R}$ its *negation*, R^+ its *transitive closure*, and R^* its *reflexive transitive closure*. The *empty relation* is denoted by O, the *universal relation* by L, and the *identity relation* by I.

A relation R is *univalent* if $R^\mathsf{T}; R \subseteq \mathsf{I}$, *total* if $R; \mathsf{L} = \mathsf{L}$, and a *mapping* if it is both univalent and total. R is *injective* if R^T is univalent and *surjective* if R^T

is total. For a mapping R from X to Y with $R_{x,y}$, we write $R(x)$ to refer to the image $y \in Y$ of $x \in X$ as usual in this case.

A *vector* v is a relation v with $v; \mathsf{L} = v$, and a *point* v is a vector with $v; v^{\mathsf{T}} \subseteq \mathsf{I}$ and $\mathsf{L}; v = \mathsf{L}$. A concrete vector on a set X is denoted by $v : X \leftrightarrow \mathbb{1}$ for any singleton set $\mathbb{1} := \{\diamond\}$. We omit $\diamond$ as subscript and write v_x instead of $v_{x,\diamond}$. Such a vector can be considered as a Boolean matrix with exactly one column, i.e., as a Boolean column vector, and *models the subset* $\{x \in X \mid v_x\}$ of X. If v is even a point, it *models an element* x of X. This means that for all $y \in X$ it holds v_y iff $x = y$. Based on these properties, the vectors $R; \mathsf{L}$ and $R^{\mathsf{T}}; \mathsf{L}$ are called the *domain* and *codomain* of R, respectively, and for a vector v we call $R \cap v; \mathsf{L}$ the *domain restriction of R through v*. Each relation R representable as $v; w^{\mathsf{T}}$, with v and w being points, is an atom of the underlying relation algebra. Moreover, for each concrete non-empty relation $R : X \leftrightarrow Y$ there exist a pair $v : X \leftrightarrow \mathbb{1}$ and $w : Y \leftrightarrow \mathbb{1}$ of points such that $v; w^{\mathsf{T}} \subseteq R$ (cf. point axiom of [14]).

The *membership* symbol $\in$ is modeled by a relation $\mathsf{M} : X \leftrightarrow 2^X$ between X and its powerset such that for all $x \in X$ and $S \in 2^X$ we have $\mathsf{M}_{x,S}$ iff $x \in S$. A *relational direct product* is a pair of relations (π, ρ) with $\pi^{\mathsf{T}}; \pi = \mathsf{I}$, $\rho^{\mathsf{T}}; \rho = \mathsf{I}$, $\pi; \pi^{\mathsf{T}} \cap \rho; \rho^{\mathsf{T}} = \mathsf{I}$, and $\pi^{\mathsf{T}}; \rho = \mathsf{L}$. This axiomatization has up to isomorphism only one model and this is in the case of concrete relations the pair consisting of the natural projections $\pi : X \times Y \leftrightarrow X$ and $\rho : X \times Y \leftrightarrow Y$ of $X \times Y$.

Finally, for each relation R, by $\mathbf{R} := (\pi; R \cap \rho); \mathsf{L}$ the *vector representation* of R is defined, and for each vector v by $\mathbf{v} := \overline{\mathsf{M}^{\mathsf{T}}; \overline{v} \cup \mathsf{M}^{\mathsf{T}}; v}$ the *point representation* of v is defined. Given a concrete relation $R : X \leftrightarrow Y$, its vector representation is of type $[X \times Y \leftrightarrow \mathbb{1}]$ and for all $x \in X$ and $y \in Y$ we have $R_{x,y}$ iff $\mathbf{R}_{\langle x,y \rangle}$. The type of the point representation $\mathbf{v}$ of a concrete vector $v : X \leftrightarrow \mathbb{1}$ is $[2^X \leftrightarrow \mathbb{1}]$ and if v models Y as subset of X, then the point $\mathbf{v}$ models Y as an element of 2^X. We say that $\mathbf{r} : 2^{X \times Y} \leftrightarrow \mathbb{1}$ is the point representation of $R : X \leftrightarrow Y$ if it is the point representation of the vector representation of R.

In Section 4 we will need the following auxiliary result on relations contained in identity relations (so-called partial identities).

Lemma 1. $S \subseteq \mathsf{I}$ *implies* $S; R = S; \mathsf{L} \cap R$.

Proof. The inclusion $S; R \subseteq S; \mathsf{L} \cap R$ follows from $S; R \subseteq S; \mathsf{L}$ and $S; R \subseteq \mathsf{I}; R = R$, and for the proof of the reverse inclusion via

$$S; \mathsf{L} \cap R \subseteq (S \cap R; \mathsf{L}^{\mathsf{T}}); (\mathsf{L} \cap S^{\mathsf{T}}; R) \subseteq S; S^{\mathsf{T}}; R \subseteq S; \mathsf{I}^{\mathsf{T}}; R = S; R$$

we use the Dedekind rule (see [14]) in the first step. $\square$

We will also apply that partial identities S are symmetric, that is, the equation $S = S^{\mathsf{T}}$ holds.

2.2   Basic Notions of Petri Nets

A Petri net (or place/transition net) is a 6-tuple $\mathcal{P} = (P, T, F, C, W, M_0)$ with disjoint sets P of *places* and T of *transitions*, $F \subseteq (P \times T) \cup (T \times P)$ as (bipartite) *flow relation*, $C : P \to \mathbb{N}$ as *capacity function*, $W : F \to \mathbb{N}$ as *weight function*, and $M_0 : P \to \mathbb{N}$ as *initial marking*. For the capacities of all places $p \in P$ and the weights of all arcs $f \in F$ the properties $C(p) \neq 0$ and $W(f) \neq 0$ are demanded and, furthermore, $M_0(p) \leq C(p)$ has to hold for all $p \in P$.

For explaining the following notions, we assume a fixed Petri net $\mathcal{P}$ with constituents P, T, F, C, W and M_0 to be given.

By $\bullet t$ we denote the set of (immediate) predecessor places of $t \in T$ w.r.t. F and by $t\bullet$ the set of (immediate) successor places of t. Furthermore, $\mathcal{M}$ denotes the set off all *possible markings* of $\mathcal{P}$, that is, the set of functions $M : P \to \mathbb{N}$ satisfying $M(p) \leq C(p)$ for all $p \in P$. Then a transition $t \in T$ is *activated under* $M \in \mathcal{M}$, abbreviated as $M \overset{t}{\leadsto}$, if $M(p) \geq W(p, t)$ for all $p \in \bullet t$ and $M(p) \leq C(p) - W(t, p)$ for all $p \in t\bullet$. In this case, t can *fire* under M. Its firing produces a marking $M' \in \mathcal{M}$, where M' is defined through

$$M'(p) := \begin{cases} M(p) - W(p, t) & \text{if } p \in \bullet t \setminus t\bullet, \\ M(p) + W(t, p) & \text{if } p \in t\bullet \setminus \bullet t, \\ M(p) - W(p, t) + W(t, p) & \text{if } p \in \bullet t \cap t\bullet, \\ M(p) & \text{otherwise.} \end{cases}$$

We say that M' is *immediately reachable from M under t* and write $M \overset{t}{\leadsto} M'$ for that relationship. Based on this notion, general reachability for markings can be defined as follows: A marking $M' \in \mathcal{M}$ is *reachable* from $M \in \mathcal{M}$, in symbols $M \overset{*}{\leadsto} M'$, if $M = M'$ or there exists a non-empty sequence $t_1, \ldots, t_n$ of transitions and a non-empty sequence $N_1, \ldots, N_n, N_{n+1}$ of markings such that $M = N_1$, $M' = N_{n+1}$ and $N_i \overset{t_i}{\leadsto} N_{i+1}$ for all $i, 1 \leq i \leq n$. In this case $t_1, \ldots, t_n$ is called a *firing sequence*. The set of markings reachable from $M_0 \in \mathcal{M}$ is called the *set of reachable markings* of $\mathcal{P}$ and denoted as $\mathcal{M}_0$. If we restrict the reachability relation to this set, we arrive at the notion of the *reachability graph* of $\mathcal{P}$. That is, the vertex set of this graph is $\mathcal{M}_0$ and there is an arc from $M \in \mathcal{M}_0$ to $M' \in \mathcal{M}_0$ iff M' is reachable from M.

A set X of places is *sufficiently marked* under a marking $M \in \mathcal{M}$ if there exist $p \in X$ and $t \in T$ such that $p \in \bullet t$ (i.e., t is a successor transition of the place p) and $M(p) \geq W(p, t)$.

We close this section with introducing some kinds of liveness, the notion we are primarily interested in this paper, and its counterpart deadness.

Assume $t \in T$ to be a transition of the Petri net $\mathcal{P}$. Then t is *weakly live* if it is activated under at least one reachable marking, that is, if there exists $M \in \mathcal{M}_0$ with $M \overset{t}{\leadsto}$. It is *live* if it is weakly live under all reachable markings. This means that for all $M \in \mathcal{M}_0$ there exists $M' \in \mathcal{M}_0$ such that both $M \overset{*}{\leadsto} M'$ and $M' \overset{t}{\leadsto}$ hold. The entire Petri net $\mathcal{P}$ is said to be *live* if all its transitions are live. A transition $t \in T$ that is not weakly live is *dead*. Deadness is also defined for reachable markings: $M \in \mathcal{M}_0$ is *dead* if there is no transition activated under M, i.e., if for all $t \in T$ the property $M \overset{t}{\leadsto}$ does not hold. In this case $\mathcal{P}$ is also called dead under M. And, finally, $\mathcal{P}$ is a *weakly live* Petri net if it does not have a dead marking.

3 Relation-Algebraic Characterization of Liveness

The relation-algebraic characterization of liveness requires a transition system given by means of suitable relations. To establish it, in Section 3.1 we first start with a relation-algebraic transcription of the definition of Petri nets. Therefore, we need a relational description of the set of natural numbers. In order to not conflict with the denotation $\mathbb{N}$, we use a set N together with an injective mapping, i.e., the successor relation $S : N \leftrightarrow N$, and a point $z : N \leftrightarrow \mathbb{1}$ modeling the number zero. Furthermore, we demand the laws $S; z = O$ and $(S^T)^*; z = L$ to hold. As shown in [3], the relational structure (N, S, z) has up to isomorphism only one model and this "standard model" consists of the natural numbers with the successor function $n \mapsto n + 1$ and the number zero. For this model, the first law says that zero is not a successor of a natural number, and the second law corresponds to Peano's induction axiom. Based on S and its converse relation $P := S^T$ (the predecessor relation), the usual linear orderings $\leq$ and $\geq$ on N are represented through S^* and P^*, respectively.

3.1 Modeling Petri Nets with Relations

Supposing $S : N \leftrightarrow N$, $P : N \leftrightarrow N$ and $z : N \leftrightarrow \mathbb{1}$, we start with the following definition. It is the direct transformation of the description of a Petri net given in Section 2.2 into the language of relations.

Definition 1. *A (place/transition) Petri net is relation-algebraically modeled by a 6-tuple $\mathcal{P} = (R, S, W^{\bullet t}, W^{t\bullet}, C, M_0)$ with relations*

$$R : P \leftrightarrow T \qquad S : T \leftrightarrow P$$

for representing the (bipartite) flow relation and mappings

$$W^{\bullet t} : P \times T \leftrightarrow N \qquad W^{t\bullet} : T \times P \leftrightarrow N \qquad C : P \leftrightarrow N \qquad M_0 : P \leftrightarrow N$$

for representing the weights of the arcs contained in R and in S, respectively, the capacities of the places, and the initial marking. Furthermore, the following four properties are required to hold:

$$(a) \quad C; z = O \qquad\qquad (b) \quad \overline{R} = \pi^T; (W^{\bullet t}; z; L \cap \rho)$$
$$(c) \quad \overline{S} = \beta^T; (W^{t\bullet}; z; L \cap \alpha)^T \qquad (d) \quad M_0 \subseteq C; P^*$$

Here the relational direct products (π, ρ) and (α, β) consist of the natural projections of $P \times T$ and $T \times P$, respectively. □

The above four relational formulae (a) to (d) exactly correspond to the restrictions of and the dependencies between the respective functions defined for Petri nets in Section 2.2. We show this in the following by using well-known correspondences between relation-algebraic terms and predicate logic and the common notation of function application in the case of relations which are mappings.

To verify that the equation (a) specifies all place capacities to be non-zero is rather trivial:

$$C; z = \mathsf{O} \iff \neg\exists p : (C; z)_p \iff \neg\exists p, n : C_{p,n} \wedge z_n \iff \neg\exists p : C(p) = 0$$

By the next two equations (b) and (c) it is specified that all weights of arcs are non-zero and in the case of a non-arc the weight of the corresponding pair is defined as zero. Namely, due to

$$\begin{aligned}
\overline{R} = \pi^\mathsf{T}; (W^{\bullet t}; z; \mathsf{L} \cap \rho) &\iff \forall p, t : \overline{R}_{p,t} \leftrightarrow (\pi^\mathsf{T}; (W^{\bullet t}; z; \mathsf{L} \cap \rho))_{p,t} \\
&\iff \forall p, t : \overline{R}_{p,t} \leftrightarrow \exists q : \pi_{q,p} \wedge (W^{\bullet t}; z; \mathsf{L} \cap \rho)_{q,t} \\
&\iff \forall p, t : \overline{R}_{p,t} \leftrightarrow \exists q : \pi_{q,p} \wedge \rho_{q,t} \wedge (W^{\bullet t}; z)_q \\
&\iff \forall p, t : \overline{R}_{p,t} \leftrightarrow \exists q : q = \langle p, t \rangle \wedge (W^{\bullet t}; z)_q \\
&\iff \forall p, t : \overline{R}_{p,t} \leftrightarrow (W^{\bullet t}; z)_{\langle p,t \rangle} \\
&\iff \forall p, t : \overline{R}_{p,t} \leftrightarrow \exists n : W^{\bullet t}{}_{\langle p,t \rangle, n} \wedge z_n \\
&\iff \forall p, t : \overline{R}_{p,t} \leftrightarrow W^{\bullet t}(p, t) = 0 \\
&\iff \forall p, t : R_{p,t} \leftrightarrow W^{\bullet t}(p, t) \neq 0
\end{aligned}$$

the equation (b) says that this holds for all pairs from $P \times T$ and, analogously, one can show that the equation (c) says that this holds for all pairs from $T \times P$. And, finally, from the derivation

$$\begin{aligned}
M_0 \subseteq C; \mathsf{P}^* &\iff \forall p, n : M_{0\,p,n} \rightarrow (C; \mathsf{P}^*)_{p,n} \\
&\iff \forall p, n : M_0(p) = n \rightarrow (C; (\mathsf{S}^*)^\mathsf{T})_{p,n} \\
&\iff \forall p, n : M_0(p) = n \rightarrow \exists n' : C_{p,n'} \wedge (\mathsf{S}^*)^\mathsf{T}{}_{n',n} \\
&\iff \forall p, n : M_0(p) = n \rightarrow \exists n' : C_{p,n'} \wedge \mathsf{S}^*{}_{n,n'} \\
&\iff \forall p, n : M_0(p) = n \rightarrow \mathsf{S}^*{}_{n,C(p)} \\
&\iff \forall p, n : M_0(p) = n \rightarrow n \leq C(p) \\
&\iff \forall p : M_0(p) \leq C(p)
\end{aligned}$$

we obtain that the last formula (d) of the definition specifies that all values of the initial marking M_0 are bounded by the respective capacities.

3.2 A Relation-Algebraic Transition System

As the descriptions in Section 2.2 show, considering liveness properties of Petri nets means to investigate which markings are reachable from each other and which transitions are activated under which markings. Thus, the Petri net's state space is under consideration and the desired information can be deduced from a transition system with rules of the form $M \xrightarrow{t} M'$ for all $M, M' \in \mathcal{M}_0$ and $t \in T$ such that $M \overset{t}{\rightsquigarrow} M'$. Section 2.2 also shows that the different notions of liveness we have introduced are not referring to the specific immediately reachable marking generated by firing an activated transition. Hence, a first step to liveness analysis is to generate the reachability graph if the state space is finite, or the coverability graph (see [5]) if the state space is infinite.

Markings are functions from P to N and, since functions are specific concrete relations[1], each marking is a relation between P and N, too. As a consequence, the type of the reachability relation $\mathcal{R}$ of a Petri net – the relation of the reachability graph if restricted to the reachable markings – can be taken as $[[P \leftrightarrow N] \leftrightarrow [P \leftrightarrow N]]$. The relation itself is component-wisely specified by

$$\mathcal{R}_{M,M'} :\Longleftrightarrow M \in \mathcal{M}_0 \wedge M' \in \mathcal{M}_0 \wedge M \overset{*}{\leadsto} M' \tag{1}$$

for all markings M and M', now taken as elements from the set $[P \leftrightarrow N]$. Besides the reachability relation of (1), a relation $\mathcal{A}$ of type $[[P \leftrightarrow N] \leftrightarrow T]$ is necessary for dealing with liveness. It relates each reachable marking to the transitions it activates. Hence, we have

$$\mathcal{A}_{M,t} :\Longleftrightarrow M \in \mathcal{M}_0 \wedge M \overset{t}{\leadsto} \tag{2}$$

for all markings $M : P \leftrightarrow N$ and transitions $t \in T$ as component-wise specification. The two relations $\mathcal{R}$ and $\mathcal{A}$ are sufficient to elegantly characterize the different notions of liveness relation-algebraically and to formally prove liveness properties of a Petri net as will be shown in the subsequent sections.

Next, we present an algorithm to generate the relations $\mathcal{R}$ and $\mathcal{A}$ component-wisely specified by (1) and (2) for a given Petri net. To this end, we assume a relational function $cn : [P \leftrightarrow N] \to [T \leftrightarrow \mathbb{1}]$ to be at hand that yields for a marking $M : P \leftrightarrow N$ the modeling of the set of transitions that are activated under M as a vector $cn(M)$ of type $[T \leftrightarrow \mathbb{1}]$, and also a relational function $irm : [P \leftrightarrow N] \times [T \leftrightarrow \mathbb{1}] \to [P \leftrightarrow N]$, that yields for a marking $M : P \leftrightarrow N$ and a transition t, modeled as a point v of type $[T \leftrightarrow \mathbb{1}]$, the immediately reachable marking M' of M under t, again as a relation $irm(M, v)$ of type $[P \leftrightarrow N]$. For the formulation of the algorithm and its understanding the explicit relation-algebraic specifications of the functions are not necessary. Therefore, we refer the interested reader to [6], where it is shown how to formally develop the relation-algebraic expressions for $cn(M)$ and $irm(M, v)$ from their component-wise definitions

$$cn(M)_t :\Longleftrightarrow M \overset{t}{\leadsto} \tag{3}$$

for all $M : P \leftrightarrow N$ and $t \in T$, and

$$irm(M, v) = M' :\Longleftrightarrow M \overset{t}{\leadsto} M' \tag{4}$$

---

[1] To distinguish between the "usual" notion of a function and the relational view as univalent and total elements of a relation algebra, we have introduced the notion "mapping" for the latter ones in Section 2.1. Although for concrete relations both notions coincide, we prefer the term "mapping" if the relation-algebraic view plays the decisive role and use the word "function" in the other cases.

for all $M : P \leftrightarrow N$, $M' : P \leftrightarrow N$, $t \in T$, and points $v : T \leftrightarrow \mathbb{1}$ such that the transition t is modeled by the point v. The paper [6] also presents the RELVIEW-implementations of cn and irm.

From (2) we get that the relation $\mathcal{A}$ thereupon relates a reachable marking M to the transitions that are activated under M. To make both the construction and the use of these relations clear, we provide a depth-first algorithm simultaneously constructing $\mathcal{R}$ and $\mathcal{A}$. Therewith, we proceed to establish the desired transition system by the following algorithm. It is formulated in pseudo code, but this immediately may be translated into the language of the RELVIEW tool.

Algorithm 1. *Let $\mathcal{P}$ be a Petri net. Then do the following:*

1. *initialize $\mathcal{R}$ with $\mathsf{O} : [P \leftrightarrow N] \leftrightarrow [P \leftrightarrow N]$*
2. *initialize $\mathcal{A}$ with $\mathsf{O} : [P \leftrightarrow N] \leftrightarrow T$*
3. *let $M := M_0$*
4. *let $\mathbf{m}$ be the point representation of M*
5. *let $\mathcal{R} := \mathcal{R} \cup \mathbf{m}; \mathbf{m}^\mathsf{T}$*
6. *let $\mathcal{A} := \mathcal{A} \cup \mathbf{m}; cn(M)^\mathsf{T}$*
7. *for each point v included in $cn(M)$ do:*
   (a) let $\mathbf{m}'$ be the point representation of $M' = irm(M, v)$
   (b) let $\mathcal{R} := \mathcal{R} \cup \mathbf{m}; \mathbf{m}'^\mathsf{T}$
   (c) if $\mathcal{R} \cap \mathbf{m}'; \mathbf{m}'^\mathsf{T} = \mathsf{O}$ restart at step 4 with $M := M'$
8. *let $\mathcal{R} := \mathcal{R}^+$*

In step 5 of the algorithm, the singleton set $\{\langle M, M \rangle\}$ is modeled by the atom $\mathbf{m}; \mathbf{m}^\mathsf{T}$. Its insertion into $\mathcal{R}$ by means of the union operation reflects the fact that each marking is immediately reachable from itself. Simultaneously, we mark the marking M "already processed" to establish the termination condition for recursive calls in step (c). Step 6 relates M to all transitions activated under it because of the meaning of the relational function cn stated in (3). Due to (4), the call of irm in step (a) constitutes M' to be reachable from M. If M' has already been reached before, we continue with the next transition activated under M. Otherwise, we depth-first execute the algorithm recursively on M'. When no further reachable marking can be calculated, the transitive closure of $\mathcal{R}$ constructed so far delivers the desired reachability relation.

Since the depth-first algorithm computes all markings reachable from the initial marking M_0, and since $\mathcal{R}_{M,M}$ holds for all such markings we have that the vector $\mathcal{R}; \mathsf{L}$ exactly models the set $\mathcal{M}_0$ of the reachable markings of $\mathcal{P}$, but now as subset of the set $[P \leftrightarrow N]$. This means that for all $M : P \leftrightarrow N$ the equivalence

$$(\mathcal{R}; \mathsf{L})_M \iff M \in \mathcal{M}_0 \tag{5}$$

holds. For the domain of $\mathcal{R}$ we even have the equation

$$\mathcal{R}; \mathsf{L} = (\mathsf{I} \cap \mathcal{R}); \mathsf{L}. \tag{6}$$

Here "$\subseteq$" follows from $\mathcal{R} \subseteq (\mathsf{I} \cap \mathcal{R}); \mathsf{L}$ which, in turn, is a consequence of

$$\mathcal{R}_{M,M'} \implies \exists M'' : \mathcal{R}_{M,M''} \wedge \mathsf{I}_{M,M''} \wedge \mathsf{L}_{M'',M'} \iff ((\mathcal{R} \cap \mathsf{I}); \mathsf{L})_{M,M'}$$

for all markings $M, M' \in [P \leftrightarrow N]$, and the converse inclusion is always true. To test whether a reachable marking M' is reachable from a reachable marking M, we only have to test $\mathbf{m}; \mathbf{m'}^\mathsf{T} \subseteq \mathcal{R}$, where $\mathbf{m}$ and $\mathbf{m'}$ are the point representations of M and M', respectively. Note that, because of (2), the domain of $\mathcal{A}$ only contains reachable markings which activate at least one transition. If we remember the notion of a dead marking M as one such that there is no transition activated under M, then we have

$$\mathcal{R} \cap \overline{\mathcal{A}; \mathsf{L}} \subseteq \mathsf{I}, \tag{7}$$

since in the relation $\mathcal{R}$ reachable dead markings are only related to themselves.

3.3 Characterization of Liveness

With a transitions system given by $\mathcal{R}$ and $\mathcal{A}$, we can now relation-algebraically characterize the different notions of liveness as introduced in Section 2.2. We prove their correctness by showing their equivalence to the respective formalized definition. In doing so, we use rigorous transformation rules between predicate logic and relation-algebraic expressions and also laws of relation algebra. We start with the liveness properties (and their counterparts) of transitions (and markings, respectively).

Theorem 1. *Let $\mathcal{P}$ be a Petri net and $\mathcal{R}$ and $\mathcal{A}$ be the relations of Section 3.2. Then the vectors*

$$\text{(a)} \quad \mathcal{A}^\mathsf{T}; \mathsf{L} \qquad \text{(b)} \quad \overline{\mathcal{A}^\mathsf{T}; \mathsf{L}} \qquad \text{(c)} \quad \overline{\mathcal{R}; \overline{\mathcal{A}^\mathsf{T}}; \mathcal{R}; \mathsf{L}} \qquad \text{(d)} \quad \mathcal{R}; \mathsf{L} \cap \overline{\mathcal{A}; \mathsf{L}}$$

of type $[T \leftrightarrow \mathbb{1}]$ in the cases (a) to (c) and $[[P \leftrightarrow N] \leftrightarrow \mathbb{1}]$ in the case (d) model the set of weakly live transitions, of all dead transitions, of all live transitions, and of all dead markings of $\mathcal{P}$, respectively.

Proof. (a) Because of (2), we have for all all transitions $t \in T$:

$$(\mathcal{A}^\mathsf{T}; \mathsf{L})_t \iff \exists M : \mathcal{A}^\mathsf{T}_{t,M} \wedge \mathsf{L}_t \iff \exists M : M \in \mathcal{M}_0 \wedge M \overset{t}{\rightsquigarrow}$$

The right-most formula of this calculation formalizes the fact that t is a weakly live transition.

(b) This proof is a trivial consequence of (a), since, by definition, each transition $t \in T$ is not weakly live iff t is dead.

(c) Using (1), (2), and (5), the claim follows from the following equivalence for all transitions $t \in T$, since the last formula of the derivation is the formalization of t to be a live transition.

$$\overline{(\mathcal{R}; \overline{\mathcal{A}^\mathsf{T}}; \mathcal{R}; \mathsf{L})_t} \iff \neg \exists M : \overline{\mathcal{R}; \overline{\mathcal{A}}_{M,t}} \wedge (\mathcal{R}; \mathsf{L})_M$$

$$\iff \neg \exists M : \neg \exists M' : (\mathcal{R}_{M,M'} \wedge \mathcal{A}_{M',t}) \wedge (\mathcal{R}; \mathsf{L})_M$$

$$\iff \forall M : (\mathcal{R}; \mathsf{L})_M \rightarrow \exists M' : (\mathcal{R}_{M,M'} \wedge \mathcal{A}_{M',t})$$

$$\iff \forall M : M \in \mathcal{M}_0 \rightarrow \exists M' : M' \in \mathcal{M}_0 \wedge M \overset{*}{\rightsquigarrow} M' \wedge M' \overset{t}{\rightsquigarrow}$$

(d) Let $M : P \leftrightarrow N$ be any marking. With the help of (2) and (5), we get the result by the following equivalence since its last formula says that M is dead.

$$(\mathcal{R}; \mathsf{L} \cap \overline{\mathcal{A}; \mathsf{L}})_M \iff (\mathcal{R}; \mathsf{L})_M \wedge \overline{\mathcal{A}; \mathsf{L}}_M \iff M \in \mathcal{M}_0 \wedge \neg \exists t : M \overset{t}{\leadsto} \qquad \square$$

The next theorem shows how to test the two liveness properties we have introduced for entire Petri nets by means of relation-algebraic formulae.

Theorem 2. *Let again $\mathcal{P}$ be a Petri net and $\mathcal{R}$ and $\mathcal{A}$ be as in Section 3.2. Then $\mathcal{P}$ is weakly live iff $\mathcal{R}; \mathsf{L} \subseteq \mathcal{A}; \mathsf{L}$ and live iff $\mathcal{R}; \mathsf{L} \subseteq \mathcal{R}; \mathcal{A}$.*

Proof. The first statement directly follows from the definition of weak liveness and Theorem 1 (d), since $\mathcal{R}; \mathsf{L} \cap \overline{\mathcal{A}; \mathsf{L}} = \mathsf{O}$ iff $\mathcal{R}; \mathsf{L} \subseteq \mathcal{A}\mathsf{L}$.

By Theorem 1 (c), liveness is equivalent to $\overline{\mathcal{R}; \overline{\mathcal{A}}^\mathsf{T}}; \mathcal{R}; \mathsf{L} = \mathsf{L}$ and, therefore, the following computation shows the second claim.

$$\overline{\mathcal{R}; \overline{\mathcal{A}}^\mathsf{T}}; \mathcal{R}; \mathsf{L} = \mathsf{L} \iff \overline{\mathcal{R}; \overline{\mathcal{A}}^\mathsf{T}}; \mathcal{R}; \mathsf{L} \subseteq \mathsf{O}$$
$$\iff \mathsf{L}; (\mathcal{R}; \mathsf{L})^\mathsf{T} \subseteq (\mathcal{R}; \mathcal{A})^\mathsf{T} \qquad \text{Schröder equivalences [14]}$$
$$\iff \mathcal{R}; \mathsf{L} \subseteq \mathcal{R}; \mathcal{A} \qquad \text{as } \mathsf{L}; (\mathcal{R}; \mathsf{L})^\mathsf{T} = (\mathcal{R}; \mathsf{L})^\mathsf{T} \qquad \square$$

4   Proving Liveness Properties with Relation Algebra

As we have relation-algebraically specified liveness in Petri nets, we can now use this for formal calculations. The following proofs make use of laws of abstract relation algebra, which, however, we refer to only in the case of a non-trivial proof step. In contrast to proofs usually found in the Petri net literature, where net-theoretic considerations are used, we base our proofs on sheer mathematical argumentation. To elaborate the main result of this section, a sufficient criterion for weak liveness saying that a Petri net is weakly live if it contains at least one live transition, we need some further properties on the relations $\mathcal{R}$ and $\mathcal{A}$ of Section 3.2. In words, the following theorem says that for all reachable markings M from the existence of a transition that is activated under M it follows for all transitions t that there is an ancestor marking of M under which t is activated.

Theorem 3. *We have $\mathcal{R}; \mathcal{A} \subseteq \mathcal{A}; \mathsf{L}$.*

Proof. Starting with the left-hand side of the inclusion to show, we calculate:

$$\mathcal{R}; \mathcal{A} = ((\mathcal{R} \cap \mathcal{A}; \mathsf{L}) \cup (\mathcal{R} \cap \overline{\mathcal{A}; \mathsf{L}})); \mathcal{A}$$
$$\subseteq ((\mathcal{R} \cap \mathcal{A}; \mathsf{L}) \cup (\mathsf{I} \cap \overline{\mathcal{A}; \mathsf{L}})); \mathcal{A} \qquad \text{due to (7)}$$
$$= (\mathcal{R} \cap \mathcal{A}; \mathsf{L}); \mathcal{A} \cup (\mathsf{I} \cap \overline{\mathcal{A}; \mathsf{L}}); \mathcal{A} \qquad \text{distributivity}$$

Now, we start from the right-hand side of the inclusion and obtain

$$\mathcal{A}; \mathsf{L} \supseteq \mathcal{A}; \mathsf{L}; \mathcal{A} \supseteq (\mathcal{R} \cap \mathcal{A}; \mathsf{L}); \mathcal{A}$$

due to $L \supseteq L$; $\mathcal{A}$ and $\mathcal{A}; L \supseteq (\mathcal{R} \cap \mathcal{A}; L)$. So, it remains to show that $(I \cap \overline{\mathcal{A}; L}); \mathcal{A}$ is empty. For this, we apply Lemma 1 and that $\overline{\mathcal{A}; L}$ is a vector and obtain then

$$(I \cap \overline{\mathcal{A}; L}); \mathcal{A} = (I \cap \overline{\mathcal{A}; L}); L \cap \mathcal{A} \subseteq \overline{\mathcal{A}; L} \cap \mathcal{A} = O. \qquad \square$$

An immediate consequence of Theorem 3 is the following corollary. In usual terminology it says that for all markings M' and transitions t, if there is an ancestor marking M of M' such that no transition is activated under M, then there is an ancestor marking M of M' such that t is not activated under all markings reachable from M.

Corollary 1. *We have* $\overline{\mathcal{A}; L}^\mathsf{T}; \mathcal{R} \subseteq \overline{\mathcal{R}; \mathcal{A}}^\mathsf{T}; \mathcal{R}$.

It is known from Petri net theory that a live Petri net is also weakly live. We formulate this fact as another immediate consequence of Theorem 3:

Corollary 2. $\mathcal{R}; L \subseteq \mathcal{R}; \mathcal{A}$ *implies* $\mathcal{R}; L \subseteq \mathcal{A}; L$.

As the last result of this section we prove the above stated criterion for weak liveness, viz. that at least one live transition exists, relation-algebraically by means of the relations $\mathcal{R}$ and $\mathcal{A}$.

Theorem 4. $\overline{\mathcal{R}; \mathcal{A}}^\mathsf{T}; \mathcal{R}; L \neq O$ *implies* $\mathcal{R}; L \subseteq \mathcal{A}; L$.

Proof. We use contraposition and assume the negation of $\mathcal{R}; L \subseteq \mathcal{A}; L$ to hold, i.e., $\mathcal{R}; L \cap \overline{\mathcal{A}; L} \neq O$. Hence, we have to show $\overline{\mathcal{R}; \mathcal{A}}^\mathsf{T}; \mathcal{R}; L = L$. From Corollary 1 we get that $\overline{\mathcal{A}; L}^\mathsf{T}; \mathcal{R}; L = L$ implies $\overline{\mathcal{R}; \mathcal{A}}^\mathsf{T}; \mathcal{R}; L = L$ and this reduces our task to the proof of $\overline{\mathcal{A}; L}^\mathsf{T}; \mathcal{R}; L = L$. To reach this goal, we calculate as follows:

$$
\begin{aligned}
\left((\overline{\mathcal{A}; L}^\mathsf{T}); \mathcal{R} \right)^\mathsf{T} &= \mathcal{R}^\mathsf{T}; \overline{\mathcal{A}; L} \\
&\supseteq (I \cap \mathcal{R})^\mathsf{T}; \overline{\mathcal{A}; L} \\
&= (I \cap \mathcal{R}); \overline{\mathcal{A}; L} && \text{since } I \cap \mathcal{R} \subseteq I \\
&= (I \cap \mathcal{R}); L \cap \overline{\mathcal{A}; L} && \text{due to Lemma 1} \\
&= \mathcal{R}; L \cap \overline{\mathcal{A}; L} && \text{due to (6)}
\end{aligned}
$$

Since both $\mathcal{R}; L$ and $\overline{\mathcal{A}; L}$ are vectors, their intersection is a vector as well. By assumption, $\mathcal{R}; L \cap \overline{\mathcal{A}; L} \neq O$. As a consequence, $L; (\mathcal{R}; L \cap \overline{\mathcal{A}; L}) = L$ due to the Tarski rule (see [14]). Combining this with the above calculation, we arrive at $L; \left((\overline{\mathcal{A}; L}^\mathsf{T}); \mathcal{R} \right)^\mathsf{T} = L$ which, in turn, is equivalent to $\overline{\mathcal{A}; L}^\mathsf{T}; \mathcal{R}; L = L$. $\qquad \square$

Certainly, mindful readers have noticed that the results of this section do not depend on the specific interpretation of $\mathcal{R}$ and $\mathcal{A}$ and, therefore, can be transformed into general relation-algebraic properties. This allows for their reuse. For instance, Theorem 3 becomes that $R; A \subseteq A; L$ for all relations R and A such that $R \cap \overline{A; L} \subseteq I$. If, for example, $R : V \leftrightarrow V$ is the relation of a directed graph with vertex set V and $A : V \leftrightarrow \mathbb{1}$ is a vector describing a subset X of the vertex set V, then this generalization expresses that if all arcs starting not in X are loops, then X is a predecessor-closed set.

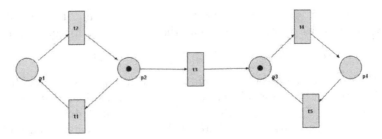

Fig. 1. A Petri net not having the deadlock/trap property

5   The Deadlock/Trap Property

A set X of places of a Petri net is called a *trap* if it is non-empty and each of its successor transitions is also a predecessor transition. The dual notation is that of a co-trap. X is a *co-trap* (or *deadlock*) if it is non-empty and each of its predecessor transitions is also a successor transition. Both notions are very useful for reasoning about liveness properties. The most important results in respect thereof use the *deadlock/trap property*. This property says that each minimal co-trap (w.r.t. set inclusion) contains a trap that is sufficiently marked under M_0. With it, in certain situations liveness can be treated without reference to reachability. For instance, a Petri net that possesses the deadlock/trap property is weakly live; a *Free-Choice net* (a net where each transition with a forward-branching predecessor place may not be backwards branching) is live iff it has the deadlock/trap property (Commoner's theorem).

Example 1. *Figure 1 shows a simple Petri net as depicted by the tool* PETRA *described in the next section. The net has four places p_1, p_2, p_3, p_4, drawn as cycles, five transitions t_1, t_2, t_3, t_4, t_5, drawn as rectangles, and an initial marking, indicated by the tokens. The latter means the following:*

$$M_0 : P \to \mathbb{N} \qquad M_0(p_2) = M_0(p_3) = 1 \qquad M_0(p_1) = M_0(p_4) = 0$$

For this net, the deadlock/trap property does not hold, since, for instance, the set $\{p_1, p_2\}$ is a minimal co-trap that contains no trap. From the flow relation we get that the net possesses the Free-Choice property. Hence, is not live due to Commoner's theorem. Indeed, using the specifications of Theorem 1, the PETRA *tool shows that the transitions t_4, t_5 are live, whereas the transitions t_1, t_2, t_3 are only weakly live.*   □

For the following, we assume that each Petri net under consideration is relation-algebraically modeled by the 6-tuple $\mathcal{P} = (R, S, W^{\bullet t}, W^{t\bullet}, C, M_0)$ as introduced in Section 3.1. Furthermore, we denote by $\mathsf{M} : P \leftrightarrow 2^P$ the membership relation between P and its powerset, and by $\pi : P \times T \leftrightarrow P$ and $\rho : P \times T \leftrightarrow T$ the projections of $P \times T$. The following theorem is the first step towards a relation-algebraic specification of the deadlock/trap property.

Theorem 5. *Let $\mathcal{P}$ be a Petri net. Then the vectors*

$$\mathcal{T} := \mathsf{M}^\mathsf{T};\mathsf{L} \cap \overline{(R^\mathsf{T};\mathsf{M} \cap \overline{S;\mathsf{M}})^\mathsf{T};\mathsf{L}} \qquad \mathcal{C} := \mathsf{M}^\mathsf{T};\mathsf{L} \cap \overline{(S;\mathsf{M} \cap \overline{R^\mathsf{T};\mathsf{M}})^\mathsf{T};\mathsf{L}}$$

of type $[2^P \leftrightarrow \mathbb{1}]$ model the set of traps and co-traps, respectively.

Proof. For all $X \in 2^P$ we have

$$(\mathsf{M}^\mathsf{T};\mathsf{L})_X \iff \exists p : \mathsf{M}_{p,X} \iff X \neq \emptyset$$

and, furthermore,

$$\overline{(R^\mathsf{T};\mathsf{M} \cap \overline{S;\mathsf{M}})^\mathsf{T};\mathsf{L}}_X \iff \neg \exists t : (R^\mathsf{T};\mathsf{M})_{t,X} \wedge \overline{S;\mathsf{M}}_{t,X} \wedge \mathsf{L}_X$$
$$\iff \forall t : (R^\mathsf{T};\mathsf{M})_{t,X} \to (S;\mathsf{M})_{t,X}$$
$$\iff \forall t : (\exists p : R_{p,t} \wedge p \in X) \to (\exists p : S_{t,p} \wedge p \in X).$$

The last formula of the second derivation says that each successor transition t of X is also a predecessor transition of X. Hence, we have shown the first claim.

The second claim follows from the above proof by simultaneously exchanging the flow relations R^T and S. □

Hence, for all $X \in 2^P$ we have $\mathcal{T}_X$ if X is a trap and $\mathcal{C}_X$ if X is a co-trap. Since $\mathsf{E} := \overline{\mathsf{M}^\mathsf{T};\overline{\mathsf{M}}}$ is the relation-algebraic specification of set inclusion on 2^P, it is very easy to model the set of all minimal co-traps by a vector $\tilde{\mathcal{C}} : 2^P \leftrightarrow \mathbb{1}$. We obtain (see e.g., [14] for a relation-algebraic specification of minimal elements)

$$\tilde{\mathcal{C}} = \mathcal{C} \cap \overline{(\mathsf{E}^\mathsf{T} \cap \overline{\mathsf{I}});\mathcal{C}}. \tag{8}$$

As the next step towards our goal we treat sufficiently marked traps relation-algebraically.

Theorem 6. *Let $\mathcal{P}$ be a Petri net. Then the vector*

$$\mathcal{S} := \mathcal{T} \cap \mathsf{M}^\mathsf{T};(R;\rho^\mathsf{T} \cap M_0;\mathsf{P}^*;(W^{\bullet t})^\mathsf{T} \cap \pi^\mathsf{T});\mathsf{L}$$

of type $[2^P \leftrightarrow \mathbb{1}]$ models the set of traps that are sufficiently marked under M_0.

Proof. For all $X \in 2^P$ we have the following equivalence.

$$(\mathsf{M}^\mathsf{T};(R;\rho^\mathsf{T} \cap M_0;\mathsf{P}^*;(W^{\bullet t})^\mathsf{T} \cap \pi^\mathsf{T});\mathsf{L})_X$$
$$\iff \exists p : \mathsf{M}_{p,X} \wedge ((R;\rho^\mathsf{T} \cap M_0;\mathsf{P}^*;(W^{\bullet t})^\mathsf{T} \cap \pi^\mathsf{T});\mathsf{L})_p$$
$$\iff \exists p : \mathsf{M}_{p,X} \wedge \exists \langle q,t \rangle : (R;\rho^\mathsf{T})_{p,\langle q,t \rangle} \wedge (M_0;\mathsf{P}^*;(W^{\bullet t})^\mathsf{T})_{p,\langle q,t \rangle} \wedge \pi^\mathsf{T}_{p,\langle q,t \rangle}$$
$$\iff \exists p : p \in X \wedge \exists \langle q,t \rangle : (R;\rho^\mathsf{T})_{p,\langle q,t \rangle} \wedge (M_0;\mathsf{P}^*;(W^{\bullet t})^\mathsf{T})_{p,\langle q,t \rangle} \wedge p = q$$
$$\iff \exists p : p \in X \wedge \exists t : (R;\rho^\mathsf{T})_{p,\langle p,t \rangle} \wedge (M_0;\mathsf{P}^*;(W^{\bullet t})^\mathsf{T})_{p,\langle p,t \rangle}$$
$$\iff \exists p : p \in X \wedge \exists t : R_{p,t} \wedge \mathsf{P}^*_{M_0(p),W(p,t)}$$
$$\iff \exists p : p \in X \wedge \exists t : R_{p,t} \wedge M_0(p) \geq W(p,t)$$

The last formula combined with the first statement of Theorem 5 yields the result. □

After these preparations, we are in the position to translate the above informal description of the deadlock/trap property into a relation-algebraic formula. This yields the following result.

Theorem 7. *A Petri net $\mathcal{P}$ possesses the deadlock/trap property iff $\tilde{C} \subseteq \mathsf{E}^\mathsf{T}; \mathcal{S}$.*

Proof. The inclusion $\tilde{C} \subseteq \mathsf{E}^\mathsf{T}; \mathcal{S}$ is equivalent to

$$\forall X : \tilde{C}_X \rightarrow \exists Y : \mathsf{E}_{Y,X} \wedge \mathcal{S}_Y$$

and from the meaning of $\tilde{C}$, E, and $\mathcal{S}$ we immediately see that the formula specifies the deadlock/trap property. □

6   Tool Support

We implemented the descriptions and characterizations presented in this paper in our Petri net tool PETRA. Based on our object-oriented Java library KURE-Java which allows us to mechanize the algebra of relations by an efficient implementation of relations via BDDs and which provides a lot of pre-defined operations and tests, PETRA is capable of executing relational characterizations and allows for both the editing and relation-algebraic analysis of Petri nets. Figure 2 gives a screen shot of the Petri net tool PETRA. A list of all reachable markings of the net shown is provided in a separate window. Selecting an entry from this list displays the respective marking within the net.

PETRA offers a flexible way to graphically draw a Petri net, to transform it into its relation-algebraic representation, to trigger analyses, and to illustrate

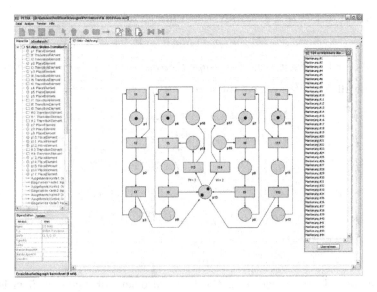

Fig. 2. A Petri net as displayed by PETRA

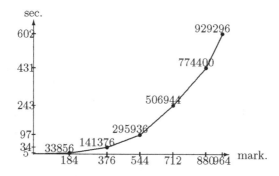

Fig. 3. Results of test runs on the net displayed in Fig. 2

within the net the results an executed relational specification delivers. The specifications where technically tuned up – we simply exploited some knowledge on the BDDs representing relations within RELVIEW and KURE-Java (see Fig. 3) – and have been executed on a 1.7 GHz Centrino with 2GB RAM. We conducted some test runs on a sample Petri net with places and transitions added stepwise to expand the state space from run to run. In the first run, we started with 17 places and 14 transitions. The net had 184 reachable markings and 33856 entries in $\mathcal{R}$, and it took 5 seconds to construct the transition system. Finally in the last run, the net consisted of 42 places and 40 transitions. The 964 markings in $\mathcal{R}$ with its 929296 entries were calculated within 10 minutes. Checking the liveness properties, however, was in all cases done in less than 1 second.

In [7] some well-known examples from computer science (viz., Petri net descriptions of readers/writers, consumer/producer, the dining philosophers and a small communication protocol) are investigated with relation-algebraic means and using PETRA. The paper also reports on the corresponding runtime tests in more detail.

7   Conclusion

In the design and implementation of large software systems usually high-level concepts, techniques, and strategies from, for instance, model-driven architecture, object-orientation, and component-oriented approaches play a prominent role. Still, source code appearing in practice is frequently very complex, and particularly parallel threads or data flows are hard to overview, to test, and to maintain. To detect livelocks or deadlocks in concurrent, distributed, or stochastic software is even today one of the most serious problems. As Petri nets allow to model such concurrent systems, analyzing liveness properties in Petri nets is a well-known approach to, for instance, detect both deadlocked threads and the reasons for such situations in the underlying system. We have presented a set of relation-algebraic specifications of liveness properties, of traps and co-traps, and of the deadlock/trap property. These specifications are correct by construction and, as another main benefit, they can immediately be executed within PETRA.

Recently, rather than stand-alone tools modern software development environment are used to cope with the technical and organizational complexity arising when large software systems are developed. Consequently, a well-suited development environments needs to be flexible and open for extensions providing powerful functionality needed to develop complex systems. As another relation-algebraic Petri net tool besides PETRA, we developed such an extension, called RELCLIPSE, for the Eclipse platform. This tool allows in an industrial development environment to model as a Petri net suitable parts of a software system under implementation and to analyze its properties by means of relation algebra. Hence, the approach makes the development of the crucial analysis features of a Petri net tool less error-prone than without a formal approach or with one that does not allow for correct-by-construction transformations. The integration into Eclipse demonstrates the practical applicability of relation-algebra within working software development environments and processes.

References

1. Berghammer, R., Fronk, A.: Applying relational algebra in 3D graphical software design. In: Berghammer, R., Möller, B., Struth, G. (eds.) RelMiCS 2003. LNCS, vol. 3051, pp. 62–73. Springer, Heidelberg (2004)
2. Berghammer, R., Karger, B., Ulke, C.: Relational-algebraic analysis of Petri nets with RELVIEW. In: Margaria, T., Steffen, B. (eds.) TACAS 1996. LNCS, vol. 1055, pp. 49–69. Springer, Heidelberg (1996)
3. Berghammer, R., Zierer, H.: Relational algebraic semantics of deterministic and nondeterministic programs. Theoretical Computer Science 43, 123–147 (1996)
4. Doberkat, E.-E.: Pipelines: Modelling a software architecture through relations. Acta Informatica 40, 37–79 (2003)
5. Finkel, A.: The minimal coverability graph for Petri nets. In: Rozenberg, G. (ed.) APN 1993. LNCS, vol. 674, pp. 210–243. Springer, Heidelberg (1993)
6. Fronk, A.: Using relation algebra for the analysis of Petri nets in a CASE tool based approach. In: 2nd IEEE International Conference on Software Engineering and Formal Methods (SEFM), Beijing, pp. 396–405. IEEE, Los Alamitos (2004)
7. Fronk, A., Kehden, B.: State space analysis of petri nets with relation-algebraic methods. Journal of Symbolic Computation 44, 15–47 (2009)
8. Fronk, A., Pleumann, J.: Relation-algebraic Analysis of Petri Nets: Concepts and Implementation. Petri Net News Letters, 61–68 (April 2005)
9. Fronk, A., Pleumann, J.: On Relational Cycles. In: MacCaull, W., Winter, M., Düntsch, I. (eds.) RelMiCS 2005. LNCS, vol. 3929, pp. 83–95. Springer, Heidelberg (2006)
10. Leoniuk, B.: ROBDD-based implementation of relational algebra with applications. Dissertation, Universität Kiel (2001) (in German)
11. Milanese, U.: On the implementation of a ROBDD-based tool for the manipulation and visualization of relations. Dissertation, Universität Kiel (2003) (in German)
12. Pastor, E., Roig, O., Cortadella, J., Badia, R.M.: Petri net analysis using Boolean manipulation. In: Valette, R. (ed.) ICATPN 1994. LNCS, vol. 815, pp. 416–435. Springer, Heidelberg (1994)
13. Reisig, W.: Petri nets: an introduction. EATCS Monographs on Theoretical Computer Science. Springer, Heidelberg (1985)
14. Schmidt, G., Ströhlein, T.: Relations and graphs. EATCS Monographs on Theoretical Computer Science. Springer, Heidelberg (1993)

∗-Continuous Idempotent Left Semirings and Their Ideal Completion

Hitoshi Furusawa and Fumiya Sanda

Department of Mathematics and Computer Science, Kagoshima University
furusawa@sci.kagoshima-u.ac.jp, k6618413@kadai.jp

Abstract. In this paper, we introduce two notions of continuity for idempotent left semirings, which are called ∗-continuity and D-continuity. Also, for a ∗-continuous idempotent left semiring, we introduce a notion of ∗-ideals. Then, we show that the set of ∗-ideals of a ∗-continuous idempotent left semiring forms a D-continuous idempotent left semiring and the construction satisfies a universal property.

1 Introduction

Kleene algebras [4] are idempotent semirings with * satisfying

$$1 + aa^* \leq a^* \tag{1}$$

$$1 + a^*a \leq a^* \tag{2}$$

$$ax \leq x \rightarrow a^*x \leq x \tag{3}$$

$$xa \leq x \rightarrow xa^* \leq x \tag{4}$$

where $\leq$ refers to the natural order in the idempotent semiring. A Kleene algebra is called ∗-continuous if it satisfies the axiom

$$ab^*c = \sum_{n \geq 0} ab^n c \tag{5}$$

where $\sum$ refers to the least upper bounds in the natural order $\leq$. In fact, (5) implies (1)-(4) in each idempotent semiring. Thus, ∗-continuous Kleene algebras might be called ∗-continuous idempotent semirings. Conway's **S**-algebras are idempotent semirings with arbitrary sums $\sum$ and their binary operator $\cdot$ distributes over $\sum$ from the both sides. Thus, **S**-algebras might be called complete idempotent semirings. In [1], Conway has given a construction that embeds every ∗-continuous Kleene algebra into an **S**-algebra. This construction can be described as an ideal completion. In [3], Kozen has shown that the construction satisfies a universal property.

This paper provides similar results to [3] on variants of lazy Kleene algebras [6] that are idempotent left semirings with a unary operator * satisfying (1) and (3). We define ∗-continuous idempotent left semirings and D-continuous

R. Berghammer et al. (Eds.): RelMiCS/AKA 2009, LNCS 5827, pp. 119–133, 2009.

idempotent left semirings. For an idempotent left semiring, we do not adopt (5) as a condition of $*$-continuity since it implies (1)-(4) even in each idempotent left semiring. Therefore, the condition of $*$-continuity is weakened, $*$-continuous idempotent left semirings satisfy (1) and (3). We give a construction that embeds every $*$-continuous idempotent left semiring into a D-continuous idempotent left semiring. As in the case of [1], this construction can be described as an ideal completion. We also show that the construction satisfies a universal property.

2   Idempotent Left Semirings

Idempotent left semirings [6] are defined as follows.

Definition 1. An *idempotent left semiring*, or briefly an *IL-semiring* is a tuple $(S, +, \cdot, 0, 1)$ with a set S, two binary operations $+$ and $\cdot$, and $0, 1 \in S$ satisfying the following properties:

- $(S, +, 0)$ is an idempotent commutative monoid.
- $(S, \cdot, 1)$ is a monoid.
- For all $a, b, c \in S$, $a \cdot c + b \cdot c = (a + b) \cdot c$, $a \cdot b + a \cdot c \leq a \cdot (b + c)$, and $0 \cdot a = 0$,

where the *natural order* $\leq$ is given by $a \leq b$ iff $a + b = b$.

We often abbreviate $a \cdot b$ to ab.

Remark 1. An IL-semiring S satisfying $ab + ac = a(b + c)$ and $a0 = 0$ for all $a, b, c \in S$ is an idempotent semiring.

Let S be an IL-semiring. For $a \in S$, a mapping $\varphi_a \colon S \to S$ is defined by $\varphi_a(x) = ax + 1$. Then the mapping preserves the natural order $\leq$ on S. For a natural number n, φ_a^n is defined by induction:

- φ_a^0 is the identity mapping,
- $\varphi_a^{n+1} = \varphi_a \circ \varphi_a^n$.

For $A \subseteq S$, $\sum A$ denotes the least upper bound of A with respect to the natural order $\leq$ on S if it exists.

Definition 2. A $*$-*continuous* IL-semiring is a tuple $(S, +, \cdot, *, 0, 1)$ with a set S, two binary operations $+$ and $\cdot$, a unary operation $*$, and $0, 1 \in S$ satisfying the following properties:

- $(S, +, \cdot, 0, 1)$ is an IL-semiring.
- For all $a, b, c \in S$, the least upper bound of $\{a\varphi_b^n(0)c \mid n \geq 0\}$ exists in S.
- $ab^*c = \displaystyle\sum_{n \geq 0} a\varphi_b^n(0)c$ holds for all $a, b, c \in S$.

Replacing a and c with 1, we have $b^* = \displaystyle\sum_{n \geq 0} \varphi_b^n(0)$. We write ILS^* for the category whose objects are $*$-continuous IL-semirings and whose arrows are homomorphisms between them.

Lemma 1. *Let S be a ∗-continuous IL-semiring and $a, b \in S$. Then, for each natural number $n \geq 0$,*

1. *$ab \leq b$ implies $\varphi_a^n(0)b \leq b$, and*
2. *$b(a + 1) \leq b$ implies $b\varphi_a^n(0) \leq b$.*

Proof. These two are proved by induction on $n \geq 0$. For $n = 0$, $\varphi_a^0(0)b = 0 \leq b$. Assume that $\varphi_a^n(0)b \leq b$. Then, the following holds.

$$
\begin{aligned}
\varphi_a^{n+1}(0)b &= (a\varphi_a^n(0) + 1)b \\
&= a\varphi_a^n(0)b + b \\
&\leq ab + b \qquad \text{(by induction hypothesis)} \\
&\leq b \qquad \text{(by } ab \leq b\text{)}
\end{aligned}
$$

For $n = 0$, $b\varphi_a^0(0) = b0 \leq b$. For $n = 1$, $b\varphi_a^1(0) = b(a0 + 1) \leq b(a + 1) \leq b$. Assume that $b\varphi_a^{n+1}(0) \leq b$ and note that $1 \leq a\varphi_a^n(0) + 1 = \varphi_a^{n+1}(0)$. Then, the following holds.

$$
\begin{aligned}
b\varphi_a^{n+2}(0) &= b(a\varphi_a^{n+1}(0) + 1) \\
&\leq b(a\varphi_a^{n+1}(0) + \varphi_a^{n+1}(0)) \text{ (by notice)} \\
&= b(a + 1)\varphi_a^{n+1}(0) \\
&\leq b\varphi_a^{n+1}(0) \qquad \text{(by } b(a + 1) \leq b\text{)} \\
&\leq b \qquad \text{(by induction hypothesis)} \qquad \square
\end{aligned}
$$

A lazy Kleene algebra is an IL-semiring with a unary operation ∗ satisfying

$$
1 + aa^* \leq a^* \quad \text{and} \quad ab \leq b \rightarrow a^*b \leq b \ .
$$

The following property shows that a ∗-continuous IL-semiring is a lazy Kleene algebra satisfying the D-axiom [7] (or a monodic tree Kleene algebra [8]).

Proposition 1. *A ∗-continuous IL-semiring S satisfies the following.*

1. *$1 + aa^* \leq a^*$.*
2. *$ab \leq b$ implies $a^*b \leq b$.*
3. *$b(a + 1) \leq b$ implies $ba^* \leq b$ (the D-axiom).*

Proof. Note that $1 \leq a^*$ since $1 \leq a0 + 1 \leq \sum_{n \geq 0} \varphi_a^n(0) = a^*$. Also $aa^* \leq a^*$ since $aa^* = \sum_{n \geq 0} a\varphi_a^n(0) \leq \sum_{n \geq 1} \varphi_a^n(0) \leq \sum_{n \geq 0} \varphi_a^n(0) = a^*$. So we have $1 + aa^* \leq a^*$. Suppose $ab \leq b$. Then we have $a^*b = \sum_{n \geq 0} \varphi_a^n(0)b \leq b$ by Lemma 1. Similarly, it holds that $b(a + 1) \leq b$ implies $ba^* \leq b$. $\qquad \square$

Definition 3. A ∗-continuous IL-semiring satisfying *the right zero law* (or *the 0-axiom*) is a ∗-continuous IL-semiring satisfying $a0 = 0$ for every element a. A ∗-continuous IL-semiring satisfying *left distributivity* (or *the +-axiom*) is a ∗-continuous IL-semiring satisfying $ab + ac = a(b + c)$ for all elements a, b, and c.

We write

- ILS_0^* for the category whose objects are ∗-continuous IL-semirings satisfying the right zero law and whose arrows are homomorphisms between them,
- ILS_+^* for the category whose objects are ∗-continuous IL-semirings satisfying left distributivity and whose arrows are homomorphisms between them, and
- $\mathsf{ILS}_{0,+}^*$ for the category whose objects are ∗-continuous IL-semirings satisfying both of the right zero law and left distributivity and whose arrows are homomorphisms between them.

The following is immediate from Proposition 1.

Corollary 1. *Let S be a ∗-continuous IL-semiring.*

- *If S satisfies the right zero law, S is a lazy Kleene algebra satisfying the D-axiom and the 0-axiom [7] (or a probabilistic Kleene algebra [5]).*
- *If S satisfies left distributivity, S is a lazy Kleene algebra satisfying the D-axiom and the +-axiom [7].*
- *If S satisfies the right zero law and left distributivity, S is a lazy Kleene algebra satisfying the D-axiom, the 0-axiom, and the +-axiom [7] (or a Kleene algebra [3,4]).*

Remark 2. If S is a ∗-continuous IL-semiring satisfying the right zero law and left distributivity, then $\varphi_a^{n+1}(0) = \sum_{0 \leq k \leq n} a^k$ holds for each $a \in S$ and natural number n. Thus, $\sum_{n \geq 0} \varphi_a^n(0) = \sum_{n \geq 0} a^n$. Therefore, S is a ∗-continuous Kleene algebra [3].

Kozen has given an example of a Kleene algebra which is not ∗-continuous in [3, Section 3]. The example shows that the converse statements of Proposition 1 and Corollary 1 need not hold.

Definition 4. A *complete* IL-semiring S is an IL-semiring satisfying the following properties: For each $A \subseteq S$,

- the least upper bound of A exists in S, and
- $(\sum A)a = \sum \{xa \mid x \in A\}$ for each $a \in S$.

A *D-continuous* IL-semiring (or a complete IL-semiring preserving *right directed joins* [7]) S is a complete IL-semiring satisfying

$$a \sum A = \sum \{ax \mid x \in A\}$$

for each $a \in S$ and each directed subset $A \subseteq S$. A D-continuous IL-semiring satisfying *the right zero law* (or preserving *the right 0* [7]) is a D-continuous IL-semiring satisfying $a0 = 0$ for each element a. A D-continuous IL-semiring satisfying *left distributivity* (or preserving *the right +* [7]) is a D-continuous IL-semiring satisfying $ab + ac = a(b + c)$ for any elements a, b, and c.

Remark 3 (Nishizawa et al. [7]). A D-continuous IL-semiring satisfies the right zero law and left distributivity iff it is an **S**-algebra or a complete idempotent semiring.

We write

- ILS^D for the category whose objects are D-continuous IL-semirings and whose arrows are completely join-preserving homomorphisms between them,
- ILS_0^D for the category whose objects are D-continuous IL-semirings satisfying the right zero law and whose arrows are completely join-preserving homomorphisms between them,
- ILS_+^D for the category whose objects are D-continuous IL-semirings satisfying left distributivity and whose arrows are completely join-preserving homomorphisms between them, and
- $\mathsf{ILS}_{0,+}^D$ for the category whose objects are D-continuous IL-semirings satisfying both of the right zero law and left distributivity and whose arrows are completely join-preserving homomorphisms between them.

3   Connection between ILS^* and ILS^D

Let Q and Q' be objects and $f\colon Q \to Q'$ an arrow of ILS^D. For each $a \in Q$, defining $a^* = \sum_{n\geq 0} \varphi_a^n(0)$, Q is a ∗-continuous IL-semiring since the set $\{\varphi_a^n(0) \mid n \geq 0\}$ is a directed subset of Q. Also, f preserves ∗ since f preserves $+, \cdot, \sum, 0, 1$, and ∗ is defined using these operators. Therefore, ILS^D is a subcategory of ILS^*. The inclusion functor from ILS^D to ILS^* is denoted by $G\colon \mathsf{ILS}^D \to \mathsf{ILS}^*$. In this section, we construct a functor from ILS^* to ILS^D which is a left adjoint to G.

3.1   ∗-Ideal

Conway [1] introduced ∗-ideals for ∗-continuous Kleene algebras. ∗-ideals for ∗-continuous IL-semirings are defined as follows.

Definition 5. Let S be a ∗-continuous IL-semiring. A *∗-ideal* is a subset A of S such that

- A is nonempty,
- A is closed under $+$,
- A is closed downward under $\leq$,
- if $a\varphi_b^n(0)c \in A$ for all $n \geq 0$, then $ab^*c \in A$.

The set of ∗-ideals of ∗-continuous IL-semiring S is denoted by $\mathcal{I}(S)$. Note that $\mathcal{I}(S)$ is closed under arbitrary intersection.

We say that a nonempty set A *generates* a ∗-ideal I if I is the smallest ∗-ideal containing A. $\langle A \rangle$ denotes the ∗-ideal generated by A. If A is a singleton $\{a\}$, we often abbreviate $\langle\{a\}\rangle$ to $\langle a \rangle$. Such a ∗-ideal is called *principal*. Note that $\langle\_\rangle$ is well-defined on nonempty subsets of S. Also note that $\langle\_\rangle$ is monotone and idempotent, i.e. $A \subseteq B$ implies $\langle A \rangle \subseteq \langle B \rangle$ and $\langle\langle A \rangle\rangle = \langle A \rangle$ for any nonempty subsets $A, B \subseteq S$.

Lemma 2. *Let S be a $*$-continuous IL-semiring. For a set $\mathcal{A}$ of nonempty subsets of S,*

$$\left\langle \bigcup \mathcal{A} \right\rangle = \left\langle \bigcup \{ \langle A \rangle \mid A \in \mathcal{A} \} \right\rangle \ .$$

Proof. The inclusion $\subseteq$ follows from monotonicity of $\langle \_ \rangle$. Again, by monotonicity of $\langle \_ \rangle$, $\langle A \rangle \subseteq \left\langle \bigcup \mathcal{A} \right\rangle$ for each $A \in \mathcal{A}$. Thus, $\bigcup \{ \langle A \rangle \mid A \in \mathcal{A} \} \subseteq \left\langle \bigcup \mathcal{A} \right\rangle$. So, we have $\left\langle \bigcup \{ \langle A \rangle \mid A \in \mathcal{A} \} \right\rangle \subseteq \left\langle \bigcup \mathcal{A} \right\rangle$ by monotonicity and idempotency of $\langle \_ \rangle$.   □

Let S be a $*$-continuous IL-semiring. For subsets $A, B \subseteq S$, define

$$A \oplus B = \{ a + b \mid a \in A, \ b \in B \}$$
$$A \odot B = \{ a \cdot b \mid a \in A, \ b \in B \}$$
$$A{\downarrow} = \{ y \mid \exists x \in A. \ y \leq x \}$$
$$A_* = \{ ab^*c \mid \forall n \geq 0. \ a\varphi_b^n(0)c \in A \} \ .$$

Note that for principal ideals,

$$\langle a \rangle = \{ a \}{\downarrow} \ .$$

Also note that

$$(A \oplus B) \odot C = (A \odot C) \oplus (B \odot C)$$
$$A \odot (B{\downarrow}) \subseteq (A \odot B){\downarrow}$$
$$(A{\downarrow}) \odot B \subseteq (A \odot B){\downarrow}$$
$$A \odot (B_*) \subseteq (A \odot B)_*$$
$$(A_*) \odot B \subseteq (A \odot B)_* \ .$$

Remark 4 (Kozen [3]). If S satisfies left distributivity, then

$$C \odot (A \oplus B) = (C \odot A) \oplus (C \odot B)$$

for any subsets $A, B, C \subseteq S$.

For a nonempty subset $A \subseteq S$, we define

$$\tau(A) = (A \oplus A) \cup A{\downarrow} \cup A_* \ .$$

Note that $A \subseteq \tau(A)$ and that τ is monotone, i.e. $A \subseteq B$ implies $\tau(A) \subseteq \tau(B)$ for any nonempty subsets $A, B \subseteq S$. Also note that A is a $*$-ideal iff it is nonempty and $\tau(A) \subseteq A$. So, if a $*$-ideal I is generated by some $A \subseteq S$, I is the least fixed point of τ containing A. Using τ, we define the transfinite sequence

$$\tau^0(A) = A$$
$$\tau^{\alpha+1}(A) = \tau(\tau^\alpha(A))$$
$$\tau^\lambda(A) = \bigcup_{\alpha < \lambda} \tau^\alpha(A) \quad \text{if } \lambda \text{ is a limit ordinal}$$
$$\tau^*(A) = \bigcup_\alpha \tau^\alpha(A) \ .$$

for each nonempty subset $A \subseteq S$. Since τ is monotone, the least ordinal κ such that $\tau^{\kappa+1}(A) = \tau^{\kappa}(A)$ exists. For such a κ, $\tau^*(A) = \tau^{\kappa}(A)$. Thus, $\tau^*(A)$ is the least fixed point of τ containing A. Therefore

$$\tau^*(A) = \langle A \rangle .$$

For ∗-ideals I and J, $I \odot J$ and $A \odot B$ generate the same ∗-ideal if B is closed under $+$, and A and B generate I and J, respectively. We use Lemma 3 to show this. Unlike in the case of [3], the assumption on B is needed in the case of ∗-continuous IL-semirings due to lack of left distributivity of $\cdot$ over $+$.

Lemma 3. *Let S be a ∗-continuous IL-semiring. The following holds for nonempty subsets $A, B \subseteq S$.*

1. *$\tau(A) \odot B \subseteq \tau(A \odot B)$.*
2. *$A \odot \tau(B) \subseteq \tau(A \odot B)$ if B is closed under $+$.*

Proof. 1 follows form

$$\begin{aligned}
\tau(A) \odot B &= ((A \oplus A) \cup A{\downarrow} \cup A_*) \odot B \\
&= ((A \oplus A) \odot B) \cup (A{\downarrow} \odot B) \cup (A_* \odot B) \\
&\subseteq ((A \odot B) \oplus (A \odot B)) \cup (A \odot B){\downarrow} \cup (A \odot B)_* \\
&= \tau(A \odot B) .
\end{aligned}$$

2 follows from

$$\begin{aligned}
A \odot \tau(B) &= A \odot ((B \oplus B) \cup B{\downarrow} \cup B_*) \\
&= A \odot (B \cup B{\downarrow} \cup B_*) \\
&= A \odot (B{\downarrow} \cup B_*) \\
&= (A \odot B{\downarrow}) \cup (A \odot B_*) \\
&\subseteq ((A \odot B) \oplus (A \odot B)) \cup (A \odot B){\downarrow} \cup (A \odot B)_* \\
&= \tau(A \odot B) .
\end{aligned}$$ □

Remark 5 (Kozen [3]). If S satisfies left distributivity, then

$$A \odot \tau(B) \subseteq \tau(A \odot B)$$

for any nonempty subsets $A, B \subseteq S$.

Lemma 4. *Let S be a ∗-continuous IL-semiring. The following holds for nonempty subsets $A, B \subseteq S$.*

1. *$\langle A \odot B \rangle = \langle \langle A \rangle \odot B \rangle$.*
2. *$\langle A \odot B \rangle = \langle A \odot \langle B \rangle \rangle$ if B is closed under $+$.*

Proof. 1. The inclusion $\subseteq$ follows from monotonicity of $\langle \_ \rangle$. For the reverse inclusion, we prove by transfinite induction that for each ordinal α,

$$\tau^\alpha(A) \odot B \subseteq \langle A \odot B \rangle \ .$$

It is clear that $\tau^0(A) \odot B = A \odot B \subseteq \langle A \odot B \rangle$. Also,

$$\begin{aligned}
\tau^{\alpha+1}(A) \odot B &= \tau(\tau^\alpha(A)) \odot B \\
&\subseteq \tau(\tau^\alpha(A) \odot B) \ \text{(by 1 of Lemma 3)} \\
&\subseteq \tau(\langle A \odot B \rangle) \quad \text{(by induction hypothesis)} \\
&= \langle A \odot B \rangle
\end{aligned}$$

since $\tau(I) = I$ for a $*$-ideal I. For limit ordinal λ,

$$\begin{aligned}
\tau^\lambda(A) \odot B &= \left(\bigcup_{\alpha < \lambda} \tau^\alpha(A) \right) \odot B \\
&= \bigcup_{\alpha < \lambda} (\tau^\alpha(A) \odot B) \\
&\subseteq \langle A \odot B \rangle \ .
\end{aligned}$$

Thus we have $\tau^*(A) \odot B \subseteq \langle A \odot B \rangle$. Therefore, by monotonicity and idempotency of $\langle \_ \rangle$,

$$\begin{aligned}
\langle \langle A \rangle \odot B \rangle &= \langle \tau^*(A) \odot B \rangle \\
&\subseteq \langle A \odot B \rangle \ .
\end{aligned}$$

Using 2 of Lemma 3 instead of 1 of it, 2 is proved similarly to 1.   □

Remark 6 (Kozen [3]). If S satisfies left distributivity, then

$$\langle A \odot B \rangle = \langle \langle A \rangle \odot \langle B \rangle \rangle$$

for any nonempty subsets $A, B \subseteq S$.

3.2   Functor from ILS* to ILS^D

Let S be a $*$-continuous IL-semiring and consider the poset $(\mathcal{I}(S), \subseteq)$. Since $\langle 0 \rangle = \{0\}$, $\langle 0 \rangle$ is the least element of $\mathcal{I}(S)$. For each subset $\mathcal{A} \subseteq \mathcal{I}(S)$, a $*$-ideal $\langle \bigcup \mathcal{A} \rangle$ is the least upper bound of $\mathcal{A}$. We write $\sum \mathcal{A}$ for $\langle \bigcup \mathcal{A} \rangle$. For any $I, J \in \mathcal{I}(S)$, we write $I + J$ for $\sum \{I, J\}$, and define $I \cdot J = \langle I \odot J \rangle$.

Proposition 2. *Let S be a $*$-continuous IL-semiring. For each $I \in \mathcal{I}(S)$ and $\mathcal{A} \subseteq \mathcal{I}(S)$, the following holds.*

1. $\langle 0 \rangle \cdot I = \langle 0 \rangle$.
2. $\langle 1 \rangle \cdot I = I = I \cdot \langle 1 \rangle$.
3. $(\sum \mathcal{A}) \cdot I = \sum \{J \cdot I \mid J \in \mathcal{A}\}$.
4. *If $\mathcal{A}$ is directed,* $I \cdot (\sum \mathcal{A}) = \sum \{I \cdot J \mid J \in \mathcal{A}\}$.

Proof. 1 follows from definition of $\odot$. The first equation of 2 follows from

$$\begin{aligned}
\langle 1 \rangle \cdot I &= \langle \langle 1 \rangle \odot I \rangle \\
&= \langle \{1\} \odot I \rangle \quad \text{(by 1 of Lemma 4)} \\
&= \langle I \rangle \\
&= I \ .
\end{aligned}$$

Since $\langle 1 \rangle = \{1\}{\downarrow}$, $\langle 1 \rangle$ is closed under $+$. So, using 2 of Lemma 4 instead of 1 of it, the second equation of 2 is proved similarly to the first equation. 3 follows from

$$\begin{aligned}
\left(\sum \mathcal{A} \right) \cdot I &= \left\langle \left(\sum \mathcal{A} \right) \odot I \right\rangle \\
&= \left\langle \left\langle \bigcup \mathcal{A} \right\rangle \odot I \right\rangle \\
&= \left\langle \left(\bigcup \mathcal{A} \right) \odot I \right\rangle \quad \text{(by 1 of Lemma 4)} \\
&= \left\langle \bigcup \{ J \odot I \mid J \in \mathcal{A} \} \right\rangle \\
&= \left\langle \bigcup \{ \langle J \odot I \rangle \mid J \in \mathcal{A} \} \right\rangle \quad \text{(by Lemma 2)} \\
&= \left\langle \bigcup \{ J \cdot I \mid J \in \mathcal{A} \} \right\rangle \\
&= \sum \{ J \cdot I \mid J \in \mathcal{A} \} \ .
\end{aligned}$$

If $\mathcal{A}$ is a directed subset of $\mathcal{I}(S)$, $\bigcup \mathcal{A}$ is closed under $+$ since

$$\begin{aligned}
x, y \in \bigcup \mathcal{A} &\iff \exists I, J \in \mathcal{A}.\, x \in I \text{ and } y \in J \\
&\implies \exists H \in \mathcal{A}.\, x, y \in H \quad (\mathcal{A} \text{ is directed}) \\
&\implies \exists H \in \mathcal{A}.\, x + y \in H \quad (H \text{ is a } \ast\text{-ideal}) \\
&\iff x + y \in \bigcup \mathcal{A} \ .
\end{aligned}$$

So, using 2 of Lemma 4 instead of 1 of it, 4 is proved similarly to 3.   □

Therefore $\mathcal{I}(S)$ forms a D-continuous IL-semiring.

Let S be a $\ast$-continuous IL-semiring. Using $\langle x \rangle = \{x\}{\downarrow}$, it is verified that the mapping

$$x \mapsto \langle x \rangle$$

from S to $\mathcal{I}(S)$ is one-to-one and preserves $+$, $\cdot$, $\ast$, 0, and 1. Thus, this mapping is an arrow from S to $\mathcal{I}(S)$ in $\mathsf{ILS}^\ast$.

Let S be a $\ast$-continuous IL-semiring and Q a D-continuous IL-semiring. Given a homomorphism $g \colon S \to G(Q)$, we define

$$g[A] = \{ g(a) \mid a \in A \}$$

for each $A \subseteq S$. Note that

$$\begin{aligned}
g[A \odot B] &= g[A] \odot g[B] \\
g[A \oplus B] &= g[A] \oplus g[B] \\
g[A{\downarrow}] &\subseteq g[A]{\downarrow} \\
g[A_\ast] &\subseteq g[A]_\ast
\end{aligned}$$

for all $A, B \subseteq S$.

Lemma 5. *For each nonempty subset $A \subseteq S$ and ordinal α, the following holds.*

$$g[\tau(A)] \subseteq \tau(g[A])$$
$$g[\tau^\alpha(A)] \subseteq \tau^\alpha(g[A])$$
$$g[\langle A \rangle] \subseteq \langle g[A] \rangle$$
$$\langle g[\langle A \rangle] \rangle = \langle g[A] \rangle \ .$$

Proof. The first inclusion follows from

$$
\begin{aligned}
g[\tau(A)] &= g[(A \oplus A) \cup A{\downarrow} \cup A_*] \\
&= g[A \oplus A] \cup g[A{\downarrow}] \cup g[A_*] \\
&\subseteq (g[A] \oplus g[A]) \cup g[A]{\downarrow} \cup g[A]_* \\
&= \tau(g[A]) \ .
\end{aligned}
$$

The second follows from the first by transfinite induction. The third follows from the second since $\langle \_ \rangle = \tau^*$. The inclusion $\langle g[\langle A \rangle] \rangle \subseteq \langle g[A] \rangle$ follows from the third, and the reverse inclusion follows from the monotonicity of g and $\langle \_ \rangle$.   □

Therefore, $g[I]$ and $g[A]$ generate the same $*$-ideal if $I \in \mathcal{I}(S)$ is generated by $A \subseteq S$. For $I \in \mathcal{I}(S)$, the least upper bound of $g[I]$ exists in Q, which is denoted by $\sum g[I]$, since the least upper bound of any subset of Q exists in Q.

Lemma 6. *If $I \in \mathcal{I}(S)$, $\sum g[I] = \sum g[A]$ for any generating set A of I.*

Proof. By Lemma 5,
$$\langle g[I] \rangle = \langle g[\langle A \rangle] \rangle = \langle g[A] \rangle$$
for any generating set A of I. Since

$$\langle g[I] \rangle = \langle g[A] \rangle \subseteq \langle \sum g[A] \rangle = \left(\sum g[A] \right){\downarrow} \ ,$$

$\sum g[A]$ is an upper bound of $g[I]$. By $g[A] \subseteq g[I]$, $\sum g[A] \leq \sum g[I]$. Thus, $\sum g[A] = \sum g[I]$ since $\sum g[I]$ is the least upper bound of $g[I]$.   □

Let S be a $*$-continuous IL-semiring and Q a D-continuous IL-semiring. For a homomorphism $g\colon S \to G(Q)$, define the map $\widehat{g}\colon \mathcal{I}(S) \to Q$ by

$$\widehat{g}(I) = \sum g[I] \ .$$

Proposition 3. *The map $\widehat{g}$ preserves $\sum$, $\cdot$, 0 and 1.*

Proof. Note that $\sum (\bigcup \mathcal{B}) = \sum \{ \sum \mathcal{B} \mid \mathcal{B} \in \mathcal{B} \}$ for a subset $\mathcal{B}$ of the powerset $\wp(Q)$. Then, for $\mathcal{A} \subseteq \mathcal{I}(S)$,

$$\hat{g}(\sum \mathcal{A}) = \sum g[\sum \mathcal{A}]$$
$$= \sum g[\langle \bigcup \mathcal{A} \rangle]$$
$$= \sum g[\bigcup \mathcal{A}] \qquad \text{(by Lemma 6)}$$
$$= \sum (\bigcup \{g[I] \mid I \in \mathcal{A}\})$$
$$= \sum \{\sum g[I] \mid I \in \mathcal{A}\}$$
$$= \sum \{\hat{g}[I] \mid I \in \mathcal{A}\} \ .$$

For $I, J \in \mathcal{I}(S)$,

$$\hat{g}(I \cdot J) = \sum g[I \cdot J]$$
$$= \sum g[\langle I \odot J \rangle]$$
$$= \sum g[I \odot J] \qquad \text{(by Lemma 6)}$$
$$= \sum g[I] \odot g[J]$$
$$= (\sum g[I]) \cdot (\sum g[J]) \quad (g[J] \text{ is directed})$$
$$= \hat{g}(I) \cdot \hat{g}(J) \ .$$

Also, we have

$$\hat{g}(\langle 0 \rangle) = \sum g[\langle 0 \rangle] = \sum g[\{0\}] = 0$$

and

$$\hat{g}(\langle 1 \rangle) = \sum g[\langle 1 \rangle] = \sum g[\{1\}] = 1 \ . \qquad \square$$

Theorem 1. *Let S be a ∗-continuous IL-semiring and Q a D-continuous IL-semiring. For a homomorphism $g \colon S \to G(Q)$, $\hat{g}$ is a unique completely join-preserving homomorphism from $\mathcal{I}(S)$ to Q such that $g = \hat{g} \circ \langle \_ \rangle$.*

Proof. For each $a \in S$, we have

$$\hat{g}(\langle a \rangle) = \sum g[\langle a \rangle] = \sum g[\{a\}] = g(a)$$

by Lemma 6. Assume that a completely join-preserving homomorphism f from $\mathcal{I}(S)$ to Q satisfies $g = f \circ \langle \_ \rangle$. Then, it holds that

$$\hat{g}(I) = \sum g[I]$$
$$= \sum \{g(a) \mid a \in I\}$$
$$= \sum \{f(\langle a \rangle) \mid a \in I\} \quad \text{(by assumption)}$$
$$= f(\sum \{\langle a \rangle \mid a \in I\})$$
$$= f(I)$$

for each $I \in \mathcal{I}(S)$. Thus, $\hat{g} = f$. $\qquad \square$

For a $*$-continuous IL-semiring S and a homomorphism $h\colon S \to S'$, we define

$$F(S) = \mathcal{I}(S) \quad \text{and} \quad F(h) = \widehat{\langle\_\rangle \circ h}\ ,$$

respectively. Then, F is a functor from ILS^* to ILS^D. It is immediate from Theorem 1 that the following holds.

Corollary 2. *The functor* $F\colon \mathsf{ILS}^* \to \mathsf{ILS}^D$ *is a left adjoint to the inclusion functor* $G\colon \mathsf{ILS}^D \to \mathsf{ILS}^*$.

4 Connections between ILS_0^* and ILS_0^D, ILS_+^* and ILS_+^D, and $\mathsf{ILS}_{0,+}^*$ and $\mathsf{ILS}_{0,+}^D$

ILS_0^*, ILS_+^*, and $\mathsf{ILS}_{0,+}^*$ are full subcategories of ILS^*. The inclusion functors from ILS_0^*, ILS_+^*, and $\mathsf{ILS}_{0,+}^*$ to ILS^* are denoted by E_0^*, E_+^*, and $E_{0,+}^*$, respectively. Also, ILS_0^D, ILS_+^D, and $\mathsf{ILS}_{0,+}^D$ are full subcategories of ILS^D. The inclusion functors from ILS_0^D, ILS_+^D, and $\mathsf{ILS}_{0,+}^D$ to ILS^D are denoted by E_0^D, E_+^D, and $E_{0,+}^D$, respectively. These six inclusion functors are visualised as follows.

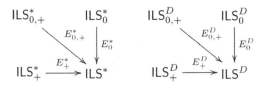

Note that

$$E_v^*(S) = S$$
$$E_v^D(Q) = Q$$
$$\mathsf{ILS}_v^*(S, S') = \mathsf{ILS}^*(E_v^*(S), E_v^*(S'))$$
$$\mathsf{ILS}_v^D(Q, Q') = \mathsf{ILS}^D(E_v^D(Q), E_v^D(Q'))$$

for any objects S, S' of ILS_v^* and Q, Q' of ILS_v^D, where v is either 0, $+$, or $0, +$.

We have already shown that ILS^D is a subcategory of ILS^*. It is immediate from this fact that ILS_0^D is a subcategory of ILS_0^*, so is ILS_+^D of ILS_+^*, and so is $\mathsf{ILS}_{0,+}^D$ of $\mathsf{ILS}_{0,+}^*$. The inclusion functors from ILS_0^D to ILS_0^*, from ILS_+^D to ILS_+^*, and from $\mathsf{ILS}_{0,+}^D$ to $\mathsf{ILS}_{0,+}^*$ are denoted by G_0, G_+, and $G_{0,+}$, respectively. Note that

$$G_v = G \circ E_v^D\ ,$$

where v is either 0, $+$, or $0, +$. In this section, we construct left adjoint functors to $G_0\colon \mathsf{ILS}_0^D \to \mathsf{ILS}_0^*$, $G_+\colon \mathsf{ILS}_+^D \to \mathsf{ILS}_+^*$, and $G_{0,+}\colon \mathsf{ILS}_{0,+}^D \to \mathsf{ILS}_{0,+}^*$.

Proposition 4. *Let S be a $*$-continuous IL-semiring.*

1. *If S satisfies the right zero law, $I \cdot \langle 0 \rangle = \langle 0 \rangle$ for each $I \in \mathcal{I}(S)$.*
2. *If S satisfies left distributivity, $I \cdot (J + J') = I \cdot J + I \cdot J'$ for all $I, J, J' \in \mathcal{I}(S)$.*

Proof. 1. Since $\langle 0 \rangle = \{0\}$ and S satisfies the right zero law, we have

$$I \cdot \langle 0 \rangle = \langle I \odot \{0\} \rangle = \langle 0 \rangle \ .$$

2. It is sufficient to show that $I \cdot (J + J') \subseteq I \cdot J + I \cdot J'$. It follows from

$$
\begin{aligned}
I \cdot (J + J') &= \langle I \odot \langle J \cup J' \rangle \rangle \\
&= \langle I \odot (J \cup J') \rangle \qquad \text{(by Remark 6)} \\
&= \langle (I \odot J) \cup (I \odot J') \rangle \\
&\subseteq \langle \langle I \odot J \rangle \cup \langle I \odot J' \rangle \rangle \\
&= I \cdot J + I \cdot J'
\end{aligned}
$$
□

Define $F_0 = F \circ E_0^*$, $F_+ = F \circ E_+^*$, and $F_{0,+} = F \circ E_{0,+}^*$. Then, by Proposition 4, F_0 is a functor from ILS_0^* to ILS_0^D, F_+ is a functor from ILS_+^* to ILS_+^D, and $F_{0,+}$ is a functor from $\mathsf{ILS}_{0,+}^*$ to $\mathsf{ILS}_{0,+}^D$.

Theorem 2.   *1. The functor $F_0 \colon \mathsf{ILS}_0^* \to \mathsf{ILS}_0^D$ is a left adjoint to the inclusion functor $G_0 \colon \mathsf{ILS}_0^D \to \mathsf{ILS}_0^*$.*
  2. The functor $F_+ \colon \mathsf{ILS}_+^* \to \mathsf{ILS}_+^D$ is a left adjoint to the inclusion functor $G_+ \colon \mathsf{ILS}_+^D \to \mathsf{ILS}_+^*$.
  3. The functor $F_{0,+} \colon \mathsf{ILS}_{0,+}^* \to \mathsf{ILS}_{0,+}^D$ is a left adjoint to the inclusion functor $G_{0,+} \colon \mathsf{ILS}_{0,+}^D \to \mathsf{ILS}_{0,+}^*$.

Proof. 1. Let S be an object of ILS_0^* and Q an object of ILS_0^D. Then, it follows from

$$
\begin{aligned}
\mathsf{ILS}_0^*(S, G_0(Q)) &= \mathsf{ILS}^*(E_0^*(S), G(E_0^D(Q))) \\
&\cong \mathsf{ILS}^D(F(E_0^*(S)), E_0^D(Q)) \quad \text{(by Corollary 2)} \\
&= \mathsf{ILS}_0^D(F_0(S), Q) \ .
\end{aligned}
$$

2 and 3 are shown analogously.   □

5   Conclusion and Outlook

In this paper, we have studied eight categories ILS^*, ILS_0^*, ILS_+^*, $\mathsf{ILS}_{0,+}^*$, ILS^D, ILS_0^D, ILS_+^D, and $\mathsf{ILS}_{0,+}^D$. Among them, there are the inclusion functors as follows.

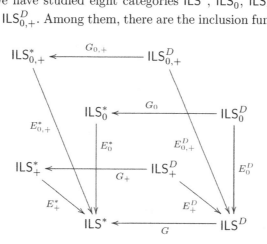

A functor $F\colon \mathsf{ILS}^* \to \mathsf{ILS}^D$ has been constructed via an ideal completion. Then, it has been shown that the functor F is a left adjoint to $G\colon \mathsf{ILS}^D \to \mathsf{ILS}^*$. Also, functors $F_0\colon \mathsf{ILS}_0^* \to \mathsf{ILS}_0^D$, $F_+\colon \mathsf{ILS}_+^* \to \mathsf{ILS}_+^D$, and $F_{0,+}\colon \mathsf{ILS}_{0,+}^* \to \mathsf{ILS}_{0,+}^D$ have been given by restriction of F to subcategories ILS_0^*, ILS_+^*, and $\mathsf{ILS}_{0,+}^*$, respectively. Then, it also has been shown that

- F_0 is a left adjoint to $G_0\colon \mathsf{ILS}_0^D \to \mathsf{ILS}_0^*$,
- F_+ is a left adjoint to $G_+\colon \mathsf{ILS}_+^D \to \mathsf{ILS}_+^*$, and
- $F_{0,+}$ is a left adjoint to $G_{0,+}\colon \mathsf{ILS}_{0,+}^D \to \mathsf{ILS}_{0,+}^*$

using that the inclusion functors E^*, E_0^*, E_+^*, $E_{0,+}^*$, E^D, E_0^D, E_+^D, and $E_{0,+}^D$ are full. These four adjunctions are visualised as follows.

$$\mathsf{ILS}^* \xrightleftharpoons[G]{F} \bot \; \mathsf{ILS}^D \qquad \mathsf{ILS}_0^* \xrightleftharpoons[G_0]{F_0} \bot \; \mathsf{ILS}_0^D$$

$$\mathsf{ILS}_+^* \xrightleftharpoons[G_+]{F_+} \bot \; \mathsf{ILS}_+^D \qquad \mathsf{ILS}_{0,+}^* \xrightleftharpoons[G_{0,+}]{F_{0,+}} \bot \; \mathsf{ILS}_{0,+}^D$$

A D-continuous IL-semiring is defined as an IL-semiring having arbitrary joins. Restricting to countable, rather than arbitrary, joins, we may consider ω-continuous IL-semirings. We have not checked whether each adjunction we gave in this paper factors into two adjunctions as follows or not

$$\mathsf{ILS}^* \xrightleftharpoons{\bot} \mathsf{ILS}^\omega \xrightleftharpoons{\bot} \mathsf{ILS}^D$$

where ILS^ω denotes the category of ω-continuous IL-semirings with ω-continuous semiring homomorphisms.

$*$-continuity introduced in this paper is too strong for lazy Kleene algebras since these need not satisfy

$$b(a+1) \le b \;\to\; ba^* \le b \;.$$

Also, in [2], it is shown that a^* need not be given by $\sum_{n\ge 0} \varphi_a^n(0)$ for each element a if we consider a lazy Kleene algebra consisting of the set of up-closed multirelations over the set of natural numbers and the first transfinite ordinal number. We need more consideration to obtain a similar relationship among lazy Kleene algebras and complete IL-semirings.

Acknowledgement

The authors have benefited from discussion with Bernhard Möller and Peter Höfner about $*$-continuity for variants of lazy Kleene algebras. Koki Nishizawa has given constructive suggestion to the previous version of this paper. We also thank to the anonymous referees for their helpful comments and suggestions.

References

1. Conway, J.H.: Regular Algebra and Finite Machines. Chapman and Hall, Boca Raton (1971)
2. Furusawa, H., Nishizawa, K., Tsumagari, N.: Multirelational Models of Lazy, Monodic Tree, and Probabilistic Kleene Algebras. Bulletin of Informatics and Cybernetics (to appear)
3. Kozen, D.: On Kleene Algebras and Closed Semirings. In: Rovan, B. (ed.) MFCS 1990. LNCS, vol. 452, pp. 26–47. Springer, Heidelberg (1990)
4. Kozen, D.: A Completeness Theorem for Kleene Algebras and the Algebra of Regular Events. Information and Computation 110, 366–390 (1994)
5. McIver, A., Weber, T.: Towards Automated Proof Support for Probabilistic Distributed Systems. In: Sutcliffe, G., Voronkov, A. (eds.) LPAR 2005. LNCS (LNAI), vol. 3835, pp. 534–548. Springer, Heidelberg (2005)
6. Möller, B.: Lazy Kleene Algebra. In: Kozen, D. (ed.) MPC 2004. LNCS, vol. 3125, pp. 252–273. Springer, Heidelberg (2004)
7. Nishizawa, K., Tsumagari, N., Furusawa, H.: The cube of Kleene algebras and the triangular prism of multirelations. In: Berghammer, R., Jaoua, A., Möller, B. (eds.) RelMiCS/AKA 2009. LNCS, vol. 5827, Springer, Heidelberg (2009)
8. Takai, T., Furusawa, H.: Monodic Tree Kleene Algebra. In: Schmidt, R.A. (ed.) RelMiCS/AKA 2006. LNCS, vol. 4136, pp. 402–416. Springer, Heidelberg (2006)

A Semiring Approach to Equivalences, Bisimulations and Control

Roland Glück[1], Bernhard Möller[1], and Michel Sintzoff[2]

[1] Universität Augsburg
[2] Université catholique de Louvain

Abstract. Equivalences, partitions and (bi)simulations are usually tackled using concrete relations. There are only few treatments by abstract relation algebra or category theory. We give an approach based on the theory of semirings and quantales. This allows applying the results directly to structures such as path and tree algebras which is not as straightforward in the other approaches mentioned. Also, the amount of higher-order formulations used is low and only a one-sorted algebra is used. This makes the theory suitable for fully automated first-order proof systems. As a small application we show how to use the algebra to construct a simple control policy for infinite-state transition systems.

1 Introduction

Semirings have turned out to be useful for algebraic reasoning about relations and graphs, for example in [3]. Even edge-weighted graphs were successfully treated in this setting by means of fuzzy relations, as shown in [5] and [6]. Hence it is surprising that up to now no treatment of equivalence relations and bisimulations in this area has taken place, although a relational-algebraic approach was given in [11]. A recent, purely lattice-theoretic abstraction of bisimulations appears in [8]. The present paper was stimulated by [10], since the set-theoretic approach of that paper lends itself to a more compact treatment using modal semirings. This motivated the treatment of subsequent work by the third author by the same algebraic tools, as presented here. In Sect. 2 we consider partitions and equivalences. Sect. 3 explores equivalences in depth, while Sect. 4 is dedicated to bisimulations. As a short application in Sect. 5 a generic construction of a simple control policy is shown.

2 Semirings, Tests and Partitions

2.1 Idempotent Semirings and Tests

Semirings abstract the operations of choice and sequential composition.

Definition 2.1
1. An *idempotent semiring* is a structure $S = (M, +, 0, \cdot, 1)$ such that $0 \neq 1$, $(M, +, 0)$ and $(M, \cdot, 1)$ are monoids, choice $+$ is commutative and idempotent, and composition $\cdot$ distributes through $+$ and is strict in both arguments.

R. Berghammer et al. (Eds.): RelMiCS/AKA 2009, LNCS 5827, pp. 134–149, 2009.

2. The *natural order* $\leq$ is given by $x \leq y \Leftrightarrow_{df} x + y = y$.
3. We call an element $x \in N$ of a subset $N \subseteq M$ *atomic in* N if $x \neq 0$ and
   $\forall y \in N : y \neq 0 \wedge y \leq x \Rightarrow y = x$.
4. A subset $N \subseteq M$ is *atomic* if every element $x \in N$ is the supremum of the
   atoms of N below x.
5. A *quantale* is an idempotent semiring that is a complete lattice under the
   natural order and in which composition distributes over arbitrary suprema.

In an idempotent semiring the element 0 is $+$-irreducible, i.e., $x + y = 0 \Rightarrow x = 0 = y$. Moreover, 0 is the least element w.r.t. $\leq$.

An example for a quantale is $(\mathsf{Rel}([0,9]), \cup, \emptyset, ;, \mathsf{id}_{[0,9]})$, where $\mathsf{Rel}([0,9])$ denotes the set of all binary relations over the interval $[0,9] \subseteq \mathbb{R}$, id_X denotes the identity relation on $X \subseteq [0,9]$ and ; denotes relational composition.

As a running example we will use a simple non-deterministic transition system the state of which is given by a single variable with values in $[0,9]$. It is described by the following relation T between input states x and output states y:

$$xTy \Leftrightarrow_{df} (x \in [0,2] \wedge y = x+4) \vee (x \in [4,6] \wedge y = x+3) \vee (x \in [4,6] \wedge y = x-4).$$

To model sets of states (or equivalently assertions about states) in the semiring setting one uses tests [7].

Definition 2.2. The set $\mathsf{test}(S)$ of *tests* of an idempotent semiring S is the maximal Boolean subalgebra of the elements below 1. The complement of a test p w.r.t. 1 is denoted by $\neg p$; it is the unique test q with $p + q = 1$ and $p \cdot q = 0$. The set of atomic tests of S, i.e., of atoms in $\mathsf{test}(S)$, is denoted by $\mathsf{atest}(S)$.

The elements 1 and 0 are tests. Moreover, for tests p, q their composition $p \cdot q$ coincides with their infimum. Hence $p \leq q \Leftrightarrow p \cdot q = p$. $\qquad(1)$

The tests in our running example are precisely the subrelations of the identity relation on $[0,9]$, including the empty relation. They correspond in an obvious manner to subsets of $[0,9]$ and thus can be used to handle these without introducing a new sort. In this example the set $\mathsf{test}(S)$ is atomic and $\mathsf{atest}(S)$ is the set $\{\{(x,x)\} \mid x \in [0,9]\}$.

For the remainder of this paper we assume $\mathsf{test}(S)$ to be atomic; an atomic test abstractly corresponds to a single state or graph node.

In the sequel we will often form sums of subsets of $\mathsf{test}(S)$. For better readability we use the abbreviation $\sum P =_{df} \sum_{p \in P} p$ for finite $P \subseteq \mathsf{test}(S)$; it coincides with the supremum of P. If the underlying semiring is a quantale we therefore denote by $\sum P$ the supremum of an arbitrary $P \subseteq \mathsf{test}(S)$.

By $+$-irreducibility of 0 we have

$$\sum P \neq 0 \Leftrightarrow \exists p \in P : p \neq 0 \qquad(2)$$

2.2   Partitions

We now define the familiar concept of partition in terms of the tests of an idempotent semiring.

Definition 2.3. A finite subset $P \subseteq \text{test}(S)$ is called a *partition* if $\sum P = 1$ and for all $p, q \in P$ the equivalence $p \cdot q = 0 \Leftrightarrow p \neq q$ holds.

Hence $\{1\}$ is a partition, as is $\text{atest}(S)$, since $\text{test}(S)$ is assumed to be atomic. Moreover, every element of a partition is a test different from 0. For a subset $P' \subseteq P$ of a partition P we have $\neg \sum P' = \sum(P - P')$. Hence the complement $\neg p$ of $p \in P$ satisfies

$$\neg p = \sum(P - \{p\}) . \tag{3}$$

In our running example $\{\text{id}_{[0,6]}, \text{id}_{]6,9]}\}$ and $\{\text{id}_{[0,9]\cap\mathbb{Q}}, \text{id}_{[0,9]-\mathbb{Q}}\}$ are partitions.

Definition 2.4. We say that partition Q *refines* partition P, in signs $Q \leq_r P$, if every element of P can be written as the sum of suitable elements of Q. When $Q \leq_r P$ we say also that P is *coarser* than Q. Clearly, $\leq_r$ is an order.

Lemma 2.5. *Assume that partition Q refines partition P. Then for all $q \in Q$ and $p \in P$ we have $q \cdot p \neq 0 \Rightarrow q \leq p$.*

Proof. By assumption there is a subset $Q' \subseteq Q$ with $p = \sum Q'$. By distributivity, $q \cdot p = q \cdot \sum Q' = \sum_{q' \in Q'} q \cdot q'$. By (2) there must be a $q' \in Q'$ with $q \cdot q' \neq 0$. Since Q is a partition this implies $q = q'$ and hence $q \cdot q' = q$ and $q \leq p$. $\square$

Lemma 2.6. *A partition Q refines a partition P iff for all $q \in Q$ there is a unique $p \in P$ with $p \cdot q \neq 0$.*

Proof. ($\Rightarrow$) For an arbitrary $q \in Q$ we show first the existence of a $p \in P$ with $p \cdot q \neq 0$. This is seen by $q = 1 \cdot q = (\sum P) \cdot q = \sum_{p \in P}(p \cdot q)$. Since $q \neq 0$, by (2) there must be a $p \in P$ with $p \cdot q \neq 0$.

To show uniqueness we assume that there are two different $p, p' \in P$ such that $p \cdot q \neq 0$ and $p' \cdot q \neq 0$ hold. By Lemma 2.5 this implies $q \leq p$ and $q \leq p'$. Hence $0 \neq q = q \cdot q \leq p \cdot p'$, contradicting $p \neq p'$.

($\Leftarrow$) Consider an arbitrary $p \in P$ and set $Q' = \{q \in Q \mid p \cdot q \neq 0\}$. We claim $p = \sum Q'$. First, $p = p \cdot \sum Q = p \cdot (\sum Q' + \sum(Q - Q')) = p \cdot \sum Q'$. By (1) this is equivalent to $p \leq \sum Q'$.

The reverse inequality $\sum Q' \leq p$ holds iff $\forall q \in Q' : q \leq p$. Suppose $q \not\leq p$ for some $q \in Q'$. By (3) this is equivalent to $0 \neq q \cdot \neg p = q \cdot \sum(P - \{p\}) = \sum_{p' \in P - \{p\}} q \cdot p'$. By (2) there must be a $p' \in P - \{p\}$ with $q \cdot p' \neq 0$. But this is a contradiction to $q \cdot p \neq 0$ and the uniqueness assumption. $\square$

Lemma 2.7. *Let P and Q be partitions with $Q \leq_r P$ and assume $p \in P$ and $p = \sum Q'$ for some $Q' \subseteq Q$. Then for all $q \in Q$ we have $p \cdot q \neq 0 \Leftrightarrow q \in Q'$.*

Proof. ($\Rightarrow$) Because of $q \cdot p = q \cdot \sum_{q' \in Q'} q' = \sum_{q' \in Q'}(q \cdot q') \neq 0$ there has to be a $q' \in Q'$ with $q \cdot q' \neq 0$. According to the definition of a partition this implies $q = q'$ and hence $q \in Q'$.

($\Leftarrow$) Let $q \in Q'$. Then $q \leq p$ and therefore $p \cdot q = q$. The claim follows from $q \neq 0$, because $q \neq 0$ holds for all $q \in Q$. $\square$

We now focus on binary relations R which do not connect different sets in a given partition P: for all $p \in P$ and for all x, y such that $x \, R \, y$ we have $x \in p \Leftrightarrow y \in p$. For example, R may be an equivalence and P may be the set of its equivalence classes. This is captured by the following abstract definition.

Definition 2.8. An element $r \in M$ *respects* a partition P if $r = \sum_{p \in P} p \cdot r \cdot p$.

Lemma 2.9. *Let $r \in M$ respect the partition P and let $p, p' \in P$ such that $p \neq p'$. Then $p \cdot r \cdot p' = 0$.*

Proof. Due to the definition of a partition we have for all $p'' \in P$ that $p \cdot p'' = 0$ or $p' \cdot p'' = 0$. Now, by respectance and distributivity,

$$p \cdot r \cdot p' = p \cdot \left(\sum_{p'' \in P} p'' \cdot r \cdot p'' \right) \cdot p' = \sum_{p'' \in P} p \cdot p'' \cdot r \cdot p'' \cdot p' = \sum_{p'' \in P} 0 = 0. \qquad \square$$

An easy consequence of this is the following.

Corollary 2.10. *Let $r \in M$ respect the partition P. Then for all $p \in P$ one has $p \cdot r \cdot \neg p = 0 = \neg p \cdot r \cdot p$.*

Proof. By (3), respectance, distributivity and Lemma 2.9,

$$p \cdot r \cdot \neg p = p \cdot r \cdot \sum (P - \{p\}) = \sum_{p' \in P - \{p\}} p \cdot r \cdot p' = \sum_{p' \in P - \{p\}} 0 = 0. \qquad \square$$

The above lemma is used to prove another important property.

Theorem 2.11. *Let partition Q refine partition P. If $r \in M$ respects Q then it respects P, too.*

Proof. For every $p \in P$ we denote by $Q_p \subseteq Q$ the unique subset of Q with $\sum Q_p = p$. Because of the partition properties $\bigcup_{p \in P} Q_p = Q$ holds. Then by definition of Q_p, distributivity, splitting the sum, Lemma 2.9 and since $\bigcup_{p \in P} Q_p = Q$ and r respects Q,

$$\begin{aligned}
\sum_{p \in P} p \cdot r \cdot p &= \sum_{p \in P} \left(\left(\sum Q_p \right) \cdot r \cdot \left(\sum Q_p \right) \right) = \sum_{p \in P} \left(\sum_{q, q' \in Q_p} (q \cdot r \cdot q') \right) \\
&= \sum_{p \in P} \left(\left(\sum_{q \in Q_p} q \cdot r \cdot q \right) + \left(\sum_{q, q' \in Q_p, q \neq q'} q \cdot r \cdot q' \right) \right) \\
&= \sum_{p \in P} \left(\sum_{q \in Q_p} q \cdot r \cdot q \right) = \sum_{q \in Q} q \cdot r \cdot q = r \qquad \square
\end{aligned}$$

2.3   Modal Semirings and Symmetry

In the sequel the concept of symmetry will play an important role. To define it we use modal operators.

Definition 2.12. A *modal (idempotent) semiring* $(M, +, 0, \cdot, 1, |\cdot\rangle, \langle\cdot|)$ consists of an idempotent semiring $S = (M, +, 0, \cdot, 1)$ and the forward and backward diamond operators $|\cdot\rangle, \langle\cdot| : M \to (\text{test}(S) \to \text{test}(S))$, characterised by the following axioms (e.g. [4]): for all $x, y \in M$ and $p, q \in \text{test}(S)$,

$$|x\rangle q \leq \neg p \Leftrightarrow p \cdot x \cdot q \leq 0 \Leftrightarrow \langle x | p \leq \neg q, \tag{4}$$

$$|x\rangle(|y\rangle q) = |x \cdot y\rangle q \qquad \langle x|(\langle y| q) = |y \cdot x\rangle q \tag{5}$$

By these definitions, $\langle x|q$ and $|x\rangle q$ abstractly describe the image and the inverse image of q under x, resp. From (4) we obtain, for all $p, q \in \mathsf{test}(S)$,

$$|p\rangle q = p \cdot q = \langle p|q . \tag{6}$$

The operators enjoy many further properties, e.g., strictness $|x\rangle 0 = 0 = \langle x|0$ and additivity $|x + y\rangle q = |x\rangle q + |y\rangle q$ and $|x\rangle(q + r) = |x\rangle q + |x\rangle r$. This also entails that they are isotone in both arguments.

In our example we have $|T\rangle \mathsf{id}_{[7,8]} = \mathsf{id}_{[4,5]}$ and $\langle T|\mathsf{id}_{[5,6[} = \mathsf{id}_{[1,2[} \cup \mathsf{id}_{[8,9[}$.

Corresponding box operators can be defined as standard de Morgan duals of the diamonds, but we do not need them for the present paper. For details see again [4].

As mentioned above, the diamonds distribute through $+$ in both arguments; in a quantale they even distribute through arbitrary sums. Moreover, by shunting we obtain from (4) that $p \cdot |x\rangle q \leq 0 \Leftrightarrow p \cdot x \cdot q \leq 0 \Leftrightarrow q \cdot \langle x|p \leq 0$. Therefore, for atomic p,

$$p \cdot x \cdot q \neq 0 \Leftrightarrow p \leq |x\rangle q . \tag{7}$$

A symmetric property holds for $\langle \cdot|$.

Frequently, reasoning can be made more compact by lifting the order on $\mathsf{test}(S)$ pointwise to functions $f, g : \mathsf{test}(S) \rightarrow \mathsf{test}(S)$ by setting

$$f \leq g \Leftrightarrow_{df} \forall p \in \mathsf{test}(S) : f(p) \leq g(p)$$

E.g., $\langle x| \leq \langle y|$ abbreviates $\forall p \in \mathsf{test}(S) : \langle x|p \leq \langle y|p$. An analogous convention applies to the equality of such functions.

In relation algebra, symmetry of a relation R is expressed as $R^{\smile} \subseteq R$ or, equivalently, as $R^{\smile} = R$, where $R^{\smile}$ is the converse of R. Since in semirings there is no converse operation, we have to find express symmetry differently.

Assuming temporarily an abstract converse $x^{\smile}$ of an element x we would certainly expect $p \cdot x^{\smile} \cdot q = 0 \Leftrightarrow q \cdot x \cdot p = 0$ for all $p, q \in \mathsf{test}(S)$. By Axiom (4) this means $|x^{\smile}\rangle = \langle x|$ and $\langle x^{\smile}| = |x\rangle$. Therefore if we consider just the behaviour of an element w.r.t. tests we can avoid the converse by passing to the respective mirror diamond. This motivates the following.

Definition 2.13. An element x of a modal semiring is *symmetric* if $\langle x| = |x\rangle$.

It is straightforward to check that the set of symmetric elements is closed under $+$. In a quantale it is even closed under arbitrary sums.

Our example relation is not symmetric: We have $|T\rangle\{(5,5)\} = \{(1,1)\}$, but $\langle T|\{(5,5)\} = \{(1,1), (8,8)\}$. If we restrict it to the relation $T' = T; (\mathsf{id}_{[0,2]} \cup \mathsf{id}_{[4,6]})$ we obtain a symmetric relation, as is easily verified.

It turns out that in a special class of semirings this notion has interesting equivalent characterisations.

Definition 2.14. Assume a modal semiring S with a greatest element $\top$. Then S satisfies the *Tarski rule* if $x \neq 0 \Leftrightarrow \top \cdot x \cdot \top = \top$.

The Tarski rule holds, for instance, in the modal semiring of binary relations. Since 0 is an annihilator for $\cdot$, this rule is equivalent to

$$\top \cdot x \cdot \top = \top \cdot y \cdot \top \;\Leftrightarrow\; (x = 0 \Leftrightarrow y = 0) \;\Leftrightarrow\; (x \leq 0 \Leftrightarrow y \leq 0) \,. \tag{8}$$

A useful consequence of the Tarski rule is

$$\top \cdot x \cdot \top \cdot y \cdot \top = 0 \;\Leftrightarrow\; (x \leq 0 \vee y \leq 0) \tag{9}$$

which, in turn, implies $\top \cdot x \cdot \top \cdot y \cdot \top = 0 \;\Leftrightarrow\; \top \cdot y \cdot \top \cdot x \cdot \top = 0$ and hence

$$\top \cdot x \cdot \top \cdot y \cdot \top = \top \cdot y \cdot \top \cdot x \cdot \top \,. \tag{10}$$

For the remainder of this section we assume that the Tarski rule holds.

Lemma 2.15. *The following statements for an element x are equivalent:*

1. $\forall p, q \in \mathsf{test}(S) : \top \cdot p \cdot x \cdot q \cdot \top = \top \cdot q \cdot x \cdot p \cdot \top$.
2. $\forall p, q \in \mathsf{test}(S) : p \cdot x \cdot q \leq 0 \Leftrightarrow q \cdot x \cdot p \leq 0$.
3. x *is symmetric.*

Proof. The equivalence of Parts 1 and 2 is just a special case of (8). Therefore it suffices to show the equivalence of Parts 2 and 3. For an arbitrary $x \in M$ we argue as follows:

$$\forall p, q \in \mathsf{test}(S) : p \cdot x \cdot q \leq 0 \Leftrightarrow q \cdot x \cdot p \leq 0$$
$$\Leftrightarrow \quad \{\!\!\{ \text{ by (4) } \}\!\!\}$$
$$\forall p, q \in \mathsf{test}(S) : |x\rangle q \leq \neg p \Leftrightarrow \langle x|q \leq \neg p$$
$$\Leftrightarrow \quad \{\!\!\{ \text{ substitution } p \mapsto \neg p, \text{ bijectivity of negation } \}\!\!\}$$
$$\forall p, q \in \mathsf{test}(S) : |x\rangle q \leq p \Leftrightarrow \langle x|q \leq p$$
$$\Leftrightarrow \quad \{\!\!\{ \text{ indirect equality } \}\!\!\}$$
$$\forall q \in \mathsf{test}(S) : |x\rangle q = \langle x|q \qquad\qquad\qquad \square$$

An immediate consequence of this lemma is the following:

Lemma 2.16. *If s is symmetric and $p \in \mathsf{test}(S)$ then $p \cdot s \cdot p$ is symmetric, too.*

Proof. For arbitrary $q, q' \in \mathsf{test}(S)$ we have, by associativity of multiplication, its commutativity on tests and symmetry of s,

$$\top \cdot q \cdot (p \cdot s \cdot p) \cdot q' \cdot \top = \top \cdot (q \cdot p) \cdot s \cdot (p \cdot q') \cdot \top = \top \cdot (p \cdot q') \cdot s \cdot (p \cdot q) \cdot \top$$
$$= \top \cdot (q' \cdot p) \cdot s \cdot (p \cdot q) \cdot \top = \top \cdot q' \cdot (p \cdot s \cdot p) \cdot q \cdot \top \qquad \square$$

This implies

Corollary 2.17. *For all $p \in \mathsf{test}(S)$ the product $p \cdot \top \cdot p$ is symmetric; in particular, $\top$ is symmetric.*

Proof. Symmetry of $\top$ is immediate from (10) and Lemma 2.15. Then symmetry of $p \cdot \top \cdot p$ is a consequence of Lemma 2.16. $\qquad \square$

Finally we show a consequence of the Tarski rule for diamonds.

Lemma 2.18

1. $|\top\rangle 1 = 1 = \langle\top|1.$
2. *If $p \neq 0$ then $|\top\rangle p = 1 = \langle\top|p.$*

Proof

1. This follows already without the Tarski rule by setting $p = q = 1$ in (6) and using isotony of the diamonds in their first argument.
2. We only show the property for the forward diamond. We have, using (6), Part 1, (5), the Tarski rule and Part 1 again,

$$|\top\rangle p = |\top\rangle|p\rangle 1 = |\top\rangle|p\rangle|\top\rangle 1 = |\top \cdot p \cdot \top\rangle 1 = |\top\rangle 1 = 1 . \qquad \square$$

3   Equivalences

3.1   Equivalences and Fixpoints

Definition 3.1. An element $x \in M$ is called *reflexive* if $1 \leq x$ and *transitive* if $x \cdot x \leq x$. A reflexive and transitive element is called a *preorder* and a symmetric preorder is an *equivalence*.

More liberally, one could define x to be reflexive and transitive if $|1\rangle \leq |x\rangle$ and $|x\rangle|x\rangle \leq |x\rangle$, resp. These conditions are equivalent to $\langle 1| \leq \langle x|$ and $\langle x|\langle x| \leq \langle x|$. As an example of the difference to the above formulation, consider the modal semiring of sets of paths in a graph under path concatenation. The element 1 there is the set of all single-node paths. The condition $1 \leq x$ hence means that the set x of paths includes all these paths, whereas $|1\rangle p \leq |x\rangle p$ means that x must for every node from p contain a path from that node to some node in p, but not necessarily a single-node one. However, for the current treatment we found it more convenient to omit the diamonds.

For an equivalence x and an atomic test p the test $|x\rangle p$ (which by symmetry of x coincides with $\langle x|p$) will play the role of the equivalence class of p under x. If p is a general test then $|x\rangle p = \langle x|p$ will correspond to the union of the equivalence classes of the elements in p. This will be made precise later.

Lemma 3.2. *Let x be a preorder.*

1. *$|x\rangle$ is a closure operator.*
2. *If p is a test then $|x\rangle p$ is a fixed point of $|x\rangle$ and $\langle x|p$ is a fixed point of $\langle x|$.*
3. *The sets of fixed points of $|x\rangle$ and $\langle x|$ each are closed under composition $\cdot$.*

Proof

1. We have to show isotony, extensivity and idempotence. Isotony holds for all diamonds. Extensivity follows from $1 \leq x$ and isotony. For idempotence we have, by transitivity of x and isotony of $|\cdot\rangle$, $|1\rangle = id$ and (5), reflexivity of x and isotony of $|\cdot\rangle$ again, that $|x\rangle|x\rangle = |x \cdot x\rangle \leq |x\rangle = |x\rangle|1\rangle \leq |x\rangle|x\rangle.$

2. First, $|x\rangle(|x\rangle p) = |x \cdot x\rangle p \leq |x\rangle p$ by (5), transitivity of x and isotony of diamonds. Second, $|x\rangle p = |1\rangle(|x\rangle p) \leq |x\rangle(|x\rangle p)$ by (6), reflexivity of x and isotony of diamonds. The statement about $\langle x|p$ is proved symmetrically.
3. The claim follows from the two previous claims as shown in more general context in [2]. □

If x is an equivalence the above lemma means that unions of equivalence classes of x are saturated w.r.t. x.

Lemma 3.3. *Let r be an equivalence and p a fixed point of the function $\langle r|$. Then $\neg p$ is a fixed point of $\langle r|$, too. (For a similar result see [9], p. 33.)*

Proof. Reflexivity of r yields $\langle r|\neg p \geq \neg p$. The reverse inequation follows using (4) twice, symmetry of r and that $\langle r|p = p$ by assumption:

$$\langle r|\neg p \leq \neg p \;\Leftrightarrow\; \neg p \cdot r \cdot p \leq 0 \;\Leftrightarrow\; |r\rangle p \leq p \;\Leftrightarrow\; \langle r|p \leq p \;\Leftrightarrow\; \text{true} . \qquad \square$$

3.2 Equivalences and Partitions

Lemma 3.4. *Let r be an equivalence. Assume that the set F of all fixed points of $\langle r|$ is atomic and let $A \subseteq F$ be the set of all its atoms. Then A is a partition.*

Proof. First we show that $\sum A = 1$. Assume $\sum A < 1$. Then by Lemma 3.3 $\neg \sum A$ is also a fixed point of f and it is different from 0. So there has to be an atomic fixed point below $\neg \sum A$, which leads to a contradiction.

For disjointness of the elements of A we consider arbitrary $p, q \in A$ with $p \neq q$. By Lemma 3.2, $p \cdot q$ is again a fixed point below p and q. Since p and q are assumed to be two different atomic fixed points of f, this implies $p \cdot q = 0$. □

Definition 3.5. For an equivalence r we call the set of the atomic fixed points of the function $\langle r|$, denoted by $Pa(r)$, the *equivalence classes of r*.

Lemma 3.6. *Assume the Tarski rule and let P be a partition. Then $Eq(P) =_{df} \sum_{p\in P} p \cdot \top \cdot p$ is an equivalence. It is called the* equivalence induced by P.

Proof. For transitivity we have, using distributivity, that $p \cdot p' = 0 \Leftarrow p \neq p'$ for $p, p' \in P$, idempotence of multiplication on tests and associativity as well as $p \in P \Rightarrow p \neq 0$ and the Tarski rule,

$$\sum_{p\in P} p \cdot \top \cdot p \cdot \left(\sum_{p\in P} p \cdot \top \cdot p\right) = \sum_{p,p'\in P} p \cdot \top \cdot p \cdot p' \cdot \top \cdot p'$$
$$= \sum_{p\in P} p \cdot \top \cdot p \cdot p \cdot \top \cdot p = \sum_{p\in P} p \cdot (\top \cdot p \cdot \top) \cdot p = \sum_{p\in P} p \cdot \top \cdot p .$$

Reflexivity can be shown, using $\top \geq 1$, idempotence of multiplication on tests and the definition of a partition, by

$$\sum_{p\in P} p \cdot \top \cdot p \geq \sum_{p\in P} p \cdot 1 \cdot p = \sum_{p\in P} p = 1 .$$

Symmetry of $\sum_{p\in P} p \cdot \top \cdot p$ follows easily from the distributivity of $|\cdot\rangle$ and $\langle\cdot|$ over summation and the symmetry of $p \cdot \top \cdot p$ for all $p \in P$ (cf. Corollary 2.17). □

Lemma 3.7. *For an equivalence r and $p, q \in Pa(r)$ we have $p \cdot r \cdot q = 0 \Leftrightarrow p \neq q$.*

Proof. Because of (4) the claim $p \cdot r \cdot q = 0$ is equivalent to $\langle r|p \leq \neg q$. Due to the fixed point property of p this is equivalent to $p \leq \neg q$. By shunting we obtain the equivalent statement $p \cdot q = 0$. From Lemma 3.4 we know that $Pa(r)$ is a partition, which gives us the equivalence to $p \neq q$. □

Lemma 3.8. *An equivalence $r \in M$ respects the partition $Pa(r)$ induced by itself.*

Proof. We have, by $Pa(r)$ being a partition, distributivity, splitting the sum and Lemma 3.7,

$$r = \left(\sum Pa(r) \right) \cdot r \cdot \left(\sum Pa(r) \right) = \sum_{p,p' \in Pa(r)} p \cdot r \cdot p'$$
$$= \left(\sum_{p \in Pa(r)} p \cdot r \cdot p \right) + \left(\sum_{p,p' \in Pa(r), p \neq p'} p \cdot r \cdot p' \right) = \sum_{p \in Pa(r)} p \cdot r \cdot p \quad \square$$

Lemma 3.9. *For a partition P and arbitrary test q we have*

$$|Eq(P)\rangle q = \sum_{p \in P \,\wedge\, p \cdot q \neq 0} p \,.$$

Proof. Using the definition of Eq, additivity of the diamond, (5), (6), strictness of the diamonds, Lemma 2.18.2 and (6) again we calculate

$$|Eq(P)\rangle q = |\sum_{p \in P} p \cdot \top \cdot p\rangle q = \sum_{p \in P} |p \cdot \top \cdot p\rangle q = \sum_{p \in P} |p\rangle|\top\rangle|p\rangle q$$
$$= \sum_{p \in P} |p\rangle|\top\rangle(p \cdot q) = \sum_{p \in P \,\wedge\, p \cdot q \neq 0} |p\rangle|\top\rangle(p \cdot q)$$
$$= \sum_{p \in P \,\wedge\, p \cdot q \neq 0} |p\rangle 1 = \sum_{p \in P \,\wedge\, p \cdot q \neq 0} p \,. \quad \square$$

Now we can show the connection between the operations Eq and Pa.

Theorem 3.10

1. *For an equivalence r we have $r \leq Eq(Pa(r))$.*
2. *For a partition P we even obtain $P = Pa(Eq(P))$.*

In particular, Pa and Eq form a Galois connection.

Proof

1. By Lemma 3.8 and isotony we have

$$r = \sum_{p \in Pa(r)} p \cdot r \cdot p \leq \sum_{p \in Pa(r)} p \cdot \top \cdot p = Eq(Pa(r)) \,.$$

2. First, by Lemma 3.9 and since P is a partition, every $p \in P$ is a fixpoint of $|Eq(P)\rangle$. Second, we show that the elements of P are atomic fixpoints of $|Eq(P)\rangle$. To this end we consider some $p \in P$ and some $q \neq 0$ with $q < p$. Then, again by Lemma 3.9, we have $|Eq(P)\rangle q = p \neq q$, i.e., q is not a fixpoint of $|Eq(P)\rangle$. Finally we show that every atomic fixpoint of $|Eq(P)\rangle$ is an element of P. Let q be a fixpoint of $|Eq(P)\rangle$. Then by Lemma 3.9

$$q = |Eq(P)\rangle q = \sum_{p \in P \,\wedge\, p \cdot q \neq 0} p \,.$$

This holds, in particular, for atomic fixpoints of $|Eq(P)\rangle$. But since atoms are sum-irreducible, the respective sums have to be singleton sums, i.e., the atomic fixpoints all coincide with elements of P.

The Galois property follows from these two properties via isotony. □

In the relational semiring Part 1 of this theorem strengthens to an equality. In general, however, it does not. Consider a graph with a single node x only and a looping arc on x. In the associated path semiring we have $1 = \{x\}$ is an equivalence and $P =_{df} \{1\}$ is the only partition possible. Then $Eq(Pa(1)) = \top \neq 1$, since $\top$ is the set of all finite constant paths $xxx \cdots$.

3.3 Atomic Tests and Equivalence Classes

Next we want to investigate the relationship between atomic tests and equivalence classes. We will see that atomic tests in a certain sense are generators of equivalence classes.

The following lemma states that two elements in the same equivalence class of r are connected to each other, whereas two elements in different equivalence classes are not connected under r.

Lemma 3.11. *Let r be an equivalence and p, q be atomic tests. Then*

$$p \cdot r \cdot q \neq 0 \Leftrightarrow |r\rangle p = |r\rangle q$$

Proof. ($\Rightarrow$) By atomicity of p and (7), isotony and transitivity of r,

$$p \leq |r\rangle q \;\Rightarrow\; |r\rangle p \leq |r\rangle |r\rangle q \;\Rightarrow\; |r\rangle p \leq |r\rangle q \;.$$

Symmetrically we obtain $\langle r|q \leq \langle r|p$, which by symmetry of r is equivalent to $|r\rangle q \leq |r\rangle p$. Now the claim follows by antisymmetry of $\leq$.

($\Leftarrow$) By reflexivity of r we have $p \leq |r\rangle p = |r\rangle q$ and hence $p \cdot r \cdot q$ by (7) and atomicity of p. $\square$

This yields an important relationship between equivalences and partitions:

Theorem 3.12. *Equivalence $r \in M$ respects partition P iff $Pa(r)$ refines P.*

Proof. ($\Rightarrow$) For the sake of contradiction we assume that $Pa(r)$ does not refine P. According to Lemma 2.6 there are $p \in Pa(r)$ and distinct elements $q, q' \in P$ with $p \cdot q \neq 0$ and $p \cdot q' \neq 0$. We consider now two atomic tests at_1 and at_2 with $at_1 \leq p \cdot q$ and $at_2 \leq p \cdot q'$. Because the equivalence classes of at_1 and at_2 under r coincide (both are p) we can apply Lemma 3.11 and obtain $at_1 \cdot r \cdot at_2 \neq 0$. Isotony yields $q \cdot r \cdot q' \neq 0$. But then r cannot respect P because of Lemma 2.9.

($\Leftarrow$) Lemma 3.8 states that r respects $Pa(r)$. According to Theorem 2.11 r respects P, too. $\square$

Now we are ready to prove the main result of this section:

Theorem 3.13. *Let r be an equivalence and p an atomic test. Then $|r\rangle p$ is an atom in the set of fixed points of $|r\rangle$. It is called the* equivalence class of p under r and is denoted by $[p]_r$.

Proof. Suppose $0 \neq |r\rangle q \leq |r\rangle p$ for some test q. By strictness of $|r\rangle$ we must have $q \neq 0$ and hence, by atomicity of $\mathsf{test}(S)$, there is a nonempty set $Q' \subseteq \mathsf{atest}(S)$ with $q = \sum Q'$. The assumption $|r\rangle q \leq |r\rangle p$ is, by distributivity of $|r\rangle$, equivalent to $\forall q' \in Q' : |r\rangle q' \leq |r\rangle p$. Reflexivity of r implies $\forall q' \in Q' : q' \leq |r\rangle p$. By (7) we get $\forall q' \in Q' : q' \cdot r \cdot p \neq 0$ and hence by lemma 3.11 $\forall q' \in Q' : |r\rangle q' = |r\rangle p$. But then also $|r\rangle q = \sum_{q' \in Q'} |r\rangle q' = |r\rangle p$ and we are done. $\qquad\square$

4 Bisimulations

A simulation for a relation $\to \subseteq X \times X$ (such as a transition relation) in the usual sense is a relation $R \subseteq X \times X$ such that

$$x\,R\,x' \wedge x \to y \Rightarrow \exists y' : y\,R\,y' \wedge x' \to y' .$$

In relation algebra this is written more compactly as $R^\smile; \to\, \subseteq\, \to\,; R^\smile$, where ; denotes relational composition.

A bisimulation for $\to$ is a simulation the converse of which is again a simulation for $\to$. Translated into relation algebra this becomes

$$R\,; \to\, \subseteq\, \to\,; R \ \wedge\ R^\smile; \to\, \subseteq\, \to\,; R^\smile .$$

Using the same method as in Sect. 2.3 we can give the following converse-free definition, where b replaces R and g replaces $\to$:

Definition 4.1. An element $b \in M$ is called a *bisimulation* for $g \in M$ iff

$$|b\rangle|g\rangle \leq |g\rangle|b\rangle \wedge \langle b||g\rangle \leq |g\rangle\langle b| .$$

For an element $g \in M$ the set of all bisimulations for g is denoted by bisim_g. Note that $0 \in \mathsf{bisim}_g$.

Lemma 4.2. *For all $g \in M$ the set bisim_g is closed under sum and product. If the underlying modal semiring is a quantale then it is closed under arbitrary sums.*

Proof. The closedness under sum follows easily from the distributivity properties of the diamonds. Closedness under product follows from Axiom (5). $\qquad\square$

For our further purposes it turns out that it is sufficient to require only the existence of a pseudoconverse.

Definition 4.3. We call $y \in M$ a *pseudoconverse* of $x \in M$ iff $|x\rangle = \langle y|$; in this case we write $\mathsf{pscon}(x,y)$.

Note that a symmetric element is a pseudoconverse of itself. We now require for all $x \in M$ the existence of a (not necessarily unique) pseudoconverse y.

Lemma 4.4. *Let $x, y \in M$ such that $\mathsf{pscon}(x,y)$. Then also $\mathsf{pscon}(y,x)$.*

Proof. We only show the inequality $\langle x| \leq |y\rangle$; the reverse inequality is shown analogously. We have, by (4) twice, the assumption $|x\rangle \leq \langle y|$, (4) twice and reflexivity of $\leq$,

$$\langle x|p \leq |y\rangle p \Leftrightarrow p \cdot x \cdot (\neg|y\rangle p) \leq 0 \Leftrightarrow |x\rangle(\neg|y\rangle p) \leq \neg p \Leftarrow \langle y|(\neg|y\rangle p) \leq \neg p$$
$$\Leftrightarrow \langle y|(\neg|y\rangle p) \cdot y \cdot p \leq 0 \Leftrightarrow |y\rangle p \leq |y\rangle p \Leftrightarrow \text{true} .\qquad \square$$

Lemma 4.5. *The sum of an element $x \in M$ and an arbitrary pseudoconverse $y \in M$ of x is symmetric.*

Proof. Let $x, y \in M$ be arbitrary with $\mathsf{pscon}(x, y)$ and let $p \in \mathsf{test}(S)$ be an arbitrary test. Then we calculate, using distributivity of $\langle \cdot |$ over $+$, $\mathsf{pscon}(x, y)$ and Lemma 4.4, distributivity of $\langle \cdot |$ over $+$ and commutativity of $+$,

$$\langle x + y|p = \langle x|p + \langle y|p = |y\rangle p + |x\rangle p = |x + y\rangle p .\qquad \square$$

Lemma 4.6. *Let $g \in M$ be arbitrary and $x \in \mathsf{bisim}_g$ and $\mathsf{pscon}(x, y)$. Then $y \in \mathsf{bisim}_g$.*

Proof. Immediate from the definition of bisimulation and pseudoconverse. $\square$

By definition of bisim_g for an arbitrary $g \in M$ it is obvious that in a quantale there is a coarsest bisimulation for g, namely $\widehat{g} =_{df} \sum_{b \in \mathsf{bisim}_g} b$. This element has another interesting property:

Theorem 4.7. *For all $g \in M$ the coarsest bisimulation $\widehat{g}$ for g is an equivalence.*

Proof. Reflexivity and transitivity follow quickly from Lemma 4.2. For symmetry we observe that for every element $b \in \mathsf{bisim}_g$ every pseudoconverse b' of b lies again in bisim_g. Due to commutativity, associativity and idempotence the equality $\sum_{b \in \mathsf{bisim}_g} b = \sum_{b \in \mathsf{bisim}_g} (b + b')$ holds. This means that $\widehat{g}$ can be written as a sum of symmetric elements of M and hence is symmetric itself. $\square$

The equivalence classes of $\widehat{g}$ have an important property wrt. to g: If from a nonempty part of an equivalence class p one can reach, via g, a second (or even the same) equivalence class q then it is possible to get from *every* nonempty part of p via g to q. This stability property is formally stated in the next theorem.

Theorem 4.8 (Stability). *Let $g \in M$ be arbitrary and $p, q \in \mathsf{atest}(S)$ be atomic tests. If $p \cdot g \cdot q \neq 0$ then for all $p' \leq [p]_{\widehat{g}}$ with $p' \neq 0$ we have $p' \cdot g \cdot [q]_{\widehat{g}} \neq 0$.*

Proof. Due to the atomicity of $\mathsf{test}(S)$ it suffices to show the claim for all atomic p'. Because $\widehat{g}$ is an equivalence (Theorem 4.7) we obtain $p' \cdot \widehat{g} \cdot p \neq 0$ from Lemma 3.11. Hence (7) shows $p' \leq |\widehat{g}\rangle p$. Similarly, the assumption $p \cdot g \cdot q \neq 0$ and atomicity of p yield by (7) that $p \leq |g\rangle q$. Now, by isotony and since $\widehat{g}$ is a symmetric bisimulation, we get

$$0 \neq p' \leq |\widehat{g}\rangle p \leq |\widehat{g}\rangle|g\rangle q \leq |g\rangle|\widehat{g}\rangle q = |g\rangle\langle\widehat{g}|q = |g\rangle[q]_{\widehat{g}}$$

and hence, by (7), $p' \cdot g \cdot [q]_{\widehat{g}} \neq 0$ as required. $\square$

By this result, $\widehat{g}$ determines the coarsest partition that is g-stable.

5   Generating Control Policies

We now sketch a generic method of control design and show how to handle it algebraically. As an illustration a simple control policy for our running example is generated.

5.1   Generic Control Synthesis Using Bisimulations

We are given a transition graph $G = (V, R)$, where V is the set of nodes and R is the transition relation, and a control objective C like cycle-freeness, transitivity or various liveness properties. As a further property we request that the desired control objective can be achieved by a suitable restriction of G. In other words, any controlled transition graph is a subgraph of the uncontrolled one. Most known algorithms generating control policies require that the transition graphs are finite. These algorithms are impracticable in the case of large-scale systems. In the case of infinite state spaces, algorithms have been developed only for a few particular control properties. We propose to construct control policies for large-scale systems by a generic method based on bisimulations; given a control objective, it is assumed that an algorithm A_{cp} generating control policies for that objective and finite systems is available. Relationships between bisimulation equivalence and logical equivalences (e.g. [1]) should help.

The idea is to reduce the huge given graph $G = (V, R)$ to a small finite graph $G_1 = (V_1, R_1)$, called the *(bisimulation) quotient* of G. The nodes in V_1 are the equivalence classes of the coarsest bisimulation for G, while the transitions in R_1 are the corresponding set-level liftings of the transitions in R. The part V_1 can be constructed by an algorithm A_{eq} (e.g. [1]). To this, hopefully finite, graph we apply algorithm A_{cp} and obtain a subgraph G_1' of G_1 with the required control property. Then a subgraph G' of G is obtained by inverting the set-level liftings.

The **crucial assumption** is that the given graph G belongs to the class of graphs for which the number of equivalence classes of its coarsest bisimulation is finite. Then the generic synthesis algorithm looks as follows:

Input   Transition Graph $G = (V, R)$, Control Objective C.

Step 1   Use algorithm A_{eq} to construct the quotient graph $G_1 = (V_1, R_1)$, where V_1 is the set of the equivalence classes of the coarsest bisimulation for G and R_1 is the quotient of R w.r.t. V_1.

Step 2   Use A_{cp} to construct the subgraph $G_1' = (V_1', R_1')$ of G_1. Hence G_1' satisfies C.

Step 3   Generate the controlled graph $G' = (V', R')$ by flattening V_1' into V' (i.e., V' is the union of the sets in V_1') and R_1' into R' (by the corresponding flattening of the transition relation).

Output   The Controlled Transition Graph $G' = (V', R')$, which satisfies C.

In each special case to which this generic algorithm is applied it remains to show that the generated graph G' satisfies the required control objective C. A generic proof of the correctness of Step 3 depends essentially on the formalisation of a significant family of control objectives.

The method elaborated in [10] for optimal control basically generates equivalence sets where states have an equal value. In fact, these sets are composed of equivalence classes of a coarsest bisimulation. So, that particular method is an instance of the proposed approach. In the present paper we illustrate the generic method of bisimulation-based control synthesis with a simple control objective.

5.2   Application to a Simple Control Objective

Now we will demonstrate the proof of the correctness of Step 3 for a simple control objective, namely a liveness property. We require that if a node has an ingoing edge it has to offer an outgoing one, too. A relational formulation could be the predicate $\forall\, x, y : x\,R\,y \Rightarrow (\exists\, z : y\,R\,z)$. In other words, if the pre-image of a node set is non-empty then its image has to be non-empty, too. This motivates the following definition in a general modal semiring $S = (M, +, 0, \cdot, 1)$:

Definition 5.1. An element $g \in M$ is called *live* iff for all $p \in \mathsf{test}(S)$ the implication $|g\rangle p \neq 0 \Rightarrow \langle g|p \neq 0$ holds. For an element $g \in M$ an element $g' \in M$ is called a *live part* of g iff g' is live and $g' \leq g$.

Obviously 0 is live. Moreover, due to distributivity of the diamonds over sums the sum of arbitrary live elements is live, too. So for a $g \in M$ there is a greatest live part, denoted by glp_g.

By atomicity of $\mathsf{test}(S)$, distributivity of the diamonds over arbitrary sums and $+$-irreducibility of 0, an element g is live iff the implication $|g\rangle p \neq 0 \Rightarrow \langle g|p \neq 0$ holds for all atomic tests $p \in \mathsf{atest}(S)$.

As an important concept we introduce a so-called *marker* $\delta_g(p, q)$ of an element $g \in M$ and tests $p, q \in \mathsf{test}(S)$. It can be understood as a sign whether g admits a transition from p to q. In this case it is a restriction of $\top$, otherwise it equals 0. The precise definition is as follows:

Definition 5.2. For an element $g \in M$ the *marker* function $\delta_g(\cdot, \cdot) : \mathsf{test}(S) \times \mathsf{test}(S) \to M$ is defined by $\delta_g(p, q) = p \cdot \top \cdot q$ if $p \cdot g \cdot q \neq 0$, and is 0 otherwise.

We will use this construction to express the above schematic algorithm in our algebraic setting. First we have to model the construction of the graph G_1 from the above description. The nodes correspond to equivalence classes, so a first idea could be to set $g_1 = \sum_{p,q \in Pa(\widehat{g})} \delta_g(p, q)$, where $Pa(\widehat{g})$ is the set of equivalence classes of the coarsest bisimulation for g. This models the property that G_1 admits a transition from one node to another iff there is a transition in G between two elements of the equivalence classes corresponding to the nodes in G_1. However, it turns out to be more convenient to abstract from this construction by means of a system of representatives (analogously to the classical use) and to reduce this quotient to a quotient witness:

Definition 5.3. A *system of representatives (SOR)* for an equivalence r is a set Rep of atomic tests such that $\sum_{p \in \mathsf{Rep}} [p]_r = 1$ and $p, q \in \mathsf{Rep} \wedge p \neq q \Rightarrow [p]_r \cdot [q]_r = 0$. For an element $g \in M$ an element $h \in M$ is called a *quotient* *witness* of g if there is a SOR Rep of $\hat{g}$ such that $h = \sum_{p,q \in \text{Rep}} p \cdot \delta_g([p]_{\hat{g}}, [q]_{\hat{g}}) \cdot q$. Rep is called the *associated* SOR of h, and the elements (p, q) from Rep^2 with $p \cdot h \cdot q \neq 0$ are called its *edges*, denoted by edges_h. The set of all quotient witnesses of g is denoted by $\text{qw}(g)$.

Let now $h \in \text{qw}(g)$ be arbitrary. Assume we can determine the glp_h. Our goal now is to construct from glp_h the greatest live part glp_g of g. To this end we set $g' = \sum_{(p,q) \in \text{edges}_{\text{glp}_h}} [p]_{\hat{g}} \cdot g \cdot [q]_{\hat{g}}$ ($\text{edges}_{\text{glp}_h}$ is defined analogously to Definition 5.3) and obtain an element $g' \in M$ with the desired property:

Theorem 5.4. *Let g' be constructed as above. Then $g' = \text{glp}_g$.*

Proof. The property $g' \leq g$ follows immediately from isotony of multiplication and $p \leq 1$ for all $p \in P$ for an arbitrary partition P. By atomicity of $\text{test}(S)$, distributivity of the diamonds over arbitrary sums and $+$-irreducibility of 0, it suffices to show the second claim for all atomic tests. So let p be an arbitrary atomic test with $|g'\rangle p \neq 0$. By q we denote the representative of $[p]_{\hat{g}}$ in Rep. Due to the construction of g' there has to be a pair $(q, q') \in \text{edges}_{\text{glp}_h}$ with $q \cdot \text{glp}_h \cdot q' \neq 0$. Because of $\text{glp}_h \leq h$ and the construction of h the inequality $q \cdot g \cdot q' \neq 0$ holds. According to Theorem 4.8 for every atomic test p' with $p' \leq [q]_{\hat{g}}$ the inequality $p' \cdot g \cdot [q']_{\hat{g}} \neq 0$ has to hold. Because p and p' are contained in the same equivalence class we have also $p \cdot g \cdot [q']_{\hat{g}} \neq 0$. But then by construction of g' also $\langle g'|p \neq 0$ holds.

Hence g' is a live part of g. Assume now that $g' < \text{glp}_g$. Then we consider the element $\tilde{h} = \sum_{p,q \in \text{Rep}} p \cdot \delta_{\text{glp}_g}([p]_{\hat{g}}, [q]_{\hat{g}}) \cdot q$. Because of $\text{glp}_g \leq g$ we have $\tilde{h} \leq h$. On the other hand, due to the construction of g' and $g' < \text{glp}_g$ there has to be $p, q \in \text{Rep}$ such that $(p, q) \notin \text{edges}_{\text{glp}_h}$ and $[p]_{\hat{g}} \cdot \text{glp}_g \cdot [q]_{\hat{g}} \neq 0$. Consider now an arbitrary $p \in \text{Rep}$ with $|\tilde{h}\rangle p \neq 0$. Then by construction $|\text{glp}_g\rangle[p]_{\hat{g}} \neq 0$ and hence $|\text{glp}_g\rangle[p]_{\hat{g}} \neq 0$. But then we have also $\langle\tilde{h}|p \neq 0$. This means that $\tilde{h}$ is live and $\tilde{h} \leq \text{glp}_h$ does not hold, which is a clear contradiction. $\square$

Let us now apply this construction to our running example. The coarsest bisimulation is here given by $[0, 2]^2 \cup [4, 6]^2 \cup (]2, 4[\cup]6, 9])^2$, and it has three equivalence classes, namely $[0, 2]$, $[4, 6]$ and $]2, 4[\cup]6, 9]$. As a quotient witness we can choose the relation $\{(1, 4), (4, 8), (4, 1)\}$. The greatest live part of this is $\{(1, 4), (4, 1)\}$. If we play this back to the original relation by means of the above construction we obtain the infinite relation $\{(x, y) \in \mathbb{R}^2 \mid (x \in [0, 2] \wedge y = x + 4) \vee (x \in [4, 6] \wedge y = x - 4)\}$, which is the greatest live part of the original relation according to Theorem 5.4.

Admittedly, the present algebraic modelling of system control is basic and primitive. The application of the generic method for other control objectives, e.g. optimality, may well require the use of labelled transition systems. For such cases, the algebraic framework needs to be refined.

6   Conclusion and Further Work

We have shown how equivalences, partitions and bisimulations can be conveniently described in the setting of modal semirings and quantales. With these tools we were also able to give a generic algorithm for constructing a simple policy for an infinite transition system.

Future work has several directions: First, we shall extend our methods to cover also labelled transition systems. Quantales for describing them are already known from the literature. The second focus will be to consider more significant goals than the simple liveness property given in Sect. 5.2. So we plan to tackle properties like acyclicity, termination or even (probabilistic) shortest path problems. A more general challenge would be to identify the subclass of control objectives for which the algorithm from Sect. 5.1 works correctly.

References

1. Baier, C., Katoen, J.-P.: Principles of Model Checking. MIT Press, Cambridge (2008)
2. Birkhoff, G.: Lattice Theory, 3rd edn., vol. XXV. Colloquium Publications, American Mathematical Society (1967)
3. Carré, B.: Graphs and Networks. Oxford Univ. Press, Oxford (1979)
4. Desharnais, J., Möller, B., Struth, G.: Kleene algebra with domain. ACM Transactions on Computational Logic 7, 798–833 (2006)
5. Glück, R., Möller, B.: Circulations, fuzzy relations and semirings. In: Audebaud, P., Paulin-Mohring, C. (eds.) MPC 2008. LNCS, vol. 5133, pp. 134–152. Springer, Heidelberg (2008)
6. Kawahara, Y.: On the cardinality of relations. In: Schmidt, R.A. (ed.) RelMiCS/AKA 2006. LNCS, vol. 4136, pp. 251–265. Springer, Heidelberg (2006)
7. Manes, E., Benson, D.: The inverse semigroup of a sum-ordered semiring. Semigroup Forum 31, 129–152 (1985)
8. Pous, D.: Complete lattices and up-to techniques. In: Shao, Z. (ed.) APLAS 2007. LNCS, vol. 4807, pp. 351–366. Springer, Heidelberg (2007)
9. Schmidt, G., Ströhlein, T.: Relations and Graphs: Discrete Mathematics for Computer Scientists. Springer, Heidelberg (1993)
10. Sintzoff, M.: Synthesis of optimal control policies for some infinite-state transition systems. In: Audebaud, P., Paulin-Mohring, C. (eds.) MPC 2008. LNCS, vol. 5133, pp. 336–359. Springer, Heidelberg (2008)
11. Winter, M.: A relation-algebraic theory of bisimulations. Fundam. Inf. 83(4), 429–449 (2008)

General Correctness Algebra

Walter Guttmann

Institut für Programmiermethodik und Compilerbau
Universität Ulm, 89069 Ulm, Germany
walter.guttmann@uni-ulm.de

Abstract. General correctness offers a finer semantics of programs than partial and total correctness. We give an algebraic account continuing and extending previous approaches. In particular, we propose axioms, correctness statements, a correctness calculus, specification constructs and a loop refinement rule. The Egli-Milner order is treated algebraically and we show how to obtain least fixpoints, used to solve recursion equations, in terms of the natural order.

1 Introduction

Relational approaches to program semantics vary in their treatment of termination according to [19].

Partial correctness does not distinguish between terminating and possibly non-terminating programs. Recursion is modelled by least fixpoints with respect to the subset order, which leads to angelic non-determinism. If the same program admits both a terminating and a non-terminating execution, the terminating one is chosen. Theories of partial correctness include Hoare logic [16], weakest liberal preconditions [9] and Kleene algebra with tests [21].

Total correctness does not distinguish between non-terminating and possibly terminating programs. Recursion is modelled by greatest fixpoints with respect to the subset order, which leads to demonic non-determinism. If the same program admits both a terminating and a non-terminating execution, the non-terminating one is chosen. Theories of total correctness include weakest preconditions [9], the Unifying Theories of Programming [17], demonic refinement algebra [26] and demonic algebra [5].

General correctness [2,4,19,3,25,10,24] distinguishes terminating and non-terminating executions. Recursion is modelled by least fixpoints with respect to the Egli-Milner order, which leads to erratic non-determinism.

Technically, partial correctness is the simplest approach, since there is no need to represent non-termination. For total and general correctness, this is done by adding a special value, predicate or variable. In total correctness, additionally, non-termination absorbs termination. This price is paid to keep the subset order, while in general correctness the more complicated Egli-Milner order must be used for fixpoints. Refinement is the subset order in all three approaches.

In this paper we focus on the algebraic treatment of general correctness. It offers a finer distinction than partial and total correctness [19,11]. We build upon a number of works, as discussed in the following.

R. Berghammer et al. (Eds.): RelMiCS/AKA 2009, LNCS 5827, pp. 150–165, 2009.

In [10] the Unifying Theories of Programming are adapted to general correctness using a restricted class of predicates called 'prescriptions'. They are generalised using matrices over semirings in [23]. While the semantics of loops is missing for prescriptions, it is given in [24] using 'commands' over modal semirings and the Egli-Milner order. Still missing, however, is the semantics of full recursion. This is contributed by Section 6 of the present paper.

Another result of [24] is that weakest preconditions are actually the weakest liberal preconditions of an appropriate modal semiring. It is used to derive a Hoare calculus for weakest preconditions. As such, the calculus is useful for total correctness claims. To this end, however, a total correctness semantics of commands, such as the one given in [14], would be more appropriate and also more simple by not having to use the Egli-Milner order. Another way to overcome the mismatch is to devise a calculus for general correctness claims. This is contributed by Section 4 of the present paper.

In [12] the absence of loop refinement rules is noted for general correctness, in contrast to total correctness. They are necessary to introduce loops when specifications are refined into programs. A general correctness loop rule is given based on prescriptions. Section 5 of the present paper contributes an algebraic statement and proof of that rule, using the calculus of Section 4.

As another ingredient of refinement, [26] discusses specifications given only by preconditions and postconditions in demonic refinement algebra. Such 'pre-post specifications' can conveniently be used to express rules like the one for loop refinement. To this end, Section 5 also contributes specifications suitable for general correctness.

All contributions are wrapped in an algebraic theory of general correctness encompassing those of [10,24,23,12] along the lines of [15]. It is based on Kleene algebra with a domain operator and developed in Section 2 of the present paper. Section 3 takes it as a guide and contributes an axiomatic description of the key constituents of general correctness, such as the Egli-Milner order. The axioms are used to derive the results announced above.

2   Semirings and Prescriptions

Prescriptions have been introduced in [10] to model general correctness in the Unifying Theories of Programming. An algebraic account using modal semirings is given in [24] and, using matrices over modal semirings, in [23]. In this section, we adapt these approaches and develop them further according to our treatment of total correctness [15].

We first recall how to extend semirings by axioms for conditions, which represent subsets of states. Based on this structure, we algebraically define prescriptions, which model programs and specifications in general correctness. To conveniently express the semantics of loops, we then introduce the Kleene star and omega operations. We finally impose further structure using the domain operation, which is necessary for our axiomatic treatment of general correctness in Section 3.

2.1   Condition Semirings

A *weak semiring* is a structure $(S, +, 0, \cdot, 1)$ such that $(S, +, 0)$ is a commutative monoid, $(S, \cdot, 1)$ is a monoid, the operation $\cdot$ distributes over $+$ in both arguments and 0 is a left annihilator, that is, $0 \cdot x = 0$. We assume $0 \neq 1$, otherwise S would be trivial. A weak semiring is *idempotent* if $+$ is, that is, if $x + x = x$. In an idempotent weak semiring the relation $x \leq y \Leftrightarrow_{\text{def}} x + y = y$ is a partial order, called the *natural order* on S, and $\cdot$ and $+$ are isotone. A *semiring* is a weak semiring in which 0 is also a right annihilator, that is, $x \cdot 0 = 0$. The $\cdot$ operation is extended elementwise to sets $A, B \subseteq S$ by $A \cdot B =_{\text{def}} \{a \cdot b \mid a \in A \land b \in B\}$ and $A \cdot b =_{\text{def}} A \cdot \{b\}$ for $b \in S$. We frequently abbreviate $a \cdot b$ with ab.

A structure $(S, T, +, 0, \cdot, 1, \sqcap, \top, \bar{})$ is a *condition semiring* if the following properties hold.

- $(S, +, 0, \cdot, 1)$ is an idempotent weak semiring having a greatest element $\top$.
- $(T, +, 0)$ is a submonoid of $(S, +, 0)$ and $T \subseteq T \cdot \top$.
- The *restriction operation* $\sqcap : T \times S \to S$ distributes over $+$, that is,
  * $\forall a \in S : \forall t, u \in T : (t + u) \sqcap a = (t \sqcap a) + (u \sqcap a)$ and
  * $\forall a, b \in S : \forall t \in T : t \sqcap (a + b) = (t \sqcap a) + (t \sqcap b)$.
- $\forall a \in S : \top \sqcap a = a$.
- $(T, +, 0, \sqcap, \top, \bar{})$ is a Boolean algebra; in particular, $0 \in T$ and $\top \in T$.

We abbreviate condition semirings with (S, T) and call the elements of T *conditions*. A condition semiring (S, T) is an *ideal condition semiring* if $S \cdot T \subseteq T$, hence T is a left ideal of S. An (ideal) condition semiring is *strict* if the underlying weak semiring is a semiring, that is, if 0 is both a left and a right annihilator.

Our notation reflects the intended, relational model (where 0, 1 and $\top$ are the empty, identity and universal relations, respectively, and $\leq$ is the subset order), so that $0 \leq 1 \leq \top$ holds, for example. To avoid confusion, it should be kept in mind that other approaches in the literature use different conventions (for example, demonic refinement algebra [26] uses the reverse order).

In relational semantics, a condition semiring (S, T) is used as follows. The state transition relation or input/output behaviour of programs is represented by elements of S. The elements of T represent subsets of states by relating each initial state in the subset to all final states. The operations $+$, $\cdot$ and $\sqcap$ model non-deterministic choice, sequential composition and input-restriction, respectively. In particular, $t \sqcap a$ restricts the transitions permitted by a to those starting in a state described by the condition t. The elements of T are also used as preconditions that represent those states from which a non-terminating execution of the program exists.

The following, basic properties are proved in [15]. In a condition semiring (S, T), the operation $\sqcap$ is associative, isotone, the greatest lower bound on $T \times S$ and satisfies the shunting rule $t \sqcap a \leq b \Leftrightarrow a \leq \bar{t} + b$ as well as $(t \sqcap a) \cdot b = t \sqcap (a \cdot b)$ for all $t \in T$ and $a, b \in S$. Reminding us that conditions represent the vectors of relation algebra, we have $t \cdot \top = t$ for all $t \in T$, and thus $T \cdot \top = T$. In an ideal condition semiring (S, T) this extends to $S \cdot T = S \cdot \top = T \cdot \top = T$.

2.2   Prescriptions

We continue with the matrix representation of prescriptions, generalised to the present axiomatisation. Let (S, T) be an ideal condition semiring. The set of *normal prescriptions* over (S, T) is

$$\mathrm{NP}(S, T) =_{\mathrm{def}} \left\{ \begin{pmatrix} a & b \\ c & d \end{pmatrix} \in S^{2 \times 2} \;\middle|\; a = \top \wedge b = 0 \wedge c \in T \right\}.$$

The components a and b are for structural purposes, making composition work as expected. The components c and d are the precondition and transition elements mentioned above. The adjective 'normal' [10] refers to the restriction $c \in T$, by which preconditions are indeed conditions on the initial states and not arbitrary relations between input and output states. For $t \in T$ and $a \in S$, we define the *normal prescription*

$$(t \Vdash a) =_{\mathrm{def}} \begin{pmatrix} \top & 0 \\ \overline{t} & a \end{pmatrix}.$$

It represents the program whose execution performs transitions allowed by a and is guaranteed to terminate when started in states described by t.

Particular normal prescriptions are $\mathsf{skip} =_{\mathrm{def}} (\top \Vdash 1)$, $\mathsf{loop} =_{\mathrm{def}} (0 \Vdash 0)$, $\mathsf{fail} =_{\mathrm{def}} (\top \Vdash 0)$, $\mathsf{havoc} =_{\mathrm{def}} (\top \Vdash \top)$ and $\mathsf{chaos} =_{\mathrm{def}} (0 \Vdash \top)$. For example, skip models the program which must terminate without changing the state, and loop the one which must not terminate, see [25].

These special prescriptions are landmarks of the structure inherent to normal prescriptions as follows. Let (S, T) be a strict ideal condition semiring, then $(\mathrm{NP}(S, T), +, \mathsf{fail}, \cdot, \mathsf{skip})$ is an idempotent weak semiring. Using the set $C =_{\mathrm{def}} \{(\overline{t} \Vdash t) \mid t \in T\}$ as conditions, $(\mathrm{NP}(S, T), C, +, \mathsf{fail}, \cdot, \mathsf{skip}, \sqcap, \mathsf{chaos}, \overline{})$ is a condition semiring. The operations $+$ and $\sqcap$ act elementwise on the matrices, $\cdot$ is the matrix product, and $\overline{}$ applies to both arguments of $\Vdash$. Both strictness and the ideal property are necessary for these results, which are proved analogously to the corresponding ones in [15] that apply to 'normal designs' modelling total correctness. The technical difference is that the matrices for normal designs satisfy $b = \top$ instead of $b = 0$ and the additional restriction $c \leq d$ that lets non-termination absorb terminating transitions (for example, $c = \top$ forces $d = \top$).

To appreciate the different structures introduced above we note the following distinctions. Relations form a strict ideal condition semiring. Normal designs over relations [15], which are the basis of the Unifying Theories of Programming, form an ideal condition semiring that is not strict. Normal prescriptions over relations, which are the basis of general correctness semantics, form a condition semiring that is not an ideal condition semiring. Every idempotent weak semiring with $\top$ forms a condition semiring with 0 and $\top$ as the only conditions, but all previous structures generally contain additional conditions.

In the remainder of this paper, we omit the adjective 'normal'. Several consequences about the natural order, sum and product of prescriptions are

- $(t \Vdash a) \leq (u \Vdash b) \Leftrightarrow u \leq t \wedge a \leq b$,
- $(t \Vdash a) + (u \Vdash b) = (t \sqcap u \Vdash a + b)$ and
- $(t \Vdash a) \cdot (u \Vdash b) = (t \sqcap \overline{a\overline{u}} \Vdash ab)$.

Hence prescriptions are equal just if both components are equal, fail is the least prescription and chaos the greatest. Moreover, $(t \Vdash 0)$ is a left annihilator for each $t \in T$. The vector property of prescriptions is derived by

$$(\bar{t} \Vdash t) \cdot (0 \Vdash \top) = (\bar{t} \sqcap \bar{t}\bar{0} \Vdash t\top) = (\bar{t} \sqcap \overline{t\top} \Vdash t) = (\bar{t} \sqcap \bar{t} \Vdash t) = (\bar{t} \Vdash t) .$$

An intuitive interpretation of the natural order is that non-terminating executions may be refined to terminating ones that do not introduce new transitions. This contrasts with designs, where any terminating execution can be introduced by such a refinement.

2.3   Kleene Algebra and Omega Algebra

A *(weak) Kleene algebra* [20,22] is a structure $(S, ^*)$ such that S is an idempotent (weak) semiring and the operation star * satisfies the unfold and induction laws

$$1 + a \cdot a^* \leq a^* \qquad\qquad b + a \cdot c \leq c \Rightarrow a^* \cdot b \leq c$$
$$1 + a^* \cdot a \leq a^* \qquad\qquad b + c \cdot a \leq c \Rightarrow b \cdot a^* \leq c$$

for $a, b, c \in S$. Hence a^*b is the least fixpoint of $\lambda x.ax + b$, denoted $\mu x.ax + b$. The star operation on prescriptions is derived using the general matrix construction presented, for example, in [13]. Let (S, T) be an ideal condition semiring such that S is a Kleene algebra, then $(\mathrm{NP}(S, T), +, \mathsf{fail}, \cdot, \mathsf{skip}, ^*)$ is a weak Kleene algebra, where

$$\begin{pmatrix} \top & 0 \\ t & a \end{pmatrix}^* = \begin{pmatrix} (\top + 0a^*t)^* & (\top + 0a^*t)^*0a^* \\ (a + t\top^*0)^*t\top^* & (a + t\top^*0)^* \end{pmatrix} = \begin{pmatrix} \top^* & 0 \\ a^*t\top & a^* \end{pmatrix} = \begin{pmatrix} \top & 0 \\ a^*t & a^* \end{pmatrix},$$

hence $(t \Vdash a)^* = (\overline{a^*\bar{t}} \Vdash a^*)$.

A *(weak) omega algebra* [6,22] is a structure $(S, ^\omega)$ such that S is a (weak) Kleene algebra and the operation omega $^\omega$ satisfies the unfold and co-induction laws

$$a^\omega = a \cdot a^\omega \qquad\qquad c \leq a \cdot c + b \Rightarrow c \leq a^\omega + a^* \cdot b$$

for $a, b, c \in S$. Hence $a^\omega + a^*b$ is the greatest fixpoint of $\lambda x.ax + b$, denoted $\nu x.ax + b$. It follows that $a^\omega \top = a^\omega = a^* a^\omega$ and $c \leq a \cdot c \Rightarrow c \leq a^\omega$. The omega operation on prescriptions cannot be derived via the matrix construction since the greatest prescription is not the matrix with four $\top$ entries. Nevertheless, a direct argument can be used to show the following result. Let (S, T) be an ideal condition semiring such that S is an omega algebra, then $(\mathrm{NP}(S, T), +, \mathsf{fail}, \cdot, \mathsf{skip}, ^*, ^\omega)$ is a weak omega algebra, where $(t \Vdash a)^\omega = (\overline{a^\omega + a^*\bar{t}} \Vdash a^\omega)$.

2.4   Tests and Domain

A *test semiring* [22] is an idempotent weak semiring $(S, +, 0, \cdot, 1)$ with a distinguished set of elements $\mathrm{test}(S) \subseteq S$ called *tests* and a *negation* operation $\neg$ such that $(\mathrm{test}(S), +, 0, \cdot, 1, \neg)$ is a Boolean algebra. By slightly generalising a proof of [15] we can show that any condition semiring $(S, T, +, 0, \cdot, 1, \sqcap, \top, \bar{})$ is a test semiring, where $\text{test}(S, T) =_{\text{def}} \{t \sqcap 1 \mid t \in T\}$ and $\neg p =_{\text{def}} \overline{p\top} \sqcap 1$ for $p \in \text{test}(S, T)$. Hence prescriptions form a test semiring with tests of the form

$$(\bar{t} \Vdash t) \sqcap (\top \Vdash 1) = \begin{pmatrix} \top & 0 \\ t & t \end{pmatrix} \sqcap \begin{pmatrix} \top & 0 \\ 0 & 1 \end{pmatrix} = \begin{pmatrix} \top & 0 \\ 0 & t \sqcap 1 \end{pmatrix} = (\top \Vdash t \sqcap 1),$$

and negation $\neg(\top \Vdash t \sqcap 1) = (\top \Vdash \bar{t} \sqcap 1)$. This allows us to represent conditional statements by either $(t \sqcap a) + (\bar{t} \sqcap b)$ or $pa + \neg pb$, using either the condition t or its corresponding test $p = t \sqcap 1$. The use of conditions (or another set satisfying the ideal property) in the underlying semiring is necessary if prescriptions are to be represented by matrices; otherwise tests can be used for the termination information as in [24].

A *domain semiring* [8] is a structure $(S, \ulcorner\urcorner)$ such that S is a test semiring and the domain operation $\ulcorner\urcorner : S \to \text{test}(S)$ satisfies the axioms

$$a \leq \ulcorner a \urcorner \cdot a \qquad \ulcorner(p \cdot a)\urcorner \leq p \qquad \ulcorner(a \cdot \ulcorner b \urcorner)\urcorner \leq \ulcorner(a \cdot b)\urcorner$$

for $a, b \in S$ and $p \in \text{test}(S)$. Useful properties for $a, b \in S$ and $p \in \text{test}(S)$ are

$$\ulcorner a \urcorner \leq p \Leftrightarrow a \leq pa \qquad a \leq b \Rightarrow \ulcorner a \urcorner \leq \ulcorner b \urcorner \qquad a = \ulcorner a \urcorner a \qquad \ulcorner p \urcorner = p$$
$$a \leq 0 \Leftrightarrow \ulcorner a \urcorner \leq 0 \qquad \ulcorner(a + b)\urcorner = \ulcorner a \urcorner + \ulcorner b \urcorner \qquad \ulcorner(pa)\urcorner = p \ulcorner a \urcorner \qquad \ulcorner(a \cdot \ulcorner b \urcorner)\urcorner = \ulcorner(a \cdot b)\urcorner$$

If a greatest element $\top$ exists, another characterisation is $\ulcorner a \urcorner \leq p \Leftrightarrow a \leq p\top$ [1]. For prescriptions we obtain $\ulcorner(t \Vdash a)\urcorner = (\top \Vdash \neg\bar{t} + \ulcorner a \urcorner)$ this way. If the test semiring is induced from an ideal condition semiring as above, we even have $\ulcorner a \urcorner = a\top \sqcap 1$.

Domain induces the operations *diamond of a* given by $\langle a \rangle p =_{\text{def}} \ulcorner(ap)\urcorner$ and its dual *box of a* given by $[a]p =_{\text{def}} \neg \langle a \rangle \neg p$. For prescriptions they amount to $\langle t \Vdash a \rangle(\top \Vdash \ulcorner u \urcorner) = (\top \Vdash \neg\bar{t} + \langle a \rangle \ulcorner u \urcorner)$ and $[t \Vdash a](\top \Vdash \ulcorner u \urcorner) = (\top \Vdash \bar{t} \cdot [a] \ulcorner u \urcorner)$.

3   Towards Axioms for General Correctness

Kleene star and omega cannot be used directly to express the general correctness semantics of loops. This is due to the fact that star and omega are taken with respect to the natural order $\leq$ that corresponds to the subset order used for partial and total correctness, but not to the Egli-Milner order.

For example, consider the endless loop while true do skip. Its partial correctness semantics is the least fixpoint $(\mu x.x) = 1^* \cdot 0 = 1 \cdot 0 = 0$. The total correctness semantics is the greatest fixpoint $(\nu x.x) = 1^\omega + 1^* \cdot 0 = 1^\omega = \top$. Instantiated to prescriptions, they are fail and chaos, respectively. However, the general correctness semantics is loop that lies properly between the least and the greatest fixpoints with respect to the natural order.

Another difference between partial, total and general correctness is observed about the term $\top \cdot 0$. For partial correctness, Kleene algebra is used where $\top \cdot 0 = 0$ (assuming $\top$ exists). For total correctness, this right annihilation axiom is dropped to obtain weak Kleene algebra, with the freedom to impose the left annihilation axiom $\top \cdot 0 = \top$ instead, as done by [26,5,15]. For general correctness, we have to drop this left annihilation axiom, too. This is easily observed from prescriptions, since the product of the greatest and the least prescription is

$$(0 \Vdash \top) \cdot (\top \Vdash 0) = \begin{pmatrix} \top & 0 \\ \top & \top \end{pmatrix} \cdot \begin{pmatrix} \top & 0 \\ 0 & 0 \end{pmatrix} = \begin{pmatrix} \top & 0 \\ \top & 0 \end{pmatrix} = (0 \Vdash 0) \, ,$$

which is neither the greatest nor the least prescription, but again loop. Since the term $\top \cdot 0$ cannot be simplified in weak Kleene or omega algebra, but is important as the intended least element of the Egli-Milner order, we call it $\mathsf{L} =_{\mathrm{def}} \top \cdot 0$.

In the following, we work towards axiomatising the structure that underlies prescriptions and their use in general correctness. We start by assuming a weak omega algebra and domain semiring S, since we have seen that prescriptions form one. Hence $\mathsf{L} = \top \cdot 0$ exists and already satisfies a number of properties.

Lemma 1. $\top \mathsf{L} = \mathsf{L}^\omega = \mathsf{L} \neq 1$ and $\mathsf{L}^* = 1 + \mathsf{L}$. Let $x \in S$, then $x\mathsf{L} \leq \ulcorner x\mathsf{L} \leq \mathsf{L} = \mathsf{L}x$ and $x0 \leq \ulcorner(x0)\mathsf{L}$.

Proof. $x\mathsf{L} \leq \top\mathsf{L} = \top\top 0 = \top 0 = \mathsf{L}$, thus $x\mathsf{L} \leq \ulcorner xx\mathsf{L} \leq \ulcorner x\mathsf{L} \leq \mathsf{L}$, and $\mathsf{L}x = \top 0 x = \top 0 = \mathsf{L}$. Hence $\mathsf{L}^* = 1 + \mathsf{L}\mathsf{L}^* = 1 + \mathsf{L}$ and $\mathsf{L}^\omega = \mathsf{L}\mathsf{L}^\omega = \mathsf{L}$. Assuming $\mathsf{L} = 1$ gives the contradiction $0 = 1 \cdot 0 = \mathsf{L} \cdot 0 = \mathsf{L} = 1$. Moreover, $x0 = x0\mathsf{L} \leq \ulcorner(x0)\mathsf{L}$. □

However, other properties which we expect to hold (since they hold for prescriptions) cannot be derived from the axioms of weak omega algebra. We therefore have to introduce further axioms.

$$\begin{array}{ll} x \leq \mathsf{L} \Rightarrow x \leq x0 & \text{(L0)} \\ \ulcorner x\mathsf{L} \leq x\mathsf{L} & \text{(L1)} \\ 1 \leq \ulcorner\mathsf{L} & \text{(L2)} \end{array}$$

Axiom (L0) is provisional and follows from axioms presented below. Its consequent can equivalently be replaced by $x = x0$. Its backward implication holds by $x \leq x0 \leq \top 0 = \mathsf{L}$. The term $x0$ represents the states which may lead to non-termination. Axioms (L1) and (L2) can equivalently be strengthened to equalities. Consequences of these axioms are recorded in the next lemma.

Lemma 2. Let $x, y \in S$ and $p, q \in \mathrm{test}(S)$. Axiom (L0) implies $x \leq \mathsf{L} \Rightarrow x = xy$ and $\mathsf{L} \neq \top$ and $px0 \leq 0 \wedge pxq \leq \mathsf{L} \Rightarrow pxq \leq 0$ and $x0 = \inf\{x, \mathsf{L}\}$. Axiom (L2) implies $\ulcorner(x\mathsf{L}) = \ulcorner x$ and $\mathsf{L} \neq 0$. Axioms (L0) and (L2) together are equivalent to $\ulcorner\mathsf{L}x \leq \mathsf{L} \Rightarrow x = x0$. Axiom (L1) implies $\ulcorner(x0)\mathsf{L} \leq x$, which is equivalent to $\ulcorner(x0)\mathsf{L} = x0$ and together with (L2) conversely implies (L1).

Proof. For $x \leq \mathsf{L}$ we have $xy \leq x0y = x0 \leq x$ and $x \leq x0 \leq xy$ by (L0). Assuming $1 \leq \mathsf{L}$ gives the contradiction $1 \leq 1 \cdot 0 = 0$. Let $px0 \leq 0$ and $pxq \leq \mathsf{L}$, then $pxq \leq pxq0 = px0 \leq 0$. Let $z \leq x$ and $z \leq \mathsf{L}$, then $z \leq z0 \leq x0$ by (L0), and $x0$ is a lower bound of x and L since $x0 \leq x1 = x$ and $x0 \leq \top 0 = \mathsf{L}$.

$\ulcorner(x\mathsf{L}) = \ulcorner(x\ulcorner\mathsf{L}) = \ulcorner(x1) = \ulcorner x$; assuming $\mathsf{L} = 0$ gives the contradiction $1 \leq \ulcorner 0 = 0$.

Let $\ulcorner\mathsf{L}x \leq \mathsf{L}$, then $x \leq \mathsf{L}$ by (L2), hence $x = x0$ by (L0). Let $\ulcorner\mathsf{L}x \leq \mathsf{L} \Rightarrow x = x0$ hold, then $x \leq \mathsf{L}$ implies $\ulcorner\mathsf{L}x \leq x \leq \mathsf{L}$, hence $x = x0$, which shows (L0). Moreover, $\ulcorner\mathsf{L}\neg\mathsf{L} = 0 \leq \mathsf{L}$ implies $\neg\mathsf{L} = \neg\mathsf{L}0 = 0$, and hence (L2).

$\ulcorner(x0)L \leq x0L = x0 \leq x$ by (L1). This implies $x0 \leq \ulcorner(x0)L = \ulcorner(x0)L0 \leq x0$ by Lemma 1. Conversely, $\ulcorner xL = \ulcorner(xL)L = \ulcorner(xL0)L \leq xL$ by (L2) and Lemma 1. $\quad\square$

Let us define the initial states of $x \in S$ from which infinite transition paths emerge as $\nabla x =_{\mathrm{def}} \ulcorner x^\omega$, and the 'convergent' states $\Delta x =_{\mathrm{def}} \neg\nabla x$. In presence of (L1) and (L2), this complies with the axiomatisation of ∇ given in [7]. To see this, observe that $\ulcorner x^\omega = \ulcorner(xx^\omega) = \langle x\rangle\ulcorner x^\omega$ by omega unfold, and $p \leq \langle x\rangle p + q$ implies $pL \leq \ulcorner(xp)L + qL \leq xpL + qL$ by (L1), hence $pL \leq x^\omega + x^*qL$ by omega co-induction, thus $p = \ulcorner(pL) \leq \ulcorner x^\omega + \langle x^*\rangle q$ by Lemma 2. Particular consequences are $\ulcorner(x^*0) \leq \ulcorner(x^*x^\omega) = \ulcorner x^\omega = \nabla x$ and $x^\omega = \ulcorner x^\omega x^\omega = \nabla x x^\omega \leq \nabla x\top$. By (L1) we also obtain $\nabla xL = \ulcorner x^\omega L = x^\omega L = x^\omega\top 0 = x^\omega 0$.

Another prescription that needs a representation is havoc. To this end, we introduce the element $H \in S$ by the following axioms that relate L and H to represent programs as pairs of termination and state transition information.

$$x \leq y + L \wedge x \leq y + H \Rightarrow x \leq y \qquad \text{(H1)}$$
$$x \leq x0 + H \qquad \text{(H2)}$$

Instantiating $y = x0$ in (H1) gives $x \leq x0 + L \Rightarrow x \leq x0$ by (H2), which implies (L0) immediately. Instantiating $x = \top$ in (H2) gives $\top = L + H$. Instantiating $x = H$ in (H1) results in $H \leq y + L \Rightarrow H \leq y$. Together we obtain $\top \leq L + y \Leftrightarrow H \leq y$, thus H is the least additive pseudo-complement of L. In particular, H is unique if it exists. An equivalent formulation of (H1) is $x + L = y + L \wedge x + H = y + H \Rightarrow x = y$. In particular, we also obtain $x \leq L \wedge x \leq H \Rightarrow x = 0$.

Lemma 3. (H1) *implies* $H0 \leq 0$ *and* (H2) *implies* $x0 \leq 0 \Rightarrow x \leq H$ *for* $x \in S$.

Proof. $H0 \leq \top 0 = L$ and $H0 \leq H1 = H$, hence $H = 0$ by (H1). Let $x0 \leq 0$, then $x \leq x0 + H \leq H$ by (H2). $\quad\square$

The two conditions shown in the previous lemma are the axioms of [26] for havoc, but in a total correctness setting. Together, they are equivalent to $x0 \leq 0 \Leftrightarrow x \leq H$, thus H is the greatest strict element. The next lemma records further consequences of our axioms.

Lemma 4. *Axiom* (H2) *implies* $1 \leq H \leq H^2$ *and* $\top H = \top = H\top = H^\omega$ *and* $HL = L$. *Axioms* (H1) *and* (H2) *together imply* $H^* = H^2 = H \neq L$. *Axioms* (H1) *and* (L2) *together imply* $H \neq \top$.

Proof. $1 \leq 1\cdot 0 + H = 0 + H = H$ by (H2). Hence $H \leq H^2$ and $\top \leq \top H$ and $\top \leq H\top$ by isotony, thus $HL = H\top L = \top L = L$ by Lemma 1, and $\top \leq H^\omega$.

$H^2 \leq H$ follows by Lemma 3 since $H^2 0 \leq H0 \leq 0$ by the same lemma using (H2) and (H1), respectively. Hence $1 + H^2 \leq H$, which implies $H^* \leq H$. Assuming $H = L$ gives the contradiction $1 \leq H = L = L0 = H0 = 0$ by (H2), Lemma 1 and Lemma 3 using (H1).

Assuming $H = \top$ gives the contradiction $0 \neq L = \top 0 = H0 = 0$ by Lemma 2 using (L2) and Lemma 3 using (H1). $\quad\square$

For prescriptions over relations we generally have $H \neq 1$, but this cannot be proved from the axioms since the underlying semiring may be such that $1 = \top$ and hence havoc is skip (the relations ≤ 1 are an example).

We are now ready to define the Egli-Milner order $\sqsubseteq$ based on our axioms:

$$x \sqsubseteq y \Leftrightarrow_{\text{def}} x \leq y + L \wedge y \leq x + \ulcorner(x0)H .$$

This definition is justified by the instance for prescriptions: We obtain the characterisation expected from [25,24,12] by calculating

$$
\begin{aligned}
&(t \Vdash a) \sqsubseteq (u \Vdash b) \\
&\Leftrightarrow (t \Vdash a) \leq (u \Vdash b) + (0 \Vdash 0) \wedge (u \Vdash b) \leq (t \Vdash a) + \ulcorner((t \Vdash a)(\top \Vdash 0))(\top \Vdash \top) \\
&\Leftrightarrow (t \Vdash a) \leq (0 \Vdash b) \wedge (u \Vdash b) \leq (t \Vdash a + \bar{t}) \\
&\Leftrightarrow a \leq b \wedge t \leq u \wedge b \leq a + \bar{t} ,
\end{aligned}
$$

since $(u \Vdash b) + (0 \Vdash 0) = (u \sqcap 0 \Vdash b + 0) = (0 \Vdash b)$ and

$$
\begin{aligned}
&(t \Vdash a) + \ulcorner((t \Vdash a)(\top \Vdash 0))(\top \Vdash \top) = (t \Vdash a) + \ulcorner(t \sqcap \overline{a0} \Vdash a0)(\top \Vdash \top) \\
&= (t \Vdash a) + \ulcorner(t \Vdash 0)(\top \Vdash \top) = (t \Vdash a) + (\top \Vdash \neg\ulcorner t)(\top \Vdash \top) \\
&= (t \Vdash a) + (\top \Vdash \neg\ulcorner t\top) = (t \Vdash a + \bar{t}) ,
\end{aligned}
$$

since $\neg\ulcorner t\top = (\overline{\ulcorner t\top} \sqcap 1)\top = \overline{\ulcorner t\top} = \overline{(t\top \sqcap 1)\top} = \overline{t\top} = \bar{t}$. The following lemma shows basic properties of $\sqsubseteq$.

Lemma 5. *Axiom* (H1) *implies that* $\sqsubseteq$ *is a partial order. Axioms* (L2) *and* (H2) *together imply that* L *is its least element. Axioms* (H1) *and* (H2) *together imply that* $\sqsubseteq$ *has no greatest element.*

Proof. Reflexivity follows immediately. For transitivity, let $x \sqsubseteq y$ and $y \sqsubseteq z$. Then $x \leq y + L$ and $y \leq z + L$, which implies $x \leq z + L + L = z + L$. Moreover $y \leq x + \ulcorner(x0)H$ and $z \leq y + \ulcorner(y0)H$, hence

$$z \leq x + \ulcorner(x0)H + \ulcorner((x + \ulcorner(x0)H)0)H = x + \ulcorner(x0)H + \ulcorner(\ulcorner(x0)H0)H = x + \ulcorner(x0)H$$

by Lemma 3. Together we have $x \sqsubseteq z$. For antisymmetry, let $x \sqsubseteq y$ and $y \sqsubseteq x$. Then $x \leq y + L$ and $y \leq x + \ulcorner(x0)H \leq x + H$ and $y \leq x + L$ and $x \leq y + \ulcorner(y0)H \leq y + H$. Hence $x \leq y$ and $y \leq x$ by (H1).

For any $x \in S$ we have $x \leq \top = L + H = L + \ulcorner LH = L + \ulcorner(L0)H$ by (H2), (L2) and Lemma 1. With $L \leq x + L$ we obtain $L \sqsubseteq x$.

Assume that $0 \sqsubseteq x$ and $1 \sqsubseteq x$, then $x \leq 0 + \ulcorner(0 \cdot 0)H = \ulcorner 0H = 0H = 0$, and therefore $1 \leq x + L \leq L$. Since $1 \leq H$ by Lemma 4, we obtain the contradiction $1 \leq 0$ by (H1). □

It can furthermore be shown that $\cdot$ and $+$ are isotone with respect to $\sqsubseteq$. We have thus derived a number of useful properties from our axioms. In the remainder of this paper we assume that (L1), (L2), (H1) and (H2) hold in S.

Least fixpoints with respect to the Egli-Milner order, denoted by ξ, are used to define the general correctness semantics of recursion. In particular, the semantics of loops is while p do $a =_{\text{def}} \xi x.pax + \neg p$.

Theorem 6. *Let $p \in \text{test}(S)$ and $a \in S$, then* while p do $a = \nabla(pa)\mathsf{L} + (pa)^*\neg p$.

A direct proof can be given using Lemmas 1 and 2. It is omitted since the result follows from our treatment of full recursion in Section 6.

4   General Correctness

Consider a domain semiring D, an element $a \in D$ and two tests $p, q \in \text{test}(D)$. Soundness of the Hoare triple $p\{a\}q$ is defined by [24] as $p \leq [a]q$, which is equivalent to $pa\neg q \leq 0$ [21]. This claims partial correctness: When started in a state satisfying p, the program a will not lead to a state satisfying $\neg q$. Thus $[a]q$ is the weakest liberal precondition of statement a and postcondition q.

The remarkable observation of [24] is that the *same* triple claims total correctness if it is interpreted in an appropriate semiring. In particular, $[a]q$ then is the weakest precondition of statement a and postcondition q. This is beneficial, since statements proved in general domain semirings automatically hold in both interpretations. For example, a calculus for weakest liberal preconditions in domain semirings yields one for weakest preconditions.

An appropriate semiring to interpret the Hoare triple is given by prescriptions. Let us verify that the Hoare triple indeed yields a total correctness claim:

$$(\top \Vdash p)(t \Vdash a)\neg(\top \Vdash q) = (\overline{pt} \Vdash pa)(\top \Vdash \neg q) = (\overline{pt} \Vdash pa\neg q) \leq (\top \Vdash 0)$$
$$\Leftrightarrow p\overline{t} \leq 0 \wedge pa\neg q \leq 0 .$$

Hence the termination claim $p\overline{t} \leq 0$ is a part of the Hoare triple. It is equivalent to $p \leq \ulcorner t$ and expresses that the starting state must be one in which the execution of $(t \Vdash a)$ is guaranteed to terminate.

Such a claim is characteristic of total correctness. Actually, the same claim is obtained for the Hoare triple interpreted in the semiring of designs [15]. Working with designs would then have the additional advantage of not having to deal with the Egli-Milner order. Instead, the semantics of recursion uses the simpler natural order of the semiring.

Another conclusion is that the Hoare triple does not express general correctness adequately. To derive a more suitable correctness claim, we again look at the concrete instance of prescriptions. The two occurrences of the precondition p in the claim above have to be separated as in $r\overline{t} \leq 0 \wedge pa\neg q \leq 0$. Now r describes the initial states from where termination has to be guaranteed, and p describes the initial states which do not lead to states satisfying $\neg q$. Partial correctness is recovered by choosing $r = 0$ and total correctness by $r = p$, but we can now make full use of the 'generality' provided by general correctness to distinguish claims about terminating and non-terminating executions.

For prescriptions we observe that the first condition is obtained by

$$(\top \Vdash r)(t \Vdash a)(\top \Vdash 0) = (\overline{rt} \Vdash ra)(\top \Vdash 0) = (\overline{rt} \Vdash 0) \leq (\top \Vdash 0) \Leftrightarrow r\overline{t} \leq 0 ,$$

and the second by

$$(\top \Vdash p)(t \Vdash a)\neg(\top \Vdash q) = (\overline{pt} \Vdash pa\neg q) \leq (0 \Vdash 0) \Leftrightarrow pa\neg q \leq 0 .$$

Generalised to our axiomatic framework of Section 3, we thus obtain the general correctness statement

$$ra0 \leq 0 \wedge pa\neg q \leq \mathsf{L} \; .$$

A notation analogous to Hoare triples would be a quadruple containing the program a, the termination precondition r, the precondition p and the postcondition q. We rather observe that the first claim $ra0 \leq 0$ is equivalent to the Hoare triple $r\{a\}1$, which can be derived using existing Hoare calculi except for constructs based on L or H. For these we can derive $0\{a\}1$ and $a \leq \mathsf{H} \Rightarrow r\{a\}1$ for any $a \in S$ and $r \in \mathrm{test}(S)$, thus in particular $0\{\mathsf{L}\}1$ and $1\{\mathsf{H}\}1$. For the while loop we calculate using Theorem 6

$$[\text{while } p \text{ do } a]1 = \neg^\ulcorner((\nabla(pa)\mathsf{L} + (pa)^*\neg p)0) = \neg(^\ulcorner(\nabla(pa)\mathsf{L}) + {}^\ulcorner((pa)^*0)) = \neg\nabla(pa)$$

to obtain the triple $\Delta(pa)\{\text{while } p \text{ do } a\}1$.

New rules are, however, necessary for the second claim $pa\neg q \leq \mathsf{L}$, which we denote by $p\left(\!\left|a\right|\!\right)q$ since it amounts to 'weak correctness' of [26]. To see this, we show $pa\neg q \leq \mathsf{L} \Leftrightarrow pa = paq$. The forward implication follows since $pa\neg q \leq \mathsf{L}$ implies $pa\neg q = pa0$ by (L0), hence $pa = paq + pa\neg q = paq + pa0 = paq$. The backward implication follows since $pa\neg q = paq\neg q = pa0 \leq \mathsf{T}0 = \mathsf{L}$. The rules for weak correctness are provided by the following theorem.

Theorem 7. *Let $a, b \in S$ and $p, q, r \in \mathrm{test}(S)$. Then*

$p\left(\!\left\|0\right\|\!\right)q$	$p\left(\!\left\|\mathsf{L}\right\|\!\right)q$	$q\left(\!\left\|1\right\|\!\right)q$
$p\left(\!\left\|a\right\|\!\right)q \wedge p\left(\!\left\|b\right\|\!\right)q \Rightarrow p\left(\!\left\|a+b\right\|\!\right)q$		
$rp\left(\!\left\|a\right\|\!\right)q \wedge \neg rp\left(\!\left\|b\right\|\!\right)q \Rightarrow p\left(\!\left\|ra+\neg rb\right\|\!\right)q$		

$$pr\left(\!\left|1\right|\!\right)q \Rightarrow p\left(\!\left|r\right|\!\right)q$$
$$p\left(\!\left|a\right|\!\right)q \wedge q\left(\!\left|b\right|\!\right)r \Rightarrow p\left(\!\left|ab\right|\!\right)r$$
$$pq\left(\!\left|a\right|\!\right)q \Rightarrow q\left(\!\left|\text{while } p \text{ do } a\right|\!\right)\neg pq$$

Proof. $p0\neg q = 0 \leq \mathsf{L}$ and $p\mathsf{L}\neg q \leq \mathsf{L}$ and $q1\neg q = 0 \leq \mathsf{L}$. The rule for tests is immediate and the rule for choice follows by $p(a+b)\neg q = pa\neg q + pb\neg q \leq \mathsf{L}$ from its premises. Composition is calculated as

$$pa\neg q \leq \mathsf{L} \wedge qb\neg r \leq \mathsf{L} \Rightarrow pab\neg r = paqb\neg r + pa\neg qb\neg r \leq pa\mathsf{L} + \mathsf{L}b\neg r \leq \mathsf{L}$$

by Lemma 1. A consequence of the rules for 1 and tests is $p\left(\!\left|q\right|\!\right)pq$. Using this and the rules for composition and choice we obtain the rule for the conditional.

To obtain the rule for the while loop, we first derive $q\left(\!\left|a\right|\!\right)q \Rightarrow q\left(\!\left|a^*\right|\!\right)q$. Assume $qa\neg q \leq \mathsf{L}$, then

$$q + (\mathsf{L} + qa^*q)a = qq + \mathsf{L}a + qa^*qaq + qa^*qa\neg q \leq qa^*q + \mathsf{L} + qa^*\mathsf{L} \leq \mathsf{L} + qa^*q$$

by Lemma 1, hence $qa^* \leq \mathsf{L} + qa^*q$ by star induction, thus

$$qa^*\neg q \leq \mathsf{L}\neg q + qa^*q\neg q = \mathsf{T}0 + qa^*0 = \mathsf{T}0 = \mathsf{L} \; .$$

Second, we derive $pq\left(\!\left|a\right|\!\right)q \Rightarrow q\left(\!\left|(pa)^*\neg p\right|\!\right)\neg pq$. This follows by the composition rule, since $q\left(\!\left|p\right|\!\right)pq \Rightarrow q\left(\!\left|pa\right|\!\right)q \Rightarrow q\left(\!\left|(pa)^*\right|\!\right)q$ and $q\left(\!\left|\neg p\right|\!\right)\neg pq$. Third, we have $q\left(\!\left|\nabla(pa)\mathsf{L}\right|\!\right)\neg pq$ by the rules for L and composition, since $q\left(\!\left|\nabla(pa)\right|\!\right)q\nabla(pa)$ holds. Apply the choice rule to these claims and Theorem 6. $\square$

5   Pre-Post Specifications

Complementary to the verification approach using correctness claims that can be derived through a calculus is the transformation approach, where specifications are refined into implementations. Specifications given by pre- and postconditions are well-known in total correctness and treated algebraically in [26]. In this section we propose specifications suitable for general correctness refinement.

Our specification $(r \mid p \rightsquigarrow q)$ consists of three components. One of them is new: The termination precondition r describes the initial states from which execution must terminate. The other two are as usual: If the precondition p holds in the initial state, the postcondition q must be established. We axiomatise the specification as the greatest element of S satisfying our general correctness claim of Section 4 for tests $r, p, q \in \text{test}(S)$:

$$r(r \mid p \rightsquigarrow q)0 = 0 \tag{G1}$$
$$p(r \mid p \rightsquigarrow q)\neg q \leq \mathsf{L} \tag{G2}$$
$$rx0 = 0 \wedge px\neg q \leq \mathsf{L} \Rightarrow x \leq (r \mid p \rightsquigarrow q) \tag{G3}$$

The greatest element leaves the greatest amount of freedom in implementation, since $x \leq y$ means that x refines y. The conjunction of (G1), (G2) and (G3) can equivalently be stated as $rx0 = 0 \wedge px\neg q \leq \mathsf{L} \Leftrightarrow x \leq (r \mid p \rightsquigarrow q)$, thus the specification is unique if it exists. These axioms are stated to show the intention, but in our algebra of Section 3 we can give an explicit characterisation.

Theorem 8. *Let $p, q, r \in \text{test}(S)$. Then $(r \mid p \rightsquigarrow q) = \neg r\mathsf{L} + \neg p\mathsf{H} + \mathsf{H}q$.*

Proof. To show that $\neg r\mathsf{L} + \neg p\mathsf{H} + \mathsf{H}q$ satisfies (G1) and (G2), we calculate

$$r(\neg r\mathsf{L} + \neg p\mathsf{H} + \mathsf{H}q)0 = r\neg r\mathsf{L}0 + r\neg p\mathsf{H}0 + r\mathsf{H}q0 = 0 + r\neg p0 + r\mathsf{H}0 = 0$$
$$p(\neg r\mathsf{L} + \neg p\mathsf{H} + \mathsf{H}q)\neg q = p\neg r\mathsf{L}\neg q + p\neg p\mathsf{H}\neg q + p\mathsf{H}q\neg q \leq \mathsf{L} + 0 + \mathsf{H}0 = \mathsf{L}$$

by Lemma 3. For (G3), let $rx0 = 0$ and $px\neg q \leq \mathsf{L}$. Then $rx \leq \mathsf{H}$ by Lemma 3 and $px = pxq$ as shown in Section 4. Therefore $x \leq \neg r\mathsf{L} + \neg p\mathsf{H} + \mathsf{H}q$ follows from the cases

$$
\begin{array}{ll}
prxq \leq rxq \leq \mathsf{H}q & prx\neg q = prxq\neg q \leq p\mathsf{H}0 = p0 = 0 \\
p\neg rx\neg q \leq \neg r\mathsf{L} & p\neg rxq \leq \neg r\mathsf{T}q = \neg r\mathsf{L}q + \neg r\mathsf{H}q \leq \neg r\mathsf{L} + \mathsf{H}q \\
\neg prx \leq \neg p\mathsf{H} & \neg p\neg rx \leq \neg p\neg r\mathsf{T} = \neg p\neg r\mathsf{L} + \neg p\neg r\mathsf{H} \leq \neg r\mathsf{L} + \neg p\mathsf{H}
\end{array}
$$

which hold by Lemma 3.   □

Thus the total correctness pre-post specification $[p, q]$ of [26] can be recovered as $(p \mid p \rightsquigarrow q)$, where both preconditions coincide. This again characterises general correctness by its separated treatment of the termination precondition. Furthermore, we can recover the special elements $0 = (1 \mid 1 \rightsquigarrow 0)$, $\mathsf{T} = (0 \mid 0 \rightsquigarrow 0)$, $\mathsf{L} = (0 \mid 1 \rightsquigarrow 0)$ and $\mathsf{H} = (1 \mid 1 \rightsquigarrow 1)$. The representation in these terms is not necessarily unique: For example, $\mathsf{T} = (0 \mid 1 \rightsquigarrow 1)$ also holds. The following two corollaries establish basic properties of our specification elements.

Corollary 9. $(r_1 \,|\, p_1 \rightsquigarrow q_1) + (r_2 \,|\, p_2 \rightsquigarrow q_2) = (r_1 r_2 \,|\, p_1 p_2 \rightsquigarrow q_1 + q_2)$. *Hence* $(\cdot \,|\, \cdot \rightsquigarrow \cdot)$ *is antitone in its first and second arguments, and isotone in its third.* *Moreover,* $q_1 \leq r_2 p_2$ *implies* $(r_1 \,|\, p_1 \rightsquigarrow q_1) \cdot (r_2 \,|\, p_2 \rightsquigarrow q_2) \leq (r_1 p_1 \,|\, p_1 \rightsquigarrow q_2)$.

Proof. Let $q_1 \leq r_2 p_2$, then

$$
\begin{aligned}
&(r_1 \,|\, p_1 \rightsquigarrow q_1) \cdot (r_2 \,|\, p_2 \rightsquigarrow q_2) \\
&= (\neg r_1 \mathsf{L} + \neg p_1 \mathsf{H} + \mathsf{H} q_1) \cdot (\neg r_2 \mathsf{L} + \neg p_2 \mathsf{H} + \mathsf{H} q_2) \\
&\leq \neg r_1 \mathsf{L} + \neg p_1 \mathsf{H} \mathsf{L} + \neg p_1 \mathsf{H} \mathsf{H} + \mathsf{H} q_1 \neg r_2 \mathsf{L} + \mathsf{H} q_1 \neg p_2 \mathsf{H} + \mathsf{H} \mathsf{H} q_2 \\
&= \neg r_1 \mathsf{L} + \neg p_1 \mathsf{L} + \neg p_1 \mathsf{H} + \mathsf{H} 0 + \mathsf{H} 0 + \mathsf{H} q_2 \\
&= \neg (r_1 p_1) \mathsf{L} + \neg p_1 \mathsf{H} + \mathsf{H} q_2 \\
&= (r_1 p_1 \,|\, p_1 \rightsquigarrow q_2)
\end{aligned}
$$

by Theorem 8 and Lemmas 1, 3 and 4. The other claims are proved similarly. □

For prescriptions, we obtain $((\top \Vdash r) \,|\, (\top \Vdash p) \rightsquigarrow (\top \Vdash q)) = (r\top \Vdash \overline{p\top} + \top q)$. Let us furthermore mention the interpretation of $(r \,|\, p \rightsquigarrow p)$ for $r \in \{0, p, 1\}$. We call such a specification an 'invariant' since it guarantees that p holds after the execution if it holds before. If $r = 0$ or $r = 1$ or $r = p$, termination is not guaranteed or always guaranteed or guaranteed if p holds, respectively.

Corollary 10. $1 \leq (r \,|\, p \rightsquigarrow p) = (r \,|\, p \rightsquigarrow p)^2$ *for* $p \in \mathrm{test}(S)$ *and* $r \in \{0, p, 1\}$.

Proof. $1 = \neg p + p \leq \neg p \mathsf{H} + \mathsf{H} p \leq (r \,|\, p \rightsquigarrow p)$ by Lemma 4 and Theorem 8. Thus,

$$(\neg p \mathsf{H} + \mathsf{H} p) \cdot (\neg p \mathsf{H} + \mathsf{H} p) \leq \neg p \mathsf{H}^2 + \mathsf{H} 0 + \mathsf{H}^2 p = \neg p \mathsf{H} + \mathsf{H} p \leq (\neg p \mathsf{H} + \mathsf{H} p)^2$$

by Lemmas 3 and 4. Moreover, $(\neg p \mathsf{H} + \mathsf{H} p) \neg r \mathsf{L} \leq \neg r \mathsf{L}$ by Lemma 1 if $r = 0$, by Lemma 3 if $r = 1$, and by both lemmas if $r = p$. Therefore,

$$
\begin{aligned}
&(r \,|\, p \rightsquigarrow p)^2 = (\neg r \mathsf{L} + \neg p \mathsf{H} + \mathsf{H} p) \cdot (\neg r \mathsf{L} + \neg p \mathsf{H} + \mathsf{H} p) \\
&= \neg r \mathsf{L} + (\neg p \mathsf{H} + \mathsf{H} p) \neg r \mathsf{L} + (\neg p \mathsf{H} + \mathsf{H} p)^2 = \neg r \mathsf{L} + \neg p \mathsf{H} + \mathsf{H} p = (r \,|\, p \rightsquigarrow p)
\end{aligned}
$$

by Theorem 8 and Lemma 1. □

The characterisation $rx0 = 0 \wedge px = pxq \Leftrightarrow x \leq (r \,|\, p \rightsquigarrow q)$ can be used to axiomatise our general correctness pre-post specifications without the use of L and H which can be added by defining them as particular specifications. While a number of properties, such as those shown in Lemmas 1 and 3, follow from this axiomatisation, the axioms (L0), (L1), (L2), (H1) and (H2) cannot be derived.

We can now use the specifications to algebraically state and prove a loop introduction rule for general correctness semantics given by [12]. Note the use of the invariant $(0 \,|\, q \rightsquigarrow q)$.

Theorem 11. *Let* $a \in S$ *and* $p, q, r \in \mathrm{test}(S)$ *such that* $pa \leq (0 \,|\, q \rightsquigarrow q)$ *and* $r \leq \Delta(pa)$. *Then* while p do $a \leq (r \,|\, q \rightsquigarrow q \neg p)$.

Proof. By (G3) it remains to show $r \,\{\text{while } p \text{ do } a\}\, 1$ and $q \,(\!\text{while } p \text{ do } a\!)\, q \neg p$. The first claim is immediate from $r \leq \Delta(pa)$ and $\Delta(pa) \,\{\text{while } p \text{ do } a\}\, 1$ derived in Section 4. The second claim follows by Theorem 7 from $pq \,(\!a\!)\, q$, which holds since $qpa \neg q \leq q(0 \,|\, q \rightsquigarrow q) \neg q \leq \mathsf{L}$ by the assumption and (G2). □

6   Recursion

In this section we generalise from loops to full recursion, an open issue of [24]. In particular, we show how to calculate least fixpoints with respect to the Egli-Milner order from fixpoints with respect to the natural order.

Throughout this section let $f : S \to S$ be isotone with respect to $\leq$ and $\sqsubseteq$, and assume that the least fixpoint μf and the greatest fixpoint νf of f with respect to $\leq$ exist. The least fixpoint of f with respect to $\sqsubseteq$ is denoted by ξf.

Theorem 12. *Let $x \in S$, then $x = \xi f \Leftrightarrow \mu f \leq x \leq \nu f \wedge x \sqsubseteq \mu f \wedge x \sqsubseteq \nu f$.*

Proof. The forward implication is immediate since ξf is the least fixpoint with respect to $\sqsubseteq$. For the backward implication, let $\mu f \leq x \leq \nu f \wedge x \sqsubseteq \mu f \wedge x \sqsubseteq \nu f$. By isotony of f we obtain $\mu f = f(\mu f) \leq f(x) \leq f(\nu f) = \nu f$ and $f(x) \sqsubseteq f(\mu f) = \mu f$ and $f(x) \sqsubseteq f(\nu f) = \nu f$. From these facts and the assumptions we obtain:

- $x \sqsubseteq f(x)$ since $x \leq \mu f + \mathsf{L} \leq f(x) + \mathsf{L}$ and $f(x) \leq \nu f \leq x + \ulcorner(x0)\urcorner\mathsf{H}$.
- $f(x) \sqsubseteq x$ since $f(x) \leq \mu f + \mathsf{L} \leq x + \mathsf{L}$ and $x \leq \nu f \leq f(x) + \ulcorner(f(x)0)\urcorner\mathsf{H}$.

Hence $x = f(x)$ by Lemma 5. Let $y \in S$ such that $y = f(y)$, hence $\mu f \leq y \leq \nu f$. Then $x \sqsubseteq y$ since $x \leq \mu f + \mathsf{L} \leq y + \mathsf{L}$ and $y \leq \nu f \leq x + \ulcorner(x0)\urcorner\mathsf{H}$.   □

As a consequence, we can give an explicit formula for ξf.

Corollary 13. *Assume ξf exists. Then $x = \xi f \Leftrightarrow x + \mathsf{L} = \mu f + \mathsf{L} \wedge x + \mathsf{H} = \nu f + \mathsf{H}$ and $\nu f \leq \mu f + \ulcorner(\nu f 0)\urcorner\mathsf{H} + \mathsf{L}$ and $\xi f = \nu f 0 + \mu f$.*

Proof. By Theorem 12 we obtain $\xi f \leq \mu f + \mathsf{L}$ and $\nu f \leq \xi f + \ulcorner(\xi f 0)\urcorner\mathsf{H} \leq \xi f + \mathsf{H}$, hence $\nu f \leq \mu f + \mathsf{L} + \ulcorner(\nu f 0)\urcorner\mathsf{H}$ using $\xi f \leq \nu f$. Let $x + \mathsf{L} = \mu f + \mathsf{L}$ and $x + \mathsf{H} = \nu f + \mathsf{H}$.

- $\mu f \leq x$ since $\mu f \leq x + \mathsf{L}$ and $\mu f \leq \nu f \leq x + \mathsf{H}$.
- $x \leq \nu f$ since $x \leq \nu f + \mathsf{H}$ and $x \leq \mu f + \mathsf{L} \leq \nu f + \mathsf{L}$.
- $\nu f \leq x + \ulcorner(x0)\urcorner\mathsf{H} + \mathsf{H}$, and $\nu f \leq x + \ulcorner(x0)\urcorner\mathsf{H} + \mathsf{L}$ since $\nu f 0 \leq (x + \mathsf{H})0 = x0$.

Hence $\mu f \leq \nu f \leq x + \ulcorner(x0)\urcorner\mathsf{H}$, yielding $x \sqsubseteq \mu f$ and $x \sqsubseteq \nu f$. The first claim follows by Theorem 12. It implies the third claim since $\nu f 0 + \mu f + \mathsf{L} = \mu f + \nu f 0 + \top 0 = \mu f + \top 0 = \mu f + \mathsf{L}$ and $\nu f + \mathsf{H} \leq \nu f 0 + \mathsf{H} \leq \nu f 0 + \mu f + \mathsf{H} \leq \nu f + \mathsf{H}$ by (H2).   □

Inspection of the proof reveals that ξf exists $\Leftrightarrow \nu f \leq \mu f + \ulcorner(\nu f 0)\urcorner\mathsf{H} + \mathsf{L}$. In particular, we prove Theorem 6 by letting $f(x) = pax + q$. Then

$$\nu f = (pa)^\omega + (pa)^* q \leq \nabla(pa)\top + \mu f \leq \mu f + \nabla(pa)\mathsf{H} + \mathsf{L} = \mu f + \ulcorner(\nu f 0)\urcorner\mathsf{H} + \mathsf{L},$$

since $\ulcorner(\nu f 0)\urcorner = \ulcorner((pa)^\omega 0 + (pa)^* q 0)\urcorner = \nabla(pa) + \ulcorner((pa)^* 0)\urcorner = \nabla(pa)$, and the least fixpoint is $\xi f = \nu f 0 + \mu f = (pa)^\omega 0 + (pa)^* q 0 + (pa)^* q = \nabla(pa)\mathsf{L} + (pa)^* q$.

We have thus established $\nu f 0 + \mu f$ as the appropriate solution to recursion in general correctness. The same term is appropriate also in partial correctness, where $\nu f 0 = 0$ vanishes. It is not appropriate in total correctness, however, since it is not equal to νf in general.

Let us finally consider the instance of prescriptions again.

Corollary 14. *Assume that ξf exists and $\mu f = (t \Vdash a)$ and $\nu f = (u \Vdash b)$. Then $u \sqcap a = u \sqcap b$ and $\xi f = (u \Vdash a)$.*

Proof. We have $u \leq t$ since $\mu f \leq \nu f$, hence $\xi f = (u \Vdash b)(\top \Vdash 0) + (t \Vdash a) = (u \Vdash 0) + (t \Vdash a) = (u \sqcap t \Vdash a) = (u \Vdash a)$ by Corollary 13. The remaining claim follows since $(u \Vdash a) \sqsubseteq (u \Vdash b)$ is equivalent to $a \leq b \wedge b \leq a + \overline{u}$.   □

A calculation shows that ξf exists $\Leftrightarrow b \leq a + \overline{u}$. It thus remains to calculate the least and greatest fixpoints for prescriptions. This can be done by the following result similar to those of [17,15] for total correctness. We omit its proof.

Proposition 15. *Let $H(t \Vdash a) = F(t \Vdash a) \Vdash G(t \Vdash a)$ be isotone with respect to $\leq$. Then $\nu H = (P_\nu(Q_\nu) \Vdash Q_\nu)$ and $\mu H = (P_\mu(Q_\mu) \Vdash Q_\mu)$, where*

$$
\begin{array}{lll}
P_\nu(a) = \mu t.F(t \Vdash a) & R_\nu(a) = G(P_\nu(a) \Vdash a) & Q_\nu = \nu R_\nu \\
P_\mu(a) = \nu t.F(t \Vdash a) & R_\mu(a) = G(P_\mu(a) \Vdash a) & Q_\mu = \mu R_\mu
\end{array}
$$

7   Conclusion

Our work shows how to treat general correctness algebraically, despite its additional complexity caused by the Egli-Milner order and the finer termination information. We have thus extended the algebraic approach already available for partial and total correctness semantics.

Future work shall further investigate the calculus and refinement, and provide operators particularly suitable for general correctness, such as the 'concert' operator of [12]. Further applications arise in the area of hybrid systems [18]. We also observe that the assumption of a weak omega algebra in Sections 3–6 is only essential for $\top$ and the results concerning while loops.

Acknowledgement. I thank the anonymous referees for their valuable remarks and helpful suggestions.

References

1. Aarts, C.J.: Galois connections presented calculationally. Master's thesis, Department of Mathematics and Computing Science, Eindhoven University of Technology (1992)
2. de Bakker, J.W.: Semantics and termination of nondeterministic recursive programs. In: Michaelson, S., Milner, R. (eds.) Automata, Languages and Programming: Third International Colloquium, pp. 435–477. Edinburgh University Press (1976)
3. Berghammer, R., Zierer, H.: Relational algebraic semantics of deterministic and nondeterministic programs. Theoretical Computer Science 43, 123–147 (1986)
4. Broy, M., Gnatz, R., Wirsing, M.: Semantics of nondeterministic and noncontinuous constructs. In: Bauer, F.L., Broy, M. (eds.) Program Construction. LNCS, vol. 69, pp. 553–592. Springer, Heidelberg (1979)

5. De Carufel, J.-L., Desharnais, J.: Demonic algebra with domain. In: Schmidt, R. (ed.) RelMiCS/AKA 2006. LNCS, vol. 4136, pp. 120–134. Springer, Heidelberg (2006)
6. Cohen, E.: Separation and reduction. In: Backhouse, R., Oliveira, J.N. (eds.) MPC 2000. LNCS, vol. 1837, pp. 45–59. Springer, Heidelberg (2000)
7. Desharnais, J., Möller, B., Struth, G.: Algebraic notions of termination. Report 2006-23, Institut für Informatik, Universität Augsburg (October 2006)
8. Desharnais, J., Möller, B., Struth, G.: Kleene algebra with domain. ACM Transactions on Computational Logic 7(4), 798–833 (2006)
9. Dijkstra, E.W.: A Discipline of Programming. Prentice Hall, Englewood Cliffs (1976)
10. Dunne, S.: Recasting Hoare and He's Unifying Theory of Programs in the context of general correctness. In: Butterfield, A., Strong, G., Pahl, C. (eds.) 5th Irish Workshop on Formal Methods. Electronic Workshops in Computing. The British Computer Society (July 2001)
11. Dunne, S., Galloway, A.: Lifting general correctness into partial correctness is *ok*. In: Davies, J., Gibbons, J. (eds.) IFM 2007. LNCS, vol. 4591, pp. 215–232. Springer, Heidelberg (2007)
12. Dunne, S., Hayes, I., Galloway, A.: Reasoning about loops in total and general correctness. In: Butterfield, A. (ed.) UTP 2008. LNCS. Springer, Heidelberg (to appear)
13. Ésik, Z., Leiß, H.: Algebraically complete semirings and Greibach normal form. Annals of Pure and Applied Logic 3(1-3), 173–203 (2005)
14. Guttmann, W., Möller, B.: Modal design algebra. In: Dunne, S., Stoddart, W. (eds.) UTP 2006. LNCS, vol. 4010, pp. 236–256. Springer, Heidelberg (2006)
15. Guttmann, W., Möller, B.: Normal design algebra. Journal of Logic and Algebraic Programming (to appear)
16. Hoare, C.A.R.: An axiomatic basis for computer programming. Communications of the ACM 12(10), 576–580/583 (1969)
17. Hoare, C.A.R., He, J.: Unifying theories of programming. Prentice Hall Europe (1998)
18. Höfner, P., Möller, B.: An algebra of hybrid systems. Journal of Logic and Algebraic Programming 78(2), 74–97 (2009)
19. Jacobs, D., Gries, D.: General correctness: A unification of partial and total correctness. Acta Informatica 22(1), 67–83 (1985)
20. Kozen, D.: A completeness theorem for Kleene algebras and the algebra of regular events. Information and Computation 110(2), 366–390 (1994)
21. Kozen, D.: On Hoare logic and Kleene algebra with tests. ACM Transactions on Computational Logic 1(1), 60–76 (2000)
22. Möller, B.: Lazy Kleene algebra. In: Kozen, D. (ed.) MPC 2004. LNCS, vol. 3125, pp. 252–273. Springer, Heidelberg (2004)
23. Möller, B.: The linear algebra of UTP. In: Uustalu, T. (ed.) MPC 2006. LNCS, vol. 4014, pp. 338–358. Springer, Heidelberg (2006)
24. Möller, B., Struth, G.: WP is WLP. In: MacCaull, W., Winter, M., Düntsch, I. (eds.) RelMiCS 2005. LNCS, vol. 3929, pp. 200–211. Springer, Heidelberg (2006)
25. Nelson, G.: A generalization of Dijkstra's calculus. ACM Transactions on Programming Languages and Systems 11(4), 517–561 (1989)
26. von Wright, J.: Towards a refinement algebra. Science of Computer Programming 51(1-2), 23–45 (2004)

Foundations of Concurrent Kleene Algebra

C.A.R. Hoare[1], Bernhard Möller[2], Georg Struth[3], and Ian Wehrman[4]

[1] Microsoft Research, Cambridge, UK
[2] Universität Augsburg, Germany
[3] University of Sheffield, UK
[4] University of Texas at Austin, USA

Abstract. A Concurrent Kleene Algebra offers two composition operators, one that stands for sequential execution and the other for concurrent execution [10]. In this paper we investigate the abstract background of this law in terms of independence relations on which a concrete trace model of the algebra is based. Moreover, we show the interdependence of the basic properties of such relations and two further laws that are essential in the application of the algebra to a Jones style rely/guarantee calculus. Finally we reconstruct the trace model in a more abstract setting based on the notion of atoms from lattice theory.

1 Introduction

A Concurrent Kleene Algebra (CKA) is one which offers two composition operators, one that stands for sequential execution and the other for concurrent execution. They are related by an inequational form of the exchange law $(a \circ b) \bullet (c \circ d) = (a \bullet c) \circ (b \bullet d)$ of two-category or bicategory theory (e.g. [16]).

The applicability of the algebra to a partially-ordered trace model of program execution semantics and to the validation familiar proof rules for sequential programs (Hoare triples) and for concurrent programming (Jones's rely/guarantee calculus) is shown in [10]. The mentioned trace model is based on a dependence relation between atomic events.

In the present paper we investigate how the laws of concurrent Kleene algebra reflect this relation; we show that two central laws are equivalent to its transitivity and acyclicity, resp. The traces obeying a generalised version of the second law are characterised in terms of convexity w.r.t the dependence relation. Moreover we introduce the notion of an event-based concurrent Kleene algebra which is a more abstract version of the concrete trace model, based on the notions of atoms and irreducible elements. We show that in such algebras the dependence relation can be recovered from the operations of sequential and concurrent composition. Most of our reasoning has been checked by computer using the system Prover9/Mace4 [17]. A collection of input files and proofs can be found under http://www.dcs.shef.ac.uk/~georg/ka/

Sect. 2 summarises the definitions of the trace model and its essential operators. In Sect. 3 we develop an abstract calculus of independence relations, both in formulas and diagrammatic rules. Sect. 4 presents quantales as a fundamental structure and gives the axiomatisation of CKAs in terms of quantales. Sect. 5

R. Berghammer et al. (Eds.): RelMiCS/AKA 2009, LNCS 5827, pp. 166–186, 2009.

gives a definition of invariants as used in the mentioned rely/guarantee calculus. In Sect. 6 we establish the equivalence of two fundamental laws with (weak) acyclicity and transitivity of the basic dependence relation. Sect. 7 presents a simplified rely/guarantee calculus. Finally, Sect. 8 develops the notion of event-based CKAS and reconstructs the trace model and the dependence relation in terms of that notion. Appendix A summarises the laws characterising the various structures involved.

2   Operations on Traces and Programs

In this section we present a concrete model of Concurrent Kleene Algebra which serves as a motivation of the abstract algebraic treatment in the later sections.

We assume a set EV of *event occurrences* together with a *dependence relation* $\rightarrow\ \subseteq EV \times EV$ between them: $e \rightarrow f$ indicates occurrence of a data flow or control flow from event e to event f.

Definition 2.1. A *trace* is a set of events; the set of all traces over EV is $TR(EV) =_{df} \mathcal{P}(EV)$. A *program* is a set of traces; the set of all programs is $PR(EV) =_{df} \mathcal{P}(TR(EV))$.

We deliberately keep the definition of traces and programs so liberal to accommodate systems with very loose coupling of events; "conventional" linear traces can, e.g., be obtained by including time stamps into the events and defining the dependence relation such that it respects time.

Examples of very simple programs are the following. The program skip, which does nothing, is defined as $\{\emptyset\}$, and the program $[e]$, which does only $e \in EV$, is $\{\{e\}\}$. The program false $=_{df} \emptyset$ has no traces, and therefore cannot be executed at all. It serves the rôle of the 'miracle' [18] in the development of programs by stepwise refinement.

Following [11] we will define four operators on programs P and Q:

$P * Q$ fine-grain concurrent composition, allowing dependences between P and Q;

$P \,;Q$ weak sequential composition, forbidding dependence of P on Q;

$P \parallel Q$ disjoint parallel composition, with no dependence in either direction;

$P \,[]\, Q$ alternation – only one of P or Q is executed, if at all; details will be given below.

To express the restrictions in this list we introduce the following independence relation.

Definition 2.2. For traces tp, tq we define the *independence relation* by

$$tp \not\leftarrow tq \Leftrightarrow_{df} \neg\exists\, p \in tp, q \in tq : q \rightarrow p .$$

Viewing tp as a set of events that should occur before all the ones in tq, one can read $tp \not\leftarrow tq$ as the requirement that tp must not depend on its "future" tq.

Now, for each operator $\circ \in \{*, ;, \|, []\}$ we define an associated binary relation $(\circ)$ between traces such that for programs P, Q we can define generically

$$P \circ Q =_{df} \{tp \cup tq \mid tp \in P \wedge tq \in Q \wedge tp\,(\circ)\,tq\} . \tag{1}$$

From this definition it is immediate that $\circ$ distributes through arbitrary unions of families of programs and hence is $\subseteq$-isotone and $false$-strict, i.e., $false \circ P = false = P \circ false$. Moreover, skip is a neutral element for $\circ$, i.e.,

$$\mathsf{skip} \circ P = P = P \circ \mathsf{skip} . \tag{2}$$

Finally, if $(\circ)$ is symmetric then $\circ$ is commutative.

Now the above informal descriptions are captured by the definitions

$$\begin{aligned}
tp\,(*)\,tq &\Leftrightarrow_{df} tp \cap tq = \emptyset , \\
tp\,(;)\,tq &\Leftrightarrow_{df} tp\,(*)\,tq \wedge tp \not\rightarrow tq , \\
tp\,(\|)\,tq &\Leftrightarrow_{df} tp\,(;)\,tq \wedge tq \not\rightarrow tp , \\
tp\,([])\,tq &\Leftrightarrow_{df} tp = \emptyset \vee tq = \emptyset .
\end{aligned}$$

It is clear that $([]) \subseteq (\|) \subseteq (;) \subseteq (*)$ and that $(*)$, $(\|)$ and $([])$ are symmetric.

Another essential operator is the union operator which again is $\subseteq$-isotone and distributes through arbitrary unions. However, it is *not* $false$-strict.

By the Tarski-Kleene fixpoint theorems hence all recursion equations involving only the operations mentioned have $\subseteq$-least solutions which can be approximated by the familiar fixpoint iteration starting from $false$. Use of union in such a recursion enables non-trivial least fixpoints.

3 Independence Calculus and Exchange Laws

To prove the most essential laws about the interaction of our operators we now give a slightly more abstract treatment. We start by observing that an equivalent relation-algebraic formulation of the independence relation $\not\rightarrow$ is $tp \not\rightarrow tq \Leftrightarrow tp \times tq \cap \rightarrow^{\smile} = \emptyset$, where $\rightarrow^{\smile}$ is the converse of $\rightarrow$. By straightforward set theory this entails

$$\begin{aligned}
(tp \cup tq) \not\rightarrow tr &\Leftrightarrow tp \not\rightarrow tr \wedge tq \not\rightarrow tr , \\
tp \not\rightarrow (tq \cup tr) &\Leftrightarrow tp \not\rightarrow tq \wedge tp \not\rightarrow tr .
\end{aligned}$$

It turns out that these bilinearity properties are the essence of the characteristic laws about the interplay of our various operators. This motivates the following definition.

Definition 3.1. An *aggregation algebra* is a structure $(A, +)$ consisting of a set A and a binary operation $+ : A \times A \rightarrow A$. An *independence relation* on an aggregation algebra is a bilinear relation $R \subseteq A \times A$, i.e.,

$$\begin{aligned}
(p + q)\,R\,r &\Leftrightarrow p\,R\,r \wedge q\,R\,r , \\
p\,R\,(q + r) &\Leftrightarrow p\,R\,q \wedge p\,R\,r .
\end{aligned}$$

In our example of traces and programs, A would be the set of traces and $+$ would be trace union. For now, we will however consider an aggregation algebra as absolutely free, i.e., $+$ need not satisfy any laws. Later we will need aggregation algebras that are (commutative) semigroups or monoids. The first condition on R says that an aggregate $p + q$ is independent of r iff both its parts p and q are independent of r. The second condition says that p is independent of the aggregate $q + r$ iff it is independent of both its parts, q and r. An independence relation on $(TR(EV), \cup)$ is $\not\vdash$.

We can visualise the independence conditions by the following diagrams.

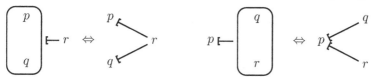

The ovals display aggregates. The letters in the ovals represent the entities that form their parts. In the first diagram, the oval around p and q denotes the aggregate $p + q$. The arrow-like symbols denote the independence relation R, where the sign $\vdash$ means that any flow of dependence is blocked there. In the leftmost diagram, the arrow relates the aggregate $p + q$ to r, hence the aggregate formed by p and q is independent of r. In its neighbour diagram, there is no aggregate and both p and q are related to r, hence independent of r. The equivalence between the diagrams visualises the first bilinearity law. Analogous remarks apply to the second pair of diagrams.

An important tool for a uniform treatment of our operators from Sect. 2 is the following lemma, whose proof is straightforward.

Lemma 3.2

1. *The set of independence relations on an aggregation algebra is closed under intersection.*
2. *The relations* $(*), (;), (\|)$ *and* $(\|)$ *are independence relations on* $(TR(EV), \cup)$.

We now consider the interplay of two independence relations R and S on an aggregation algebra A.

Lemma 3.3. *Let R and S be independence relations on an aggregation algebra $(A, +)$ such that $R \subseteq S$. Then*

1. $(p + q)\, R\, r \wedge p\, S\, q \;\Rightarrow\; p\, S\, (q + r) \wedge q\, R\, r.$
2. $p\, R\, (q + r) \wedge q\, S\, r \;\Rightarrow\; (p + q)\, S\, r \wedge p\, R\, q.$

Proof.

$$(p + q)\, R\, r \wedge p\, S\, q \;\Leftrightarrow\; p\, R\, r \wedge q\, R\, r \wedge p\, S\, q$$
$$\Rightarrow\; p\, S\, r \wedge q\, R\, r \wedge p\, S\, q$$
$$\Leftrightarrow\; q\, R\, r \wedge p\, S\, (q + r).$$

$$p R (q + r) \wedge q S r \Leftrightarrow p R q \wedge p R r \wedge q S r$$
$$\Rightarrow p R q \wedge p S r \wedge q S r$$
$$\Leftrightarrow p R q \wedge (p + q) S r. \qquad \square$$

Obviously, we now must introduce two different kinds of arrows in diagrams. The first law can be visualised as

The second law looks similar. A diagrammatic proof of the first law is

A simple consequence is the following.

Corollary 3.4. *Let R be an independence relation on an aggregation algebra $(A, +)$. Then*

$$(p + q) R r \wedge p R q \Leftrightarrow q R r \wedge p R (q + r).$$

Proof. Set $S = R$ in Lemma 3.3. $\qquad \square$

We can also prove the following *exchange laws* which are crucial for concurrent Kleene algebra.

Theorem 3.5. *Let R and S be independence relations on an aggregation algebra $(A, +)$ such that $R \subseteq S$ and S is symmetric. Then*

$$(p + q) R (r + s) \wedge p S q \wedge r S s \Rightarrow p R r \wedge q R s \wedge (p + r) S (q + s) .$$

Proof.

$$(p + q) R (r + s) \wedge p S q \wedge r S s$$
$$\Leftrightarrow p R r \wedge q R r \wedge p R s \wedge q R s \wedge p S q \wedge r S s$$
$$\Rightarrow p R r \wedge q S r \wedge p S s \wedge q R s \wedge p S q \wedge r S s$$
$$\Rightarrow p R r \wedge q R s \wedge r S q \wedge (p + r) S (s) \wedge p S q$$
$$\Rightarrow p R r \wedge q R s \wedge (p + r) S (q) \wedge (p + r) S (s)$$
$$\Leftrightarrow p R r \wedge q R s \wedge (p + r) S (q + s) . \qquad \square$$

The diagrammatic statement of the exchange law (neglecting hypotheses) is

A diagrammatic proof is

As immediately obvious from the diagrams, the hypotheses also entail

$$(p+q)\,R\,(r+s) \wedge p\,S\,q \wedge r\,S\,s \Rightarrow p\,R\,s \wedge q\,R\,r \wedge (p+s)\,S\,(q+r)\,. \quad (3)$$

The proofs in this section, logical and diagrammatical, are only intended to give a flavour of the approach. In fact, the former ones have all been automated, hence formally verified, with Prover9 [17] and could as well have been omitted. Proofs at this level of complexity present no obstacle to ATP systems.

We now apply our results to our special aggregation algebra $(TR(EV), \cup)$.

Lemma 3.6. *Let* $\circ, \bullet \in \{*, ;, \|, \, []\, \}$.

1. *If* $(\bullet) \subseteq (\circ)$ *and* $(\circ)$ *is symmetric then* $(P \circ Q) \bullet (R \circ S) \subseteq (P \bullet R) \circ (Q \bullet S)$.
2. *If* $(\bullet) \subseteq (\circ)$ *then* $(P \circ Q) \bullet R \subseteq P \circ (Q \bullet R)$ *and* $P \bullet (Q \circ R) \subseteq (P \bullet Q) \circ R$.
3. $\circ$ *is associative.*

Proof.

1. For traces $tp \in P, tq \in Q, tr \in R, ts \in S$ we have by (1) and Theorem 3.5

$$\begin{aligned}
& (tp \cup tq) \cup (tr \cup ts) \in (P \circ Q) \bullet (R \circ S) \\
\Leftrightarrow\ & tp(\circ)tq \wedge tr(\circ)ts \wedge (tp \cup tq))(\bullet)(tr \cup ts) \\
\Rightarrow\ & tp(\bullet)tr \wedge tq(\bullet)ts \wedge (tp \cup tr))(\circ)(tq \cup ts) \\
\Leftrightarrow\ & (tp \cup tr) \cup (tq \cup ts) \in (P \bullet R) \circ (Q \bullet S)\,.
\end{aligned}$$

Since $(tp \cup tq) \cup (tr \cup ts) = (tp \cup tr) \cup (tq \cup ts)$, we are done.
2. Similar to the proof of Part 1, using Lemma 3.3.
3. Use the two previous laws with $\bullet = \circ$. $\qquad \square$

A particularly important special case of Part 1 is the exchange law

$$(P * Q)\,;\,(R * S) \subseteq (P\,;\,R) * (Q\,;\,S)\,. \quad (4)$$

In the remainder of this paper we shall no longer deal with the less interesting operators $\|$ and $[]$.

4   Quantales and Concurrent Kleene Algebras

We now abstract further from the concrete example of traces and programs.

Definition 4.1. A *semiring* is a structure $(S, +, 0, \cdot, 1)$ such that $(S, +, 0)$ is a commutative monoid, $(S, \cdot, 1)$ is a monoid, multiplication distributes over addition in both arguments and 0 is a left and right annihilator with respect to multiplication $(a \cdot 0 = 0 = 0 \cdot a)$. A semiring is *idempotent* if its addition is.

In an idempotent semiring, the relation $\leq$ defined by $a \leq b \Leftrightarrow_{df} a + b = b$ is a partial ordering, in fact the only partial ordering on S for which 0 is the least element and for which addition and multiplication are isotone in both arguments. It is therefore called the *natural ordering* on S. This makes S into a semilattice with addition as join and least element 0.

Definition 4.2. A *quantale* [19] or *standard Kleene algebra* [5] is an idempotent semiring that is a complete lattice under the natural order and in which composition distributes over arbitrary suprema. The infimum and the supremum of a subset T are denoted by $\sqcap T$ and $\sqcup T$, respectively. Their binary variants are $x \sqcap y$ and $x \sqcup y$ (the latter coinciding with $x + y$).

Let now $PR(EV)$ denote the set of all programs over the event set EV. From the observations in Sect. 2 the following is immediate:

Lemma 4.3. $(PR(EV), \cup, false, *, \mathsf{skip})$ *and* $(PR(EV), \cup, false, ;, \mathsf{skip})$ *both are quantales.*

In a quantale S, finite and infinite iteration * and $^\omega$ are defined by

$$a^* = \mu x \,.\, 1 + a \cdot x \,, \qquad\qquad a^\omega = \nu x \,.\, a \cdot x \,,$$

where μ and ν denote the least and greatest fixpoint operators. The star used here is not to be confused with the separation operator $*$ above; it should also be noted that a^ω in [1] corresponds to $a^* + a^\omega$ in the quantale setting.

It is well known that then $(S, +, \cdot, 0, 1, ^*)$ forms a Kleene algebra [14]. From this we obtain many useful laws for free. As instances we mention

$$a^* \cdot a^* = (a^*)^* = a^* \,, \qquad (a \cdot b)^* \cdot a = a \cdot (b \cdot a)^* \,, \qquad (a + b)^* = a^* \cdot (b \cdot a^*)^* \,.$$

Since in a quantale the function defining star is continuous, Kleene's fixpoint theorem shows that $a^* = \bigsqcup_{i \in \mathbb{N}} a^i$. Moreover, we have the star induction rules

$$b + a \cdot x \leq x \Rightarrow a^* \cdot b \leq x \,, \qquad b + x \cdot a \leq x \Rightarrow b \cdot a^* \leq x \,. \tag{5}$$

Hence in $(PR(EV), \cup, false, *, \mathsf{skip})$ and $(PR(EV), \cup, false, ;, \mathsf{skip})$ the program P^* consists of all finite disjoint unions and all finite sequential compositions of traces in P, resp. In the latter case P^* is denoted by P^∞ in [11].

If, in addition, the complete lattice $(S, \leq)$ in a quantale is completely distributive, i.e., if $+$ distributes over arbitrary infima, then $(S, +, \cdot, 0, 1, *, {}^\omega)$ forms an omega algebra [4]. Again this entails many useful laws, e.g.,

$$(a \cdot b)^\omega = a \cdot (b \cdot a)^\omega , \qquad (a + b)^\omega = a^\omega + a^* \cdot b \cdot (a + b)^\omega .$$

Since $PR(EV)$ is a power set lattice, it is completely distributive. Hence both program quantales also admit infinite iteration with all its laws. The infinite iteration ${}^\omega$ in $(PR(EV), \cup, false, *, \mathsf{skip})$ is similar to the unbounded parallel spawning $!P$ in the π-calculus [22]. The abstract combination of two quantales leads to the following definition [10].

Definition 4.4. A *concurrent Kleene algebra (CKA)* is a structure $(S, +, 0, *, ;, 1)$ such that $(S, +, *, 0, 1)$ and $(S, +, ;, 0, 1)$ are quantales linked by the exchange axiom

$$(a * b) ; (c * d) \leq (b ; c) * (a ; d) .$$

Compared with the original exchange law (4) this one has its free variables in a different order. This does no harm, since the concrete $*$ operator on programs is commutative and hence satisfies the above law as well. Hence we have

Corollary 4.5. $(PR(EV), \cup, false, *, ;, \mathsf{skip})$ *is a CKA.*

The reason for our formulation of the exchange axiom here is that this form of the law implies commutativity of $*$ as well as $a ; b \leq a * b$ and hence saves two axioms. We list some important consequences of this axiomatisation; the proofs are given in [10].

Lemma 4.6. *In a CKA the following laws hold.*

1. $a * b = b * a$.
2. $(a * b) ; (c * d) \leq (a ; c) * (b ; d)$.
3. $a ; b \leq a * b$.
4. $(a * b) ; c \leq a * (b ; c)$.
5. $a ; (b * c) \leq (a ; b) * c$.

5 Invariants

For the further development we need to deal with the set of events a program may use.

Definition 5.1. A *power invariant* is a program R of the form $R = \mathcal{P}(E)$ for a set $E \subseteq EV$ of events.

It consists of all possible traces that can be formed from events in E and hence is the most general program using only those events. The smallest power invariant is $\mathsf{skip} = \mathcal{P}(\emptyset)$. The term "invariant" expresses that often a program relies (whence the name R) on the assumption that its environment only uses events from a particular subset, i.e., preserves the invariant of staying in that set. We will now investigate the properties of power invariants and, later on, of their abstract counterparts.

To this end we want to define a function that forms from a program the smallest power invariant containing it. We denote, for a program P, by $|P| =_{df} \bigcup P$ the set of all events occurring in traces of P; when convenient, $|P|$ can also be considered as a trace.

It is straightforward to check that $|\_|$ distributes through arbitrary unions. Hence it has an upper adjoint F, defined by the Galois connection

$$P \subseteq F(X) \Leftrightarrow |P| \subseteq X .$$

This entails $F(X) = \mathcal{P}(X)$ and $|\mathcal{P}(X)| = X$. Moreover, as adjoints of a Galois connection, $\mathcal{P}(\_)$ and $|\_|$ are $\subseteq$-isotone. Setting $X = |P|$ we obtain $P \subseteq \mathcal{P}(|P|)$. Finally, for $X, Y \subseteq EV$ we have $\mathcal{P}(X) \subseteq \mathcal{P}(Y) \Leftrightarrow X \subseteq Y$.

Motivated by the above remarks we now define $\mathsf{INV}(P) =_{df} \mathcal{P}(|P|)$. Then $\mathsf{INV}(P)$ is the most general program that can be formed from the events of P. As a composition of isotone functions, INV is isotone again.

We now prepare for our abstract notion of invariants. An *invariant* is a program R with $R = \mathsf{INV}(R)$. In particular, every invariant in our concrete CKA of programs is a power invariant. In general CKAs we will replace INV by a suitable abstract operator the properties of which will be discussed below. The definition implies that invariants are fixpoints of an isotone function and hence form a complete lattice under the inclusion order.

The operation ∇ from [11] and INV are interrelated. To this end we set $\mathsf{SINGLES}(P) =_{df} \{\{e\} : \{e\} \in P\}$. Then

$$\mathsf{INV}(\mathsf{SINGLES}(Q)) = Q\nabla Q , \qquad Q\nabla R = \mathsf{INV}(\mathsf{SINGLES}(Q \cup R)) .$$

We shall use INV, since it leads to simpler and more intuitive formulations.

We give a few useful properties of INV.

Theorem 5.2

1. $\mathsf{INV}(P)$ *is the smallest invariant containing P.*
2. $\mathsf{INV}(\mathsf{INV}(P)) = \mathsf{INV}(P)$*; hence* $\mathsf{INV}(P)$ *is an invariant.*
3. INV *is a closure operator.*
4. $\mathsf{skip} \subseteq \mathsf{INV}(P)$.
5. $\mathsf{INV}(P * Q) \subseteq \mathsf{INV}(P \cup Q)$.

Proof.

1. We have already seen above that $P \subseteq \mathsf{INV}(P)$. Let S be another invariant with $P \subseteq S$. Then, by isotony of INV and the definition of invariants, $\mathsf{INV}(P) \subseteq \mathsf{INV}(S) = S$.
2. Since, as remarked above, $|\mathcal{P}(X)| = X$, we have

$$\mathsf{INV}(\mathsf{INV}(P)) = \mathcal{P}(|\mathcal{P}(|P|)|) = \mathcal{P}(|P|) = \mathsf{INV}(P) .$$

3. By Part 1 INV is extensive. By the Galois connection it is isotone and by Part 2 it is idempotent.

4. Immediate from the definition of INV.
5. By the definition of $*$ we have $|P * Q| \subseteq |P \cup Q|$ and the property follows by isotony of $\mathcal{P}$.   □

Since INV is a closure operator we have the following (see e.g. [3]).

Corollary 5.3. *For a set $\mathcal{R}$ of power invariants, $\bigcap \mathcal{R}$ and $INV(\bigcup \mathcal{R})$ are the meet and join of $\mathcal{R}$ in the complete lattice of invariants, resp.*

We now again abstract from the concrete case of programs and use the fact that INV is a closure and Parts 4 and 5 of Theorem 5.2 as the characteristics of abstract invariants, since these properties suffice to prove the results about the rely/guarantee calculus in Section 7 we are after.

Definition 5.4. A *CKA with invariants* is a structure $(S, +, 0, *, ;, 1, \iota)$ such that $(S, +, 0, *, ;, 1)$ is a CKA and $\iota : S \rightarrow S$ is a closure operator that additionally satisfies, for all $a, b \in S$,

$$1 \leq \iota\, a\,, \qquad\qquad \iota\,(a * b) \leq \iota\,(a + b)\,.$$

An *invariant* is an element $a \in S$ with $\iota\, a = a$.

In [10] a more specific view of invariants is taken: there an invariant is an element r with $1 \leq r$ and $r * r \leq r$, which are the properties of invariants shown in Theorem 5.6 below. This entails that the invariants are precisely the fixpoints of the finite iteration operator $*$ w.r.t. concurrent composition $*$. This still allows proving many of the above properties, but does not characterise power invariants and hence is not adequate for all purposes. However, we have the following connection.

Lemma 5.5. *Defining in a CKA $\iota\, a =_{df} a^*$ makes it a CKA with invariants.*

Proof. By standard Kleene algebra $*$ is a closure operator with $1 \leq a^*$. The remaining axiom is shown by star induction (5) and isotony as follows:

$$(a * b)^* \leq (a + b)^* \Leftarrow 1 + a * b * (a + b)^* \leq (a + b)^* \Leftarrow$$
$$1 \leq (a + b)^* \wedge (a + b) * (a + b) * (a + b)^* \leq (a + b)^* \Leftrightarrow \text{TRUE}\,.   □$$

Again it is clear that the invariants in the abstract sense form a complete lattice with properties analogous to those of Corollary 5.3. Moreover, one has the usual Galois connection for closures (e.g. [7]):

$$a \leq \iota\, b \Leftrightarrow \iota\, a \leq \iota\, b\,. \tag{6}$$

With this definition we can give a uniform abstract proof of idempotence of operators on invariants.

Theorem 5.6. *Consider a CKA S with invariants. Let $\circ$ be an isotone binary operation on S that has 1 as neutral element and satisfies $\forall a, b : \iota\,(a \circ b) \subseteq \iota\,(a + b)$. Then for invariant r we have $r \circ r = r$.*

Proof. We first show $r \circ r \leq r$. By extensivity of ι, the assumption and $r + r = r$ as well as invaraince of r we have $r \circ r \subseteq \iota\,(r \circ r) \subseteq \iota\,r = r$. The converse inclusion is shown by $r = r \circ 1 \leq r \circ r$, using neutrality of 1, the axiom $1 \leq \iota\,a$ and isotony of $\circ$. □

The original motivation for discussing invariants was that they should allow guaranteeing that a program only uses events from a given admissible set. To this end we define a *guarantee relation*, slightly more liberally than [11], by

$$a \text{ guar } b \Leftrightarrow_{df} \iota\,a \leq \iota\,b \,.$$

Since ι as a closure is extensive, isotone and idempotent, the right hand side is equivalent to $a \leq \iota\,b$. If b is an invariant, i.e., $b = \iota\,b$, we obtain by (6)

$$a \text{ guar } b \Leftrightarrow \iota\,a \leq \iota\,b \Leftrightarrow a \leq \iota\,b \Leftrightarrow a \leq b \,.$$

We have the following properties.

Theorem 5.7

1. *If g is an invariant then 1 guar g.*
2. *If g, g' are invariants and $\circ$ is again an isotone binary operation satisfying*
   $\forall a, b : \iota\,(a \circ b) \leq \iota\,(a + b)$ *then*

$$b \text{ guar } g \;\wedge\; b' \text{ guar } g' \;\Rightarrow\; (b \circ b') \text{ guar } (g + g') \,.$$

3. *For the concrete case of programs, $[e]$ guar $G \Leftrightarrow e \in |G|$.*

Proof.

1. Immediate from the axioms and the above remark on guar.
2.
$$b \text{ guar } g \wedge b' \text{ guar } g'$$
$$\Leftrightarrow \quad \{\!| \text{ above remark on guar } |\!\}$$
$$\iota\,b \leq g \wedge \iota\,b' \leq g'$$
$$\Rightarrow \quad \{\!| \text{ isotony of } + |\!\}$$
$$\iota\,b + \iota\,b' \leq g + g'$$
$$\Rightarrow \quad \{\!| \text{ isotony of } \iota |\!\}$$
$$\iota\,(b + b') \leq g + g'$$
$$\Rightarrow \quad \{\!| \text{ assumption about } \circ |\!\}$$
$$\iota\,(b \circ b') \leq g + g'$$
$$\Leftrightarrow \quad \{\!| \text{ extensivity of } \iota |\!\}$$
$$\iota\,(b \circ b') \leq \iota\,(g + g')$$
$$\Leftrightarrow \quad \{\!| \text{ definition } |\!\}$$
$$(b \circ b') \text{ guar } (g + g') \,.$$

3. By the definitions and the Galois connection for $|\_|$,

$$[e] \text{ guar } G \Leftrightarrow \mathsf{INV}([e]) \subseteq \mathsf{INV}(G) \Leftrightarrow \{e\} \subseteq \mathsf{INV}(G) \Leftrightarrow e \in |G| \,. \qquad □$$

6   Characterising Dependence

In [10] it is shown that the definitions of $*$ and ; for concrete programs in terms of the transitive closure $\to^+$ of the dependence relation $\to$ entail two important further laws that are essential for the rely/guarantee calculus to be presented below:

Theorem 6.1. *Let* $R = \mathcal{P}(E)$ *be a power invariant in* $PR(EV)$.

1. *If* $\to$ *is acyclic and* $e \in EV$ *then*

$$R * [e] \subseteq R \,; [e] \,; R \ .$$

2. *If* $\to$ *is transitive then for all* $P, Q \in PR(EV)$ *we have*

$$R * (P \,; Q) \subseteq (R * P) \,; (R * Q) \ .$$

This means that the two properties of Theorem 6.1 hold if $\to$ is a strict-order. But in fact, in a sense also the reverse implication holds. To formulate it we need a further notion.

Definition 6.2. *We call* $\to$ *weakly acyclic if for all events* e, f,

$$e \to^+ f \to^+ e \ \Rightarrow \ f = e \ ,$$

and *weakly transitive if*

$$e \to f \to g \ \Rightarrow \ (e = g \lor e \to g) \ .$$

Weak acyclicity means that $\to$ may at most have immediate self-loops (which cannot be "detected" by the ; operator, since it is defined in terms of distinct events only).

Theorem 6.3. *Let* $[e]$ *be again the single-event program* $\{\{e\}\}$.

1. *If* $R * [e] \subseteq R \,; [e] \,; R$ *is valid for all power invariants* R *and events* e *then* $\to$ *is weakly acyclic.*
2. *If* $R * (P \,; Q) \subseteq (R * P) \,; (R * Q)$ *is valid for all power invariants* R *and programs* P, Q *then* $\to$ *is weakly transitive.*

Proof of Part 2.
Assume events p, q, r with $q \to r$ and $r \to p$ but $q \not\to p$. This implies $q \neq r$ and $r \neq p$. Assume now $p \neq q$ and set $P =_{df} [p], Q =_{df} [q]$ and $R =_{df} [\,] \cup [r]$. Then $P \,; Q = [p, q]$ and $R * (P \,; Q) = [p, q] \cup [r, p, q]$. Moreover, $R * P = [p] \cup [r, p]$ and $R * Q = [q] \cup [r, q]$, hence $(R * P) \,; (R * Q) = [p, q]$ contradicting the assumed property. Therefore we must have $p \leftarrow q$.　　　　□

We abstract this into the following

Definition 6.4. *A CKA* S *with invariants is* $*$-distributive *if all invariants* r *and all* $a, b \in S$ *satisfy*

$$r * (a \,; b) \leq (r * a) \,; (r * b) \ .$$

We still have to prove Part 1. Rather than doing this directly we investigate a slightly more general property which is equivalent to an interesting property of traces more general than single-event ones.

Definition 6.5. A trace tp is *convex* if for all events $p, q \in tp$ and arbitrary event f we have

$$p \rightarrow^+ f \rightarrow^+ q \Rightarrow f \in tp \ .$$

A convex trace can be considered as "closed" under dependence.

For the following lemma we introduce the auxiliary function $dep(tp) =_{df} \{q \mid \exists p \in tp : q \rightarrow^+ p\}$ on traces tp. Hence $dep(tp)$ consists of all events on which some event of tp depends. Then we have

Lemma 6.6. Let tp be a trace and assume that $R * \{tp\} \subseteq R \ ; \{tp\} \ ; R$ holds for all power invariants R.

1. Dependence between a trace and any event outside occurs at most in one direction, i.e., for any event $f \notin tp$ we have

$$tp \cap dep(\{f\}) = \emptyset \ \vee \ \{f\} \cap dep(tp) = \emptyset \ .$$

2. As a consequence, tp is convex.

Proof.

1. Set $R =_{df} \mathcal{P}(\{f\})$. By assumption the trace $tr = \{f\} \in R$ can be split as $tr = tr' \ ; tr''$ such that $tr * tp = tr' \ ; tp \ ; tr''$.
   Case 1: $tr' = \{f\} \wedge tr'' = \emptyset$. Hence $tr * tp = \{f\} \ ; tp$. This implies $\{f\} \cap dep(tp) = \emptyset$.
   Case 2: $tr' = \emptyset \wedge tr'' = \{f\}$. Hence $tr * tp = tp \ ; \{f\}$. This implies $tp \cap dep(\{f\}) = \emptyset$.
2. Suppose $f \notin tp$. The premise $p \rightarrow^+ f$ implies $p \in tp \cap dep(\{f\})$ while $f \rightarrow^+ q$ implies $f \in \{f\} \cap dep(tp)$. In particular, both sets are non-empty, contradicting Part 1. $\qquad\square$

The case of singleton traces is covered as follows:

Lemma 6.7. All traces $\{e\}$ are convex iff $\rightarrow$ is weakly acyclic.

Proof. ($\Rightarrow$) Assume $e \rightarrow^+ f \rightarrow^+ e$. Then by the assumed convexity of $\{e\}$ we get $f \in \{e\}$, i.e., $f = e$.

($\Leftarrow$) Assume $p \rightarrow^+ f \rightarrow^+ q$ for $p, q \in \{e\}$, i.e., $e \rightarrow^+ f \rightarrow^+ e$. Then by the assumed weak acyclicity $f = e$, i.e., $f \in \{e\}$. $\qquad\square$

We now want to show that also the reverse of Lemma 6.6 holds.

Lemma 6.8. Let tp be convex. Then for all power invariants R the formula $R * \{tp\} \subseteq R \ ; \{tp\} \ ; R$ is valid.

Proof. Consider some $tr \in R$. We need to show $\{tr\} * \{tp\} \subseteq R \,;\, \{tp\} \,;\, R$. The claim holds vacuously if $tp \cap tr \neq \emptyset$. Hence assume that $tp \cap tr = \emptyset$ and set

$$tr' =_{df} tr \cap dep(tp) \,, \qquad tr'' =_{df} tr - dep(tp) \,.$$

In particular, $tp \cap tr' = \emptyset$. From Lemma 6.3 of [10] we know

$$tr'' \cap dep(tp) = tr'' \cap dep(tr') = \emptyset \,.$$

If we can show that also $tp \cap dep(tr') = \emptyset$ we have $\{tr\} * \{tp\} = \{tr'\} \,;\, \{tp\} \,;\, \{tr''\}$ and are done. Therefore, suppose $p \in tp \cap dep(tr')$, say $p \rightarrow^+ r$ for some $r \in tr'$. By definition of tr' there is a $q \in tp$ with $r \rightarrow^+ q$. Since tp is assumed to be convex, this implies $r \in tp$, a contradiction to $r \in tr'$ and $tp \cap tr' = \emptyset$.   □

Next, we consider general programs.

Definition 6.9. A program is *convex* if all its traces are.

Lemma 6.10. *P is convex iff it satisfies for all power invariants R*

$$R * P \subseteq R \,;\, P \,;\, R \,.$$

Proof. ($\Rightarrow$) Immediate from the definition and Lemma 6.8.

($\Leftarrow$) Consider traces $tp \in P$ and $tr \in R$. We need to show $\{tr\} * \{tp\} \subseteq R;\{tp\};R$. The claim holds vacuously if $tp \cap tr \neq \emptyset$. Hence let $tp \cap tr = \emptyset$. By the assumption there are traces $tp' \in P$ and $tr', tr'' \in tr$ with $tp' \cap tr' = tp' \cap tr'' = tr' \cap tr'' = \emptyset$ and $tr' \not\leftarrow tp' \wedge tp' \not\leftarrow tr'' \wedge tr' \not\leftarrow tr''$ such that $tp \cup tr = tr' \cup tp' \cup tr''$. But by disjointness this implies $tp' = tp$ and we are done.   □

These results motivate the following abstraction.

Definition 6.11. An element a of a CKA with invariants is called *convex* iff for all invariants r we have $r * a \leq r \,;\, a \,;\, r$.

By $b \,;\, c \leq b * c$, commutativity of $*$ and idempotence of invariants (Theorem 5.6) this inequation strengthens to an equality. This means that convex elements behave like "atoms" w.r.t. sequentialisation. Convexity will be important for one of the rules presented in the next section.

7   A Simplified Rely/Guarantee-Calculus

In [13] Jones has presented a calculus that considers properties of the environment on which a program wants to rely and the ones it, in turn, guarantees for the environment. The basis of this calculus are quintuples of the form $P\,R\,\{Q\}\,S\,G$ which express that after a pre-history modelled by program P the program Q when run in concurrent composition with a program satisfying the invariant R will achieve overall history S and guarantee invariant G. In general CKAs these quintuples can be formalised by

$$a\,r\,\{b\}\,s\,g \quad \Leftrightarrow_{df} \quad a\,\{r * b\}\,s \;\wedge\; b\,\mathsf{guar}\,g$$

when r and g are invariants. They are based on the following Hoare triples:

$$c\{d\}e \Leftrightarrow_{df} c;d \le e .$$

In [10] it is shown that all the standard rules for Hoare triples also hold for this abstract version.

However, in the setting of the present paper the following type of quadruples with an invariant r works just as well:

$$a\ r\{b\}s \Leftrightarrow_{df} a\{r*b\}s .$$

If information about the events of a program b is needed (the rôle of g in the original quintuples of the Jones calculus is, to a certain extent, to carry this information), one can use the smallest invariant ιb containing b, since b guar ιb.

We give the simplified versions of the original rely/guarantee-properties; the proofs result in a straightforward way from the ones shown in [10] by omitting the guarantee parts.

For concurrent composition we obtain

Theorem 7.1. *For invariants r, r',*

$$a\ r\{b\}s \ \wedge \ a'\ r'\{b'\}s' \ \wedge \ b' \le r \ \wedge b \le r' \ \Rightarrow \\ (a \sqcap a')\ (r \sqcap r')\{b*b'\}(s \sqcap s') .$$

For sequential composition one has

Theorem 7.2. *Let, for CKA elements a, b, denote $a \sqcap b$ their meet (which exists, since CKAs are quantales). For invariants r, r',*

$$a\ r\{b\}s \wedge s\ r'\{b'\}s' \Rightarrow a\ (r \sqcap r')\{b;b'\}s'$$

provided $b;b'$ is protected from $r \sqcap r'$, i.e.,

$$(r \sqcap r') * (b;b') \le (r*b);(r'*b') .$$

The protectedness assumption holds in particular if the underlying CKA is *-distributive, since $r \sqcap r'$ is again an invariant and hence

$$(r \sqcap r') * (b;b') \le ((r \sqcap r') * b);(r \sqcap r') * b') \le (r*b);(r'*b') .$$

Next we give rules for 1, union and convex programs.

Theorem 7.3

1. $a\ r\{1\}s \ \Leftrightarrow \ a\{r\}s.$
2. $a\ r\{b+b'\}s \ \Leftrightarrow \ a\ r\{b\}s \wedge a\ r\{b'\}s.$
3. If b is convex then $a\ r\{b\}s \ \Leftrightarrow \ a\{r;b;r\}s.$

Part 3 has only been given for concrete single-event programs in [10]; therefore we give a quick proof for the abstract form here:

$$a\ r\{b\}s \Leftrightarrow a;(r*b) \subseteq s \Leftrightarrow a;(r;b;r) \subseteq s \Leftrightarrow a\{r;b;r\}s . \qquad \square$$

8   Event-Based Algebras

The definition of a CKA does not mention the dependence relation anymore. However, in the next section, when we establish a sufficient condition for protectedness, we shall need it, even at the level of single events. Therefore we now give algebraic characterisations of traces and events.

Throughout this section we assume a CKA S with $1 \neq 0$. A *subatom* is an element a such that $b \leq a \Rightarrow b = 0 \vee b = a$. A subatom different from 0 is called an *atom*.

Definition 8.1. An element $t \in S$ is called a *trace* if it is a subatom and join-prime, i.e., if

$$\forall a \in S : a \leq t \Rightarrow a = 0 \vee a = t ,$$
$$\forall T \subseteq S : T \neq \emptyset \wedge t \leq \bigsqcup T \Rightarrow \exists a \in T : t \leq a .$$

The set of all traces is denoted by $TR(S)$. For b in S the set of traces of b is

$$TR(b) =_{df} \{a \in TR(S) \,|\, a \leq b\} .$$

By this definition 0 is a trace. The traces different from 0 would be called atoms in lattice theory (e.g. [3]). Admitting also 0 as a trace saves a number of case distinctions. It is immediate that every trace a is $+$-irreducible, i.e.,

$$a = b + c \Rightarrow b = a \vee c = a .$$

Moreover, if a is a trace and $b \leq a$ then b is a trace again. In particular, if $a * b$ is a trace then by Lemma 4.6(3) also $a \,;\, b$ is a trace.

In our concrete model the abstract traces different from 0 correspond to singleton programs.

Definition 8.2. In a CKA S we define a relation $\sqsubseteq$ by

$$a \sqsubseteq b \Leftrightarrow_{df} \exists c : b = a * c .$$

To investigate its properties we need

Definition 8.3. A subset $E \subseteq S$ is *well behaved* if the following conditions hold (for $a, b, c \in E$):
  (a) $1 \in E$.
  (b) $E * E \subseteq E$.
  (c) $*$ is cancellative on E, i.e., $a * b \neq 0 \wedge a * b = a * c \Rightarrow b = c$.
  (d) 1 is $*$-irreducible in E, i.e., $1 = a * b \Rightarrow a = 1 \vee b = 1$.

Lemma 8.4

1. $\sqsubseteq$ is a preorder, i.e., reflexive and transitive.

Assume now that $E \subseteq S$ is well behaved. Then we have the following additional properties.

2. $\sqsubseteq$ *is antisymmetric on* E.
3. 1 *is the* $\sqsubseteq$-*least element of* E.
4. *If* $0 \in E$ *then it is the* $\sqsubseteq$-*greatest element of* E.

Proof.

1. Reflexivity follows by choosing $c = 1$ in the definition of $\sqsubseteq$.
   For transitivity assume $a \sqsubseteq b$ and $b \sqsubseteq c$, say $b = a * d$ and $c = b * e$. Then $c = (a * d) * e = a * (d * e)$.
2. Assume $a \sqsubseteq b$ and $b \sqsubseteq a$. If $a = 0$ then $b = 0$ follows form the definition of $a \sqsubseteq b$, since 0 is an annihilator for $*$. Otherwise let $b = a * c$ and $a = b * d$. Then $a * 1 = a = b * d = a * c * d$, hence $1 = c * d$ by cancellativity. Now, irreducibility of 1 implies $c = 1 \lor d = 1$ and hence $c = 1 = d$, showing $a = b$.
3. and (4) are straightforward from the definition of $\sqsubseteq$, neutrality of 1 and annihilation of 0. $\square$

In our concrete model, the set E of singleton programs is well behaved and the relation $\sqsubseteq$ is isomorphic to the subset relation on concrete traces.

Assume now that E is well behaved and hence $\sqsubseteq$ is a partial order on E. The supremum of a subset $D \subseteq E$ w.r.t. $\sqsubseteq$, if existent, is denoted by $\circledast D$.

Lemma 8.5. *If* $0 \in D \subseteq E$ *then* $0 = \circledast D$.

This is immediate from the definition of $\sqsubseteq$ and suprema.

Definition 8.6. Assume that E is well behaved. Then $e \in E$ is called an E-*event* if it is subatomic and join-prime w.r.t. $\sqsubseteq$, i.e., if

$$\forall d \in E : d \sqsubseteq e \Rightarrow d = 1 \lor d = e ,$$
$$\forall D \subseteq E : D \neq \emptyset \land \circledast D \text{ exists} \Rightarrow (t \sqsubseteq \circledast D \Rightarrow \exists d \in D : t \sqsubseteq d) .$$

By this definition 1 is an E-event, as is 0 if $0 \in E$. The E-events different from $0, 1$ are atoms w.r.t. $\sqsubseteq$ in E. Clearly, every E-event a is $*$-irreducible in E:

$$a = b * c \Rightarrow b = a \lor c = a .$$

To put things into perspective, we note that the order $\sqsubseteq$ corresponds to the well-known divisibility order on the natural numbers and E-events play the same rôle as the prime numbers.

Definition 8.7. A CKA S is *event-based* if the following properties hold:

(a) 1 is a trace.
(b) Every element is the supremum of its traces, i.e., for all $a \in S$ we have $a = \sqcup TR(a)$.
(c) The set $TR(S)$ of traces is well behaved. By $EV(S)$ we denote the set of $TR(S)$-events and call them the *events* of S. The set of events of trace t is

$$EV(t) =_{df} \{e \in EV(S) \mid e \sqsubseteq t\} .$$

(d) The set $TR(S)$ of traces is a complete lattice w.r.t. $\sqsubseteq$ and every trace is the supremum of its events, i.e., for all $t \in TR(S)$ we have $t = \circledast EV(t)$.

(e) For all events e we have $e * e = 0$ and hence $e\,;e = 0$.

For an arbitrary $a \in S$ we then set $EV(a) =_{df} \bigcup_{t \in TR(a)} EV(t)$.

Hence our concrete model of programs forms an event-based CKA. Event-based CKAs are quite similar to the feature algebras developed in [12] for the description of product families.

The definition of an event-based CKA S immediately yields

Lemma 8.8

1. $EV(0) = EV(S)$.
2. $EV(1) = \{1\}$.
3. For traces a, b with $a * b \neq 0$ we have $EV(a * b) = EV(a) \cup EV(b)$ and hence $a * b = \circledast\{a, b\}$.

8.1   Abstract Dependence and Protection

In this section we define an abstract counterpart to the dependence relation and use it to give an intuitive sufficient criterion for protectedness. For it we need an abstract formulation of the dependence relation.

Definition 8.9. We call element a *sequentially independent* of element b, in signs $a \not\leftarrow b$, if $a * b \leq a\,;b$.

The following properties are shown by straightforward calculation and, in the last case, by Theorem 5.6:

Lemma 8.10

1. $0 \not\leftarrow a$ and $a \not\leftarrow 0$.
2. $1 \not\leftarrow a$ and $a \not\leftarrow 1$.
3. $a \not\leftarrow c \wedge b \not\leftarrow c \Rightarrow (a + b) \not\leftarrow c$.
4. $a \not\leftarrow b \wedge a \not\leftarrow c \Rightarrow a \not\leftarrow (b + c)$.
5. If r is an invariant then $r \not\leftarrow r$.

Part 5 shows that for general programs this notion behaves in an unexpected way. However, in our concrete model it works fine for singleton programs:

$$\{tp\} \not\leftarrow \{tq\} \quad \Leftrightarrow \quad \forall p \in tp, q \in tq : \neg(p \leftarrow q) \ .$$

In particular, $[p] \not\leftarrow [q] \Leftrightarrow \neg(p \leftarrow q)$. This motivates the following

Definition 8.11. In an event-based CKA we define the dependence relation between events e, f by

$$e \rightarrow f \Leftrightarrow_{df} \neg(f \not\leftarrow e) \Leftrightarrow f\,;e \neq f * e \ .$$

We denote the converse of $\rightarrow$ by $\leftarrow$. We say that the algebra *respects dependence* if $e \leftarrow f \Rightarrow e\,;f = 0$.

Lemma 8.12. *Consider traces tp, tq of an event-based CKA that respects dependence.*

1. *If $p \to q$ for some $p \in EV(tp)$ and $q \in EV(tq)$ then $tp \,;\, tq = 0$.*
2. *If $tp * tq \neq 0$ then*

$$tp \not\rightarrow tq \quad \Leftrightarrow \quad \forall p \in EV(tp), q \in EV(tq) : p \not\rightarrow q \;.$$

Proof.

1. By additivity of ; we have $tp \,;\, tq = \circledast \{u \,;\, v \mid u \in EV(tp), v \in EV(tq)\}$ and the claim follows from Lemma 8.5.
2. ($\Leftarrow$) Immediate from event-basedness and additivity of $*$ and ; .
   ($\Rightarrow$) By Part 1 we have $p \,;\, q \neq 0$ for all $p \in EV(tp)$ and $q \in EV(tq)$. Since $TR(S)$ is assumed to be well behaved, also $p * q$ is a trace, and from $p \,;\, q \leq p * q$ it follows that $p \,;\, q = p * q$.   □

With these prerequisites it is now possible to completely replay the proof of Theorem 6.1 in the abstract setting of event-based CKAs; we omit the details.

9 Related Work and Outlook

Our basic model and its algebraic abstraction by CKAs reflect a non-interleaving view of concurrency and hence rather falls into the class of partial-order models for true concurrency, which is also shown by the discussion after Theorem 6.1. Nevertheless, as detailed in [10], there are certain connections to interleaving-based process algebras such as ACP, CCS, CSP, mCRL2 and the π-calculus. Moreover, [10] provides a discussion of the relation of our approach to some to the more prominent representatives of partial order-semantics.

Recently, Prisacariu has proposed *synchronous Kleene algebra (SKA)* [21]. Conceptually, it seems to be an interesting special case of ours, useful when communication between threads is synchronised. Semantically, Prisacariu's model is based on languages formed by strings of sets of letters; e.g., a, b, cad, eb is a string. The letters in each set model actions that are executed in parallel. Besides the usual regular operations, an operation of synchronous parallel composition on strings is defined similarly to a zip in functional programming, and lifted to a complex product at the language level. It is shown that the language models are the free algebras in the class of SKAs. At the axiomatic level, parallel composition interacts with sequential composition via an *equational* strengthening of the exchange law (4). Our model cannot satisfy such an equational form of the exchange law, because our sequential and concurrent compositions have the same unit, and the exchange equation would make the compositions identical. CKAs are more general also in that they cover also asynchronous models and interleaving semantics. An additional minor difference is that the axiomatisation of SKAs is purely based on Kozen's first-order axiomatisation[14,15], and not on quantales. Hence it can be fully treated with first-order theorem provers.

Although CKA is not a direct abstraction of the familiar concurrency calculi, we envisage that many of them can be mapped into our basic model of programs and its abstraction CKA. A first experiment along these lines is a trace model of a core subset of the π-calculus in [11]. Moreover, the study [9] shows how to apply the trace model in a unified description of the various phenomena arising in concurrent programs operating on shared/private and weakly/strongly consistent memory, communicating in a synchronised or buffered way, and using dynamic/nested and disposed/collected resources. Further studies will concern the elaboration of these ideas.

Acknowledgement. We are grateful for valuable comments by J. Desharnais, H.-H. Dang, R. Glück, W. Guttmann, P. Höfner, P. O'Hearn, H. Yang and by the anonymous referees of RelMiCS/AKA09.

References

1. Back, R., von Wright, J.: Refinement calculus — a systematic introduction. Springer, Heidelberg (1998)
2. Benabou, J.: Introduction to bicategories. In: Reports of the Midwest Category Seminar. Lecture Notes in Math., vol. 47, pp. 1–77. Springer, Heidelberg (1967)
3. Birkhoff, G.: Lattice Theory, 3rd edn. Amer. Math. Soc. (1967)
4. Cohen, E.: Separation and reduction. In: Backhouse, R., Oliveira, J. (eds.) MPC 2000. LNCS, vol. 1837, pp. 45–59. Springer, Heidelberg (2000)
5. Conway, J.: Regular Algebra and Finite Machines. Chapman & Hall, Boca Raton (1971)
6. Desharnais, J., Möller, B., Struth, G.: Kleene Algebra with domain. Trans. Computational Logic 7, 798–833 (2006)
7. Erné, M., Koslowski, J., Melton, A., Strecker, G.E.: A primer on Galois connections. In: Proc. 1991 Summer Conference on General Topology and Applications in Honor of Mary Ellen Rudin and Her Work. Annals of the New York Academy of Sciences, vol. 704, pp. 103–125 (1993)
8. Hoare, C.A.R.: An axiomatic basis for computer programming. Commun. ACM 12, 576–585 (1969)
9. Hoare, C.A.R.: Unifying models of execution history (manuscript) (June 2009)
10. Hoare, C.A.R., Möller, B., Struth, G., Wehrman, I.: Concurrent Kleene Algebra. In: Bravetti, M., Zavattaro, G. (eds.) CONCUR 2009. LNCS, vol. 5710, pp. 399–414. Springer, Heidelberg (2009)
11. Hoare, C.A.R., Wehrman, I., O'Hearn, P.: Graphical models of separation logic. In: Proc. Marktoberdorf Summer School 2008 (2008) (forthcoming)
12. Höfner, P., Khedri, R., Möller, B.: Feature algebra. In: Misra, J., Nipkow, T., Sekerinski, E. (eds.) FM 2006. LNCS, vol. 4085, pp. 300–315. Springer, Heidelberg (2006)
13. Jones, C.: Development methods for computer programs including a notion of interference. PhD Thesis, University of Oxford. Programming Research Group, Technical Monograph 25 (1981)
14. Kozen, D.: A completeness theorem for Kleene algebras and the algebra of regular events. Information and Computation 110, 366–390 (1994)
15. Kozen, D.: Kleene algebra with tests. Trans. Programming Languages and Systems 19, 427–443 (1997)
16. Mac Lane, S.: Categories for the working mathematician, 2nd edn. Springer, Heidelberg (1998)

17. McCune, W.: Prover9 and Mace4, `http://www.prover9.org/` (accessed March 1, 2009)
18. Morgan, C.: Programming from Specifications. Prentice-Hall, Englewood Cliffs (1990)
19. Mulvey, C.: Rendiconti del Circolo Matematico di Palermo 12, 99–104 (1986)
20. O'Hearn, P.: Resources, concurrency, and local reasoning. Theor. Comput. Sci. 375, 271–307 (2007)
21. Prisacariu, C.: Extending Kleene lgebra with synchrony — technicalities. University of Oslo, Department of Informatics, Research Report No. 376 (October 2008)
22. Sangiorgi, D., Walker, D.: The π-calculus — A theory of mobile processes. Cambridge University Press, Cambridge (2001)

A   Axiom Systems

For ease of reference we summarise the algebraic structures employed in the paper.

1. A *semiring* is a structure $(S, +, 0, \cdot, 1)$ such that $(S, +, 0)$ is a commutative monoid, $(S, \cdot, 1)$ is a monoid, multiplication distributes over addition in both arguments and 0 is a left and right annihilator with respect to multiplication $(a \cdot 0 = 0 = 0 \cdot a)$. A semiring is *idempotent* if its addition is.

2. A *quantale* [19] or *standard Kleene algebra* [5] is an idempotent semiring that is a complete lattice under the natural order and in which composition distributes over arbitrary suprema. The infimum and the supremum of a subset T are denoted by $\sqcap T$ and $\sqcup T$, respectively. Their binary variants are $x \sqcap y$ and $x \sqcup y$ (the latter coinciding with $x + y$).

3. A *concurrent Kleene algebra* (CKA) is a structure $(S, +, 0, *, ;, 1)$ such that $(S, +, 0, *, 1)$ and $(S, +, 0, ;, 1)$ are quantales linked by the exchange axiom

$$(a * b) ; (c * d) \leq (b ; c) * (a ; d) .$$

4. A *CKA with invariants* $(S, +, 0, *, ;, 1, \iota)$ consists of a CKA $(S, +, 0, *, ;, 1)$ and a closure operator $\iota : S \to S$ that additionally satisfies, for all $a, b \in S$,

$$1 \leq \iota a , \qquad\qquad \iota (a * b) \leq \iota (a + b) .$$

An *invariant* is an element $a \in S$ with $\iota a = a$.

5. A *rely/guarantee-CKA* [10] is a pair (S, I) such that S is a CKA and $I \subseteq I(S)$ is a set of invariants, i.e. of elements r satisfying $r = r^*$, such that $1 \in I$ and for all $r, r' \in I$ also $r \sqcap r' \in I$ and $r * r' \in I$. Moreover, all $r \in I$ and $a, b \in S$ have to satisfy

$$r * (a ; b) \leq (r * a) ; (r * b) .$$

Armstrong's Inference Rules in Dedekind Categories

Toshikazu Ishida[1], Kazumasa Honda[2], and Yasuo Kawahara[2]

[1] Center for Fundamental Education, The University of Kitakyushu 4-2-1, Kitagata,
Kokuraminami-ku, Kitakyushu, 802-8577, Japan
t-ishida@kitakyu-u.ac.jp
[2] Department of Informatics, Kyushu University, Fukuoka, 819-0395, Japan
kawahara@i.kyusu-u.ac.jp

Abstract. It is well-known that Armstrong's inference rules are sound and complete for functional dependencies of relational data bases and for implication in the theory of formal concepts by Wille and Ganter. In this paper the authors treat Armstrong's inference rules and the implication as (binary) relations in an upper semi lattice in a Dedekind category, and give a relation algebraic proof of the completeness theorem for Armstrong's inference rules in a Schröder category.

1 Introduction

Functional dependency, initiated by Codd[1], is the best known class of dependencies, related to logical constraints, on relational databases. It is well-known [2] that Armstrong's inference rules [3] are sound and complete for functional dependencies (Cf. [4]). On the other hand Wille and Ganter [5] introduced a notion of implication for formal contexts (i.e. boolean valued functional data bases), slightly different from functional dependency, and they showed that Armstrong's inference rules are sound and complete for implication, too.

This paper attempts to extend Armstrong's inference rules and the implication into appropriate relations in a Dedekind category, and to give a unified proof of the completeness theorem for Armstrong's inference rules in a Schröder category.

This paper is organized as follows. In section 2 we will review some basic facts on Armstrong's inference rules, which are main subject of the paper. In section 3 we review the definitions and basic properties of Dedekind category [6] and Schröder category [7][8]. In section 4 an inference relation in a Dedekind category is defined as an extension of dependency relations satisfying Armstrong's inference rules on an upper semi-lattice. In section 5 we construct a closure system in the sense of [9] from an arbitrary relation on a complete upper semi-lattice in a Dedekind category. In section 6 we define two relations: the first one is an implication relation which extends Wille and Ganter's implication relation, and the second one is a closure operation generalizing usual closure operations used in defining formal concepts. Moreover we show a basic relationship between them. In the final section we give a relation algebraic proof of the completeness theorem for Armstrong's inference rules in a Schröder category with unit satisfying the point axiom.

R. Berghammer et al. (Eds.): RelMiCS/AKA 2009, LNCS 5827, pp. 187–198, 2009.
© Springer-Verlag Berlin Heidelberg 2009

2   Armstrong's Inference Rules

Armstrong's inference rules give a basic framework to treat the logical structure of dependencies in an attribute set. Let A and B be subsets of an attribute set Y. A formal expression $A \to B$, i.e. an ordered pair of subsets A and B of Y, is called a *dependency* on Y.

Armstrong's Inference Rules:

$$[\text{A1}] \; \frac{A \supseteq B}{A \to B} \quad [\text{A2}] \; \frac{A \to B \quad B \to C}{A \to C} \quad [\text{A3}] \; \frac{A \to B \quad A \to C}{A \to B \cup C}$$

Let $\mathcal{L}$ be a set of dependencies. A derivation from $\mathcal{L}$ is a nonempty sequence

$$\{A_0 \to B_0, A_1 \to B_1, \ldots, A_n \to B_n\}$$

of dependencies such that, for all $k = 0, 1, \ldots, n$, one of the following holds:

(i) $A_k \supseteq B_k$ ([A1]) or $A_k \to B_k$ is in $\mathcal{L}$ (assumption) ([A0]),
(ii) $\exists i, j < k$ such that

$$[\text{A2}] \quad \frac{A_i \to B_i \quad A_j \to B_j}{A_k \to B_k}, \quad (B_i = A_j, A_k = A_i, B_k = B_j)$$

(iii) $\exists i, j < k$ such that

$$[\text{A3}] \quad \frac{A_i \to B_i \quad A_j \to B_j}{A_k \to B_k}. \quad (A_i = A_j, B_k = B_i \cup B_j)$$

A dependency $A \to B$ is *provable* from $\mathcal{L}$, written $\mathcal{L} \vdash A \to B$, if there is a derivation $\{A_0 \to B_0, A_1 \to B_1, \ldots, A_n \to B_n\}$ from $\mathcal{L}$ such that $A = A_n$ and $B = B_n$.

For a set $\mathcal{L}$ of dependencies we define for every $A \subseteq Y$ a subset $A_{\mathcal{L}}$ of Y by $A_{\mathcal{L}} = \{y \in Y \mid \mathcal{L} \vdash A \to y\}$.

Lemma 1. *Let B be a finite subset of Y. Then $\mathcal{L} \vdash A \to B$ iff $B \subseteq A_{\mathcal{L}}$.*

Proof. ($\to$) Assume that $\mathcal{L} \vdash A \to B$ and let $y \in B$. Then $\vdash B \to y$ by [A1] and so $\mathcal{L} \vdash A \to y$ by [A2]. Hence $y \in A_{\mathcal{L}}$.

($\leftarrow$) Assume that $B \subseteq A_{\mathcal{L}}$. Then, for all $y \in B$, we have $\mathcal{L} \vdash A \to y$ by the definition of $A_{\mathcal{L}}$ and hence $\mathcal{L} \vdash A \to B$ by the union rule [A3], because B is finite.   □

3   Dedekind Category, Schröder Category

In this section, we recall the definition of a kind of relation category which we will call Dedekind categories following Olivier and Serrato [6]. Dedekind categories are equivalent to locally complete division allegories introduced in Freyd and Scedrov [7].

Throughout this paper, a morphism α from an object X into an object Y in a Dedekind category (which will be defined below) will be denoted by a half arrow, and the composite of a morphism $\alpha : X \rightharpoonup Y$ followed by a morphism $\beta : Y \rightharpoonup Z$ will be written as $\alpha\beta : X \rightharpoonup Z$. Also we will denote the identity morphism on X as id_X.

Definition 1. A Dedekind category $\mathcal{D}$ is a category satisfying the following:

D1. [Complete Heyting Algebra]: For all pairs of objects X and Y the homset $\mathcal{D}(X,Y)$ consisting of all morphisms of X into Y is a complete Heyting algebra (namely, a complete distributive lattice) with the least morphism 0_{XY} and the greatest morphism ∇_{XY}. Its algebraic structure will be denoted by

$$\mathcal{D}(X,Y) = (\mathcal{D}(X,Y), \sqsubseteq, \sqcup, \Rightarrow, \sqcap, 0_{XY}, \nabla_{XY}).$$

D2. [Converse]: There is given a converse operation $^\sharp : \mathcal{D}(X,Y) \to \mathcal{D}(Y,X)$. That is, for all morphisms $\alpha, \alpha' : X \rightharpoonup Y$, $\beta : Y \rightharpoonup Z$, the following converse laws hold:

(a) $(\alpha\beta)^\sharp = \beta^\sharp \alpha^\sharp$,
(b) $(\alpha^\sharp)^\sharp = \alpha$,
(c) $\alpha \sqsubseteq \alpha'$, then $\alpha^\sharp \sqsubseteq \alpha'^\sharp$.

D3. [Dedekind Formula]: For all morphisms $\alpha : X \rightharpoonup Y$, $\beta : Y \rightharpoonup Z$ and $\gamma : X \rightharpoonup Z$, the Dedekind formula $\alpha\beta \sqcap \gamma \sqsubseteq \alpha(\beta \sqcap \alpha^\sharp \gamma)$ holds.

D4. [Residual Composition]: For all morphisms $\alpha : X \rightharpoonup Y$ and $\beta : Y \rightharpoonup Z$, the residual composite $\alpha \ominus \beta : X \rightharpoonup Z$ is a morphism such that $\gamma \sqsubseteq \alpha \ominus \beta \iff \alpha^\sharp \gamma \sqsubseteq \beta$ for all morphisms $\gamma : X \rightharpoonup Z$. □

For all morphisms α, β and γ, the following hold.

Proposition 1. If $\alpha^\sharp \alpha \sqsubseteq \mathrm{id}_Y$ then $\alpha(\beta \sqcap \gamma) = \alpha\beta \sqcap \alpha\gamma$. □

For a set of relation $\{\alpha_j : X \rightharpoonup Y | j \in J\}$, we define two relations $\sqcup_{j \in J}\alpha_j$ and $\sqcap_{j \in J}\alpha_j$ as follows:

$$\sqcup_{j \in J}\alpha_j = \{(a,b) \in X \times Y | \exists j \in J, (a,b) \in \alpha_j\},$$

$$\sqcap_{j \in J}\alpha_j = \{(a,b) \in X \times Y | \forall j \in J, (a,b) \in \alpha_j\}.$$

A morphism $f : X \to Y$ such that $f^\sharp f \sqsubseteq \mathrm{id}_Y$ (univalent) and $\mathrm{id}_X \sqsubseteq ff^\sharp$ (total) is called a *function* and may be introduced as $f : X \to Y$. In what follows, the word *relation* is a synonym for morphism of a Dedekind category.

A function $f : X \to Y$ is called a *surjection* if $f^\sharp f = \mathrm{id}_Y$, and f is called a *injection* if $ff^\sharp = \mathrm{id}_X$.

Next, we review some fundamental properties of residual composition [10] [11].

Proposition 2. Let $\alpha, \alpha' : A \rightharpoonup B$, $\beta, \beta' : B \rightharpoonup C$, $\gamma : C \rightharpoonup D$, $\mu : V \rightharpoonup B$ and $\rho, \rho', \rho_j : V \rightharpoonup A (j \in J)$ be relations in $\mathcal{D}$. Then the following hold:

(1) If $\alpha' \sqsubseteq \alpha$ and $\beta \sqsubseteq \beta'$ then $\alpha \ominus \beta \sqsubseteq \alpha' \ominus \beta'$.

(2) $\mathrm{id}_A \sqsubseteq \alpha \ominus \alpha^\sharp$.

(3) $\alpha \ominus (\beta \ominus \gamma) = \alpha\beta \ominus \gamma$.

(4) If α is a function then $\alpha \ominus \beta = \alpha\beta$.

(5) If $\mu \sqsubseteq \rho \ominus \alpha$ then $\rho \sqsubseteq \mu \ominus \alpha^\sharp$.   (Galois connection)

(6) $\rho \sqsubseteq (\rho \ominus \alpha) \ominus \alpha^\sharp$.

(7) $((\rho \ominus \alpha) \ominus \alpha^\sharp) \ominus \alpha = \rho \ominus \alpha$.

(8) If α and α' are functions such that $\alpha \sqsubseteq \alpha'$, then $\alpha = \alpha'$.   □

We define four relations $\max(\rho; \zeta)$, $\min(\rho; \zeta)$, $\sup(\rho; \zeta)$ and $\inf(\rho; \zeta) : V \rightharpoonup X$ for two relations $\rho : V \rightharpoonup X$ and $\zeta : X \rightharpoonup X$ as following:

maximum $\max(\rho; \zeta) = \rho \sqcap (\rho \ominus \zeta)$,

minimum $\min(\rho; \zeta) = \rho \sqcap (\rho \ominus \zeta^\sharp)$,

supremum $\sup(\rho; \zeta) = (\rho \ominus \zeta) \sqcap ((\rho \ominus \zeta) \ominus \zeta^\sharp)$,

infimum $\inf(\rho; \zeta) = (\rho \ominus \zeta^\sharp) \sqcap ((\rho \ominus \zeta^\sharp) \ominus \zeta)$.

The relations satisfy the following proposition.

Proposition 3. *Let $\alpha : X \to Y$ be a function, and $\beta : X \rightharpoonup Y$ and $\zeta : Y \rightharpoonup Y$ relations. Then the following holds.*

(a) $\max(\alpha\beta; \zeta) = \alpha \max(\beta; \zeta)$ *and* $\min(\alpha\beta; \zeta) = \alpha \min(\beta; \zeta)$,*function*

(b) $\sup(\alpha\beta; \zeta) = \alpha \sup(\beta; \zeta)$ *and* $\inf(\alpha\beta; \zeta) = \alpha \inf(\beta; \zeta)$.   □

A relation $\zeta : X \rightharpoonup X$ is called an *order* if $\mathrm{id}_X \sqsubseteq \zeta$(reflexive), $\zeta\zeta \sqsubseteq \zeta$(transitive) and $\zeta \sqcap \zeta^\sharp \sqsubseteq \mathrm{id}_X$(antisymmetric). A relation $\zeta : X \rightharpoonup X$ is *complete* if $\sup(\rho; \zeta)$ is a function for any relation $\rho : V \rightharpoonup X$.

A Schröder category are a particular Dedekind category whose hom-set are complete Boolean algebras.

Definition 2. *A Schröder category $\mathcal{S}$ is a category satisfying the following two conditions SC1 and SC2 in addition to D2 and D3 in definition 1:*

SC1.[Complete Boolean Algebra]: *For all pairs of objects X and Y the hom-set $\mathcal{S}(X, Y)$ consisting of all relations of X into Y is a complete Boolean algebra with the least relation 0_{XY} and the greatest relation ∇_{XY}. Its algebraic structure will be denoted by*

$$\mathcal{S}(X, Y) = (\mathcal{S}(X, Y), \sqsubseteq, \sqcup, \sqcap, ^-, 0_{XY}, \nabla_{XY}),$$

where $\sqsubseteq$, $\sqcup$, $\sqcap$ and $^-$ denote the inclusion order, the join, the meet and the complement of relations, respectively.

SC2.[Zero Relation]: *The least relation 0_{XY} is a zero relation, that is,*
$$\alpha 0_{YZ} = 0_{XZ}.$$

4   Inference Relation

In this section, we introduce inference relations in a Dedekind category [6][11], correspond to Armstrong's inference rules.

Definition 3. *Let* $\xi : X \rightharpoonup X$ *be a complete order and* $\gamma : X \rightharpoonup X$ *a relation in a Dedekind category* $\mathcal{D}$. *Define a relation* $\lambda = \lambda_\gamma : X \rightharpoonup X$ *as the least relation satisfying the following four conditions:*

(A0) $\gamma \sqsubseteq \lambda$,
(A1) $\xi^\sharp \sqsubseteq \lambda$,
(A2) $\lambda\lambda \sqsubseteq \lambda$,
(A3) $\sup(\lambda; \xi) \sqsubseteq \lambda$. (union rule)

Define an ascending chain of relations

$$\lambda_0 \sqsubseteq \lambda_1 \sqsubseteq \lambda_2 \sqsubseteq \cdots : X \rightharpoonup X$$

by

$$\lambda_0 = \gamma \sqcup \xi^\sharp, \quad \lambda_{n+1} = \lambda_n \sqcup \lambda_n\lambda_n \sqcup \sup(\lambda_n; \xi) \quad (n \geq 0).$$

Suppose that the hom-set $\mathcal{D}(X, X)$ satisfies the ascending chain condition. Then there exists a natural number N such that

$$\lambda_{N+1} = \lambda_N.$$

It is trivial that $\lambda = \lambda_N$ satisfies (A0) - (A3).

Proposition 4. $\lambda = \lambda_N$ *is the least relation which satisfies* (A0) - (A3).

Proof. Assume that $\lambda' : X \rightharpoonup X$ is another relation satisfying (A0) - (A3). By induction we will prove that $\lambda_n \sqsubseteq \lambda'$ for all natural numbers n. It is trivial that $\lambda_0 = \gamma \sqcup \xi^\sharp \sqsubseteq \lambda'$ by (A0) and (A1). Assume $\lambda_n \sqsubseteq \lambda'$. Then we have $\lambda_n\lambda_n \sqsubseteq \lambda'$ by (A2), and

$$
\begin{aligned}
\sup(\lambda_n; \xi) &\sqsubseteq (\lambda_n \ominus \xi) \ominus \xi^\sharp && \{ \ Def \ of \ sup \ \} \\
&\sqsubseteq (\lambda' \ominus \xi) \ominus \xi^\sharp && \{ \ \lambda_n \sqsubseteq \lambda' \ \} \\
&= \sup(\lambda'; \xi)\xi \ominus \xi^\sharp \\
&= \sup(\lambda'; \xi)(\xi \ominus \xi^\sharp) && \{ \ \sup(\lambda'; \xi) : function \ \} \\
&= \sup(\lambda'; \xi)\xi^\sharp && \{ \ \xi \ominus \xi^\sharp = \xi^\sharp \ \} \\
&\sqsubseteq \lambda'\lambda' && \{ \ (A3) \ and \ (A1) \ \} \\
&\sqsubseteq \lambda'. && \{ \ (A2) \ \}
\end{aligned}
$$

This proves $\lambda_{n+1} \sqsubseteq \lambda'$ and consequently $\lambda \sqsubseteq \lambda'$. □

Proposition 5. $\sup(\lambda; \xi) \sqsubseteq \xi$, $\sup(\lambda; \xi) = \max(\lambda; \xi)$ *and* $\lambda = \sup(\lambda; \xi)\xi^\sharp$.

Proof

$$\sup(\lambda;\xi) \sqsubseteq \lambda \ominus \xi \qquad \{\ Def\ of\ sup\ \}$$
$$\sqsubseteq id_X \ominus \xi \qquad \{\ (A1)\ \xi^\sharp \sqsubseteq \lambda\ \}$$
$$= \xi,$$

$$\sup(\lambda;\xi) \sqsubseteq \lambda \sqcap (\lambda \ominus \xi) \ \{\ (A3)\ \sup(\lambda;\xi) \sqsubseteq \lambda\ (union\ rule)\ \}$$
$$= \max(\lambda;\xi) \quad \{\ Def\ of\ max\ \}$$
$$\sqsubseteq \sup(\lambda;\xi), \quad \{\ \lambda \sqsubseteq (\lambda \ominus \xi) \ominus \xi^\sharp\ \}$$

$$\lambda \sqsubseteq \sup(\lambda;\xi)\xi^\sharp \ \{\ Prop\ 2\ \}$$
$$\sqsubseteq \lambda\lambda \qquad \{\ (A3),\ (A1)\ \}$$
$$\sqsubseteq \lambda. \qquad \{\ (A2)\ \} \qquad\qquad \square$$

5   Closure System

In this section, we define a relation ν and prove the relation is a closure system [9].

Definition 4. *Define a relation* $\nu = \nu_\gamma : I \rightharpoonup X$ *by*

$$\nu = \nabla_{IX}[id_X \sqcap (\xi^\sharp \gamma \ominus \xi)].$$

For this relation the following propositions hold. In the following proofs, we abbreviate the expression "Dedekind formula" by the symbol "DF".

Proposition 6. *Let* $\alpha : X \rightharpoonup Y$ *and* $\beta : Y \rightharpoonup X$ *be relations. Then the identity*

$$\nabla_{IX}[id_X \sqcap (\alpha \ominus \beta)] = \nabla_{IY} \ominus (\alpha^\sharp \Rightarrow \beta)$$

holds.

Proof
(1) $\rho \sqsubseteq \nabla_{IX}[id_X \sqcap (\alpha \ominus \beta)] \rightarrow \rho \sqsubseteq \nabla_{IY} \ominus (\alpha^\sharp \Rightarrow \beta)$:
Assume $\rho \sqsubseteq \nabla_{IX}[id_X \sqcap (\alpha \ominus \beta)]$. Then

$$\alpha^\sharp \sqcap \nabla_{YI}\rho \sqsubseteq \alpha^\sharp \sqcap \nabla_{YI}\nabla_{IX}[id_X \sqcap (\alpha \ominus \beta)]$$
$$= \alpha^\sharp[id_X \sqcap (\alpha \ominus \beta)] \qquad \{\ DF\ \}$$
$$\sqsubseteq \alpha^\sharp(\alpha \ominus \beta) \qquad \{\ id_X \sqcap (\alpha \ominus \beta) \sqsubseteq \alpha \ominus \beta\ \}$$
$$\sqsubseteq \beta \qquad \{\ D4\ \}.$$

Hence $\rho \sqsubseteq \nabla_{IY} \ominus (\alpha^\sharp \Rightarrow \beta)$.
(2) $\rho \sqsubseteq \nabla_{IY} \ominus (\alpha^\sharp \Rightarrow \beta) \rightarrow \rho \sqsubseteq \nabla_{IX}[id_X \sqcap (\alpha \ominus \beta)]$:

$$\rho \sqsubseteq \nabla_{IY} \ominus (\alpha^\sharp \Rightarrow \beta) \leftrightarrow \alpha^\sharp \sqcap \nabla_{YI}\rho \sqsubseteq \beta \qquad \{\ D4\ and\ Prop\ of \Rightarrow\ \}$$
$$\leftrightarrow \alpha^\sharp \sqcap \nabla_{YX}(id_X \sqcap \rho^\sharp\rho) \sqsubseteq \beta \ \{\ \nabla_{YI}\rho = \nabla_{YI}(id_X \sqcap \rho^\sharp\rho)\ \}$$
$$\leftrightarrow \alpha^\sharp(id_X \sqcap \rho^\sharp\rho) \sqsubseteq \beta$$
$$\leftrightarrow id_X \sqcap \rho^\sharp\rho \sqsubseteq \alpha \ominus \beta \qquad \{\ D4\ \}$$
$$\rightarrow \rho = \rho(id_X \sqcap \rho^\sharp\rho) \sqsubseteq \nabla_{IX}[id_X \sqcap (\alpha \ominus \beta)]. \qquad \square$$

Corollary 1. *If $\gamma = p^\sharp q$ for a pair of functions $p, q : R \to X$, then*

$$\nu = \nabla_{RI} \ominus (p\xi \Rightarrow q\xi).$$

Proof

$$
\begin{aligned}
\nu &= \nabla_{IX}[\mathrm{id}_X \sqcap (\xi^\sharp \gamma \ominus \xi)] &&\{ \text{ Def of } \nu \} \\
&= \nabla_{IX}[\mathrm{id}_X \sqcap (\xi^\sharp p^\sharp q \ominus \xi)] &&\{ \gamma = p^\sharp q \} \\
&= \nabla_{IX}[\mathrm{id}_X \sqcap (\xi^\sharp p^\sharp \ominus q\xi)] &&\{ q : function \} \\
&= \nabla_{IX} \ominus (p\xi \Rightarrow q\xi). &&\{ \text{ Prop 6 } \}
\end{aligned}
$$

$\square$

Proposition 7. ν *is a closure system.*

Proof. For all relations $\rho \sqsubseteq \nu$ we will show $\inf(\rho; \xi) \sqsubseteq \nu$. Assume $\rho \sqsubseteq \nu$ and set $a = \inf(\rho; \xi)$. Then we have $\mathrm{id}_X \sqcap \rho^\sharp \rho \sqsubseteq \xi^\sharp \gamma \ominus \xi$ by Proposition 6, and $a\xi^\sharp = \rho \ominus \xi^\sharp$ and $\rho \sqsubseteq a\xi$ since ξ is a complete order.

$$
\begin{aligned}
a^\sharp \rho &= a^\sharp \rho(\mathrm{id}_X \sqcap \rho^\sharp \rho) &&\{ \rho = \rho(\mathrm{id}_X \sqcap \rho^\sharp \rho) \} \\
&\sqsubseteq a^\sharp a\xi(\xi^\sharp \gamma \ominus \xi) &&\{ \rho \sqsubseteq a\xi, \ \rho \sqsubseteq \nu \} \\
&\sqsubseteq \xi(\xi^\sharp \xi^\sharp \gamma \ominus \xi) &&\{ a : function, \ \xi = \xi\xi \} \\
&= \xi[\xi^\sharp \ominus (\xi^\sharp \gamma \ominus \xi)] &&\{ \text{ Prop 2 } \} \\
&\sqsubseteq \xi^\sharp \gamma \ominus \xi,
\end{aligned}
$$

$$
\begin{aligned}
\mathrm{id}_X \sqcap a^\sharp a = a^\sharp a &&\{ a : function \} \\
\sqsubseteq (a^\sharp a \ominus \xi^\sharp) \ominus \xi &&\{ \text{ Prop 2 } \} \\
= (a^\sharp \ominus a\xi^\sharp) \ominus \xi &&\{ a : function \} \\
= [a^\sharp \ominus (\rho \ominus \xi^\sharp)] \ominus \xi &&\{ a\xi^\sharp = \rho \ominus \xi^\sharp \} \\
= (a^\sharp \rho \ominus \xi^\sharp) \ominus \xi &&\{ \text{ Prop 2 } \} \\
\sqsubseteq [(\xi^\sharp \gamma \ominus \xi) \ominus \xi^\sharp] \ominus \xi &&\{ a^\sharp \rho \sqsubseteq \xi^\sharp \gamma \ominus \xi \} \\
= \xi^\sharp \gamma \ominus \xi.
\end{aligned}
$$

Hence, this proves $a \sqsubseteq \nu$ by Proposition 6.

$\square$

6   Closure Operation

In this section, we define a closure operation and show some propositions.

Definition 5. *We define a function $f : X \to X$ by*

$$f = \inf(\xi \sqcap \nabla_{XI}\nu; \xi).$$

Then f is called a closure operation [9].

For this closure operation the following propositions hold.

Proposition 8. *Let* $p, q : R \to X$ *be functions. Then*

$$p^\sharp q \sqsubseteq f\xi^\sharp \;\leftrightarrow\; \nabla_{RI}\nu \sqsubseteq p\xi \Rightarrow q\xi.$$

Proof

$$
\begin{aligned}
p^\sharp q \sqsubseteq f\xi^\sharp &\leftrightarrow q \sqsubseteq pf\xi^\sharp && \{\, p : \text{function} \,\} \\
&\leftrightarrow q \sqsubseteq (p\xi \sqcap p\nabla_{XI}\nu) \ominus \xi^\sharp && \{\, f\xi^\sharp = (\xi \sqcap \nabla_{XI}\nu) \ominus \xi^\sharp \,\} \\
&\leftrightarrow q \sqsubseteq (p\xi \sqcap \nabla_{RI}\nu) \ominus \xi^\sharp && \{\, p\nabla_{XI} = \nabla_{RI} \,\} \\
&\leftrightarrow p\xi \sqcap \nabla_{RI}\nu \sqsubseteq q\xi && \{\, \text{Galois connection} \,\} \\
&\leftrightarrow \nabla_{RI}\nu \sqsubseteq p\xi \Rightarrow q\xi. && \{\, \text{Prop of } \Rightarrow \,\}
\end{aligned}
$$

$\qquad\qquad\qquad\qquad\qquad\qquad\qquad\qquad\qquad\qquad\qquad\qquad\square$

Lemma 2. $\lambda \sqsubseteq f\xi^\sharp$.

Proof. By Proposition 4 we will show that $f\xi^\sharp$ satisfies (A0) - (A3).
(A0)

$$
\begin{aligned}
\xi \sqcap \nabla_{XI}\nu &= \xi \sqcap \nabla_{XI}\nabla_{IX}[\mathrm{id}_X \sqcap (\xi^\sharp\gamma \ominus \xi)] && \{\, \text{Def of } \nu \,\} \\
&\sqsubseteq \xi[\mathrm{id}_X \sqcap (\xi^\sharp\gamma \ominus \xi)] && \{\, \text{DF} \,\} \\
&\sqsubseteq \xi(\xi^\sharp\gamma \ominus \xi) \\
&\sqsubseteq \gamma \ominus \xi, && \{\, \text{Prop 2 and D4} \,\}
\end{aligned}
$$

$$
\begin{aligned}
\gamma &\sqsubseteq (\gamma \ominus \xi) \ominus \xi^\sharp && \{\, \text{Prop 2} \,\} \\
&\sqsubseteq (\xi \sqcap \nabla_{XI}\nu) \ominus \xi^\sharp && \{\, \xi \sqcap \nabla_{XI}\nu \sqsubseteq \gamma \ominus \xi \,\} \\
&= f\xi^\sharp. && \{\, f = \inf(\xi \sqcap \nabla_{XI}\nu; \xi) \,\}
\end{aligned}
$$

(A1) $\xi^\sharp = \xi \ominus \xi^\sharp \sqsubseteq (\xi \sqcap \nabla_{XI}\nu) \ominus \xi^\sharp = f\xi^\sharp$.
(A2) As f is a closure operation, the monotonic law $\xi \sqsubseteq f\xi f^\sharp$ holds. Then we have

$$
\begin{aligned}
\xi \sqsubseteq f\xi f^\sharp &\leftrightarrow f^\sharp\xi \sqsubseteq \xi f^\sharp && \{\, f : \text{function} \,\} \\
&\leftrightarrow \xi^\sharp f \sqsubseteq f\xi^\sharp,
\end{aligned}
$$

$$
\begin{aligned}
f\xi^\sharp f\xi^\sharp &\sqsubseteq ff\xi^\sharp\xi^\sharp && \{\, \xi^\sharp f \sqsubseteq f\xi^\sharp \,\} \\
&\sqsubseteq f\xi^\sharp. && \{\, ff = f,\ \xi\xi \sqsubseteq \xi \,\}
\end{aligned}
$$

(A3) $\sup(f\xi^\sharp; \xi) = f\sup(\xi^\sharp; \xi) = f\,\mathrm{id}_X \sqsubseteq f\xi^\sharp$. $\qquad\qquad\qquad\square$

Proposition 9. $\sup(\lambda; \xi) \sqsubseteq \nabla_{XI}\nu$.

Proof. Set $s = \sup(\lambda; \xi)$. Then $s^\sharp\lambda \sqsubseteq \xi^\sharp$ holds because of $\lambda \sqsubseteq s\xi^\sharp$. Also recall that (A0) $\gamma \sqsubseteq \lambda$, (A1) $\xi^\sharp \sqsubseteq \lambda$ and (A3) $s \sqsubseteq \lambda$.

$$
\begin{aligned}
\gamma^\sharp\xi s^\sharp s &\sqsubseteq \lambda^\sharp\lambda^\sharp\lambda^\sharp s && \{\, \gamma \sqsubseteq \lambda,\ \xi^\sharp \sqsubseteq \lambda,\ s \sqsubseteq \lambda \,\} \\
&\sqsubseteq \lambda^\sharp s && \{\, \lambda\lambda \sqsubseteq \lambda \,\} \\
&\sqsubseteq \xi. && \{\, s^\sharp\lambda \sqsubseteq \xi^\sharp \,\}
\end{aligned}
$$

Hence it follows that

$$
\begin{aligned}
s &= ss^\sharp s \\
&\sqsubseteq \nabla_{XX}[\mathrm{id}_X \sqcap (\xi^\sharp\gamma \ominus \xi)] && \{\, s \sqsubseteq \nabla_{XX},\ \gamma^\sharp\xi s^\sharp s \sqsubseteq \xi \,\} \\
&\sqsubseteq \nabla_{XI}\nu. && \{\, \nabla_{XX} = \nabla_{IX}\nabla_{IX} \,\}
\end{aligned}
$$

$\qquad\qquad\qquad\qquad\qquad\qquad\qquad\qquad\qquad\qquad\qquad\qquad\square$

Lemma 3. $f\xi^\sharp \sqsubseteq \lambda$.

Proof

$$
\begin{aligned}
f\xi^\sharp &= (\xi \sqcap \nabla_{XI}\nu) \ominus \xi^\sharp \quad \{\ f : \text{function}\ \} \\
&\sqsubseteq \sup(\lambda; \xi) \ominus \xi^\sharp \quad \{\ \textit{Prop 5 and prop 9}\ \} \\
&= \sup(\lambda; \xi)\xi^\sharp \quad \{\ \sup(\lambda; \xi) : \text{function}\ \} \\
&\sqsubseteq \lambda\lambda \quad \{\ (A3), (A1)\ \} \\
&\sqsubseteq \lambda. \quad \{\ (A2)\ \} \qquad\qquad\qquad \square
\end{aligned}
$$

Therefore $\lambda = f\xi^\sharp$ and $\sup(\lambda; \xi) = f$ hold. Hence, the supremum of the inference relation, introduced by Armstrong's inference rules, and the closure operation f are equivalent.

Next, we introduce an implication relation, and show the following propositions.

Definition 6. *Let* $\tau : I \to X$ *be a relation and* $u_\tau = \mathrm{id}_X \sqcap \tau^\sharp \tau$. *Define an implication relation* $\delta_\tau : X \to X$ *and a closure operation* $f_\tau : X \to X$ *by*

$$
\delta_\tau = \xi u_\tau \ominus \xi^\sharp \quad \text{and} \quad f_\tau = \inf(\xi \sqcap \nabla_{XI}\tau; \xi).
$$

For this implication relation and closure operation the following propositions hold.

Proposition 10. $\delta_\tau = f_\tau \xi^\sharp$ *and* $f_\tau = \max(\delta_\tau; \xi)$.

Proof

$$
\begin{aligned}
f_\tau \xi^\sharp &= (\xi \sqcap \nabla_{XI}\tau) \ominus \xi^\sharp \\
&= \xi u_\tau \ominus \xi^\sharp \quad \{\ \xi \sqcap \nabla_{XI}\tau = \xi u_\tau\ \} \\
&= \delta_\tau, \quad \{\ \textit{Def of } \delta_\tau\ \}
\end{aligned}
$$

and

$$
\begin{aligned}
\max(\delta_\tau; \xi) &= \max(f_\tau \xi^\sharp; \xi) \quad \{\ \delta_\tau = f_\tau \xi^\sharp\ \} \\
&= f_\tau \max(\xi^\sharp; \xi) \quad \{\ f_\tau : \text{function}\ \} \\
&= f_\tau. \quad \{\ \max(\xi^\sharp; \xi) = \mathrm{id}_X\ \} \qquad \square
\end{aligned}
$$

Proposition 11. *Let* $0, 1, a : I \to X$ *be I-points and* $\delta_a : X \to X$ *an implication relation.*

(a) *If* $0\xi = \nabla_{IX}$, *then* $0\delta_a = a\xi^\sharp$,
(b) *If* $1\xi^\sharp = \nabla_{IX}$ *and* $a \sqcap 1 = 0_{IX}$, *then* $1\delta_a^\sharp = a\xi^{\sharp -}$.

Proof
(a)

$$
\begin{aligned}
0\delta_a &= 0(\xi a^\sharp a \ominus \xi^\sharp) \quad \{\ u_a = a^\sharp a \text{ by } a : \text{function}\ \} \\
&= 0\xi a^\sharp a \ominus \xi^\sharp \quad \{\ 0 : \text{function}\ \} \\
&= \nabla_{IX} a^\sharp a \ominus \xi^\sharp \quad \{\ 0\xi = \nabla_{IX}\ \} \\
&= a \ominus \xi^\sharp \quad \{\ a\nabla_{XI} = \mathrm{id}_I\ \} \\
&= a\xi^\sharp. \quad \{\ a : \text{function}\ \}
\end{aligned}
$$

(b)

$$1\delta_a^{\sharp} = 1(\xi a^{\sharp} a \ominus \xi^{\sharp})^{\sharp}$$
$$= (\xi a^{\sharp} a \ominus \xi^{\sharp} 1^{\sharp})^{\sharp}$$

$$1\xi = 1\xi \sqcap \nabla_{IX} = 1\xi \sqcap 1\xi^{\sharp} \{ 1\xi^{\sharp} = \nabla_{IX} \}$$
$$= 1(\xi \sqcap \xi^{\sharp})$$

By $\mathrm{id}_X \sqsubseteq \xi$ and $\xi \sqcap \xi^{\sharp} \sqsubseteq \mathrm{id}_X$, $1\xi = 1$.
   Therefore

$$1\delta_a^{\sharp} = (\xi a^{\sharp} a \ominus 1^{\sharp})^{\sharp} \quad \{ 1\xi = 1 \text{ by } 1\xi^{\sharp} = \nabla_{IX} \}$$
$$= (\xi a^{\sharp} a 1^{\sharp -})^{\sharp -} \quad \{ \text{Schröder category} \}$$
$$= (\xi a^{\sharp})^{\sharp -} \quad\quad \{ a 1^{\sharp -} = \mathrm{id}_I \text{ by } a \sqcap 1 = 0_{IX} \}$$
$$= (a\xi^{\sharp})^{-}$$
$$= a\xi^{\sharp -}. \quad\quad \{ a : \text{I-point} \} \qquad\qquad \square$$

7   Completeness Theorem for Armstrong's Inference Rules

In this section, we give a relation algebraic proof of the completeness theorem for Armstrong's inference rules in a Schröder category with unit satisfying the point axiom.

Theorem 1 (Completeness). *Assume that there exist I-points $0, 1 : I \to X$ such that $0\xi = \nabla_{IX}$ and $1\xi = 1$. Then the identity*

$$\lambda = \sqcap_{\gamma \sqsubseteq \delta_\tau} \delta_\tau$$

holds.

Proof. First note that δ_τ satisfies (A1) - (A3) for all relations τ.
(A1) $\xi^{\sharp} = \xi \ominus \xi^{\sharp} \sqsubseteq \xi u_\tau \ominus \xi^{\sharp} = \delta_\tau$.
(A2)

$$\delta_\tau \delta_\tau = f_\tau \xi^{\sharp} f_\tau \xi^{\sharp} \{ \text{Prop 10} \}$$
$$\sqsubseteq f_\tau f_\tau \xi^{\sharp} \xi^{\sharp} \{ \xi \sqsubseteq f_\tau \xi f_\tau^{\sharp} \}$$
$$= f_\tau \xi^{\sharp}$$
$$= \delta_\tau. \quad \{ \text{Prop 10} \}$$

(A3) $\sup(\delta_\tau; \xi) = \sup(f_\tau \xi^{\sharp}; \xi) = f_\tau \sup(\xi^{\sharp}; \xi) = f_\tau \mathrm{id}_X \sqsubseteq f_\tau \xi^{\sharp} = \delta_\tau$.

Thus for all relations τ such that (A0) $\gamma \sqsubseteq \delta_\tau$ the inclusion $\lambda \sqsubseteq \delta_\tau$ holds by Proposition 4 and consequently we have $\lambda \sqsubseteq \sqcap_{\gamma \sqsubseteq \delta_\tau} \delta_\tau$ holds. Next we will see the converse $\sqcap_{\gamma \sqsubseteq \delta_\tau} \delta_\tau \sqsubseteq \lambda$.
   By using the point axiom $\mathrm{id}_X = \sqcup_{x \in X} x^{\sharp} x$ we show

$$x(\sqcap_{\gamma \sqsubseteq \delta_\tau} \delta_\tau) \sqsubseteq x\lambda$$

for all I-points $x : I \to X$. (Set $\hat{x} = x \sup(\lambda; \xi)$. Obviously $\hat{x}$ is an I-point.)
(I) $\hat{x} = x \sup(\lambda; \xi) = 1$:

$$
\begin{aligned}
x\lambda &= x \sup(\lambda; \xi)\xi^\sharp \quad \{\ Prop\ 5\ \} \\
&= 1\xi^\sharp \qquad\qquad \{\ xsup(\lambda; \xi) = 1\ \} \\
&= \nabla_{IX} \qquad\quad \{\ 1\xi^\sharp = \nabla_{IX}\ \} \\
&\sqsupseteq x(\sqcap_{\gamma \sqsubseteq \delta_\tau} \delta_\tau).
\end{aligned}
$$

(II) $\hat{x} = x \sup(\lambda; \xi) \neq 1$:
We now assume the totality condition: All nonzero relations $\rho : I \to X$ are total.
Then $\hat{x} \sqcap 1 = 0_{IX}$ holds. First we will prove $\gamma \sqsubseteq \delta_{\hat{x}}$.

$$
\begin{aligned}
\gamma &= (\mathrm{id}_X \sqcap \nabla_{XI}\nabla_{IX})\gamma &&\{\ \nabla_{XX} = \nabla_{XI}\nabla_{IX}\ \} \\
&= (\mathrm{id}_X \sqcap \nabla_{XI}\hat{x}\nabla_{XX})\gamma &&\{\ \hat{x} : \text{function}\ \} \\
&= (\mathrm{id}_X \sqcap \nabla_{XI}\hat{x}\xi^\sharp)\gamma \sqcup (\mathrm{id}_X \sqcap \nabla_{XI}\hat{x}\xi^{\sharp-})\gamma &&\{\ \nabla_{XX} = \xi^\sharp \sqcup \xi^{\sharp-}\ \} \\
&\sqsubseteq \delta_{\hat{x}}. &&\{\ \text{(ii) and (iii)}\ \}
\end{aligned}
$$

(i) $x\lambda = 0\delta_{\hat{x}}$
$x\lambda = x \sup(\lambda; \xi)\xi^\sharp = \hat{x}\xi^\sharp = 0\delta_{\hat{x}}$ by Proposition 5 and Proposition 11.
(ii)

$$
\begin{aligned}
\nabla_{XI}\hat{x}\xi^\sharp\gamma &= \nabla_{XI}x\lambda\gamma &&\{\ \sup(\lambda; \xi)\xi^\sharp = \lambda\ \} \\
&\sqsubseteq \nabla_{XI}x\lambda &&\{\ \gamma \sqsubseteq \lambda,\ \lambda\lambda \sqsubseteq \lambda\ \} \\
&= \nabla_{XI}0\delta_{\hat{x}} &&\{\ \text{(i)}\ \} \\
&\sqsubseteq \nabla_{XI}(\nabla_{IX} \ominus \xi^\sharp)\delta_{\hat{x}} &&\{\ 0 \sqsubseteq \nabla_{IX} \ominus \xi^\sharp \text{ by } \nabla_{IX} = 0\xi\ \} \\
&\sqsubseteq \xi^\sharp\delta_{\hat{x}} &&\{\ \nabla_{XI}(\nabla_{IX} \ominus \xi^\sharp) \sqsubseteq \xi^\sharp\ \} \\
&\sqsubseteq \delta_{\hat{x}}. &&\{\ \xi^\sharp \sqsubseteq \delta_{\hat{x}},\ \delta_{\hat{x}}\delta_{\hat{x}} \sqsubseteq \delta_{\hat{x}}\ \}
\end{aligned}
$$

(iii)

$$
\begin{aligned}
&(\mathrm{id}_X \sqcap \nabla_{XI}\hat{x}\xi^{\sharp-})\gamma \\
&= (\mathrm{id}_X \sqcap \nabla_{XI}1\delta_{\hat{x}}^\sharp)\gamma \sqcap \nabla_{XI}\nabla_{IX} &&\{\ Prop\ 11\ \} \\
&\sqsubseteq (\mathrm{id}_X \sqcap \nabla_{XI}1\delta_{\hat{x}}^\sharp)\gamma \sqcap \nabla_{XI}1\delta_{\hat{x}} &&\{\ \nabla_{IX} = 1\xi^\sharp \sqsubseteq 1\delta_{\hat{x}}\ \} \\
&\sqsubseteq (\mathrm{id}_X \sqcap \nabla_{XI}1\delta_{\hat{x}}^\sharp)(\gamma \sqcap \delta_{\hat{x}}1^\sharp\nabla_{IX}\nabla_{XI}1\delta_{\hat{x}}) &&\{\ DF\ \} \\
&\sqsubseteq \delta_{\hat{x}}\delta_{\hat{x}} &&\{\ \nabla_{XI}1 : \text{function}\ \} \\
&\sqsubseteq \delta_{\hat{x}}.
\end{aligned}
$$

Hence

$$
\begin{aligned}
x(\sqcap_{\gamma \sqsubseteq \delta_\tau} \delta_\tau) &\sqsubseteq x\delta_{\hat{x}} &&\{\ \gamma \sqsubseteq \delta_{\hat{x}}\ \} \\
&\sqsubseteq x\lambda\delta_{\hat{x}} &&\{\ \mathrm{id}_X \sqsubseteq \xi^\sharp \sqsubseteq \lambda\ \} \\
&= 0\delta_{\hat{x}}\delta_{\hat{x}} &&\{\ \text{(i)}\ \} \\
&\sqsubseteq 0\delta_{\hat{x}} &&\{\ \delta_{\hat{x}}\delta_{\hat{x}} \sqsubseteq \delta_{\hat{x}}\ \} \\
&= x\lambda. &&\{\ \text{(i)}\ \} \qquad\qquad \square
\end{aligned}
$$

8   Conclusion

We present relational formulations for Armstrong's inference rules and implication in a Dedekind category. We treat Armstrong's inference rules and the implication as (binary) relations in an upper semi lattice in a Dedekind category, and give a relation algebraic proof of the completeness theorem for Armstrong's inference rules in a Schröder category. In the future, relation calculus might demonstrate more definitions and theorems of formal concepts. Really, it could completely represent an algorithm of the analytic method.

References

1. Codd, E.: A Relational Model of Data for Large Shared Data Banks. Communications of the ACM 13, 377–387 (1970)
2. Beeri, C., Fagin, R., Howard, J.: A Complete Axiomatization for Functional and Multivalued Dependencies in Database Relations. In: Proceedings of the 1977 ACM SIGMOD International Conference on Management of Data, Toronto, Canada, pp. 47–61 (1977)
3. Armstrong, W.W.: Dependency Structures in Database Relationships. In: IFIP Congress, pp. 580–583 (1974)
4. Ishida, T., Honda, K., Kawahara, Y.: Implication and Functional Dependency in Intensional Contexts. Bulletin of Informatics and Cybernetics 40, 101–111 (2008)
5. Ganter, B., Wille, R.: Formal Concept Analysis. Springer, Heidelberg (1999)
6. Olivier, J., Serrato, D.: Catégories de Dedekind. Morphismes dans les Catégories de Schröder. C. R. Acad. Sci. Paris 260, 939–941 (1980)
7. Freyd, P., Scedrov, A.: Categories, Allegories. North-Holland, Amsterdam (1990)
8. Yasuo, K.: Urysohn's lemma in Schröder categories. Bulletin of Informatics and Cybernetics 39, 69–81 (2007)
9. Kawahara, Y., Tanaka, K.: Closure Systems and Closure Operations in Dedekind Categories. Bulletin of Informatics and Cybernetics 40, 89–99 (2008)
10. Kawahara, Y.: Theory of Relations. Kyushu University (2007)
11. Ishida, T., Honda, K., Kawahara, Y.: Formal Concepts in Dedekind Categories. In: Berghammer, R., Möller, B., Struth, G. (eds.) RelMiCS/AKA 2008. LNCS, vol. 4988, pp. 221–233. Springer, Heidelberg (2008)

Data Mining, Reasoning and Incremental Information Retrieval through Non Enlargeable Rectangular Relation Coverage

Ali Jaoua[1], Rehab Duwairi[1], Samir Elloumi[2], and Sadok Ben Yahia[2]

[1] Computer Science and Engineering Department College of Engineering,
Qatar University, Doha, Qatar

[2] Faculté des Sciences de Tunis, Département des Sciences de l'Informatique,
Tunis, Tunisia

Abstract. Association rules extraction from a binary relation as well as reasoning and information retrieval are generally based on the initial representation of the binary relation as an adjacency matrix. This presents some inconvenience in terms of space memory and knowledge organization. A coverage of a binary relation by a minimal number of non enlargeable rectangles generally reduces memory space consumption without any loss of information. It also has the advantage of organizing objects and attributes contained in the binary relation into a conceptual representation. In this paper, we propose new algorithms to extract association rules (i.e. data mining), conclusions from initial attributes (i.e. reasoning), as well as retrieving the total objects satisfying some initial attributes, by using only the minimal coverage. Finally we propose an incremental approximate algorithm to update a binary relation organized as a set of non enlargeable rectangles. Two main operations are mostly used during the organization process: First, separation of existing rectangles when we delete some pairs. Second, join of rectangles when common properties are discovered, after addition or removal of elements from a binary context. The objective is the minimization of the number of rectangles and the maximization of their structure. The article also raises the problems of equational modeling of the minimization criteria, as well as incrementally providing equations to maintain them.

Keywords: Formal Concept Analysis, Minimal representation, Galois Connection.

1 Introduction

Decomposition of binary relations into non enlargeable rectangles emerges in several contexts, including formal concept analysis, relational algebra and graph theory [1]. However, relational algebra offers more accurate relational equations to describe optimal decompositions. A binary relation is also central in the definition of a formal concept analysis [2,3], where the triplet $(\mathcal{D}, \mathcal{R}, \mathcal{T})$, built from a set of objects $\mathcal{D}$, a set of attributes $\mathcal{T}$, and a relation $\mathcal{R}$ linking objects to

R. Berghammer et al. (Eds.): RelMiCS/AKA 2009, LNCS 5827, pp. 199–210, 2009.

attributes, is very useful for data analysis and classification. For several years, studies of minimal decomposition with non enlargeable rectangles have been done concurrently with minimal knowledge representation of a binary context done in the field of formal concept analysis. These two views are first related in the papers of Belohlavek and Vychodil [4], and Kcherif et. al. [5]. Minimal rectangular representation of a binary relation $\mathcal{R}$ has several advantages. First, it offers the possibility of representing a binary relation $\mathcal{R}$ with less memory space. Second, each non enlargeable rectangle represents a class of maximal number of objects (i.e. the domain of the rectangle) sharing a maximal number of attributes (i.e. the range of the rectangle), all together these rectangles represent a pyramidal classification of the relation $\mathcal{R}$. Third, the minimal representation represents a coverage of the initial binary relation, therefore it is always possible to generate any other non enlargeable rectangle from the minimal coverage without reusing the initial relation. In this paper, we propose an exact algorithm to map a Galois Connection [6] through any representation of a binary relation with non enlargeable rectangles. We also propose an exact algorithm to find additional attributes starting from initial subset of attributes, and to find all objects satisfying some query through the minimal rectangular decomposition [7,8,9]. Finally, we propose an incremental approximate algorithm for restructuring the minimal representation when we add or remove some pairs, attributes or objects. Working on a rectangular representation is also useful for experimenting a new model for information representation, similar to semantic and neural networks, but original because here, we will try to minimize the number of non enlargeable rectangles. We show that two operations are necessary to maintain the equilibrium of the rectangular system: separation, and join of rectangles. These operations allow us to maintain approximately some minimal representation by non enlargeable rectangles.

In the next section, we define some relational algebraic background; afterwards, we present the definition of minimal rectangular decomposition of a binary relation. In section 5, we make the link between information retrieval and knowledge engineering through maximal rectangular representation. We then develop an algorithm implementing Galois connection through a minimal rectangular coverage, later, in section 6, we derive an approximate algorithm for incremental organization of the rectangular coverage.

2   Preliminary Concepts

2.1   Binary Relations

A binary relation $\mathcal{R}$ between two finite sets $\mathcal{D}$ and $\mathcal{T}$ is a subset of the Cartesian product $\mathcal{D} \times \mathcal{T}$. An element in $\mathcal{R}$ is denoted by (e, e'), where e' denotes an image of e by $\mathcal{R}$. For a binary relation, we associate the following subsets [10]:

- The set of images of e is defined by : $e.\mathcal{R} = \{e' | (e, e') \in \mathcal{R}\}$,
- The set of antecedents of e' is defined by : $\mathcal{R}.e' = \{e | (e, e') \in \mathcal{R}\}$,
- The domain of $\mathcal{R}$ is defined by: $Dom(\mathcal{R}) = \{e | \exists e' : (e, e') \in \mathcal{R}\}$,

- The range of $\mathcal{R}$ is defined by: $Ran(\mathcal{R}) = \{e'|\exists e : (e, e') \in \mathcal{R}\}$,
- The cardinality of $\mathcal{R}$ (Card($\mathcal{R}$)) is the number of pairs belonging to $\mathcal{R}$.

Let $\mathcal{R}$ and $\mathcal{R}'$ be two binary relations. We define the composition of $\mathcal{R}$ and $\mathcal{R}'$ as the relation given by: $\mathcal{R}; \mathcal{R}'=\{(e,e')|\exists\, t : (e,t) \in \mathcal{R} \wedge (t,e') \in \mathcal{R}'\}$.

- The inverse of a relation is: $\mathcal{R}^T=\{(e,e')|(e',e) \in \mathcal{R}\}$
- The complement of a relation is: $\overline{\mathcal{R}}=\{(e,e')|(e,e') \notin \mathcal{R}\}$
- The relation $\mathcal{I}$, identity over a set A, is given by : $\mathcal{I}(A)= \{(e,e)|e \in A\}$.

Relations on finite sets can be represented by tables or matrices, as shown in Figure 1 for the relation x congruent y modulo 3 on the set $\{1, 2, 3, 4, 5, 6, 7\}$.

	1	2	3	4	5	6	7
1	1	0	0	1	0	0	1
2	0	1	0	0	1	0	0
3	0	0	1	0	0	1	0
4	1	0	0	1	0	0	1
5	0	1	0	0	1	0	0
6	0	0	0	0	0	1	0
7	1	0	0	1	0	1	1

Fig. 1. A relation represented by a Boolean Matrix

Notion of Rectangle and Non Enlargeable Rectangle (or Maximal Rectangle)

Let $\mathcal{R}$ be a binary relation defined between $\mathcal{D}$ and $\mathcal{T}$: A rectangle of $\mathcal{R}$ is a Cartesian product of two non empty sets $A \subseteq \mathcal{D}$ and $B \subseteq \mathcal{T}$ and $A \times B \subseteq \mathcal{R}$ where A is the domain (also called *objects*), and B is the range (also called *attributes*) of the rectangle. The rectangular closure of a binary relation is: $\mathcal{R}^*= Dom(\mathcal{R}) \times Cod(\mathcal{R})$.

A Rectangle $A \times B \subseteq \mathcal{R}$ is called non enlargeable if:

$A \times B \subseteq A' \times B' \subseteq \mathcal{R} \rightarrow (A = A')\wedge(B = B')$.

In terms of formal concept analysis a non enlargeable rectangle is called a formal concept, but it is defined as a pair (A, B) , enabling us to include (ϕ, B) and (A, ϕ), where ϕ represents the empty set.

3   Formal Concept Analysis

Here, we recall some basic notions from Formal Concept Analysis (FCA) [2,11,12]. Let D and T be sets, called the set of objects and attributes, respectively, and let R be a relation included in $D \times T$: for g belonging to D; m belonging to T, gRm holds if the object g has the attribute m. The triple $K = (D; T; R)$ is called a formal context.

3.1   Definitions

The operators u and v, from $2^T \rightarrow 2^D$ (respectively, from $2^D \rightarrow 2^T$) where $u(A) = \{m \in T | gRm, \forall g \in A\}$, and $v(B) = \{g \in D | gRm, \forall m \in B\}$, represent a specific Galois connection [1,2,8].

We notice that $u(A)$ represents the set of all attributes shared by all objects in A, and that $v(B)$ represents the set of all objects sharing all properties in B.

By definition we call $v(u(A))$ the closure of A, and $u(v(B))$ the closure of B.

The pair $(A; B)$, where A is included in D, B is included in T, $u(A) = B$, and $v(B) = A$ is called a **formal concept** of the context K with extent A and intent B (in this case we have also $v(u(A)) = A$ and $u(v(B)) = B$).

$u(v(B)) - B$ is by definition the set of associated attributes to the set B. $u(v(B)) - B$ contains all attributes shared by all objects satisfying all attributes in the set B. We may also conclude the association rule: $B \rightarrow u(v(B)) - B$.

3.2   Association Rules Extraction

If $\mathcal{B}$ and $\mathcal{C}$ are subsets of $\mathcal{T}$, the implication $\mathcal{B} \rightarrow \mathcal{C}$ holds if $v(\mathcal{B})$ is included in $v(\mathcal{C})$. We can also say that $\mathcal{A} \rightarrow u(v(\mathcal{A}))$ for any subset $\mathcal{A}$ of $\mathcal{T}$.

In the context of this paper, $\mathcal{T}$ is the set of attributes (or propositions). In Table 1, we find a representation of a context $\mathcal{K} = (\mathcal{D}; \mathcal{T}; \mathcal{R})$, where $\mathcal{D} = \{o_1, o_2, o_3\}$, $\mathcal{T} = \{p, q\}$ and $\mathcal{R} = \{(o_1, p), (o_1, q), (o_2, q)\}$.

Example 1. We associate to each implication: $p \rightarrow q$ the relation $\mathcal{R}(p, q)$ in Table 1. Notice that p and q are attributes (or propositions).

Table 1. Binary relation $R(p, q)$ associated to implication $p \rightarrow q$

	p	q
o_1	1	1
o_2	0	1
o_3	0	0

$u(v(\{p\})) = \{p, q\}$. This means that the closure of set $\{p\}$ equals the set $\{p, q\}$. Using the definition given in the previous subsection, we can represent the closure of set $\{p\}$ by the association rule $p \rightarrow q$. We also may conclude that $p \rightarrow q$ because $v(\{p\}) = \{o_1\}$ is included in $v(\{q\}) = \{o_1, o_2\}$.

4   Minimal Coverage of a Binary Relation by Non Enlargeable Rectangles

It is proved in the theory of formal concept analysis that the set of non enlargeable rectangles (i.e. concepts) in a binary relation $\mathcal{R}$ has a structure of lattice with respect to two operators (join and meet). However, the number of concepts in the lattice is generally exponential in terms of the size of the binary relation $\mathcal{R}$, while we are able to have different possible coverages of $\mathcal{R}$ with a minimal number of non enlargeable rectangles.

Example 2. Let $\mathcal{R}$ be the binary relation given in Table 2, relating four objects $\{o_1, o_2, o_3, o_4\}$ to five attributes $\{a_1, a_2, a_3, a_4, a_5\}$.

Table 2. Binary relation

$\mathcal{R}$	a_1	a_2	a_3	a_4	a_5
o_1	1	1	0	0	0
o_2	1	1	0	0	1
o_3	1	1	1	1	0
o_4	1	0	0	0	1

We can extract three non enlargeable rectangles, covering $\mathcal{R}$:
$\rho_1 = \{o_1, o_2, o_3\} \times \{a_1, a_2\}$.
$\rho_2 = \{o_3\} \times \{a_1, a_2, a_3, a_4\}$.
$\rho_3 = \{o_2, o_4\} \times \{a_1, a_5\}$.

$R = \rho_1 \cup \rho_2 \cup \rho_3$

Remark 1. In this paper, we don't discuss the different algorithms and heuristics to find the minimum coverage of $\mathcal{R}$. These questions are discussed in several papers [13,14]. However, it is easy to prove that from any binary relation coverage with minimal number of rectangles, we can associate the rectangular coverage of each rectangle of the coverage, and therefore find an equivalent minimal coverage with non enlargeable rectangles.

In the following section, starting from a minimal coverage of $\mathcal{R}$, we propose an algorithm to extract additional knowledge starting from initial facts or attributes. It is also a way to extract association rules.

5   Reasoning and Object Retrieval through a Minimal Coverage by Non Enlargeable Rectangles

Instead of searching through an initial binary relation, we can often profit from searching through non enlargeable rectangles. This is the case for reasoning (*i.e.* searching for conclusions starting from some initial attributes), finding association rules, as well as retrieving information (*i.e.* objects satisfying some attributes).

Here, we first explain the idea of the algorithm using Table 2 as the running example. Assume that you want to derive all associated attributes corresponding to a_1 with respect to the relation $\mathcal{R}$. If we use the closure of $\{a_1\}$ applied on $\mathcal{R}$ we obtain $\{a_1\}$ only. If we apply it to $\{a_2\}$ we obtain $\{a_1, a_2\}$ which means that $a_2 \rightarrow a_1$.

5.1 An Exact Algorithm for Reasoning, Information Retrieval, and Association Rules Extraction

Let $\rho_1, \rho_2, \rho_3, \ldots, \rho_n$ be n non enlargeable rectangles covering some binary relation $\mathcal{R}$ with a minimal way, then $R = \rho_1 \cup \rho_2 \cup \rho_3 \cup \ldots \cup \rho_n$.

The set of all objects satisfying simultaneously some subset of attributes $A = \{a_1, a_2, \ldots, a_p\}$ is by definition equal to $v(A)$, where v is the second operator of the Galois connection:

$$v(A) = a_1.R^T \cap a_2.R^T \cap \ldots \cap a_p.R^T$$
$$= (a_1.\rho_1{}^T \cup \ldots \cup a_1.\rho_n{}^T) \cap (a_2.\rho_1{}^T \cup \ldots \cup a_2.\rho_n{}^T) \cap \ldots \cap (a_p.\rho_1{}^T \cup \ldots \cup a_p.\rho_n{}^T)$$

Using a similar procedure but from a set of objects $O = \{o_1, o_2, \ldots, o_q\}$, we may in a similar way find all attributes satisfying simultaneously all objects in O, ($i.e$ $u(O)$).

$$u(O) = o_1.R \cap o_2.R \cap \ldots \cap o_q.R$$
$$= (o_1.\rho_1 \cup \ldots \cup o_1.\rho_n) \cap (o_2.\rho_1 \cup \ldots \cup o_2.\rho_n) \cap \ldots \cap (o_q.\rho_1 \cup \ldots \cup o_q.\rho_n)$$

If we use the coverage with three rectangles ρ_1, ρ_2, ρ_3, extracted from the binary relation given in Table 2, as a base for reasoning instead of the initial context, we first find all antecedents of a_1 ($i.e.v(\{a_1\})$), through the three rectangles to obtain $O = \{o_1, o_2, o_3, o_4\}$, as retrieved objects through the different existing rectangles, then through the same rectangles we look for all images of elements of O separately with respect to each non enlargeable rectangle, consecutively the images of O are:

$$S_1 = o_1.\rho_1$$
$$= \{a_1, a_2\},$$
$$S_2 = o_2.\rho_1 \cup o_2.\rho_3$$
$$= \{a_1, a_2, a_5\},$$
$$S_3 = o_3.\rho_1 \cup o_3.\rho_2$$
$$= \{a_1, a_2, a_3, a_4\},$$
$$S_4 = o_4.\rho_3$$
$$= \{a_1, a_5\}.$$

Therefore, we find $u(O) = S_1 \cap S_2 \cap S_3 \cap S_4$ equal to $\{a_1\}$ which means that $\{a_1\}$ is a closed set. If we do the same for $\{a_2\}$ we obtain $\{a_1, a_2\}$, which means that $a_2 \rightarrow a_1$. We may of course repeat the same process for as any attribute, that we can generalize to any subset of attributes.

By using these new relational expressions, of $v(A)$, $u(O)$ and $u(v(A))$ we now get respectively the set of objects $v(A)$ satisfying attributes A, the set of attributes $u(O)$ satisfying objects O, and association rules: $A \rightarrow (u(v(A)) - A)$. The advantage of the algorithm is that it runs on the base of the rectangles instead of the original relation. In Algorithm 1 of the appendix, a represents only one attribute, C is the set of all non enlargeable rectangles represented as an array of n records, where each element C[i] is itself composed of two fields C[i].antecedents as the set of objects, and C[i].images as the set of attributes of the rectangle shared by all objects belonging to C[i].antecedents. The function

Association-attributes in algorithm 1, implements $u(v(a))$. A generalization of the algorithm to several attributes is straightforward.

Association-attributes(Coverage of $\mathcal{R}$: C[], Initial attribute: a)
//Algorithm calculating $u(v(A))$ through the coverage C
//where A is only composed of one attribute
//C[i] is the ith non enlargeable rectangle
//n is the number of non enlargeable rectangles.
begin
   int n = C.length();
   Sequence_of_objects Antecedents = {};
   // First we build the set of all objects (*i.e.* antecedents of a) satisfying attribute a.
   for *(i=1; i<n;i++)* **do**
      //n is the number of non enlargeable rectangles.
      if *(a belongs to C[i].images)* **then**
         Antecedents= Antecedents ∪ C[i].antecedents;

   // Then we calculate the images of each object OB belonging to Antecedents,
   //into Attributes[OB],
   //we obtain all associated attributes in Conclusions as the conjunction of all
   //Attributes[OB] .
   Conclusions={}; // initially empty
   foreach *(object OB ∈ Antecedents)* **do**
      Attributes[OB]= {};
      for *(i=1; i<n; i++)* **do**
         if *(OB ∈ (C[i].antecedents))* **then**
            Attributes[OB]=Attributes[OB] ∪ C[i].Images;

   //Initialize by the set of all attributes of the first object OB0
   //belonging to Antecedents. Conclusion = Attributes[OB0] ;
   foreach *(object OB ∈ Antecedents)* **do**
      Conclusions= Conclusions ∩ Attributes[OB];
end

Algorithm 1. Algorithm for extracting conclusions from initial facts through a rectangular structure

The time complexity of the algorithm depends on the number of rectangles, and their size.

6 Approximate Incremental Non Enlargeable Rectangular Re-organization

In order to update the minimal rectangular coverage, after each addition of a new pair (ob, at), we analyze the following different cases: if ob , or at is new then, we necessarily need to create a new non enlargeable rectangle only containing the new pair. Otherwise, if ob and at are both already used in a current rectangle ρ, the new link (ob,at) is added to ρ only if the $Dom(\rho)$ contains only ob, or $Ran(\rho)$ contains only at, else we need to create the closure of ob with respect to each rectangle (i.e. separating pairs into two non enlargeable rectangles). Using regularly join operation, we guaranty the maximality of the rectangles and we minimize their number.

In case we delete an object, we only remove it from the domain of all existing non enlargeable rectangles. If we remove an attribute (at) for a specific object (ob) then we remove it from the ranges of all non enlargeable rectangles such that object ob belongs to their domain. A rectangle may become enlargeable if and only if its domain or (its range) is equal to the domain or (the range) of another rectangle; in that case by merging the two rectangles, we obtain a new one. The process should be continued until no new updated rectangles have the same domain or same range as any other one. By calling function join we maintain the maximality of the rectangles, and decrease their number. As a matter of fact, maintaining the minimum number of non enlargeable rectangle is still an open problem, for which we should find efficient algorithms.

Example 3. In order to explain the reorganization process, we use the binary relation $\mathcal{R}$ of example 2, including its minimal rectangular coverage with the three non enlargeable rectangles: ρ_1, ρ_2, ρ_3 :
where:
$\rho_1 = \{o_1, o_2, o_3\} \times \{a_1, a_2\}.$
$\rho_2 = \{o_3\} \times \{a_1, a_2, a_3, a_4\}.$
$\rho_3 = \{o_2, o_4\} \times \{a_1, a_5\}.$

$$\mathcal{R} = \rho_1 \cup \rho_2 \cup \rho_3.$$

If we add the pair (o_1, a_4) then a new non enlargeable rectangle ρ_4 is created (i.e. we call it separation action): $\rho_4 = \{o_1\} \times \{a_1, a_2, a_4\}.$

If we still add the pair (o_2, a_4) then a new non enlargeable rectangle ρ_5 is created: $\rho_5 = \{o_2\} \times \{a_1, a_2, a_4, a_5\}.$

If we add the pair (o_3, a_4) then no new non enlargeable rectangle is created, because the pair already belongs to ρ_2. By applying the Galois connection on each one of the rectangles, we make them non enlargeable. By this way, ρ_1 is updated to: $\rho_1 = \{o_1, o_2, o_3\} \times \{a_1, a_2, a_4\}$, and ρ_4 becomes equal to ρ_1. If we add $\{o_5, a_1\}, \{o_5, a_5\}, \{o_5, a_2\}, \{o_5, a_4\}$ to $\mathcal{R}$ then ρ_1 is update to: $\rho_1 = \{o_1, o_2, o_3, o_5\} \times \{a_1, a_2, a_4\}$, and ρ_3 is updated to $\rho_3 = \{o_2, o_4, o_5\} \times \{a_1, a_5\}.$

In case we add a new pair (o_6, a_7) then a new non enlargeable rectangle $\rho_4 = \{o_6\} \times \{a_7\}$ is created.

In case we add the new pair (o_6, a_4) we create the new non enlargeable rectangle containing all objects chairing attribute a_4: $\rho_5 = \{o_3, o_6\} \times \{a_4\}.$

If we remove the following pair: (o_3, a_4), then initial non enlargeable rectangle ρ_1 is split into two non enlargeable ones:

$\rho_5 = \{o_1, o_2, o_3, o_5\} \times \{a_1, a_2\};$ and $\rho_6 = \{o_1, o_2, o_5\} \times \{a_4\}.$

```
function adding-a-pair(Coverage of R: C[], attribute: at, object: ob)
//C[i] is the ith non enlargeable rectangle
// n is the current number of non enlargeable rectangles
begin
    int n = C.length();
    for (i=1 to n) do
        add (at,ob,i) // where i mentions non enlargeable rectangle C[i].
end

function delete-a-pair(Coverage of R: C[], attribute: at, object: ob, int: n)
//C[i] is the ith non enlargeable rectangle.
// n is the number of non enlargeable rectangles.
begin
    for (i=1 to n) do
        remove (at,ob,i); //Remove the pair (ob,at) from rectangle C[i].
        if (C[i] is empty) then
            // the rectangle i becomes empty.
            swap (i,n); // swap rectangle i with rectangle n.
            n=n-1;//decrement the number of rectangles.
end

function add(attribute: at, object: ob, int: i)
begin
    if ((ob is new) or (at is new)) then
        (ob,at) belongs to only one new non enlargeable rectangle (n+1);
        that we can build by applying Galois Connection on object ob if not
        new, else if attribute at is not new then we apply Galois Connection
        on attribute at , else we create the new rectangle containing only the
        new pair (ob,at)
    else if ((cardinal(C[i].domain)==1)and(ob ∈ C[i].domain)) then
        add (at) to C[i].attribute;
    else if ((cardinal(C[i].range==1))and(at ∈ C[i].range)) then
        add ob to C[i].domain;
    else if ((ob ∈ C[i].domain)and(cardinal(C[i].range)==1)) then
        // Creation of a new non enlargeable rectangle (n+1).
        C[n+1].domain=ob; C[n+1].range= at;
        for (j=1 to n) do
            if (at ∈ C[j].range) then
                C[n+1].domain =C[n+1].domain C[j].domain;
        for (j=1 to n) do
            Join (C[j], C[n+1]);
    else
        n=n+1;
        // Create a new rectangle containing only the new pair (ob,at).
        C[n].attribute=at; C[n].object=ob;
        for (i=1 to n) do
            Join(C[i], C[n]);
end
```

Algorithm 2. Approximate Incremental reorganization of a rectangular structure (Part I)

```
function remove(attribute: at, object: ob, int: i)
// i is a non enlargeable rectangle number.
begin
    if ( (cardinal(C[i])==1)and(ob ∈ C[i].domain)and(at ∈ C[i].range))
    then
        // rectangle i becomes empty.
        C[i].attribute=;
        C[i].Object=;
    else if (ob ∈ C[i].domain) then
        // Separation of concept i into two concepts (updated i, and (n+1)).
        C[i].range=C[i].range-at;
        Create (C[n+1]);
        C[n+1].domain=C[i].domain - ob;
        C[n+1].range=C[i].range ∪ at;
        n=n+1;
        // joining new obtained rectangles with others for (j=1 to (n-1))
        do
            Join(C[j], C[ n]);
        for (j=1 to n) do
            Join (C[i], C[j]);
end

function join (Concepts C[i], C[j])
// i and j are two non enlargeable rectangle numbers.
begin
    if ( C[i].domain==C[j].domain) then
        merge rectangles i and j into a new rectangle k;
        C[k].domain=C[i].domain;
        C[k].range= C[i].range ∪ C[j].range;
        remove the two rectangles i and j;
        Use operations u and v to make rectangle k non enlargeable;
    else if (C[i].range==C[j].range) then
        merge rectangles i and j into a new rectangle k;
        C[k].range=C[i].range;
        C[k].domain= C[i].domain ∪ C[j].domain;
        remove the two rectangles i and j;
        Use operations u and v to make rectangle k non enlargeable;
end
```

Algorithm 3. Approximate Incremental reorganization of a rectangular structure (Part II)

Finally, in the appendix, we may find approximate algorithms 2 and 3, for updating a set of non enlargeable rectangles, by adding or removing different kinds of pairs, adding or removing objects or attributes. The idea is to maintain as much as possible a minimal representation. However, here we propose a method suitable to be a starting point for more accurate ones.

The function Join, after transforming the two rectangles to their closure, merges them into only one, if they have either the same domain or the same range. If the two rectangles have the same domain, and different ranges, then the function join merges them to only one rectangle with the same domain, and as a range the union of the ranges of the two rectangles. If the two rectangles have the same range, the function merges them into only one with the same range, and as a domain the union of the domains of the two rectangles. In future research work, we will try to improve its efficiency. We will also try to minimize the number of join operations.

The complexity of the algorithm in terms of n (i.e. the number of rectangles) is quadratic, where we mainly count the number of times basic operation join is executed.

7   Conclusion

In this paper, we investigate the main incremental functions we need to maintain a rectangular decomposition of a binary relation. We discover that two important operations are the most frequently done, join and separation of non enlargeable rectangles. The present work should be the base for rectangular database, or documentary database or internet space. It should open the door for a uniform structuring approach, suitable to be generalized and improved. Even if the objective is the minimization of the number of rectangles and the maximization of their structure, open problems are raised in this paper, on how to put in equation minimization criteria in general, and how to find the incremental equations to maintain it.

Acknowledgements

Special thanks to Qatar University of financing the present research work, under research grant #05013CS. We also thank the reviewers for their valuable remarks that we considered for the improvement of the present paper.

References

1. Winter, M.: Decomposing relations into orderings. In: Berghammer, R., Möller, B., Struth, G. (eds.) RelMiCS 2003. LNCS, vol. 3051. Springer, Heidelberg (2004)
2. Ganter, B., Wille, R.: Formal Concept Analysis. Springer, Heidelberg (1999)
3. Ganter, B., Stumme, G., Wille, R.: Formal Concept Analysis, Foundations and Applications. LNCS. Springer, Heidelberg (2005)
4. Belohlavek, R., Vychodil, V.: Discovery of optimal factors in binary data via a novel method of matrix decomposition. Journal of Computer and System Sciences (to appear, 2009)
5. Kcherif, R., Gammoudi, M., Jaoua, A.: Using difunctional relations in information organization. Information Science, 153–166 (2000)

6. Jaoua, A., Elloumi, S.: Galois connection, formal concepts and galois lattice in real relations: Application in a real classifier. The Journal of Systems and Software, 149–163 (2002)
7. Riguet, J.: Relations binaires, fermetures et correspondances de Galois. Bulletin de la société mathématique de France, 114–155 (1944)
8. Riguet, J.: Deductibility and exactness. In: Proceedings of the Second International Seminar on Relational Methods in Computer Sciences, PARATY, Brazil (1995)
9. Schmidt, G.: About fringes. Private correspondance with Pr. A. Jaoua (2009)
10. Schmidt, G., Stroehlein, T.: Relations and Graphs: Discrete Mathematics for Computer Scientits. Springer, Heidelberg (1993)
11. Carpineto, C., Romano, G.: Exploiting the Potential of Concept Lattices for Information Retrieval with CREDO. Journal of Universal Computing, 985–1013 (2004)
12. Godin, R., Gecsei, J., Pichet, C.: Design of browsing interface for information retrieval. In: Proceedings of the International Conference SIGIR 1989, pp. 32–39 (1989)
13. Hasnah, A., Jaoua, A., Jaam, J.: Computer Aided Intelligent Recognition Techniques and Applications, Conceptual Data Classification: Application for Knowledge Extraction. Wiley, Chichester (2005)
14. Belkhiter, N., Bourhfir, C., Gammoudi, M., Jaoua, A., Thanh, N.L., Reguig, M.: Décomposition rectangulaire optimale d'une relation binaire: Application aux bases de données documentaires. Information Systems and Operational Research Journal, 33–54 (1994)

Collagories for Relational Adhesive Rewriting*

Wolfram Kahl

McMaster University, Hamilton, Ontario, Canada
kahl@cas.mcmaster.ca

Abstract. We define collagories essentially as "distributive allegories without zero morphisms", and show that they are sufficient for accommodating the relation-algebraic approach to graph transformation. Collagories closely correspond to the adhesive categories important for the categorical DPO approach to graph transformation. but thanks to their relation-algebraic flavour provide a more accessible and more flexible setting.

1 Introduction

One of the hallmarks of the relation-algebraic approach to graph transformation [Kaw90, Kah01, Kah04] is that it allows an abstract characterisation of the gluing condition for the double pushout approach. Nevertheless, the categorical approach to graph transformation has continued to use the node-and-edge-based formulation of the gluing condition even in the handbook chapter [CMR+97]. Recently, the literature of the categorical approach, starting essentially with [EPPH06] has adopted the "adhesive categories" of Lack and Sobociński [LS04], where however the details of the gluing condition are completely sidestepped.

The relational categories used in the relational approach so far arise from the toposes of graph structures, which are adhesive categories by virtue of being toposes. However, adhesive categories also include, for example, categories of pointed sets, which do not give rise to distributive allegories due to the failure of the zero law.

In this paper we show that dropping the zero law still produces a relational formalism that can accommodate the necessary tools for graph transformation, and furthermore relates nicely with adhesive categories.

We first re-develop, in sections 2–4, the fundamentals of the relation-algebraic approach to graph transformation using our new *bi-tabular collagories*, and show in Sect. 5 that these provide adhesive categories. Sections 6–8 are then devoted to constructing concrete bi-tabular collagories of semi-unary algebras.

2 Categories, Allegories

This section only serves to fix notation and terminology for standard concepts, see [FS90, SS93, Kah04]. Like Freyd and Scedrov and a slowly increasing number of categorists, we use denote composition in "diagram order" not only in

---

* This research is supported by the National Science and Engineering Research Council of Canada (NSERC).

R. Berghammer et al. (Eds.): RelMiCS/AKA 2009, LNCS 5827, pp. 211–226, 2009.

relation-algebraic contexts, where this is customary, but also in the context of categories. We will always use the infix operator ";" to make composition explicit: $R \mathbin{;} S = \mathcal{A} \xrightarrow{R} \mathcal{B} \xrightarrow{S} \mathcal{C}$.

Definition 2.1. A *category* $\mathbf{C}$ is a tuple $(\mathsf{Obj_C}, \mathsf{Mor_C}, \mathsf{src}, \mathsf{trg}, \mathbb{I}, \mathbin{;})$ where
- $\mathsf{Obj_C}$ is a collection of *objects*.
- $\mathsf{Mor_C}$ is a collection of *arrows* or *morphisms*.
- src (resp. trg) maps each morphism to its source (resp. target) object. Instead of $\mathsf{src}(f) = \mathcal{A} \wedge \mathsf{trg}(f) = \mathcal{B}$ we write $f : \mathcal{A} \to \mathcal{B}$. The collection of all morphisms f with $f : \mathcal{A} \to \mathcal{B}$ is denoted as $\mathsf{Mor_C}[\mathcal{A}, \mathcal{B}]$ and also called a *homset*.
- ";" is the binary *composition* operator, and composition of two morphisms $f : \mathcal{A} \to \mathcal{B}$ and $g : \mathcal{B}' \to \mathcal{C}$ is defined iff $\mathcal{B} = \mathcal{B}'$, and then $(f \mathbin{;} g) : \mathcal{A} \to \mathcal{C}$; composition is associative.
- $\mathbb{I}$ associates with every object $\mathcal{A}$ a morphism $\mathbb{I}_\mathcal{A}$ which is both a right and left unit for composition.   □

Definition 2.2. An *ordered category* is a category $\mathbf{C}$ such that
- for each two objects $\mathcal{A}$ and $\mathcal{B}$, the relation $\sqsubseteq_{\mathcal{A},\mathcal{B}}$ is a partial order on $\mathsf{Mor_C}[\mathcal{A}, \mathcal{B}]$ (the indices will usually be omitted), and
- composition is monotonic with respect to $\sqsubseteq$ in both arguments.   □

Definition 2.3. An *upper-semilattice category* is an ordered category where
- each homset is a distributive lattice with binary join $\sqcup$,
- composition distributes over binary joins from both sides.   □

For homsets that have least or greatest elements, we introduce corresponding notation:

Definition 2.4. In an ordered category, for each two objects $\mathcal{A}$ and $\mathcal{B}$ we introduce the following notions:
- If the homset $\mathsf{Mor_C}[\mathcal{A}, \mathcal{B}]$ contains a greatest element, this is denoted $\top_{\mathcal{A},\mathcal{B}}$.
- If the homset $\mathsf{Mor_C}[\mathcal{A}, \mathcal{B}]$ contains a least element, this is denoted $\bot_{\mathcal{A},\mathcal{B}}$.   □

For these extremal morphisms and for identities we frequently omit indices where these can be induced from the context.

Definition 2.5. An *allegory* is an ordered category such that
- each morphism $R : \mathcal{A} \to \mathcal{B}$ has a *converse* $R^\smile : \mathcal{B} \to \mathcal{A}$,
- the *involution equations* hold for all $R : \mathcal{A} \to \mathcal{B}$ and $S : \mathcal{B} \to \mathcal{C}$:

$$(R^\smile)^\smile = R \qquad \mathbb{I}_\mathcal{A}^\smile = \mathbb{I}_\mathcal{A} \qquad (R \mathbin{;} S)^\smile = S^\smile \mathbin{;} R^\smile$$

- conversion is monotonic with respect to $\sqsubseteq$.
- each homset is a lower semilattice with binary meet $\sqcap$.
- for all $Q : \mathcal{A} \to \mathcal{B}$, $R : \mathcal{B} \to \mathcal{C}$, and $S : \mathcal{A} \to \mathcal{C}$, the *modal rule* holds:

$$Q \mathbin{;} R \sqcap S \sqsubseteq (Q \sqcap S \mathbin{;} R^\smile) \mathbin{;} R .$$

    □

Many standard properties of relations can be characterised in the context of allegories:

Definition 2.6. A morphism $R : \mathcal{A} \to \mathcal{B}$ in an allegory is called:
- *univalent* iff $R^\smile ; R \sqsubseteq \mathbb{I}_\mathcal{B}$,
- *total* iff $\mathbb{I}_\mathcal{A} \sqsubseteq R ; R^\smile$,
- *injective* iff $R ; R^\smile \sqsubseteq \mathbb{I}_\mathcal{A}$,
- *surjective* iff $\mathbb{I}_\mathcal{B} \sqsubseteq R^\smile ; R$,
- a *mapping* iff it is univalent and total,
- *bijective* iff it is injective and surjective,
- *difunctional* iff $R ; R^\smile ; R \sqsubseteq R$. (See [SS93, 4.4] for more about difunctionality). $\square$

For an allegory $\mathbf{A}$, we write $\mathsf{Map}\,\mathbf{A}$ for the sub-category of $\mathbf{A}$ that contains only the mappings as arrows.

Definition 2.7. For a morphism $R : \mathcal{A} \to \mathcal{B}$ in an allegory, we define its *difunctional closure* $R^\boxplus : \mathcal{A} \to \mathcal{B}$ as the least difunctional morphism containing R (if this exists), and we further define $R^\rhd : \mathcal{A} \to \mathcal{A}$ and $R^\lhd : \mathcal{B} \to \mathcal{B}$ as:

$$R^\rhd := \mathbb{I} \sqcup R^\boxplus ; (R^\boxplus)^\smile \quad \text{and} \quad R^\lhd := \mathbb{I} \sqcup (R^\boxplus)^\smile ; R^\boxplus . \qquad \square$$

For endomorphisms, there are a few additional properties of interest:

Definition 2.8. A morphism $R : \mathcal{A} \to \mathcal{A}$ in an allegory is called:
- *reflexive* iff $\mathbb{I} \sqsubseteq R$,
- *transitive* iff $R ; R \sqsubseteq R$,
- *co-reflexive* or a *sub-identity* iff $R \sqsubseteq \mathbb{I}_\mathcal{A}$,
- *symmetric* iff $R^\smile \sqsubseteq R$,
- an *equivalence* iff it is symmetric, reflexive and transitive. $\square$

Definition 2.9. [FS90, 2.15] An object $\mathcal{U}$ in an allegory is a *partial unit* if $\mathbb{I}_\mathcal{U} = \mathbb{T}_{\mathcal{U},\mathcal{U}}$. The object $\mathcal{U}$ is a *unit* if, further, every object is the source of a total morphism targeted at $\mathcal{U}$. An allegory is said to be *unitary* if it has a unit. $\square$

We use the symbol "$\mathbb{1}$" for an arbitrary but fixed unit object.

3   Collagories

$$\kappa\acute{o}\lambda\lambda\alpha\text{: glue}$$

In Freyd and Scedrov's treatment, although allegories do not require zero-ary meets, distributive allegories do require zero-ary joins (least elements) together with distributivity of composition over them, that is, the zero law $\bot ; R = \bot$. We define an intermediate concept that does not assume anything about zero-ary joins:

Definition 3.1. A *collagory* is an allegory that is also an upper-semilattice category. $\qquad\square$

For Kleene star, we use Kozen's axioms [Koz94]:

Definition 3.2. A *Kleene collagory* is a collagory where, on homsets of endomorphisms, there is an additional unary operation $\_^*$ which satisfies the following axioms for all $R : \mathcal{A} \to \mathcal{A}$, $Q : \mathcal{B} \to \mathcal{A}$, and $S : \mathcal{A} \to \mathcal{C}$:

$$R^* \;=\; \mathbb{I}_A \sqcup R \sqcup R^* \,\mathring{,}\, R^* \qquad \text{recursive star definition}$$
$$Q \,\mathring{,}\, R \sqsubseteq Q \;\Rightarrow\; Q \,\mathring{,}\, R^* \sqsubseteq Q \qquad \text{right induction}$$
$$R \,\mathring{,}\, S \sqsubseteq S \;\Rightarrow\; R^* \,\mathring{,}\, S \sqsubseteq S \qquad \text{left induction} \qquad\square$$

Proposition 3.3. In a Kleene collagory, all difunctional closures exist, and:

$$R^{\triangleright} = (R \,\mathring{,}\, R^{\smile})^* \,, \qquad R^{\triangleleft} = (R^{\smile} \,\mathring{,}\, R)^* \,, \qquad R^{\boxtimes} = R^{\triangleright} \,\mathring{,}\, R = R \,\mathring{,}\, R^{\triangleleft} \,. \qquad\square$$

Alternatively, we also can fore-go the Kleene star and directly axiomatise difunctional closure:

Definition 3.4. A *difunctionally closed collagory* is a collagory where, there is an additional unary operation $\_^{\boxtimes}$ which satisfies the following axioms for all $R : \mathcal{A} \to \mathcal{B}$, $Q : \mathcal{C} \to \mathcal{A}$, and $S : \mathcal{A} \to \mathcal{C}$: $Q' : \mathcal{C} \to \mathcal{B}$, and $S' : \mathcal{B} \to \mathcal{C}$:

$$R^{\boxtimes} \;=\; R \sqcup R^{\boxtimes} \,\mathring{,}\, (R^{\boxtimes})^{\smile} \,\mathring{,}\, R^{\boxtimes} \qquad \text{recursive definition}$$
$$Q \,\mathring{,}\, R \sqsubseteq Q' \wedge Q' \,\mathring{,}\, R^{\smile} \,\mathring{,}\, R \sqsubseteq Q' \;\Rightarrow\; Q \,\mathring{,}\, R^{\boxtimes} \sqsubseteq Q' \qquad \text{right induction}$$
$$R \,\mathring{,}\, S \sqsubseteq S' \wedge R \,\mathring{,}\, R^{\smile} \,\mathring{,}\, S' \sqsubseteq S' \;\Rightarrow\; R^{\boxtimes} \,\mathring{,}\, S \sqsubseteq S' \qquad \text{left induction} \qquad\square$$

Proposition 3.5. In a difunctionally closed collagory, the operation $\_^{\boxtimes}$ produces difunctional closures.

PROOF: Containment $R \sqsubseteq R^{\boxtimes}$ and difunctionality $R^{\boxtimes} \,\mathring{,}\, (R^{\boxtimes})^{\smile} \,\mathring{,}\, R^{\boxtimes} \sqsubseteq R^{\boxtimes}$ follow directly from the recursive definition.

For minimality, assume that C is difunctional with $R \sqsubseteq C$. Then we have $\mathbb{I} \,\mathring{,}\, R \sqsubseteq C$ and $C \,\mathring{,}\, R^{\smile} \,\mathring{,}\, R \sqsubseteq C \,\mathring{,}\, C^{\smile} \,\mathring{,}\, C \sqsubseteq C$ and therefore, with the right induction rule, $R^{\boxtimes} = \mathbb{I} \,\mathring{,}\, R^{\boxtimes} \sqsubseteq C$. $\qquad\square$

4   Tabulations and Co-tabulations

Central to the connection between pullbacks and pushouts in categories of mappings on the one hand and constructions in relational theories on the other hand is the fact that a square of mappings commutes iff the "relation" induced by the source span is contained in that induced by the target co-span. The proof of this does not need the modal rule.

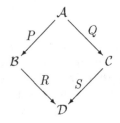

Lemma 4.1. [FS90, 2.146] Given a square of mappings in an allegory as drawn to the right, we have $P \,\mathring{,}\, R = Q \,\mathring{,}\, S$ iff $P^{\smile} \,\mathring{,}\, Q \sqsubseteq R \,\mathring{,}\, S^{\smile}$. $\qquad\square$

This provides a first hint that in the relational setting, the identity of the two mappings P and Q does not matter when looking for a pushout of the span $B \xleftarrow{P} A \xrightarrow{Q} C$ — we only need to consider the diagonal $P^\smile \,\sharp\, Q$. Dually, when looking for a pullback of the co-span $B \xrightarrow{R} D \xleftarrow{S} C$, only $R \,\sharp\, S^\smile$ needs to be considered. The gap between the two ways of calculating the horizontal diagonal can be significant since $R \,\sharp\, S^\smile$ is always difunctional.

Producing the result span of a pullback (respectively the result co-span of a pushout) from the horizontal diagonal alone is, in some sense, a generalisation of Freyd and Scedrov's splitting of idempotents; [Kah04] contains more discussion of this aspect.

Definition 4.2. [FS90, 2.14] In an allegory, let a morphism $V : B \to C$ be given. The span $B \xleftarrow{P} A \xrightarrow{Q} C$ of *mappings* P and Q is called a *tabulation of V* iff $P^\smile \,\sharp\, Q = V$ and $P \,\sharp\, P^\smile \sqcap Q \,\sharp\, Q^\smile = \mathbb{I}_A$. □

The following equivalent characterisation provided by [Kah04] has the advantage that it is fully equational, without the implicit inclusion conditions in the requirement that P and Q are mappings. This frequently facilitates calculations. Notice that $\mathbb{I} \sqcap V \,\sharp\, V^\smile = \mathsf{dom}\, V$; we use the expanded form to emphasise the duality with Proposition 4.6 below.

Proposition 4.3. In an allegory, the span $B \xleftarrow{P} A \xrightarrow{Q} C$ is a tabulation of $V : B \to C$ if and only if the following equations hold:

$$P^\smile \,\sharp\, Q = V \qquad \begin{array}{c} P^\smile \,\sharp\, P = \mathbb{I} \sqcap V \,\sharp\, V^\smile \\ Q^\smile \,\sharp\, Q = \mathbb{I} \sqcap V^\smile \,\sharp\, V \end{array} \qquad P \,\sharp\, P^\smile \sqcap Q \,\sharp\, Q^\smile = \mathbb{I}_A \ . \qquad \square$$

Tabulations in an allegory are unique up to isomorphism (this uses the modal rule), and include the following special cases:

- In a tabulation of a sub-identity, both tabulation morphisms are the induced *sub-object* injection [FS90, 2.145].
- We can define a *direct product* of A and B to be a tabulation of a $\top_{A,B}$, provided that greatest morphism exists.
- If a co-span $B \xrightarrow{R} D \xleftarrow{S} C$ of mappings is given, then its *pullback* in Map $\mathbf{A}$ is obtained as a tabulation of $R \,\sharp\, S^\smile$ [FS90, 2.147].

If an allegory in known to have all direct products and subobjects, then these can be used to construct a tabulation for each morphism.

Lemma 4.4. If a co-span $B \xrightarrow{R} D \xleftarrow{S} C$ of mappings is given with R injective, and $B \xleftarrow{P} A \xrightarrow{Q} C$ is a tabulation for $R \,\sharp\, S^\smile$, then Q is injective, too.

PROOF: First we use Proposition 4.3 to show $Q \,\sharp\, Q^\smile \sqsubseteq P \,\sharp\, P^\smile$:

$$\begin{aligned} Q \,\sharp\, Q^\smile &= Q \,\sharp\, Q^\smile \,\sharp\, Q \,\sharp\, Q^\smile &= Q \,\sharp\, (\mathbb{I} \sqcap (R \,\sharp\, S^\smile)^\smile \,\sharp\, R \,\sharp\, S^\smile) \,\sharp\, Q^\smile \\ &\sqsubseteq Q \,\sharp\, S \,\sharp\, R^\smile \,\sharp\, R \,\sharp\, S^\smile \,\sharp\, Q^\smile &= P \,\sharp\, R \,\sharp\, R^\smile \,\sharp\, R \,\sharp\, R^\smile \,\sharp\, P^\smile &= P \,\sharp\, P^\smile \end{aligned}$$

Together with the tabulation condition, this implies $Q \,\!; Q^\smile = P \,\!; P^\smile \sqcap Q \,\!; Q^\smile = \mathbb{I}_A$, that is, injectivity of Q. $\qquad\qquad\Box$

While a tabulation can be seen as a certain kind of decomposition of an arbitrary morphism in an allegory into a span, the dual of a tabulation is then a certain kind of decomposition of a difunctional morphism in a collagory into a co-span. Although the formal material here is dual to that above, we still spell it out in full detail for reference and better intuition.

Definition 4.5. [Kah04] In a collagory, let a morphism $W : \mathcal{B} \to \mathcal{C}$ be given. The co-span $\mathcal{B} \xrightarrow{R} \mathcal{D} \xleftarrow{S} \mathcal{C}$ of *mappings* R and S is called a *co-tabulation of* W iff $R \,\!; S^\smile = W$ and $R^\smile \,\!; R \sqcup S^\smile \,\!; S = \mathbb{I}_\mathcal{D}$. $\qquad\Box$

The first equation implies $W \,\!; W^\smile \,\!; W = R \,\!; S^\smile \,\!; S \,\!; R^\smile \,\!; R \,\!; S^\smile \sqsubseteq R \,\!; S^\smile$ (using univalence of R and S), so if W has a co-tabulation, it has to be difunctional.

   This also has an equivalent characterisation that is perfectly "bi-dual" to the one in Proposition 4.3:

Proposition 4.6. In a collagory, the span $\mathcal{B} \xrightarrow{R} \mathcal{D} \xleftarrow{S} \mathcal{C}$ is a co-tabulation of $W : \mathcal{B} \to \mathcal{C}$ iff the following equations hold:

$$R \,\!; S^\smile = W \qquad \begin{array}{l} R \,\!; R^\smile = \mathbb{I} \sqcup W \,\!; W^\smile \\ S \,\!; S^\smile = \mathbb{I} \sqcup W^\smile \,\!; W \end{array} \qquad R^\smile \,\!; R \sqcup S^\smile \,\!; S = \mathbb{I}_\mathcal{D} \ . \qquad\Box$$

In a collagory, co-tabulations are unique up to isomorphism [SS93, 4.4.10], and we have the following special cases:

– In a co-tabulation of an equivalence, both R and S are the induced *quotient* projections.

– We can define a *direct sum* of $\mathcal{A}$ and $\mathcal{B}$ to be a co-tabulation of $\perp\!\!\!\perp_{\mathcal{A},\mathcal{B}}$, if that least morphism exists.

If direct sums and quotients are available, then a co-tabulation can be constructed for each difunctional morphism.

   Central to applications to graph transformation is the following:

Proposition 4.7. In a collagory $\mathbf{C}$, if a span $\mathcal{B} \xleftarrow{P} \mathcal{A} \xrightarrow{Q} \mathcal{C}$ of mappings is given, then a co-tabulation for $(P^\smile \,\!; Q)^\boxplus$ is a pushout for that span in $\mathsf{Map}\,\mathbf{C}$.

PROOF: The proof of [Kah01, Theorem 5.3.5] is easily adapted. $\qquad\qquad\Box$

A co-tabulation for $U^\boxplus$ satisfies the following equations:

$$R \,\!; S^\smile = U^\boxplus \qquad R \,\!; R^\smile = U^\boxright \qquad S \,\!; S^\smile = U^\boxleft \qquad R^\smile \,\!; R \sqcup S^\smile \,\!; S = \mathbb{I}_\mathcal{D} \ .$$

This was introduced as a *gluing for* U in [Kah01]. Kawahara is the first to have characterised pushouts relation-algebraically in essentially this way [Kaw90].

   For pushouts along injective mappings, the co-tabulated morphism already is difunctional:

Lemma 4.8. If a span $\mathcal{B} \xleftarrow{P} \mathcal{A} \xrightarrow{Q} \mathcal{C}$ of mappings is given with Q injective, then $P^{\smile} \,\stretchstretch\, Q$ is difunctional (and therefore $(P^{\smile} \,\stretchstretch\, Q)^{\boxtimes} = P^{\smile} \,\stretchstretch\, Q$).

PROOF: Since P, as a mapping, is difunctional, we have

$$P^{\smile} \,\stretchstretch\, Q \,\stretchstretch\, Q^{\smile} \,\stretchstretch\, P \,\stretchstretch\, P^{\smile} \,\stretchstretch\, Q = P^{\smile} \,\stretchstretch\, P \,\stretchstretch\, P^{\smile} \,\stretchstretch\, Q = P^{\smile} \,\stretchstretch\, Q \ . \qquad \square$$

Furthermore, co-tabulations preserve injectivity:

Lemma 4.9. If a span $\mathcal{B} \xleftarrow{P} \mathcal{A} \xrightarrow{Q} \mathcal{C}$ of mappings is given with Q injective, and $\mathcal{B} \xrightarrow{R} \mathcal{D} \xleftarrow{S} \mathcal{C}$ is a co-tabulation for $P^{\smile} \,\stretchstretch\, Q$, then R is injective, too.

PROOF: Using injectivity of Q and univalence of P in one of the equations from Proposition 4.6 gives us injectivity of R:

$$R \,\stretchstretch\, R^{\smile} = \mathbb{I} \sqcup P^{\smile} \,\stretchstretch\, Q \,\stretchstretch\, (P^{\smile} \,\stretchstretch\, Q)^{\smile} = \mathbb{I} \sqcup P^{\smile} \,\stretchstretch\, Q \,\stretchstretch\, Q^{\smile} \,\stretchstretch\, P = \mathbb{I} \sqcup P^{\smile} \,\stretchstretch\, P = \mathbb{I} \ . \quad \square$$

With that, we can show that, essentially, a pushout over an injective mapping is also a pullback:

Lemma 4.10. If a span $\mathcal{B} \xleftarrow{P} \mathcal{A} \xrightarrow{Q} \mathcal{C}$ of mappings is given with Q injective, and $\mathcal{B} \xrightarrow{R} \mathcal{D} \xleftarrow{S} \mathcal{C}$ is a co-tabulation for $P^{\smile} \,\stretchstretch\, Q$, then $\mathcal{B} \xleftarrow{P} \mathcal{A} \xrightarrow{Q} \mathcal{C}$ is also a tabulation for $R \,\stretchstretch\, S^{\smile}$.

PROOF: Cross-commutativity $R \,\stretchstretch\, S^{\smile} = P^{\smile} \,\stretchstretch\, Q$ is already contained in the co-tabulation conditions. Since Q is injective and P is total, we also obtain

$$P \,\stretchstretch\, P^{\smile} \sqcap Q \,\stretchstretch\, Q^{\smile} = P \,\stretchstretch\, P^{\smile} \sqcap \mathbb{I}_{\mathcal{A}} = \mathbb{I}_{\mathcal{A}} \ . \qquad \square$$

Definition 4.11. If a collagory has a tabulation for each morphism and a co-tabulation for each difunctional morphism, then we call it *bi-tabular*. $\qquad \square$

5 Maps in Collagories form Adhesive Categories

Adhesive categories as a more specific setting for double-pushout graph rewriting have been introduced by Lack and Sobociński [LS04, LS05]; the following two definitions are taken from there:

Definition 5.1. A *van Kampen square* (i) is a pushout which satisfies the following condition: given a commutative cube (ii) of which (i) forms the bottom face and the back faces are pullbacks (where $\mathcal{C}$ is considered to be in the back), the front faces are pullbacks if and only if the top face is a pushout.

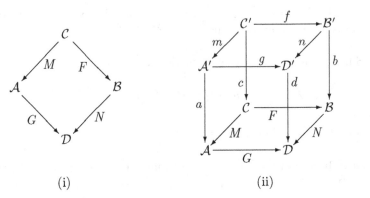

(i)   (ii)   □

Definition 5.2. A category **C** is said to be *adhesive* if
1. **C** has pushouts along monomorphisms;
2. **C** has pullbacks;
3. pushouts along monomorphisms are van Kampen squares.   □

For more concise formulations, we define:

Definition 5.3. A *van Kampen setup* in a category **C** for a square as in Def.
5.1(i) is a commuting cube in **C** as in Def. 5.1(ii) where the bottom square is a
pushout and the two back squares are pullbacks.   □

For reference, we expand this into the collagory setting:

Lemma 5.4. In a collagory **C**, a van Kampen setup in Map **C** means that the
following hold:

Bottom pushout:

$$G \,; N^\smile = (M^\smile \,; F)^\boxplus \qquad G^\smile \,; G \sqcup N^\smile \,; N = \mathbb{I}_D \qquad G \,; G^\smile = (M^\smile \,; F)^\vartriangleright$$
$$N \,; N^\smile = (M^\smile \,; F)^\vartriangleleft$$

Back pullbacks:

$$c^\smile \,; m = M \,; a^\smile \qquad c \,; c^\smile \sqcap m \,; m^\smile = \mathbb{I}_{C'} \qquad c^\smile \,; c = \mathbb{I}_C \sqcap M \,; a^\smile \,; a \,; M^\smile$$
$$m^\smile \,; m = \mathbb{I}_{A'} \sqcap a \,; M^\smile \,; M \,; a^\smile$$
$$c^\smile \,; f = F \,; b^\smile \qquad c \,; c^\smile \sqcap f \,; f^\smile = \mathbb{I}_{C'} \qquad c^\smile \,; c = \mathbb{I}_C \sqcap F \,; b^\smile \,; b \,; F^\smile$$
$$f^\smile \,; f = \mathbb{I}_{B'} \sqcap b \,; F^\smile \,; F \,; b^\smile$$

Remaining commutative squares:

$$m \,; g = f \,; n \qquad\qquad g \,; d = a \,; G \qquad n \,; d = b \,; N \qquad\qquad □$$

These equations are used in the long version of this paper [Kah09] to prove the
following:

Theorem 5.5. In the category of maps Map **C** over a collagory **C**, *pushouts
along injective maps are van Kampen squares.*   □

The main result of this section is an immediate consequence of this theorem —
note that, because of Lemma 4.8, we do not need difunctional (or transitive)
closure for this:

Corollary 5.6. For a bi-tabular collagory **C** where all monos in Map **C** are injective in **C**, the mapping category Map **C** is adhesive. ☐

(The restriction on monic mappings is necessary since there might, for example, be an object $\mathcal{A}$ in **C** for which the only mapping with target $\mathcal{A}$ is $\mathbb{I}_{\mathcal{A}}$; in that case, all mappings $f : \mathcal{A} \to \mathcal{B}$ would automatically be monos in Map **C** regardless whether they are injective in **C**. Note that f (together with identities) itself forms a tabulation and a co-tabulation for f.)

This result immediately makes the rewriting concepts and results introduced for adhesive categories in [LS04], including the local Church-Rosser theorem and the concurrency theorem, available for DPO rewriting defined via tabulations and co-tabulations in the context of collagories.

6   Collagories of Semi-unary Algebras and Bisimulations

In [Kah01, Kah04], relational homomorphisms between unary algebras have been shown to form a distributive allegory. In this section we generalise this result to collagories by allowing constant symbols and in turn dropping the zero law requirement.

Most of the mathematical content of this section has been presented and proven in more detail in [Kah01, Kah04]. Besides the proof of Theorem 6.6, also the reformulation using the sort-indexed product category and the forgetful functor $\mathcal{U}_{\Sigma}$ is new.

Definition 6.1. A *signature* is a tuple $(\mathcal{S}, \mathcal{F}, \mathsf{src}, \mathsf{trg})$ consisting of
- a set $\mathcal{S}$ of *sorts*,
- a set $\mathcal{F}$ of *function symbols*,
- a mapping $\mathsf{src} : \mathcal{F} \to \mathcal{S}^{*}$ associating with every function symbol the list of its *source sorts*, and
- a mapping $\mathsf{trg} : \mathcal{F} \to \mathcal{S}$ associating with every function symbol its *target sort*.

Such a signature is called *semi-unary* if $\mathsf{length}(\mathsf{src}(f)) \leq 1$ for each $f : \mathcal{F}$, and *unary* if $\mathsf{length}(\mathsf{src}(f)) = 1$ for each $f : \mathcal{F}$. ☐

For a function symbol $f : \mathcal{F}$, we usually employ the shorthand "$f : s_1 \times \cdots \times s_n \to t$" instead of the rather verbose "$\mathsf{src}(f) = \langle s_1, \ldots, s_n \rangle$ and $\mathsf{trg}(f) = t$". For a zero-ary function symbol, also called *constant symbol*, we write "$f : \mathbb{1} \to t$".

The following example signatures will be used for discussion and results in sections 7 and 8:

$$\mathsf{sigGraph} := \langle\ \textbf{sorts: } \mathsf{V}, \mathsf{E}$$
$$\textbf{ops: } \mathsf{s}, \mathsf{t} : \mathsf{E} \to \mathsf{V}$$
$$\rangle$$

$$\mathsf{sigPointedSet} := \langle\ \textbf{sorts: } \mathsf{S}$$
$$\textbf{ops: } \mathsf{point} : \mathbb{1} \to \mathsf{S}$$
$$\rangle$$

$$\mathsf{sigPoint} := \langle\ \textbf{sorts: } \mathsf{P}$$
$$\textbf{ops:}$$
$$\rangle$$

$$\mathsf{sigPointed} := \langle\ \textbf{sorts: } \mathsf{P}, \mathsf{O}$$
$$\textbf{ops: } \mathsf{p} : \mathsf{P} \to \mathsf{O}$$
$$\rangle$$

$$\text{sigType} := \langle \; \textbf{sorts:} \; \mathsf{T}$$
$$\textbf{ops:}$$
$$\rangle$$

$$\text{sigTyped} := \langle \; \textbf{sorts:} \; \mathsf{O}, \mathsf{T}$$
$$\textbf{ops:} \; \mathsf{t} : \mathsf{O} \to \mathsf{T}$$
$$\rangle$$

$$\text{sigNELabels} := \langle \; \textbf{sorts:} \; \mathsf{NL}, \mathsf{EL}$$
$$\textbf{ops:}$$
$$\rangle$$

$$\text{sigLGraph} := \langle \; \textbf{sorts:} \; \mathsf{N}, \mathsf{E}, \mathsf{NL}, \mathsf{EL}$$
$$\textbf{ops:} \; \mathsf{s}, \mathsf{t} : \mathsf{E} \to \mathsf{N},$$
$$\mathsf{n} : \mathsf{N} \to \mathsf{NL},$$
$$\mathsf{e} : \mathsf{E} \to \mathsf{EL}$$
$$\rangle$$

Definition 6.2. For a set $\mathcal{S}$ (of *sorts*) and a category $\mathbf{C}$, we define $\mathbf{C}^{\mathcal{S}}$, the $\mathcal{S}$-*indexed product category of* $\mathbf{C}$, as follows:

– an object $\mathcal{A}$ of $\mathbf{C}^{\mathcal{S}}$ consists of $\mathbf{C}$-objects $s^{\mathcal{A}}$ for every $s : \mathcal{S}$;
– a morphism $\varPhi : \mathcal{A} \to \mathcal{B}$ of $\mathbf{C}^{\mathcal{S}}$ is an $\mathcal{S}$-indexed family of $\mathbf{C}$-morphisms $\varPhi = (\varPhi_s)_{s:\mathcal{S}}$ such that $\varPhi_s : s^{\mathcal{A}} \to s^{\mathcal{B}}$ for every sort $s : \mathcal{S}$.
– composition $;^{\mathcal{S}}$ and identities $\mathbb{I}^{\mathcal{S}}$ are defined component-wise;
– if $\mathbf{C}$ is an allegory, then inclusion $\sqsubseteq^{\mathcal{S}}$, meet $\sqcap^{\mathcal{S}}$ and converse are defined component-wise;
– if $\mathbf{C}$ is collagory, then join $\sqcup^{\mathcal{S}}$ is defined component-wise. □

One easily verifies that the resulting $\mathcal{S}$-indexed product categories, allegories, and collagories all satisfy the respective axioms.

When defining $\varSigma$-algebras in the presence of binary function symbols, we need several technical conditions on direct products [Kah01, Def. 3.1.12]; for the current study, we can do without direct products (at the cost of some duplication of formalisation for unary and zero-ary function symbols), but we still need allegories for the characterisation of mappings:

Definition 6.3. Given a signature $\varSigma = (\mathcal{S}, \mathcal{F}, \mathsf{src}, \mathsf{trg})$ and an allegory $\mathbf{C}$, which has to have a unit $\mathbb{1}$ if $\varSigma$ contains constant symbols, an *abstract $\varSigma$-algebra* over $\mathbf{C}$ consists of the following items:

– an object $\mathcal{A}$ of $\mathbf{C}^{\mathcal{S}}$,
– for each function symbol $f : \mathcal{F}$ with $f : s \to t$ a mapping $f^{\mathcal{A}} : s^{\mathcal{A}} \to t^{\mathcal{A}}$ in $\mathbf{C}$.
– for each constant symbol $c : \mathcal{F}$ with $c : \mathbb{1} \to t$ a mapping $c^{\mathcal{A}} : \mathbb{1} \to t^{\mathcal{A}}$ in $\mathbf{C}$. □

It is important to note that, where we use sets as carriers, we have no restriction to non-empty sets — unlike most of the universal algebra literature.

Since we use this definition to construct an allegory with abstract $\varSigma$-algebras as objects, the generality of discussing *abstract $\varSigma$-algebras* over allegories allows us to stack this construction at no cost at all, with possibly different signatures at every level, building for example graphs where the nodes and edges are hypergraphs and hypergraph morphisms.

The morphisms in allegories of $\varSigma$-algebras have to behave "essentially like relations", and so it is only natural that we consider a relational generalisation of conventional (functional) $\varSigma$-homomorphisms. For arbitrary signatures, this has been presented in [Kah01]. For unary signatures, one naturally starts with defining L-simulations satisfying $\Phi_s^{\smile} ; f^{\mathcal{A}} \sqsubseteq f^{\mathcal{B}} ; \Phi_t^{\smile}$ according to de Roever and Engelhardt [dRE98], and then proceeds to L-simulations for which their converse is an L-simulation, too; these are called "bisimulations" in [Kah04].

Definition 6.4. Let a signature $\Sigma = (\mathcal{S}, \mathcal{F}, \mathsf{src}, \mathsf{trg})$, an allegory $\mathbf{C}$, and two abstract Σ-algebras $\mathcal{A}$ and $\mathcal{B}$ over $\mathbf{C}$ be given.

A Σ-*bisimulation from* $\mathcal{A}$ *to* $\mathcal{B}$ is a $\mathbf{C}^{\mathcal{S}}$-morphisms from $\mathcal{A}$ to $\mathcal{B}$ such that for every function symbol $f \in \mathcal{F}$ with $f : s \to t$ and every constant symbol $c \in \mathcal{F}$ with $c : \mathbb{1} \to t$ the following inclusions hold:

$$\Phi_s ; f^{\mathcal{B}} \sqsubseteq f^{\mathcal{A}} ; \Phi_t, \qquad \text{and} \qquad c^{\mathcal{B}} \sqsubseteq c^{\mathcal{A}} ; \Phi_t . \qquad \Box$$

Using Σ-algebras over $\mathbf{C}$ as objects and Σ-bisimulations as morphisms defines a category $\mathbf{C}^{\Sigma}$ with an obvious "underlying" functor $\mathcal{U}_{\Sigma} : \mathbf{C}^{\Sigma} \to \mathbf{C}^{\mathcal{S}}$.

This "forgetful" functor $\mathcal{U}_{\Sigma}$ is faithful. If $\mathbf{C}$ is an allegory, then $\mathcal{U}_{\Sigma}$ reflects inclusion, meets and converse in the sense that these can be defined for $\mathbf{C}^{\Sigma}$ via their $\mathcal{U}_{\Sigma}$ images. Therefore, $\mathbf{C}^{\Sigma}$ is an allegory, too [Kah01, Thm. 3.2.6].

Conventional Σ-algebra homomorphisms are just mappings in the allegory Rel^{Σ} of concrete Σ-algebras over the allegory Rel of sets and concrete relations.

If Σ contains a constant symbol, then even if the allegory $\mathbf{C}$ has least morphisms, then least homomorphisms in $\mathbf{C}^{\mathcal{S}}$ are not generally in the range of $\mathcal{U}_{\Sigma}$, and even if $\mathbf{C}^{\Sigma}$ does have least morphisms, the zero law will in general not hold for them, no matter whether it holds in $\mathbf{C}$.

If Σ contains a function symbol of arity at least 2, then even if $\mathbf{C}$ is an upper-semilattice category, then $\mathcal{U}_{\Sigma}$ does not reflect joins, in the sense that $\mathcal{U}_{\Sigma}(\Phi) \sqcup^{\mathcal{S}} \mathcal{U}_{\Sigma}(\Psi)$ is not necessarily in the range of $\mathcal{U}_{\Sigma}$. Furthermore, even if $\mathbf{C}^{\Sigma}$ has joins, composition will, in presence of function symbols of arity at least 2, in general not distribute over these joins (since non-empty joins do not distribute over the product $\times$ occurring in the homomorphism condition) so $\mathbf{C}^{\Sigma}$ will not be an upper-semilattice category.

For semi-unary signatures, however, $\mathcal{U}_{\Sigma}$ does reflect joins:

Lemma 6.5. If $\mathbf{C}$ is an upper-semilattice category, Σ is a semi-unary signature, and $\Phi, \Psi : \mathcal{A} \to \mathcal{B}$ are two Σ-bisimulations, then $\Phi \sqcup^{\mathcal{S}} \Psi$ is a Σ-bisimulation, too, and is the join in $\mathbf{C}^{\Sigma}$ of Φ and Ψ, that is, $\Phi \sqcup^{\Sigma} \Psi = \Phi \sqcup^{\mathcal{S}} \Psi$.

PROOF: We need to check the bisimulation conditions for unary function symbols $f : s \to t$ and for constant symbols $c : \mathbb{1} \to t$:

$$(\Phi \sqcup^{\mathcal{S}} \Psi)_s ; f^{\mathcal{B}} = (\Phi_s \sqcup \Psi_s) ; f^{\mathcal{B}} = \Phi_s ; f^{\mathcal{B}} \sqcup \Psi_s ; f^{\mathcal{B}}$$
$$\sqsubseteq f^{\mathcal{A}} ; \Phi_t \sqcup f^{\mathcal{A}} ; \Psi_t = f^{\mathcal{A}} ; (\Phi_t \sqcup \Psi_t) = f^{\mathcal{A}} ; (\Phi \sqcup^{\mathcal{S}} \Psi)_t$$
$$c^{\mathcal{B}} \sqsubseteq c^{\mathcal{A}} ; \Phi_t \sqcup c^{\mathcal{A}} ; \Psi_t = c^{\mathcal{A}} ; (\Phi_t \sqcup \Psi_t) = c^{\mathcal{A}} ; (\Phi \sqcup^{\mathcal{S}} \Psi)_t$$

The equation $\Phi \sqcup^{\Sigma} \Psi = \Phi \sqcup^{\mathcal{S}} \Psi$ follows from the reflection of inclusion by $\mathcal{U}_{\Sigma}$.$\Box$

Given the closure of Σ-bisimulations under the converse, meet, and join operations in $\mathbf{C}^{\mathcal{S}}$, properties of $\mathbf{C}$-morphisms for these operations are inherited by Σ-bisimulations because of the component-wise definitions, and we obtain:

Theorem 6.6. If Σ is a semi-unary signature and $\mathbf{C}$ is a collagory, then $\mathbf{C}^{\Sigma}$ is a collagory, too. □

If $\mathbf{C}$ has tabulations (respectively co-tabulations), the sort-indexed product category $\mathbf{C}^{\mathcal{S}}$ obviously has tabulations (respectively co-tabulations), too, and they can be calculated component-wise. Perhaps surprisingly, these can be extended to the collagory $\mathbf{C}^{\Sigma}$ of bisimulations between Σ-algebras without problems; we just need to provide definitions for the function symbols of the "new" objects, and verify all relevant conditions (proofs in [Kah09]):

Theorem 6.7. If $\Sigma = (\mathcal{S}, \mathcal{F}, \mathsf{src}, \mathsf{trg})$ is a semi-unary signature and $\mathbf{C}$ is an allegory, and $\mathcal{B} \xleftarrow{P} \mathcal{A} \xrightarrow{Q} \mathcal{C}$ is a tabulation in $\mathbf{C}^{\mathcal{S}}$ of the Σ-bisimulation $V : \mathcal{B} \to \mathcal{C}$, i.e., for each sort $s : \mathcal{S}$, $\mathcal{B} \xleftarrow{P_s} \mathcal{A} \xrightarrow{Q_s} \mathcal{C}$ is a tabulation of $V_s : s^{\mathcal{B}} \to s^{\mathcal{C}}$, then we define for each function symbol $f : s \to t$ and each constant symbol $c : \mathbb{1} \to t$ in Σ:

$$f^{\mathcal{A}} := P_s \,\text{;}\, f^{\mathcal{B}} \,\text{;}\, P_t^{\smile} \sqcap Q_s \,\text{;}\, f^{\mathcal{C}} \,\text{;}\, Q_t^{\smile}$$
$$c^{\mathcal{A}} := c^{\mathcal{B}} \,\text{;}\, P_t^{\smile} \sqcap c^{\mathcal{C}} \,\text{;}\, Q_t^{\smile}$$

Then $\mathcal{A}$ turns into a Σ-algebra and P and Q are Σ-bisimulations, too, so $\mathcal{B} \xleftarrow{P} \mathcal{A} \xrightarrow{Q} \mathcal{C}$ is a tabulation in $\mathbf{C}^{\Sigma}$. □

Theorem 6.8. If $\Sigma = (\mathcal{S}, \mathcal{F}, \mathsf{src}, \mathsf{trg})$ is a semi-unary signature and $\mathbf{C}$ is a collagory, and $\mathcal{B} \xrightarrow{R} \mathcal{D} \xleftarrow{S} \mathcal{C}$ is a co-tabulation in $\mathbf{C}^{\mathcal{S}}$ of the Σ-bisimulation $W : \mathcal{B} \to \mathcal{C}$, i.e., for each sort $s : \mathcal{S}$, $\mathcal{B} \xrightarrow{R_s} \mathcal{D} \xleftarrow{S_s} \mathcal{C}$ is a tabulation of $W_s : s^{\mathcal{B}} \to s^{\mathcal{C}}$, then we define for each function symbol $f : s \to t$ and each constant symbol $c : \mathbb{1} \to t$ in Σ:

$$f^{\mathcal{D}} := R_s^{\smile} \,\text{;}\, f^{\mathcal{B}} \,\text{;}\, R_t \sqcup S_s^{\smile} \,\text{;}\, f^{\mathcal{C}} \,\text{;}\, S_t$$
$$c^{\mathcal{D}} := c^{\mathcal{B}} \,\text{;}\, R_t \sqcup c^{\mathcal{C}} \,\text{;}\, S_t$$

Then $\mathcal{D}$ turns into a Σ-algebra and R and S are Σ-bisimulations, too, so $\mathcal{B} \xrightarrow{R} \mathcal{D} \xleftarrow{S} \mathcal{C}$ is a co-tabulation in $\mathbf{C}^{\Sigma}$. □

7   Reducts along Signature Homomorphisms

While the concept of Σ-algebra is sufficient to capture, for example, unlabelled graphs as sigGraph-algebras, categories of labelled graphs are frequently considered as having *fixed* label sets, which means that only certain sub-categories of $Set^{\mathsf{sigLGraph}}$ are considered.

We use the concept of *reducts* to formalise this in a general way. In the example, we consider the reduct of $Set^{\mathsf{sigLGraph}}$ to the sub-signature sigNELabels. The fixed label sets under consideration form a one-object sub-category $\mathbf{K}$ of $Set^{\mathsf{sigNELabels}}$, and in order to obtain graphs labelled over these label sets, we restrict attention to objects in $Set^{\mathsf{sigLGraph}}$ for which the reduct lies in that sub-category $\mathbf{K}$.

The current section introduces and studies the reduct relator. This is employed in Sect. 8 to implement the restriction of Σ-algebra collagories via reduct-side sub-categories. This single construction principle for generating concrete bi-tabular collagories corresponds, as shown in Corollary 8.7, to several categorical constructions that are known for adhesive categories.

Definition 7.1. Let $\Sigma = (\mathcal{S}, \mathcal{F}, \mathsf{src}, \mathsf{trg})$ and $\Sigma_R = (\mathcal{S}_R, \mathcal{F}_R, \mathsf{src}_R, \mathsf{trg}_R)$ be two signatures, and let $\sigma : \Sigma_R \to \Sigma$ be a signature homomorphism.

For any Σ-algebra $\mathcal{A}$, such a signature homomorphism $\sigma : \Sigma_R \to \Sigma$ induces a Σ_R-algebra $\mathcal{A}|\sigma$, the σ-*reduct of* $\mathcal{A}$, in the following way:

– For every sort $r : \mathcal{S}_R$, its carrier is $r^{\mathcal{A}|\sigma} = (\sigma\ r)^{\mathcal{A}}$;
– for every function symbol $f \in \mathcal{F}_R$, its interpretation is $f^{\mathcal{A}|\sigma} = (\sigma\ f)^{\mathcal{A}}$.   □

It is easy to verify that $\mathcal{A}|\sigma$ is indeed a Σ_R-algebra.

If $\sigma : \Sigma_R \to \Sigma$ is a sub-signature embedding, then we also call $\mathcal{A}|\sigma$ the Σ_R-reduct of $\mathcal{A}$ and write also $\mathcal{A}\lfloor\Sigma_R$.

Since our signatures are a special case of sketches [BW99, Chapters 4,7,8,10], $\lfloor\sigma$ is a special case of what Barr and Wells call "model category functor". We complete the definition and show that is is a relator:

Definition 7.2. For a signature homomorphism $\sigma : \Sigma_R \to \Sigma$, the σ-reduct of a $\mathbf{C}^{\mathcal{S}}$-morphism $\Phi = (\Phi_s)_{s:\mathcal{S}}$ is the $\mathbf{C}^{\mathcal{S}_R}$-morphism $\Phi|\sigma = ((\Phi|\sigma)_r)_{r:\mathcal{S}_R}$ with $(\Phi|\sigma)_r := \Phi_{\sigma\ r}$ for every $r : \mathcal{S}_R$.   □

The straight-forward proof of the following is elaborated in [Kah09]:

Proposition 7.3. For a signature homomorphism $\sigma : \Sigma_R \to \Sigma$, the σ-reduct of a Σ-bisimulation is a Σ_R-bisimulation.

Furthermore, the reduct operation $\lfloor\sigma$ is an allegory relator from $\mathbf{C}^{\Sigma}$ to $\mathbf{C}^{\Sigma_R}$ and therefore also a functor from $\mathsf{Map}(\mathbf{C}^{\Sigma})$ to $\mathsf{Map}(\mathbf{C}^{\Sigma_R})$.   □

Obviously, the reduct relator is in general not full if σ is not injective on sorts. If σ is injective, we can "replace in $\mathcal{A}$ its reduct part along a morphism to $\mathcal{A}|\sigma$"; this will be used to show the main theorems of the next section (proof in [Kah09]):

Theorem 7.4. If $\sigma : \Sigma_R \to \Sigma$ is an injective signature homomorphism, then the reduct functor $\lfloor\sigma$ is a fibration [BW99, 12.1].   □

8   Reduct-Restricted Σ-Algebra Categories

In the following, let $\sigma : \Sigma_R \to \Sigma$ be an arbitrary but fixed signature homomorphism, and $\mathbf{K}$ a sub-category of $\mathbf{C}^{\Sigma_R}$. We will further assume that $\mathbf{K}$ is contained in the image of $\lfloor\sigma$ — this restriction is not essential, but frequently allows more concise formulations.

Definition 8.1. The $\sigma, \mathbf{K}$-restriction of $\mathbf{C}^{\Sigma}$ contains exactly those objects and morphisms for which the image under $\lfloor\sigma$ is in $\mathbf{K}$.

We denote this restriction as $\mathbf{C}^{\sigma|\mathbf{K}}$.   □

Because relators preserve identities and composition, and $\mathbf{K}$ is a category, the restriction $\mathbf{C}^{\sigma|\mathbf{K}}$ is a category again.

The technical importance of the assumption on **K** is that it provides surjectivity on homsets for the reduct relator:

Proposition 8.2. If **K** is contained in the image of $\lfloor\sigma$, then the restriction of $\lfloor\sigma$ to $\mathbf{C}^{\sigma}\lfloor_{\mathbf{K}}$ is a full relator. □

If σ is a sub-signature embedding, we also write $\mathbf{C}^{\Sigma}\lfloor_{\mathbf{K}}$ instead of $\mathbf{C}^{\sigma}\lfloor_{\mathbf{K}}$. If, in addition, the restriction category **K** contains only one object $\mathcal{L}$ and its identity, we also write $\mathbf{C}^{\Sigma}\lfloor_{\mathcal{L}}$. This latter case covers in particular the situation where Σ_{R} contains only label sorts and $\mathcal{L}$ fixes the label interpretations, producing for example a category of labelled graphs with fixed label sets.

Note that every one-object-one-morphism category has all limits and colimits and is not only an allegory, but even a (trivial) relation algebra, and also a bitabular collagory. This therefore provides an important special case for many of the properties in the remainder of this paper.

Proposition 8.3. If **K** is a sub-allegory of $\mathbf{C}^{\Sigma_{\mathrm{R}}}$, then $\mathbf{C}^{\sigma}\lfloor_{\mathbf{K}}$ is an allegory.

PROOF: Assume that $\Phi\lfloor\sigma$ and $\Psi\lfloor\sigma$ are in **K**. Since **K** is closed under converse and meets, $\Phi^{\smile}\lfloor\sigma = (\Phi\lfloor\sigma)^{\smile}$ and $(\Phi\sqcap\Psi)\lfloor\sigma = (\Phi\lfloor\sigma)\sqcap(\Psi\lfloor\sigma)$ are in **K**, too.

Therefore, $\mathbf{C}^{\sigma}\lfloor_{\mathbf{K}}$ is closed under converse and meets, too, and therefore is a sub-allegory of $\mathbf{C}^{\Sigma}$. □

Proposition 8.4. For semi-unary Σ, if **K** is a sub-collagory of $\mathbf{C}^{\Sigma_{\mathrm{R}}}$, then $\mathbf{C}^{\sigma}\lfloor_{\mathbf{K}}$ is a collagory.

PROOF: Assume that $\Phi\lfloor\sigma$ and $\Psi\lfloor\sigma$ are in **K**. With Lemma 6.5 and since **K** is closed under joins, the join $(\Phi\sqcup^{\Sigma}\Psi)\lfloor\sigma = (\Phi\lfloor\sigma)\sqcup^{\Sigma_{\mathrm{R}}}(\Psi\lfloor\sigma)$ is in **K**, too.

So $\mathbf{C}^{\sigma}\lfloor_{\mathbf{K}}$ is closed under joins, too, and therefore is a sub-collagory of $\mathbf{C}^{\Sigma}$. □

This join preservation works in particular in the case where **K** is a one-object-one-morphism category, since in that case, non-empty joins in **K** are still inherited (trivially) from $\mathbf{C}^{\Sigma_{\mathrm{R}}}$.

Empty joins, i.e., least morphisms, however, are generally *not* inherited in the one-object-one-morphism category, since identity morphisms are rarely least morphisms in $\mathbf{C}^{\Sigma_{\mathrm{R}}}$. Therefore the zero law does in general not hold in $\mathbf{C}^{\sigma}\lfloor_{\mathbf{K}}$. A simple example for this arises in $Set^{\mathrm{sigPointed}}\lfloor_{\{\bullet\}}$, i.e., the allegory of relational homomorphisms between pointed sets: The presence of the point induces exactly the same counterexamples as the presence of a zero-ary function symbol, for example if $O^{\mathcal{A}} = \{0,1\}$, and the point (respectively the value of the constant) in $\mathcal{A}$ is 1, then $\mathbb{\bot}_{O^{\mathcal{A}},O^{\mathcal{A}}} = \{(1,1)\}$ is a non-trivial closure of the non-inherited least element of **K**, and with $R := \{(0,1),(1,1)\}$ we have $R\mathbin{;}\mathbb{\bot} = R \neq \mathbb{\bot}$.

Since the reduct relator $\lfloor\sigma$ distributes over all relevant operations, it also preserves (co-)tabulations, i.e.:

- If the span $\mathcal{B}\xleftarrow{P}\mathcal{A}\xrightarrow{Q}\mathcal{C}$ is a tabulation for $V : \mathcal{B} \to \mathcal{C}$ in $\mathbf{C}^{\Sigma}$, then the span $\mathcal{B}\lfloor\sigma\xleftarrow{P\lfloor\sigma}\mathcal{A}\lfloor\sigma\xrightarrow{Q\lfloor\sigma}\mathcal{C}\lfloor\sigma$ is a tabulation for $V\lfloor\sigma$ in $\mathbf{C}^{\Sigma_{\mathrm{R}}}$.
- If the co-span $\mathcal{B}\xrightarrow{R}\mathcal{D}\xleftarrow{S}\mathcal{C}$ is a co-tabulation for $W : \mathcal{B} \to \mathcal{C}$ in $\mathbf{C}^{\Sigma}$, then the co-span $\mathcal{B}\lfloor\sigma\xrightarrow{R\lfloor\sigma}\mathcal{D}\lfloor\sigma\xleftarrow{S\lfloor\sigma}\mathcal{C}\lfloor\sigma$ is a co-tabulation for $W\lfloor\sigma$ in $\mathbf{C}^{\Sigma_{\mathrm{R}}}$.

Theorem 8.5. For semi-unary Σ, injective $\sigma : \Sigma_R \to \Sigma$, and $\mathbf{K}$ a sub-collagory of $\mathbf{C}^{\Sigma_R}$, if $V : \mathcal{B} \to \mathcal{C}$ has a tabulation $\mathcal{B} \xleftarrow{P} \mathcal{A} \xrightarrow{Q} \mathcal{C}$ in $\mathbf{C}^\Sigma$, and $V|\sigma$ has a tabulation $\mathcal{B}|\sigma \xleftarrow{P_0} \mathcal{A}_0 \xrightarrow{Q_0} \mathcal{C}|\sigma$ in $\mathbf{K}$, then V also has a tabulation in $\mathbf{C}^{\sigma|\mathbf{K}}$.

PROOF: Since tabulations in $\mathbf{C}^{\Sigma_R}$ are unique up to isomorphism, there must be an isomorphism $\phi : \mathcal{A}_0 \to \mathcal{A}|\sigma$. According to Theorem 7.4, we obtain a cartesian morphism $\psi : \mathcal{A}_1 \to \mathcal{A}$ for ϕ and $\mathcal{A}$, and since this is also an isomorphism, $\mathcal{B} \xleftarrow{\psi;P} \mathcal{A}_1 \xrightarrow{\psi;Q} \mathcal{C}$ is a tabulation for V in $\mathbf{C}^{\sigma|\mathbf{K}}$. □

The corresponding statement for co-tabulations is shown in the same way, so we obtain as result:

Theorem 8.6. For semi-unary Σ and an injective signature homomorphism $\sigma : \Sigma_R \to \Sigma$, if $\mathbf{C}$ is a bi-tabular collagory and if $\mathbf{K}$ is bi-tabular sub-collagory of $\mathbf{C}^{\Sigma_R}$, then $\mathbf{C}^{\sigma|\mathbf{K}}$ is a bi-tabular collagory, too. □

This includes all the systematically constructed examples for adhesive categories provided by Lack and Sobociński [LS04], in particular the following uses of a one-object-one-morphism collagory $\mathbf{K}$:

Corollary 8.7. If $\mathbf{C}$ is a bi-tabular collagory, then the following are bi-tabular collagories, too:

– $\mathbf{C}^{\mathsf{sigPointed}}|_\mathcal{C}$ for any object $\mathcal{C}$ (conflating $\mathcal{C}$ in $\mathbf{C}$ with the sigPoint-algebra that assigns $\mathcal{C}$ to the sort P) — this is equivalent to the co-slice category $\mathcal{C}/\mathbf{C}$,
– $\mathbf{C}^{\mathsf{sigTyped}}|_\mathcal{C}$ for any object — this is equivalent to the slice category $\mathbf{C}/\mathcal{C}$,
– node- and edge-labelled graphs considered as sigLGraph-algebras with fixed node and edge label sets. □

9   Conclusion

We have streamlined the axiomatic basis of the relation-algebraic approach to graph structure transformation by introducing *collagories*, which, in comparison to earlier approaches, remove consideration of the zero-law and, to a certain extent, of difunctional closure defined via the Kleene star. We showed that the concepts of tabulation and co-tabulation, which are essential for the relation-algebraic rewriting approach, can be formalised in collagories, and that the category of mappings in a bi-tabular collagory forms an *adhesive category*, thus establishing a powerful connection to the categorical approach to graph structure transformation. We showed that all the important examples of adhesive categories can also be obtained as special cases of powerful collagory constructions; future work will investigate whether (respectively when) the category of relations [FS90, 1.412] in an adhesive category forms a collagory. Another interesting goal would be to identify a nicer collagory-level formulation of the van Kampen property, and establish connections with the characterisation as bicolimits in the bicategory of spans given by Heindel and Sobociński [HS09].

Further investigations will explore different variations of adhesive categories in a collagory setting, including the quasiadhesive categories of [LS05], and their applications.

References

[BW99]   Barr, M., Wells, C.: Category Theory for Computing Science, 3rd edn. Centre de recherches mathématiques (CRM), Université de Montréal (1999)

[CMR⁺97]   Corradini, A., Montanari, U., Rossi, F., Ehrig, H., Heckel, R., Löwe, M.: Algebraic Approaches to Graph Transformation, Part I: Basic Concepts and Double Pushout Approach. In: Rozenberg, G. (ed.) Handbook of Graph Grammars and Computing by Graph Transformation. Foundations, ch. 3, vol. 1, pp. 163–245. World Scientific, Singapore (1997)

[EPPH06]   Ehrig, H., Padberg, J., Prange, U., Habel, A.: Adhesive High-Level Replacement Systems: A New Categorical Framework for Graph Transformation. Fund. Inform. 74(1), 1–29 (2006)

[FS90]   Freyd, P.J., Scedrov, A.: Categories, Allegories. North-Holland Mathematical Library, vol. 39. North-Holland, Amsterdam (1990)

[HS09]   Heindel, T., Sobociński, P.: Van Kampen Colimits as Bicolimits in Span. In: Kurz, A., Tarlecki, A. (eds.) CALCO 2009. LNCS, vol. 5728. Springer, Heidelberg (to appear 2009)

[Kah01]   Kahl, W.: A Relation-Algebraic Approach to Graph Structure Transformation 2001. Habil. Thesis, Fakultät für Informatik, Univ. der Bundeswehr München, Techn. Report 2002-03 (2001),
            http://sqrl.mcmaster.ca/~kahl/Publications/RelRew/

[Kah04]   Kahl, W.: Refactoring Heterogeneous Relation Algebras around Ordered Categories and Converse. J. Relational Methods in Comp. Sci. 1, 277–313 (2004)

[Kah09]   Kahl, W.: Collagories for Relational Adhesive Rewriting. SQRL Report 56, Software Quality Research Laboratory, Department of Computing and Software, McMaster University (2009),
            http://sqrl.mcmaster.ca/sqrl_reports.html

[Kaw90]   Kawahara, Y.: Pushout-Complements and Basic Concepts of Grammars in Toposes. Theoretical Computer Science 77, 267–289 (1990)

[Koz94]   Kozen, D.: A Completeness Theorem for Kleene Algebras and the Algebra of Regular Events. Inform. and Comput. 110(2), 366–390 (1994)

[LS04]   Lack, S., Sobociński, P.: Adhesive Categories. In: Walukiewicz, I. (ed.) FOSSACS 2004. LNCS, vol. 2987, pp. 273–288. Springer, Heidelberg (2004)

[LS05]   Lack, S., Sobociński, P.: Adhesive and quasiadhesive categories. RAIRO Inform. Théor. Appl. 39(3), 511–545 (2005)

[dRE98]   de Roever, W.-P., Engelhardt, K.: Data Refinement: Model-Oriented Proof Methods and their Comparison. Cambridge Tracts Theoret. Comput. Sci, vol. 47. Cambridge Univ. Press, Cambridge (1998)

[SS93]   Schmidt, G., Ströhlein, T.: Relations and Graphs, Discrete Mathematics for Computer Scientists. EATCS-Monographs on Theoretical Computer Science. Springer, Heidelberg (1993)

Cardinal Addition in Distributive Allegories

Yasuo Kawahara[1] and Michael Winter[2],[*]

[1] Department of Informatics,
Kyushu University,
Fukuoka, Japan
kawahara@i.kyushu-u.ac.jp
[2] Department of Computer Science,
Brock University,
St. Catharines, Ontario, Canada, L2S 3A1
mwinter@brocku.ca

Abstract. In this paper we want to extend the abstract approach to the size of a relation based on a cardinality function. Assuming suitable extra structure on the underlying distributive allegory we are going to define addition on cardinalities and investigate its basic properties.

1 Introduction

The calculus of binary relations has been investigated since the middle of the nineteenth century. It plays an important rôle in the development of logic and algebra. In recent years it has become clear that this calculus, and its categorical versions in particular, is a fundamental conceptual and methodological tool in computer science just as much as in logic. In addition to its usage in modeling programming languages, classical and non-classical logics and different methods of data mining, relations have been a fundamental tool in formal program development [1,3,4,11,12].

A solution of a given problem, in particular an optimal solution, is often characterized as a minimal element among a set of candidates. In this context, minimality may refer to some underlying ordering or to the size of the solution, i.e. its cardinality. For example, one might be interested in a minimum vertex cover of a graph G, i.e. a subset C of vertices of minimum cardinality so that every edge of G is incident to some element in C. Instead of computing an optimal solution one is often satisfied with a good approximation. The previous example is an NP-hard problem so that finding optimal solutions is not always feasible. There exists a simple greedy approximation algorithm for this problem independently discovered by Fanica Gavril and Mihalis Yannakakis [10] computing a vertex cover with at most twice the minimum number of vertices. In [2] relations were used to formally develop a relational version of this algorithm. The correctness of the algorithm, excluding size considerations, was specified and verified

---

[*] The author gratefully acknowledges support from the Natural Sciences and Engineering Research Council of Canada.

R. Berghammer et al. (Eds.): RelMiCS/AKA 2009, LNCS 5827, pp. 227–241, 2009.

using relation algebra. Since a formal introduction of a notion of cardinality of a relation was not available, the additional size property was shown outside the theory.

The current paper is a continuation of [8,9]. Those papers introduced three different notions of the cardinality of a relation based on three different preorders on objects. In this paper we use the most general version. Given such an allegory with a cardinality function and some suitable extra structure we are going to define addition and study its basic properties. In the example mentioned above our theory of the cardinality of a relation and cardinal addition could be used to specify and verify all aspects of the algorithm within that unified framework.

2   Categories of Relations

Throughout this paper we assume that the reader is familiar with the basic notions from category and lattice theory. For notions not defined here we refer to [5,6].

We denote by $\text{Mor}_\mathcal{C}$ the collection of morphisms of a category $\mathcal{C}$. If a morphism f has source A and target B we usually write $f : A \to B$. The collection of all morphisms between A and B is denoted by $\mathcal{C}[A, B]$. We use ; for composition with the convention that it has to be read from left to right, i.e. $f; g$ means first f then g. The identity morphism on the object A is written as $\mathbb{I}_A$.

Definition 1. *An allegory $\mathcal{R}$ is a category satisfying the following:*

1. *For all objects A and B the class $\mathcal{R}[A, B]$ is a lower semi-lattice. Meet and the induced ordering are denoted by $\sqcap, \sqsubseteq$, respectively. The elements in $\mathcal{R}[A, B]$ are called relations.*
2. *There is a monotone operation $\check{\ }$ (called the converse operation) such that for all relations $Q, R : A \to B$ and $S : B \to C$ the following holds*

$$(Q; S)^\smile = S^\smile; Q^\smile \quad \text{and} \quad (Q^\smile)^\smile = Q.$$

3. *For all relations $Q : A \to B$, $R, S : B \to C$ we have $Q; (R \sqcap S) \sqsubseteq Q; R \sqcap Q; S$.*
4. *For all relations $Q : A \to B$, $R : B \to C$ and $S : A \to C$ the following modular law holds $Q; R \sqcap S \sqsubseteq Q; (R \sqcap Q^\smile; S)$.*

Within an allegory functions are of particular interest. Therefore, we call a relation $R : A \to B$ univalent (or a partial function) iff $R^\smile; R \sqsubseteq \mathbb{I}_B$ and total iff $\mathbb{I}_A \sqsubseteq R; R^\smile$. Functions are total and univalent relations and are usually denoted by lowercase letters. Furthermore, R is called injective iff $R^\smile$ is univalent and surjective iff $R^\smile$ is total. Among all the properties valid for relations in an allegory we have summarized those used in this paper in the following lemma. A proof can be found in [5,11,12].

Lemma 1. *Let $\mathcal{R}$ be an allegory. Then we have:*

1. *$Q; R \sqcap S \sqsubseteq (Q \sqcap S; R^\smile); (R \sqcap Q^\smile; S)$ for all relations $Q : A \to B$, $R : B \to C$ and $S : A \to C$ (Dedekind formula);*

2. *If $Q : A \to B$ is univalent, then $Q; (R \sqcap S) = Q; R \sqcap Q; S$ for all relations $R, S : B \to C$;*
3. *If $R : B \to C$ is univalent, then $Q; R \sqcap S = (Q \sqcap S; R^\smile); R$ for all relations $Q : A \to B$ and $S : A \to C$.*

Two functions $f : C \to A$ and $g : C \to B$ with common source are said to tabulate a relation $R : A \to B$ iff $R = f^\smile; g$ and $f; f^\smile \sqcap g; g^\smile = \mathbb{I}_C$. If for all relations of an allegory $\mathcal{R}$ there is tabulation, then $\mathcal{R}$ is called tabular. If $\mathcal{R}[A, B]$ has a greatest element $\top_{AB}$, the tabulation of $\top_{AB}$ is called a relational product [4,11,12]. In this case the the object and the two functions tabulating $\top_{AB}$ are denoted by $A \times B$, $\pi : A \times B \to A$ and $\rho : A \times B \to B$, respectively.

Later we will need the following technical lemma which relates tabulations of different relations. A proof was already given in [8,9].

Lemma 2. *Let $\mathcal{R}$ be an allegory, and $R : A \to B$ a relation that is tabulated by $f : C \to A$ and $g : C \to B$. Furthermore, let $h : D \to A$ and $k : D \to B$ be functions with $h^\smile; k \sqsubseteq R$, and define $l := h; f^\smile \sqcap k; g^\smile : D \to C$. Then we have the following:*

1. *l is the unique function with $h = l; f$ and $k = l; g$.*
2. *If $h^\smile; k = R$, then l is surjective.*
3. *If $h : D \to A$ and $k : D \to B$ is a tabulation, i.e. $h; h^\smile \sqcap k; k^\smile = \mathbb{I}_D$, then l is injective.*
4. *If R is a partial identity, i.e. $A = B$ and $R \sqsubseteq \mathbb{I}_A$, then f (or g) is a tabulation of R, i.e. $R = f^\smile; f$ and $f; f^\smile = \mathbb{I}_C$.*

Definition 2. *An allegory $\mathcal{R}$ is called distributive if:*

1. *For every pair of objects A and B the class $\mathcal{R}[A, B]$ is a distributive lattice with smallest element $\bot\!\!\bot_{AB}$. Union is denoted by $\sqcup$.*
2. *For all relations $Q : A \to B$, $R, S : B \to C$ we have $Q; \bot\!\!\bot_{BC} = \bot\!\!\bot_{AC}$ and $Q; (R \sqcup S) = Q; R \sqcup Q; S$.*

Since the dual of a distributive allegory is again a distributive allegory one might be interested in dual notions. For example, the dual notion of a relation product is given by relational sums.

Definition 3. *Let $\mathcal{R}$ be a distributive allegory and A and B be objects of $\mathcal{R}$. Then an object $A+B$ together with two relations $\iota : A \to A+B$ and $\kappa : B \to A+B$ is called a relational sum if it satisfies the following:*

$$\iota; \iota^\smile = \mathbb{I}_A, \qquad \kappa; \kappa^\smile = \mathbb{I}_B, \qquad \iota; \kappa^\smile = \bot\!\!\bot_{AB}, \qquad \iota^\smile; \iota \sqcup \kappa^\smile; \kappa = \mathbb{I}_{A+B}.$$

The next lemma is a particular interest if R and S are the injections ι and κ of a relational sum.

Lemma 3. *Let $\mathcal{R}$ be an allegory, $R : A \to C$ and $S : B \to C$ with $R; S^\smile = \bot\!\!\bot_{AB}$. The we have for all $Q_1 : D \to A, Q_2 : D \to B$ and $T_1 : A \to E, T_2 : B \to E$:*

1. $Q_1; R \sqcap Q_2; S = \amalg_{DC}$,
2. $(Q_1; R \sqcup Q_2; S); (R^\smile; T_1 \sqcup S^\smile; T_2) = Q_1; T_1 \sqcup Q_2; T_2$.

Proof

1. follows immediately from $Q_1; R \sqcap Q_2; S \sqsubseteq (Q_1; R; S^\smile \sqcap Q_2); S = \amalg_{CD}$.
2. was already shown several times, e.g. [11]. □

A subset M of a set N may be described by the canonical injection $f : M \rightarrow N$. Furthermore, the set of equivalence classes of an equivalence relation is fully determined by the function mapping each element to its equivalence class. Combining both concept we aim at the notion of a splitting.

Definition 4. *Let $Q : A \rightarrow A$ be a symmetric idempotent relation, i.e. $Q^\smile = Q$ and $Q; Q = Q$. An object B together with a relation $R : B \rightarrow A$ is called a splitting of Q (or R splits Q) iff $R; R^\smile = \mathbb{I}_B$ and $R^\smile; R = Q$.*

A splitting is unique up to isomorphism. If Q is a partial identity, the object B of the splitting corresponds to the subset given by Q. Analogously, if Q is an equivalence relation, B corresponds to the set of equivalence classes.

Some properties of cardinal numbers and their operations in the context of set theory rely on the axiom of choice. One of the possible versions of this axiom in the context of allegories is as follows:

(AC) For all total relations $R : A \rightarrow B$ there is a function $f : A \rightarrow B$ with $f \sqsubseteq R$.

3   Cardinality Function

In [8,9] three different notions of a cardinality function on an allegory based on three preorders on objects were introduced. The first notion was based on the existence of an injective function from the smaller to the bigger object, and the second notion was based on the existence of a surjective function from the bigger to the smaller object. The third notion generalized both versions and was based on the existence of a total and injective relation from the smaller to the bigger object. The relational version of the axiom of choice introduced above relates the three different notions. In this paper we want to use the most general version of a cardinality function. For a detailed motivation of axioms below we refer to [8,9].

Definition 5. *Let $\mathcal{R}$ be an allegory. A cardinality structure $(\mathcal{C}, |.|, \leq)$ on $\mathcal{R}$ consists of a (partially) ordered class $(\mathcal{C}, \leq)$ and a function $|.| : \text{Mor}_\mathcal{R} \rightarrow \mathcal{C}$ satisfying*

C0: $|R^\smile| = |R|$ *for all relations R;*
C1: $|.|$ *is monotonic, i.e. $R \sqsubseteq S$ implies $|R| \leq |S|$ for all relations $R, S : A \rightarrow B$;*
C2: *If $Q; Q^\smile \sqcap S; S^\smile \sqsubseteq \mathbb{I}_B$ for relations $Q : A \rightarrow B$ and $S : A \rightarrow C$, then for all $R : B \rightarrow C$*

$$|Q; R \sqcap S| \leq |R \sqcap Q^\smile; S|.$$

The structure is called strong iff $|.|$ is surjective as a function and $|\mathbb{I}_A| \leq |\mathbb{I}_B|$ implies that there is a total and injective relation $R : A \to B$.

Throughout this paper we will always assume that a cardinality function is surjective (whether it is strong or not). This is not a restriction since we can always restrict the class $\mathcal{C}$ to the image of $|.|$.

In [9] it was shown that a cardinal structure is strong iff it is initial with respect to all cardinal structure. Therefore strong cardinal structures are unique up to isomorphism.

The next lemma was already given in [8,9].

Lemma 4. *Let $(\mathcal{C}, |.|, \lesssim)$ be a cardinal structure on the allegory $\mathcal{R}$. Then:*

1. *If $R : A \to B$ is total and injective, then $|\mathbb{I}_A| \leq |\mathbb{I}_B|$.*
2. *If $U : C \to A$ and $V : C \to B$ are univalent with $U; U^\smile \sqcap V; V^\smile \sqsubseteq \mathbb{I}_C$, then*

$$|U^\smile; V| = |U; U^\smile \sqcap V; V^\smile|.$$

3. *If $R : A \to B$ has a tabulation $f : C \to A$ and $g : C \to B$, then $|R| = |\mathbb{I}_C|$.*

Property (1) and (3) of the previous lemma provide a hint how to define a cardinality structure if the underlying allegory is tabular.

Definition 6. *Let $\mathcal{R}$ be an allegory. Then the canonical cardinality structure $(\mathcal{OB}, |.|^*, \lesssim)$ is defined by*

1. *$\lesssim$ is a preordering on the class of objects of $\mathcal{R}$ defined by $A \lesssim B$ iff there is a total and injective relation $R : A \to B$,*
2. *$\mathcal{OB}$ is the class of equivalence classes $[A]$ of objects from $\mathcal{R}$ with respect to the equivalence relation induced by $\lesssim$,*
3. *$|.|^*$ is defined by $|R|^* := [C]$ where $R : A \to B$ has a tabulation $f : C \to A$ and $g : C \to B$.*

In [9] it was shown that the canonical cardinality structure is a strong cardinality function.

Lemma 5. *Let $(\mathcal{C}, |.|, \lesssim)$ be a cardinal structure on the allegory $\mathcal{R}$, and be $Q : A \to B$. Then:*

1. *If Q is injective, then $|Q; R| \leq |R|$ for all $R : B \to C$.*
2. *If Q is surjective and $Q^\smile; Q \sqcap R; R^\smile \sqsubseteq \mathbb{I}_B$, then $|R| \leq |Q; R|$.*
3. *If $i : C \to A$ and $j : D \to B$ are injections, then $|Q| = |i^\smile; Q; j|$.*

Proof

1. We immediately conclude

$$|Q; R| = |Q; R \sqcap Q; R|$$
$$\leq |R \sqcap Q^\smile; Q; R| \quad \text{C2 since } Q; Q^\smile \sqcap Q; R; (Q; R)^\smile \sqsubseteq Q; Q^\smile \sqsubseteq \mathbb{I}_A$$
$$\leq |R|. \qquad \text{C1}$$

2. We have

$$|R| = |R \sqcap R|$$
$$\leq |Q^\smile; Q; R \sqcap R| \qquad Q \text{ surjective}$$
$$\leq |Q; R \sqcap Q; R| \qquad \text{C2 since } Q^\smile; Q \sqcap R; R^\smile \sqsubseteq \mathbb{I}_B$$
$$= |Q; R|.$$

3. Consider the following computation

$$|Q| = |i^\smile; Q| \qquad \text{by (1) and (2) since } i \text{ is an injection}$$
$$= |Q^\smile; i| \qquad \text{C0}$$
$$= |j^\smile; Q^\smile; i| \qquad \text{by (1) and (2) since } j \text{ is an injection}$$
$$= |i^\smile; Q; j|. \qquad \text{C0}$$

This completes the proof. $\qquad\qquad\qquad\qquad\qquad\qquad\qquad\qquad\qquad\quad$ □

A distributive allegory adds an additional operation to the theory of allegories. It can be expected that a suitable notion of a cardinal structure on such an allegory has to satisfy additional axioms.

Definition 7. *Let $\mathcal{R}$ be a distributive allegory. An admissible cardinality structure $(\mathcal{C}, |.|, \leq)$ on $\mathcal{R}$ is a cardinal structure satisfying*

1. *If $|Q| \leq |Q'|$ and $(Q \sqcup Q') \sqcap R = \perp\!\!\!\perp_{AB}$, then $|Q \sqcup R| \leq |Q' \sqcup R|$.*

We want to provide an example of a non-admissible cardinal structure.

Example 1. It is easy to verify that every lower semi-lattice with a greatest element constitutes an one object allegory by defining $R^\smile := R$ and $Q; R := Q \sqcap R$. On such an allegory axioms C0 and C2 become trivial so that every monotone function $|.|$ from the allegory to an arbitrary ordering is a cardinality function. Notice that $|.|$ is strong if it is surjective since the allegory has just one object, and hence just one identity function.

Consider the example provided in Figure 1. The Boolean lattice on the left-hand side with the definitions above constitutes an allegory. The function $|.|$ is monotone so that it is a strong cardinality function on that allegory. The three relations Q, Q' and R are a counterexample to the admissibility property of $|.|$.

Considering sets and their cardinality one might expect that the assumption $(Q \sqcup Q') \sqcap R = \perp\!\!\!\perp_{AB}$ of an admissible cardinal structure can be weakened to $Q' \sqcap R = \perp\!\!\!\perp_{AB}$, i.e. that we can strengthen the admissibility property to

(*) $|Q| \leq |Q'|$ and $Q' \sqcap R = \perp\!\!\!\perp_{AB}$ implies $|Q \sqcup R| \leq |Q' \sqcup R|$.

The next example shows that this presumption is wrong.

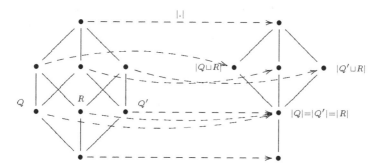

Fig. 1. A strong cardinality structure that is not admissible

Example 2. As in the previous example we are going to work with an one-object distributive allegory that is based on a lattice. Consider the example provided in Figure 2. The function $|.|$ is monotone so that it is a strong cardinality function on the allegory given on the left-hand side. Furthermore, it is easy to verify that this structure is also admissible since Q' and R are the only distinct non-zero relations with $Q' \sqcap R = \bot\!\bot$. On the other hand, the three relations Q, Q' and R are a counterexample to the stronger property $(*)$.

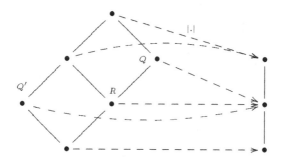

Fig. 2. A strong and admissible cardinality structure that does not satisfy $(*)$

In the next lemma we want to show that the canonical cardinal structure is admissible. Notice that the proof of the lemma requires the stronger assumption $(Q \sqcup Q') \sqcap R = \bot\!\bot_{AB}$.

Lemma 6. *Let $\mathcal{R}$ be a tabular distributive allegory. Then the canonical cardinal structure $(\mathcal{OB}, |.|^*, \lesssim)$ is admissible.*

Proof. Suppose $Q_i : A \to B$ $(i = 1, 2, 3)$ are relations with $|Q_1|^* \le |Q_2|^*$ and $(Q_1 \sqcup Q_2) \sqcap Q_3 = \bot\!\bot_{AB}$, i.e. $Q_1 \sqcap Q_3 = \bot\!\bot_{AB}$ and $Q_2 \sqcap Q_3 = \bot\!\bot_{AB}$. Furthermore, assume that $f_i : C_i \to A$ and $g_i : C_i \to B$ tabulates Q_i, and that $h_i : D_i \to A$

and $k_i : D_i \rightarrow B$ tabulates $Q_i \sqcup Q_3$ for $i = 1, 2$. From Lemma 2 we obtain injections $l_1 = f_1; h_1^{\smile} \sqcap g_1; k_1^{\smile} : C_1 \rightarrow D_1$, $l_3 = f_3; h_1^{\smile} \sqcap g_3; k_1^{\smile} : C_3 \rightarrow D_1$, $m_2 = f_2; h_2^{\smile} \sqcap g_2; k_2^{\smile} : C_2 \rightarrow D_2$ and $m_3 = f_3; h_2^{\smile} \sqcap g_3; k_2^{\smile} : C_3 \rightarrow D_2$ with the following properties:

$$l_1; h_1 = f_1, \qquad l_1; k_1 = g_1, \qquad l_3; h_1 = f_3, \qquad l_3; k_1 = g_3,$$
$$m_2; h_2 = f_2, \qquad m_2; k_2 = g_2, \qquad m_3; h_2 = f_3, \qquad m_3; k_2 = g_3.$$

From Lemma 4(3) we get $|\mathbb{I}_{C_1}|^* = |Q_1|^* \leq |Q_2|^* = |\mathbb{I}_{C_2}|^*$. Since the canonical

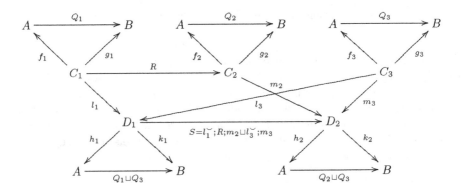

cardinal structure is strong by definition we obtain a total and injective relation $R : C_1 \rightarrow C_2$. Now, define $S = l_1^{\smile}; R; m_2 \sqcup l_3^{\smile}; m_3 : D_1 \rightarrow D_2$. We want to show that S is total and injective since this implies $|Q_1 \sqcup Q_3|^* = |\mathbb{I}_{D_1}|^* \leq |\mathbb{I}_{D_2}|^* = |Q_2 \sqcup Q_3|^*$ using Lemma 4(1). To this end consider the following computation

$$\mathbb{\,}\!\mathbb{L}_{C_2 C_3} = f_2; (Q_2 \sqcap Q_3); g_3^{\smile} \qquad\qquad\qquad Q_2 \sqcap Q_3 = \mathbb{\,}\!\mathbb{L}_{AB}$$
$$= f_2; (f_2^{\smile}; g_2 \sqcap f_3^{\smile}; g_3); g_3^{\smile}$$
$$= f_2; f_2^{\smile}; g_2; g_3^{\smile} \sqcap f_2; f_3^{\smile}; g_3; g_3^{\smile} \qquad\qquad \text{Lemma 1(2)}$$
$$= g_2; g_3^{\smile} \sqcap f_2; f_3^{\smile} \qquad\qquad\qquad\qquad\qquad f_2, g_3 \text{ injections}$$
$$= m_2; k_2; g_3^{\smile} \sqcap m_2; h_2; f_3^{\smile} \qquad\qquad\qquad \text{Properties above}$$
$$= m_2; (k_2; g_3^{\smile} \sqcap h_2; f_3^{\smile}) \qquad\qquad\qquad\quad \text{Lemma 1(2)}$$
$$= m_2; m_3^{\smile}. \qquad\qquad\qquad\qquad\qquad\qquad\quad \text{Definition } m_3$$

Analogously, we get $l_1; l_3^{\smile} = \mathbb{\,}\!\mathbb{L}_{C_1 C_3}$ using $Q_1 \sqcap Q_3 = \mathbb{\,}\!\mathbb{L}_{AB}$ and the definition of l_3. We conclude

$$S; S^{\smile} = (l_1^{\smile}; R; m_2 \sqcup l_3^{\smile}; m_3); (m_2^{\smile}; R^{\smile}; l_1 \sqcup m_3^{\smile}; l_3)$$
$$= l_1^{\smile}; R; m_2; m_2^{\smile}; R^{\smile}; l_1 \sqcup l_1^{\smile}; R; m_2; m_3^{\smile}; l_3$$
$$\sqcup l_3^{\smile}; m_3; m_2^{\smile}; R^{\smile}; l_1 \sqcup l_3^{\smile}; m_3; m_3^{\smile}; l_3$$

$$= l_1^{\smile}; R; m_2; m_2^{\smile}; R^{\smile}; l_1 \sqcup l_3^{\smile}; m_3; m_3^{\smile}; l_3 \qquad \text{Property above}$$

$$= l_1^{\smile}; R; R^{\smile}; l_1 \sqcup l_3^{\smile}; l_3 \qquad\qquad m_2, m_3 \text{ injections}$$

$$= l_1^{\smile}; l_1 \sqcup l_3^{\smile}; l_3 \qquad\qquad R \text{ injective}$$

so that it remains to show that $l_1^{\smile}; l_1 \sqcup l_3^{\smile}; l_3 = \mathbb{I}_{D_1}$. Therefore, consider the following computation

$$l_1^{\smile}; f_1 \sqcup l_3^{\smile}; f_3 = (h_1; f_1^{\smile} \sqcap k_1; g_1^{\smile}); f_1 \sqcup (h_1; f_3^{\smile} \sqcap k_1; g_3^{\smile}); f_3$$

$$= (h_1 \sqcap k_1; g_1^{\smile}; f_1) \sqcup (h_1 \sqcap k_1; g_3^{\smile}; f_3) \qquad \text{Lemma 1(3)}$$

$$= h_1 \sqcap (k_1; g_1^{\smile}; f_1 \sqcup k_1; g_3^{\smile}; f_3)$$

$$= h_1 \sqcap k_1; (g_1^{\smile}; f_1 \sqcup g_3^{\smile}; f_3) \qquad \text{Lemma 1(2)}$$

$$= h_1 \sqcap k_1; (Q_1^{\smile} \sqcup Q_3^{\smile})$$

$$= h_1 \sqcap k_1; (Q_1 \sqcup Q_3)^{\smile}$$

$$= h_1 \sqcap k_1; k_1^{\smile}; h_1$$

$$= h_1.$$

Analogously, we get $l_1^{\smile}; g_1 \sqcup l_3^{\smile}; g_3 = k_1$. We obtain

$$\mathbb{I}_{D_1} = h_1; h_1^{\smile} \sqcap k_1; k_1^{\smile} \qquad\qquad h_1, k_1 \text{ tabulation}$$

$$= (l_1^{\smile}; f_1 \sqcup l_3^{\smile}; f_3); h_1^{\smile} \sqcap (l_1^{\smile}; g_1 \sqcup l_3^{\smile}; g_3); k_1^{\smile} \qquad \text{Properties above}$$

$$= (l_1^{\smile}; f_1; h_1^{\smile} \sqcup l_3^{\smile}; f_3; h_1^{\smile}) \sqcap (l_1^{\smile}; g_1; k_1^{\smile} \sqcup l_3^{\smile}; g_3; k_1^{\smile}) \qquad \text{Lemma 1(2)}$$

$$= (l_1^{\smile}; f_1; h_1^{\smile} \sqcap l_1^{\smile}; g_1; k_1^{\smile}) \sqcup (l_1^{\smile}; f_1; h_1^{\smile} \sqcap l_3^{\smile}; g_3; k_1^{\smile})$$

$$\quad \sqcup (l_3^{\smile}; f_3; h_1^{\smile} \sqcap l_1^{\smile}; g_1; k_1^{\smile}) \sqcup (l_3^{\smile}; f_3; h_1^{\smile} \sqcap l_3^{\smile}; g_3; k_1^{\smile})$$

$$= (l_1^{\smile}; f_1; h_1^{\smile} \sqcap l_1^{\smile}; g_1; k_1^{\smile}) \sqcup (l_3^{\smile}; f_3; h_1^{\smile} \sqcap l_3^{\smile}; g_3; k_1^{\smile}) \qquad \text{Lemma 3}$$

$$= l_1^{\smile}; (f_1; h_1^{\smile} \sqcap g_1; k_1^{\smile}) \sqcup l_3^{\smile}; (f_3; h_1^{\smile} \sqcap g_3; k_1^{\smile}) \qquad l_1, l_3 \text{ injective}$$

$$= l_1^{\smile}; l_1 \sqcup l_3^{\smile}; l_3.$$

This completes the proof.   □

4 Cardinal Addition

Cardinal addition will be based on the cardinality of the disjoint union. The first lemma shows that this will establish a monotone operation.

Lemma 7. *Let $\mathcal{R}$ be a distributive allegory with relational sums, and $(\mathcal{C}, |.|, \leq)$ be an admissible cardinal structure on $\mathcal{R}$. For $Q : A \to B$, $Q' : C \to D$ with $|Q| \leq |Q'|$ we have $|\iota^{\smile}; Q; \iota \sqcup \kappa^{\smile}; R; \kappa| \leq |\iota^{\smile}; Q'; \iota \sqcup \kappa^{\smile}; R; \kappa|$.*

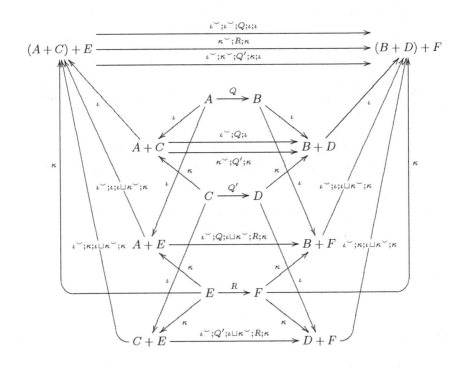

Proof. From Lemma 5(3) we conclude $|\iota^{\smile};\iota^{\smile};Q;\iota;\iota| = |\iota^{\smile};Q;\iota| = |Q| \leq |Q'| = |\kappa^{\smile};Q';\kappa| = |\kappa^{\smile};\kappa^{\smile};Q';\kappa;\kappa|$. Furthermore, we have

$$(\iota^{\smile};\iota^{\smile};Q;\iota;\iota \sqcup \iota^{\smile};\kappa^{\smile};Q';\kappa;\iota) \sqcap \kappa^{\smile};R;\kappa$$
$$= \iota^{\smile};(\iota^{\smile};Q;\iota;\iota \sqcup \kappa^{\smile};Q';\kappa);\iota \sqcap \kappa^{\smile};R;\kappa \qquad \text{Lemma 1(2)}$$
$$= \sqcup\!\!\sqcup_{(A+C)+E\,(B+D)+F} \qquad\qquad\qquad\qquad \text{Lemma 3(1)}$$

Since the cardinal structure is admissible we obtain $|\iota^{\smile};\iota^{\smile};Q;\iota;\iota \sqcup \kappa^{\smile};R;\kappa| \leq |\iota^{\smile};\kappa^{\smile};Q';\kappa;\iota \sqcup \kappa^{\smile};R;\kappa|$. Consider the following computation

$$(\iota^{\smile};\iota;\iota \sqcup \kappa^{\smile};\kappa);(\iota^{\smile};\iota;\iota \sqcup \kappa^{\smile};\kappa)^{\smile}$$
$$= (\iota^{\smile};\iota;\iota \sqcup \kappa^{\smile};\kappa);(\iota^{\smile};\iota^{\smile};\iota \sqcup \kappa^{\smile};\kappa)$$
$$= \iota^{\smile};\iota;\iota^{\smile};\iota \sqcup \kappa^{\smile};\kappa \qquad\qquad\qquad \text{Lemma 3(2)}$$
$$= \iota^{\smile};\iota \sqcup \kappa^{\smile};\kappa \qquad\qquad\qquad\qquad \iota \text{ injection}$$
$$= \mathbb{I}_{A+E},$$
$$(\iota^{\smile};\iota;\iota \sqcup \kappa^{\smile};\kappa)^{\smile};(\iota^{\smile};\iota;\iota \sqcup \kappa^{\smile};\kappa)$$
$$= (\iota^{\smile};\iota^{\smile};\iota \sqcup \kappa^{\smile};\kappa);(\iota^{\smile};\iota;\iota \sqcup \kappa^{\smile};\kappa)$$

$$= \iota^\smile; \iota^\smile; \iota; \iota \sqcup \kappa^\smile; \kappa \qquad\qquad \text{Lemma 3(2)}$$

$$\sqsubseteq \iota^\smile; \iota \sqcup \kappa^\smile; \kappa$$

$$= \mathbb{I}_{(A+C)+E},$$

which shows that $\iota^\smile; \iota; \iota \sqcup \kappa^\smile; \kappa$ is an injection. Analogously it follows that $\iota^\smile; \kappa; \iota \sqcup \kappa^\smile; \kappa$ is an injection. We conclude

$$|\iota^\smile; Q; \iota \sqcup \kappa^\smile; R; \kappa|$$

$$= |(\iota^\smile; \iota; \iota \sqcup \kappa^\smile; \kappa)^\smile; (\iota^\smile; Q; \iota \sqcup \kappa^\smile; R; \kappa); (\iota^\smile; \iota; \iota \sqcup \kappa^\smile; \kappa)| \qquad \text{Lemma 5(3)}$$

$$= |(\iota^\smile; \iota^\smile; \iota \sqcup \kappa^\smile; \kappa); (\iota^\smile; Q; \iota \sqcup \kappa^\smile; R; \kappa); (\iota^\smile; \iota; \iota \sqcup \kappa^\smile; \kappa)|$$

$$= |\iota^\smile; \iota^\smile; Q; \iota; \iota \sqcup \kappa^\smile; R; \kappa| \qquad\qquad \text{Lemma 3(2)}$$

$$\le |\iota^\smile; \kappa^\smile; Q'; \kappa; \iota \sqcup \kappa^\smile; R; \kappa| \qquad\qquad \text{see above}$$

$$= |(\iota^\smile; \kappa^\smile; \iota \sqcup \kappa^\smile; \kappa); (\iota^\smile; Q'; \iota \sqcup \kappa^\smile; R; \kappa); (\iota^\smile; \kappa; \iota \sqcup \kappa^\smile; \kappa)| \qquad \text{Lemma 3(2)}$$

$$= |(\iota^\smile; \kappa; \iota \sqcup \kappa^\smile; \kappa)^\smile; (\iota^\smile; Q'; \iota \sqcup \kappa^\smile; R; \kappa); (\iota^\smile; \kappa; \iota \sqcup \kappa^\smile; \kappa)|$$

$$= |\iota^\smile; Q'; \iota \sqcup \kappa^\smile; R; \kappa|. \qquad\qquad \text{Lemma 5(3)}$$

This completes the proof. $\qquad\qquad\qquad\qquad\qquad\qquad\qquad\qquad\qquad\qquad$ □

We are now ready to define addition for the cardinality of relations.

Definition 8. *Let $\mathcal{R}$ be a distributive allegory with relational sums, and $(\mathcal{C}, |.|, \le)$ be an admissible cardinal structure on $\mathcal{R}$. For $x, y \in \mathcal{C}$ with $x = |Q|$ and $y = |R|$ we define $x + y := |\iota^\smile; Q; \iota \sqcup \kappa^\smile; R; \kappa|$.*

Lemma 7 also shows that $+$ is well-defined, i.e. that the definition is independent of the choice of a relation Q with $|Q| = x$.

Lemma 8. *Let be $Q, R : A \to B$. Then:*

1. $|Q \sqcup R| \le |Q| + |R|$.
2. If $Q \sqcap R = \bot\!\!\!\bot_{AB}$, then $|Q| + |R| = |Q \sqcup R|$.
3. If Q (or R) has a complement, then $|Q| + |R| = |Q \sqcup R| + |Q \sqcap R|$.

Proof

1. First of all, we have

$$(\iota \sqcup \kappa); (\iota \sqcup \kappa)^\smile = (\iota \sqcup \kappa); (\iota^\smile \sqcup \kappa^\smile)$$

$$= \iota; \iota^\smile \sqcup \kappa; \kappa^\smile \qquad\qquad \text{Lemma 3(2)}$$

$$= \mathbb{I}_{A+A}$$

showing that $\iota \sqcup \kappa$ is injective. We conclude

$$|Q \sqcup R| = |(\iota \sqcup \kappa); (\iota^\smile; Q; \iota \sqcup \kappa^\smile; R; \kappa); (\iota^\smile \sqcup \kappa^\smile)| \qquad \text{Lemma 3(2)}$$

$$\le |\iota^\smile; Q; \iota \sqcup \kappa^\smile; R; \kappa| \qquad\qquad \text{Lemma 5(1)}$$

$$= |Q| + |R|.$$

2. Consider the following computation

$$\iota^\smile; \kappa \sqcap (\iota^\smile; Q^\smile; Q; \iota \sqcup \kappa^\smile; R^\smile; R; \kappa)$$
$$= (\iota^\smile; \kappa \sqcap \iota^\smile; Q^\smile; Q; \iota) \sqcup (\iota^\smile; \kappa \sqcap \kappa^\smile; R^\smile; R; \kappa)$$
$$= \amalg_{B+B, B+B} \qquad\qquad \text{Lemma 3(1)}$$

Analogously, we get $\kappa^\smile; \iota \sqcap (\iota^\smile; Q^\smile; Q; \iota \sqcup \kappa^\smile; R^\smile; R; \kappa) = \amalg_{B+B, B+B}$. This implies

$$(\iota \sqcup \kappa)^\smile; (\iota \sqcup \kappa) \sqcap (\iota^\smile; Q; \iota \sqcup \kappa^\smile; R; \kappa)^\smile; (\iota^\smile; Q; \iota \sqcup \kappa^\smile; R; \kappa)$$
$$= (\iota \sqcup \kappa)^\smile; (\iota \sqcup \kappa) \sqcap (\iota^\smile; Q^\smile; Q; \iota \sqcup \kappa^\smile; R^\smile; R; \kappa) \qquad \text{Lemma 3(2)}$$
$$= (\iota^\smile; \iota \sqcup \iota^\smile; \kappa \sqcup \kappa^\smile; \iota \sqcup \kappa^\smile; \kappa) \sqcap (\iota^\smile; Q^\smile; Q; \iota \sqcup \kappa^\smile; R^\smile; R; \kappa)$$
$$= (\mathbb{I}_{B+B} \sqcup \iota^\smile; \kappa \sqcup \kappa^\smile; \iota) \sqcap (\iota^\smile; Q^\smile; Q; \iota \sqcup \kappa^\smile; R^\smile; R; \kappa)$$
$$= (\mathbb{I}_{B+B} \sqcap (\iota^\smile; Q^\smile; Q; \iota \sqcup \kappa^\smile; R^\smile; R; \kappa))$$
$$\quad \sqcup (\iota^\smile; \kappa \sqcap (\iota^\smile; Q^\smile; Q; \iota \sqcup \kappa^\smile; R^\smile; R; \kappa))$$
$$\quad \sqcup (\kappa^\smile; \iota \sqcap (\iota^\smile; Q^\smile; Q; \iota \sqcup \kappa^\smile; R^\smile; R; \kappa))$$
$$= \mathbb{I}_{B+B} \sqcap (\iota^\smile; Q^\smile; Q; \iota \sqcup \kappa^\smile; R^\smile; R; \kappa)) \qquad \text{see above}$$
$$\sqsubseteq \mathbb{I}_{B+B}$$

and

$$(\iota \sqcup \kappa)^\smile; (\iota \sqcup \kappa) \sqcap (\iota^\smile; Q; \iota \sqcup \kappa^\smile; R; \kappa); (\iota \sqcup \kappa)^\smile; (\iota \sqcup \kappa); (\iota^\smile; Q; \iota \sqcup \kappa^\smile; R; \kappa)^\smile$$
$$= (\iota \sqcup \kappa)^\smile; (\iota \sqcup \kappa) \sqcap (\iota^\smile; Q \sqcup \kappa^\smile; R); (Q^\smile; \iota \sqcup R^\smile; \kappa) \qquad \text{Lemma 3(2)}$$
$$= (\iota^\smile; \iota \sqcup \iota^\smile; \kappa \sqcup \kappa^\smile; \iota \sqcup \kappa^\smile; \kappa) \sqcap (\iota^\smile; Q \sqcup \kappa^\smile; R); (Q^\smile; \iota \sqcup R^\smile; \kappa)$$
$$= (\mathbb{I}_{A+A} \sqcup \iota^\smile; \kappa \sqcup \kappa^\smile; \iota) \sqcap (\iota^\smile; Q \sqcup \kappa^\smile; R); (Q^\smile; \iota \sqcup R^\smile; \kappa)$$
$$= (\mathbb{I}_{A+A} \sqcap (\iota^\smile; Q \sqcup \kappa^\smile; R); (Q^\smile; \iota \sqcup R^\smile; \kappa))$$
$$\quad \sqcup (\iota^\smile; \kappa \sqcap (\iota^\smile; Q \sqcup \kappa^\smile; R); (Q^\smile; \iota \sqcup R^\smile; \kappa))$$
$$\quad \sqcup (\kappa^\smile; \iota \sqcap (\iota^\smile; Q \sqcup \kappa^\smile; R); (Q^\smile; \iota \sqcup R^\smile; \kappa))$$
$$= \mathbb{I}_{A+A} \sqcap (\iota^\smile; Q \sqcup \kappa^\smile; R); (Q^\smile; \iota \sqcup R^\smile; \kappa) \qquad \text{see above}$$
$$\sqsubseteq \mathbb{I}_{A+A}.$$

Using the two properties above together with Lemma 5(2) we conclude

$$|Q| + |R| = |\iota^\smile; Q; \iota \sqcup \kappa^\smile; R; \kappa|$$
$$= |\iota^\smile; Q^\smile; \iota \sqcup \kappa^\smile; R^\smile; \kappa| \qquad \text{C0}$$
$$\leq |(\iota \sqcup \kappa); (\iota^\smile; Q^\smile; \iota \sqcup \kappa^\smile; R^\smile; \kappa)| \qquad \text{Lemma 5(2)}$$
$$= |(\iota^\smile; Q; \iota \sqcup \kappa^\smile; R; \kappa); (\iota \sqcup \kappa)^\smile| \qquad \text{C0}$$
$$\leq |(\iota \sqcup \kappa); (\iota^\smile; Q; \iota \sqcup \kappa^\smile; R; \kappa); (\iota \sqcup \kappa)^\smile| \qquad \text{Lemma 5(2)}$$
$$= |Q \sqcup R|. \qquad \text{Lemma 3(2)}$$

3. We will denote the complement of Q by $\overline{Q}$ and compute

$$
\begin{aligned}
|Q| + |R| &= |Q| + |(Q \sqcap R) \sqcup (\overline{Q} \sqcap R)| \\
&= |Q| + |Q \sqcap R| + |\overline{Q} \sqcap R| & \text{by (2)} \\
&= |Q \sqcup (\overline{Q} \sqcap R)| + |Q \sqcap R| & \text{by (2)} \\
&= |Q \sqcup R| + |Q \sqcap R|.
\end{aligned}
$$

This completes the proof. $\qquad\square$

The following computation

$$
\begin{aligned}
|\amalg_{AB}| &= |\amalg_{AC}; R; \amalg_{DB}| \\
&\leq |R| & \text{Lemma 5(1) since } \amalg \text{ is injective}
\end{aligned}
$$

shows that $|\amalg_{AB}| \leq |R|$ for every relation $R : C \to D$. In particular, we have $|\amalg_{AB}| = |\amalg_{CD}|$ for all objects A, B, C and D. Therefore, we use the notation $0 := |\amalg_{AB}|$.

The next theorem establishes the main result about cardinal addition.

Theorem 1. *Let $\mathcal{R}$ be a distributive allegory with relational sums, and $(\mathcal{C}, |\cdot|, \leq)$ be an admissible cardinal structure on $\mathcal{R}$. Then $(\mathcal{C}, \leq, +, 0)$ is an ordered commutative monoid, i.e. $+$ is monotonic, associative and commutative with 0 as neutral element.*

Proof. Monotonicity follows immediately from Lemma 7. The relation $\iota^\smile; \kappa \sqcup \kappa^\smile; \iota$ is a bijective function from $A + B$ to $B + A$ so that we conclude

$$
\begin{aligned}
|Q| + |R| &= |\iota^\smile; Q; \iota \sqcup \kappa^\smile; R; \kappa| \\
&= |(\iota^\smile; \kappa \sqcup \kappa^\smile; \iota); (\iota^\smile; Q; \iota \sqcup \kappa^\smile; R; \kappa); (\iota^\smile; \kappa \sqcup \kappa^\smile; \iota)| & \text{Lemma 5(3)} \\
&= |\iota^\smile; R; \iota \sqcup \kappa^\smile; Q; \kappa| & \text{Lemma 3(2)} \\
&= |R| + |Q|.
\end{aligned}
$$

Associativity follows analogously using the bijective function $\iota^\smile; \iota^\smile; \iota \sqcup (\iota^\smile; \kappa^\smile; \iota \sqcup \kappa^\smile; \kappa); \kappa : (A + B) + C \to A + (B + C)$. Finally, we have

$$
\begin{aligned}
|Q| + 0 &= |Q| + |\amalg_{AB}| \\
&= |\iota^\smile; Q; \iota \sqcup \kappa^\smile; \amalg_{AB}; \kappa| \\
&= |\iota^\smile; Q; \iota| \\
&= |Q|. & \text{Lemma 5(3)}
\end{aligned}
$$

This completes the proof. $\qquad\square$

Finally, we want to concentrate on the well-known equivalence of cardinal numbers based on set theory

$$
\exists y : x + y = z \iff x \leq z.
$$

Notice that even in set theory the implication $\Leftarrow$ requires the axiom of choice.

Lemma 9. *Let $\mathcal{R}$ be a distributive allegory with relational sums, and $(\mathcal{C}, |.|, \leq)$ be an admissible cardinal structure on $\mathcal{R}$. Then we have:*

1. *If $|Q| + |R| = |S|$, then $|Q| \leq |S|$.*
2. *If $\mathcal{R}$ is Boolean, tabular, has splittings and satisfies (AC) and $|.|$ is strong, then $|Q| \leq |S|$ implies that there is a relation R with $|Q| + |R| = |S|$.*

Proof

1. This follows immediately from $|Q| = |Q| + 0 \leq |Q| + |R| = |S|$.
2. Assume $Q : A \to B$ and $S : C \to D$ are tabulated by $f : E \to A, g : E \to B$ and $h : F \to C, k : F \to D$, respectively. Then we have $|\mathbb{I}_E| = |Q| \leq |S| = |\mathbb{I}_F|$ by Lemma 4(3). Since $|.|$ is strong there is a total and injective relation $T : E \to F$. By (AC) we obtain an injection $i \sqsubseteq T$. Suppose $u : G \to F$ splits the symmetric and idempotent relation $\mathbb{I}_F \sqcap \overline{i^\smile ; i}$. Then u is an injection and we have

$$
\begin{aligned}
|Q| + |u| &= |Q| + |u^\smile ; u| & \text{Lemma 5(3)} \\
&= |Q| + |\mathbb{I}_F \sqcap \overline{i^\smile ; i}| & \\
&= |\mathbb{I}_E| + |\mathbb{I}_F \sqcap \overline{i^\smile ; i}| & \text{Lemma 4(3)} \\
&= |i^\smile ; i| + |\mathbb{I}_F \sqcap \overline{i^\smile ; i}| & \text{Lemma 5(3)} \\
&= |i^\smile ; i \sqcup (\mathbb{I}_F \sqcap \overline{i^\smile ; i})| & \text{Lemma 8(2)} \\
&= |\mathbb{I}_F| & \\
&= |S|. & \text{Lemma 4(3)}
\end{aligned}
$$

This completes the proof.   □

Notice that the assumption that $\mathcal{R}$ is Boolean from 2. of the previous lemma is redundant since this follows the remaining properties [13].

5   Conclusion and Outlook

In this paper we extended the notion of a cardinal function to distributive allegories, and we defined a suitable notion of addition of cardinal numbers. The next step in the development of this theory will be an abstract treatment of multiplication. This notion will be based on relational products, of course. Therefore, multiplication can just be defined in allegories that permit relational products. It is well known that such an allegory is representable, i.e. multiplication, unlike addition, can just be defined in representable allegories.

Another area of further study will be the application of this theory to theorems including cardinality statements and the development of algorithms. As already mentioned in the introduction certain algorithm are formulated and/or developed within the relational framework using cardinality properties. Even though the correctness of the core algorithm is shown using relation algebra, any cardinality consideration is usually treated separately. Our theory provides the theoretical background for both tasks within one framework.

References

1. Berghammer, R.: Computation of Cut Completions and Concept Lattices Using Relational Algebra and RelView. JoRMiCS 1, 50–72 (2004)
2. Berghammer, R., Müller-Olm, M.: Formal Development and Verification of Approximations Algorithms Using Auxiliary Variables. In: Bruynooghe, M. (ed.) LOPSTR 2004. LNCS, vol. 3018, pp. 59–74. Springer, Heidelberg (2004)
3. Bird, R., de Moor, O.: Algebra of Programming. Prentice-Hall, Englewood Cliffs (1997)
4. Brink, C., Kahl, W., Schmidt, G. (eds.): Relational Methods in Computer Science. Advances in Computer Science. Springer, Vienna (1997)
5. Freyd, P., Scedrov, A.: Categories, Allegories. North-Holland, Amsterdam (1990)
6. Grätzer, G.: General lattice theory, 2nd edn. Birkhäuser, Basel (2003)
7. Kawahara, Y.: On the Cardinality of Relations. In: Schmidt, R.A. (ed.) RelMiCS/AKA 2006. LNCS, vol. 4136, pp. 251–265. Springer, Heidelberg (2006)
8. Kawahara, Y., Winter, M.: Cardinality in Allegories. In: Berghammer, R., Möller, B., Struth, G. (eds.) RelMiCS/AKA 2008. LNCS, vol. 4988, pp. 274–288. Springer, Heidelberg (2008)
9. Kawahara, Y., Winter, M.: Cardinality Functions in Allegories. JLAP (submitted)
10. Papadimitriou, C.H., Steiglitz, K.: Combinatorial Optimization: Algorithms and Complexity. Dover, New York (1998)
11. Schmidt, G., Ströhlein, T.: Relationen und Graphen. Springer, Heidelberg (1989); English version: Relations and Graphs. Discrete Mathematics for Computer Scientists. EATCS Monographs on Theoret. Comput. Sci. Springer, Heidelberg (1993)
12. Winter, M.: Goguen Categories. A Categorical Approach to L-Fuzzy Relations. Trends in Logic, vol. 25. Springer, Heidelberg (2007)
13. Winter, M.: Complements in Distributive Allegories Submitted to RelMiCS 11/AKA 6

Relational Methods in the Analysis of While Loops: Observations of Versatility

Asma Louhichi[1], Olfa Mraihi[1], Lamia Labed Jilani[1],
Khaled Bsaies[2], and Ali Mili[3]

[1] Institut Superieur de Gestion Bardo 2000 Tunisia
lamia.labed@isg.rnu.tn
[2] Faculty of Science of Tunis Tunis El Manar 2092 Tunisia
khaled.bsaies@fst.rnu.tn
[3] NJIT Newark NJ 07102-1982 USA
mili@cis.njit.edu

Abstract. Despite much progress in the design of programming languages, the vast majority of software being written and deployed nowadays remains written in languages where iteration is the main inductive construct, and the main source of algorithmic complexity. For the past four decades, the analysis of iterative constructs has been dominated, not undeservedly, by the concept of invariant assertions. In this paper we submit relation-based alternatives, namely invariant relations and invariant functions, and show how these can provide complementary perspectives, and can enrich the analysis of iterations. Whereas loop invariants can be used to establish the correctness of iterative programs in Hoare logics, invariant relations and invariant functions are used to derive program functions in Mills' logic. In keeping with the conference format, we do not delve too much into theoretical results, and focus instead on the applied aspects of our relation-theoretic approach.

Keywords: Function extraction, loop functions, invariant assertions, invariant relations, invariant functions, relational calculus, refinement calculus, computing loop behavior.

1 Deriving Loop Functions

Despite several decades of evolution, most programming languages in use today are Pascal-like languages, whose most important construct is the loop. Also, despite several decades of research, the analysis of programs for the purpose of understanding them, verifying their correctness or maintaining them remains an unfulfilled challenge. Few methods in existence today scale up to industrial size software products, and most support tools require extensive human intervention, which makes them both less effective and less dependable. This combination has led to a renewed interest in the analysis of while loops, dominated to a large extent by the concept of invariant assertions [1,2,3,4,5,6,7,8,9,10] (to cite only a few). In this paper we explore relation-based concepts that offer alternative perspectives to invariant assertions, namely *invariant relations* [11] and *invariant*

R. Berghammer et al. (Eds.): RelMiCS/AKA 2009, LNCS 5827, pp. 242–259, 2009.

functions [12]. Our focus in this paper is not on theoretical results, but rather on exploring practical applications of these concepts to the analysis of loops. Also, we are not interested to compare and contrast the merits of invariant assertions versus invariant relations and invariant functions as much as we are interested to show how these two families of approaches (logic-based, relation-based) offer complementary perspectives and insights. As usual, the law of diminishing returns advocates the use of a diverse set of methods and tools, to maximize the impact of our effort.

In the following section we briefly present some mathematical background, which we use in section 3 to define and illustrate the concepts of invariant relations and invariant functions. In section 4 we present a number of theorems that form the basis of our approach to the analysis of loops. In section 4.2 we briefly discuss the design and implementation of the tool, and in section 5 we show examples of application of the proposed method and tool. Finally, in section 6 we briefly summarize our results and assess them with respect to competing approaches and tools.

2   Mathematical Background

2.1   Programs and Specifications

Adopting Mills' logic [13,14], we use functions to represent program semantics and relations to represent program specifications. Given a program P on variables x, y, z (for example), we let the set S defined by these variables be the *space* of the program, and we define the function of program P as

$$[P] = \{(s, s')| \text{ if } P \text{ starts execution in state } s \text{ then it terminates in state } s'\}.$$

From this definition, we infer:

$$dom([P]) = \{s| \text{ if } P \text{ starts execution in state } s \text{ then it terminates}\}.$$

Given a space S, we represent program specifications by binary relations on S. A specification R on S represents all acceptable input/output pairs, in the following sense: the domain of R represents all the initial states that the user may submit to the program; given an element s in $dom(R)$, the image set of s by R represents all the final states that are considered correct for s.

2.2   Refinement Ordering

We define an ordering relation on relational specifications under the name *refinement ordering*:

Definition 1. *A relation R is said to* refine *a relation R' if and only if*

$$RL \cap R'L \cap (R \cup R') = R'.$$

In set theoretic terms, this equation means that the domain of R is a superset of (or equal to) the domain of R', and that for any element s in the domain of R', the image set of s by R is a subset of (or equal to) the image set of s by R'. This is similar to, but different from, refining a pre/postcondition specification by weakening its precondition and/or strengthening its postcondition [15,16]. We abbreviate this property by $R \sqsupseteq R'$ or $R' \sqsubseteq R$. We admit that, modulo traditional definitions of total correctness [17,15,18], the following propositions hold.

- A program P is correct with respect to a specification R if and only if $[P] \sqsupseteq R$.
- $R \sqsupseteq R'$ if and only if any program correct with respect to R is correct with respect to R'.

Intuitively, R refines R' if and only if R represents a stronger requirement than R'.

2.3   Refinement Lattice

We admit without proof that the refinement relation is a partial ordering. In [19] Boudriga et al. analyze the lattice properties of this ordering and find the following results:

- Any two relations R and R' have a greatest lower bound, which we refer to as the *meet*, denote by $\sqcap$, and define by:

$$R \sqcap R' = RL \cap R'L \cap (R \cup R').$$

- Two relations R and R' have a least upper bound if and only if they satisfy the following condition:

$$RL \cap R'L = (R \cap R')L.$$

Under this condition, their least upper bound is referred to as the *join*, denoted by $\sqcup$, and defined by:

$$R \sqcup R' = \overline{RL} \cap R' \cup \overline{R'L} \cap R \cup (R \cap R').$$

- Two relations R and R' have a least upper bound if and only if they have an upper bound; this property holds in general for lattices, but because the refinement ordering is not a lattice (since the existence of the join is conditional), it bears checking for this ordering specifically.
- The lattice of refinement admits a *universal lower bound*, which is the empty relation.
- The lattice of refinement admits no *universal upper bound*.
- Maximal elements of this lattice are total deterministic relations.

3   Relational Invariants

In this section, we briefly define, discuss and illustrate these concepts, in preparation for our subsequent discussions. We consider a while statement of the form $w = $ **while** t **do** B on some space S, and we assume that w terminates normally for any initial state s in S. For the sake of illustration, we consider the following while loop on array variables a and b, real variables x and y, and index variables i and j:

$$\textbf{while } (i \neq N + 1)\{x := x + a[i]; y := y + b[j]; i := i + 1; j := j - 1; \}.$$

We submit that the following predicate is an invariant assertion:

$$x = \sum_{k=1}^{i-1} a[k].$$

3.1   Invariant Relations

Invariant relations are relations that contain pairs of states (s, s') such that s' follows from s by application of an arbitrary number of iterations. We define them as follows.

Definition 2. *Given a while loop of the form* w = while t do B *on some space* S, *and given a relation* R *on* S, *we say that* R *is an* invariant relation *for* w *if and only if* R *is reflexive, transitive, and satisfies the following conditions (where* T *is the vector defined by predicate* t:

- *The Invariance Condition:*

$$T \cap [B] \subseteq R.$$

- *The Convergence Condition:*

$$R \circ \overline{T} = L.$$

It is possible to interpret the invariant condition in many ways, of which we choose the following:

- The function of the while loop is given by the following expression [20]:

$$[w] = (I(t) \circ [B])^* \circ I(\neg t),$$

  where $(I(t) \circ [B])^*$ is the reflexive transitive closure of $(I(t) \circ [B])$ (which can, in turn, be written as $(T \cap [B])$., where T is the vector defined by t)
- The reflexive transitive closure of a relation is the smallest reflexive transitive superset of the relation.
- As an arbitrary (not necessarily smallest) reflexive transitive superset of $(I(t) \circ [B])$, an invariant relation is an approximation of the the reflexive transitive closure of $(I(t) \circ [B])$, hence can be used to approximate the function of the loop. This will be confirmed, and formalized, in section 4, by Theorem 3.

As for the *convergence condition*, it provides that any state in S can be mapped by an invariant relation R onto a state in S that satisfies $\neg t$. Given that R represents the effect of applying the loop body an arbitrary number of times, this condition ensures that these applications eventually produce a final state, i.e. a state that causes the loop to terminate.

To illustrate this concept, we consider the example of the array sum, presented above, and propose the following invariant relation for it, inviting the reader to compare it with (contrast it to) the invariant assertion presented above.

$$R = \{(s, s') | x + \sum_{k=i}^{N} a[k] = x' + \sum_{k=i'}^{N} a'[k] \wedge i \leq i'\}.$$

This relation is clearly reflexive and transitive; we leave it to the reader to check that it satisfies the invariance condition and the convergence condition.

3.2 Invariant Functions

Whereas invariant relations are used to approximate loop functions, invariant functions are used to generate a broad class of invariant relations, namely those that are symmetric (in addition to being reflexive and transitive).

Definition 3. *Let w be a while statement of the form* while t do B *that terminates normally for all initial states in S. We say that a function F on S is an invariant function if and only if it is total and*

$$T \cap [B] \circ F = T \cap F.$$

Invariant functions are due to [12]. They are relevant to our work because if F is an invariant function then its nucleus $F\widehat{F}$ is an invariant relation; we have further shown in [11] that if R is a symmetric invariant relation then it is the nucleus of some invariant function.

To illustrate the concept of invariant function, we consider the array program, and submit the following function:

$$F \begin{pmatrix} a \\ x \\ i \\ b \\ y \\ j \end{pmatrix} = \begin{pmatrix} a \\ \frac{x + \sum_{k=i}^{N} a[k]}{N+1} \\ a \\ \frac{x + \sum_{k=i}^{N} a[k]}{N+1} \end{pmatrix}.$$

We briefly verify that this function is invariant with respect to application of the loop body (assuming s satisfies condition t).

$$F \left([B] \begin{pmatrix} a \\ x \\ i \\ b \\ y \\ j \end{pmatrix} \right) = F \begin{pmatrix} a \\ x + a[i] \\ i+1 \\ b \\ y + b[j] \\ j-1 \end{pmatrix} = \begin{pmatrix} a \\ \frac{x + a[i] + \sum_{k=i+1}^{N} a[k]}{N+1} \\ a \\ \frac{x + a[i] + \sum_{k=i+1}^{N} a[k]}{N+1} \end{pmatrix} = F \begin{pmatrix} a \\ x \\ i \\ b \\ y \\ j \end{pmatrix}.$$

4   Successive Approximations

We now discuss how invariant relations and invariant functions are used in our effort to derive loop functions automatically. As a discipline of separation of concerns, we resolve to derive the function of the loop by successive approximations. To this effect, we resolve to use the lattice of refinement, and we formulate approximations as inequalties that involve the function of the loop. Because the loops we are considering are deterministic, and because we are assuming that they terminate for all states in S, the loops define total, deterministic relations. According to the lattice structure we have found for the refinement ordering, total deterministic relations are maximal; hence the only meaningful inequalities we can write with loop functions are of the form

$$[w] \sqsupseteq T,$$

for some relation T, to which we refer as a *lower bound* of the loop function. The central idea of our approach is that we derive the function of the loop ($[w]$) from a collection of statements of the form $[w] \sqsupseteq T_i$, for a number of lower bunds T_i. If we can analyze the loop and derive lower bounds T_1, T_2, ... T_k of $[w]$, we write

$$[w] \sqsupseteq T_1 \wedge [w] \sqsupseteq T_2 \wedge ... \wedge [w] \sqsupseteq T_k,$$

from which we infer (by lattice theory)

$$[w] \sqsupseteq T_1 \sqcup T_2 \sqcup ... \sqcup T_k.$$

If $T_1 \sqcup T_2 \sqcup ... \sqcup T_k$ is total and deterministic, then (according to the lattice structure) it is maximal in the lattice, hence we infer

$$[w] = T_1 \sqcup T_2 \sqcup ... \sqcup T_k.$$

If not, then the join of lower bounds forms a comprehensive lower bound (approximation) of the loop function. In the next section we present a number of theorems that provide lower bounds for the loop function; proofs of these theorems are given in [20].

4.1   Lower Bound Theorems

We distinguish between two types of theorems: Theorems that provide *constructive lower bounds*, i.e. lower bounds that are explicit expression of loop parameters (t, B); and theorems that provide *creative lower bounds*, i.e. lower bounds that must be derived creatively, then checked against specific conditions. Our algorithm for deriving loop functions uses these classes of lower bounds differently: constructive lower bounds are generated systematically for all loops; whereas creative lower bounds are generated by pattern matching against pre-catalogued code patterns.

4.1.1   Constructive Lower Bounds

The following theorem provides that the final state of an execution satisfies $\neg t$, and that it has a antecedent by the loop body that satisfies t; in other words, the final state is the first state in the sequence of iterants that fails to satisfy t. This is useful in many cases in practice, for example when the loop condition is an inequality ($\geq, \leq, <, >$).

Theorem 1. *Let w be a while loop defined by* while t do B. *If $t \not\equiv$ false then the following relation T is a lower bound of $[B]$:*

$$T = I(t) \circ L \circ I(t) \circ [B] \circ I(\neg t) \cup I(\neg t).$$

The following theorem provides that if the loop body is executed at least once, then the final state will be in the range of $[B]$. This is useful in cases when the function of the loop body is not surjective.

Theorem 2. *Given a while statement w of the form* while t do B *on space S, such that w terminates for all initial states in S, and that $t \not\equiv$ false . Then the following specification is a lower bound of the function of the loop:*

$$T = (L \circ [B] \cup I) \circ I(\neg t).$$

4.1.2   Creative Lower Bounds

The following theorems provide lower bounds that require a creative step; these lower bounds are not expressed exclusively in terms of the loop parameters, but also in terms of quantities that must be derived creatively. Our algorithm will generate them by means of a pattern matching step, which we discuss in the next section.

Theorem 3. *We consider a while loop w on space S, defined by $w =$* while t do B. *If R is an invariant relation for w then T is a lower bound for $[w]$, where*

$$T = R \circ I(\neg t).$$

This theorem allows us to derive a lower bound of $[w]$ from any invariant relation we may know about the loop. Because invariant relations are supersets of the loop body's function, it may be advantageous to structure the function of the loop body as an intersection, so as to facilitate deriving its supersets: If the function B of the loop body is written as an intersection, say

$$B = B_0 \cap B_1 \cap B_2 \cap ... \cap B_k,$$

then any superset of B_0 is a seuperset of B, any superset of $B_0 \cap B_1$ is a superset of B, any superset of $B_0 \cap B_1 \cap B_2$ is a seuperset of B, etc; this property can be used as a basis for a separation of concerns discipline, whereby we derive lower bounds of the loop function ($[w]$) by considering arbitrarily few terms of the intersection. But we do not always get to choose how to structure the function of the loop body: if the loop body has an if-then-else, or nested if-then-else's, or sequenced if-then-else's, then the outermost structure of its function is a union, not an intersection. The following theorem provides a tentative solution for such cases.

Theorem 4. *We consider a while statement of the form* `while t do B` *where the function of* `B` *can be written as the union of two relations, say P and Q. If R and R' are reflexive transitive relations such that*

$$P \subseteq R,$$

$$Q \circ R \subseteq R'$$

and

$$R \circ R' \circ I(\neg t) \circ L = L$$

then T is a lower bound of $[w]$, where

$$T = R \circ R' \circ I(\neg t).$$

In the next section we briefly discuss how these theorems are deployed in practice to build a tool that computes the function of a loop from an analysis of its source code in a variety of common programming languages (C, C++, Java).

4.2   A Tool for Loop Analysis

We use the theorems of the previous section as a basis in the design of an automated tool that derives the function of a while loop by analyzing its source code. The tool proceeds in three steps:

- *From Source Code to Internal Notation.* We map source code from the relevant languages to a unified language-independent notation. This allows us to keep the subsequent steps language-independent, and also to prepare the loop for application of theorems 3 and 4. To accomodate Theorem 3, we aim to structure the function of the loop body as an intersection; if the outermost structure of this function is a union, we structure it as a union of intersections. The notation we use to this effect is that of *Conditional Concurrent Assignments*, or CCA for short; files that contain the representation of the loop in this notation will have the extention .cca.
- *From Internal Notation to Mathematica Equations.* We analyze the CCA notation for the purpose of applying the lower bound theorems; the lower bounds we generate take the form of equations involving initial values and final values of the program variables. For constructive lower bounds, we generate systematically all the equations that stem from Theorems 1 and 2. For creative lower bounds, we proceed by matching CCA statements or combinations of statements against a library of code patterns and inferring appropriate lower bounds.
- *From Mathematica Equations to Loop Functions.* Solve the equations generated above in the final values of the program variables, as a function of the initial values; this gives, in effect, the function of the loop. This step is carried out by a combination of a computer algebra system (Mathematica, ©Wolfram Research) with a theorem prover (Otter is currently under consideration).

The success of our approach depends critically on the second step, hence we discuss it in some detail in this section. We generate creative lower bounds by matching CCA statements or combinations of statements against pre-catalogued code patterns for which we have predefined lower bounds. To this effect we define a database of patterns called *recognizers*, where each recognizer is defined by the following items:

- A space declaration, given in terms of variables and constants; unlike other approaches that work exclusively on executable code, our approach can be applied to constants whose value is left unspecified. If the function of the loop takes different forms depending on the value of the constants, our approach will highlight them.
- A code pattern that includes one, two or three CCA statements. In order to keep combinatorics under control, we have resolved to limit our pattern matching to no more than three statements; depending on the number of CCA statements in the pattern, we talk about 1-recognizers, 2-recognizers, or 3-recognizers.
- An invariant relation that is known to be maintained by the relevant statements.
- The lower bound that stems from the invariant relation (by post-restricting it to $\neg t$).

When a CCA statement or combination of statements matches the code pattern of a recognizer, we instantiate the lower bound given by the recognizer in the form of a Mathematica equation involving primed variables (referring to final values) and unprimed variables (referring to initial values of state variables). Solving the combination of these equations in the primed variables using the unprimed variables as parameters yields an explicit expression of the function of the loop.

5   Examples of Application

To illustrate our relational approach, we present below a few simple examples for which our tool generates the loop function. Along with each example, we present a sampling of the recognizers that are used to analyze it.

5.1   Numeric Example

The first example involves numeric computations. We consider the following while loop on a number of integer constants a, b and c, and integer variables x, y, z, t and i:

while $(i \neq 0)\{t := t + a * y; z := z + b * x; x := x + a; y =: y * c; i := i - 1; \}.$

We are interested in deriving the function that this loop defines between its initial states (values of x, y, z, t and i prior to the execution of the loop) and its final states (values of these variables when execution terminates). Some of the recognizers that we use to analyse this loop include the following:

ID	State Space	Code Pattern	Invariant Relation $R =$	Lower Bound $T =$
1R1	x: int const c: int >0	x=x+c	$\{(s, s') \mid x \bmod c = x' \bmod c\}$	$\{(s, s') \mid x \bmod c \wedge \neg t(s')\}$
2R1:	x, y: int const a, b: int	x = x+a y = y+b	$\{(s, s') \mid ay - bx = ay' - bx'\}$	$\{(s, s') \mid ay - bx = ay' - bx' \wedge \neg t(s')\}$
2R2:	x, y: int const a: int	x = x*a y = y+x	$\{(s, s') \mid y(1-a) + x = y'(1-a) + x'\}$	$\{(s, s') \mid y(1-a) + x = y'(1-a) + x' \wedge \neg t(s')\}$
2R3:	x, y: int const a, b: int	x = x+a y = y*b	$\{(s, s') \mid \frac{y}{b^{x/a}} = \frac{y'}{b^{x'/a}}\}$	$\{(s, s') \mid \frac{y}{b^{x/a}} = \frac{y'}{b^{x'/a}} \wedge \neg t(s')\}$

The pattern matching step produces the following Mathematica equations:

```
Reduce[
  {Reduce[
    {t+a*y/(1-c) == tP+a*yP/(1-c),
     z-b*x*(x-a)/(2*a) == zP-b*xP*(xP-a)/(2*a),
     y/c^(x/a) == yP/c^(xP/a),
     a*x+1*i == a*xP+1*iP,
     y/c^(i/-1) == yP/c^(iP/-1),
     Reduce[
       {(iP == 0),
        Exists [
          {iPP, tPP, xPP, yPP, zPP},
          (iPP == 0) && tP == tPP+a*yPP &&
          zP == zPP+b*xPP && xP == xPP+a &&
          yP == yPP*c && iP == iPP-1] } ]
    } ],
  i >= iP} {iP, tP, xP, yP, zP},
  Backsubstitution -> True]
```

We make multiple nested calls to *Reduce* in order to sequentialize the resolution of these equations; the inner call stems from application of Theorem 2, and specifies that the final state has an antecedent by the loop body that satisfies the loop condition. Mathematica solves the equations in xP, yP, zP, tP and iP using x, y, z, t, and i as variables and a, b and c as constants. Because $(1-c)$, c and a appear in the denominator in the Mathematica expression above, Mathematica assumes that $a \times c \times (c-1)$ is non-zero. Singular values of these constants produce simpler expressions of the loop function, which we do not discuss here.

$$a \times (c^2 - c) \neq 0 \land y \neq 0 \Rightarrow \begin{pmatrix} iP = 0 \\ tP = \frac{-t+ct-ay+ac^i y}{c-1} \\ xP = a \times i + x \\ yP = c^i y \\ zP = \frac{-abi+abi^2+2bix+2z}{2} \end{pmatrix}$$

$$\land$$

$$a \times (c^2 - c) \neq 0 \land y = 0 \Rightarrow \begin{pmatrix} iP = 0 \\ tP = t \\ xP = a \times i + x \\ yP = 0 \\ zP = \frac{-abi+abi^2+2bix+2z}{2} \end{pmatrix}$$

To enhance our confidence in the correctness of the generated function, we deploy a test driver that produces randomly generated test data (varying the variables and the constants, subject to the cited condition), and tests the loop against an oracle derived from the loop function; all the tests we ran were successful.

Interest of this example: This example shows how we can process numeric examples of arbitrary size, with fairly little overhead, provided we have the appropriate recognizers. This example also highlights that we can handle symbolic constants, since we deal with source code, whereas systems that deal with program execution (e.g. Daikon [5]) can only handle instantiated constants.

5.2 Array Manipulations

We consider the following while loop on array variables a and b, integer variables x and y, integer constant N and index variables i and j:

while $(i < N)\{x := x + a[i]; y := y + b[j]; i := i + 1; j := j - 1; \}$

We are interested in deriving the function of this loop; we are also interested to determine how this function is affected by minor changes to the source code, such as: changing the condition to $(i \leq N)$; to $(i \neq N)$; to $(j > 0)$; to $(j \geq 0)$; if we permute the index updates and the sum updates. Table 1 summarizes the outcomes of various combinations of the proposed modifications, with the associated functions. The table below presents two of the recognizers that we needed to analyze the example at hand.

ID	State Space	Code Pattern	Invariant Relation $R =$	Lower Bound $T =$
3R1	x: int a[N]: int i: int	i=i+1, x = x+a[i] a=a	$\{(s, s')\|a' = a$ $\land x + \sum_{k=i}^{N} a[k] =$ $x' + \sum_{k=i}^{N} a'[k]\}$	$\{(s, s')\|a' = a$ $\land x + \sum_{k=i}^{N} a[k] =$ $x' + \sum_{k=i'}^{N} a'[k] \land \neg t(s')\}$
3R2	x: int a[N]: int i: int	i=i-1, x = x+a[i] a=a	$\{(s, s')\|a' = a$ $\land x + \sum_{k=1}^{i} a[k] =$ $x' + \sum_{k=1}^{i'} a'[k]\}$	$\{(s, s')\|a' = a$ $\land x + \sum_{k=1}^{i} a[k] =$ $x' + \sum_{k=1}^{i'} a'[k] \land \neg t(s')\}$

Table 1. Array Loops and their Functions

Program	Function
`while (i<N) { x=x+a[i];` `y=y+b[j]; i=i+1; j=j-1;}`	$I(i \geq N) \cup \{(s,s') \mid i \leq N \wedge i' = N \wedge j' = i - N + j \wedge$ $x' = x + \sum_{k=i}^{N-1} a[k] \wedge y' = y + \sum_{k=i-N+j+1}^{j} b[k]$ $\wedge a' = a \wedge b' = b\}$
`while (i<N) { j=j-1;` `x=x+a[i]; y=y+b[j]; i=i+1;}`	$I(i \geq N) \cup \{(s,s') \mid i \leq N \wedge i' = N \wedge j' = i - N + j \wedge$ $x' = x + \sum_{k=i}^{N-1} a[k] \wedge y' = y + \sum_{k=i-N+1}^{j-1} b[k]$ $\wedge a' = a \wedge b' = b\}$
`while (i<N) { i=i+1;` `x=x+a[i]; y=y+b[j]; j=j-1;}`	$I(i \geq N) \cup \{(s,s') \mid i \leq N \wedge i' = N \wedge j' = i - N + j \wedge$ $x' = x + \sum_{k=i+1}^{N} a[k] \wedge y' = y + \sum_{k=i-N+j+1}^{j} b[k]$ $\wedge a' = a \wedge b' = b\}$
`while (i<N) { i=i+1;` `j=j-1; x=x+a[i]; y=y+b[j];}`	$I(i \geq N) \cup \{(s,s') \mid i \leq N \wedge i' = N \wedge j' = i - N + j \wedge$ $x' = x + \sum_{k=i+1}^{N} a[k] \wedge y' = y + \sum_{k=i-N+j}^{j-1} b[k]$ $\wedge a' = a \wedge b' = b\}$
`while (i!=N) { i=i+1;` `j=j-1; x=x+a[i]; y=y+b[j];}`	$\{(s,s') \mid i \leq N \wedge i' = N \wedge j' = i - N + j \wedge$ $x' = x + \sum_{k=i+1}^{N} a[k] \wedge y' = y + \sum_{k=i-N+j}^{j-1} b[k]$ $\wedge a' = a \wedge b' = b\}$
`while (i<=N) { x=x+a[i];` `y=y+b[j]; i=i+1; j=j-1; }`	$I(i > N) \cup \{(s,s') \mid i \leq N \wedge i' = N + 1 \wedge$ $x' = x + \sum_{k=i}^{N} a[k] \wedge y' = y + \sum_{k=i-N+j}^{j} b[k]$ $\wedge j' = i - N + j - 1 \wedge a' = a \wedge b' = b\}$
`while (i<=N) { j=j-1;` `x=x+a[i]; y=y+b[j]; i=i+1; }`	$I(i > N) \cup \{(s,s') \mid i \leq N \wedge i' = N + 1 \wedge$ $x' = x + \sum_{k=i}^{N} a[k] \wedge y' = y + \sum_{k=i-N+j-1}^{j-1} b[k]$ $\wedge j' = i - N + j - 1 \wedge a' = a \wedge b' = b\}$
`while (i<=N) { i=i+1;` `x=x+a[i]; y=y+b[j]; j=j-1; }`	$I(i > N) \cup \{(s,s') \mid i \leq N \wedge i' = N + 1 \wedge$ $x' = x + \sum_{k=i+1}^{N+1} a[k] \wedge y' = y + \sum_{k=i-N+j}^{j} b[k]$ $\wedge j' = i - N + j - 1 \wedge a' = a \wedge b' = b\}$
`while (i<=N) { i=i+1;` `j=j-1; x=x+a[i]; y=y+b[j]; }`	$I(i > N) \cup \{(s,s') \mid i \leq N \wedge i' = N + 1 \wedge$ $x' = x + \sum_{k=i+1}^{N+1} a[k] \wedge y' = y + \sum_{k=i-N+j-1}^{j-1} b[k]$ $\wedge j' = i - N + j - 1 \wedge a' = a \wedge b' = b\}$
`while (j>0) { x=x+a[i];` `y=y+b[j]; i=i+1; j=j-1;}`	$I(j \leq 0) \cup \{(s,s') \mid i \leq N \wedge i' = i + j \wedge j' = 0 \wedge$ $x' = x + \sum_{k=i}^{i+j-1} a[k] \wedge y' = y + \sum_{k=1}^{j} b[k]$ $\wedge a' = a \wedge b' = b\}$
`while (j>0) { j=j-1;` `x=x+a[i]; y=y+b[j]; i=i+1;}`	$I(j \leq 0) \cup \{(s,s') \mid i \leq N \wedge i' = i + j \wedge j' = 0 \wedge$ $x' = x + \sum_{k=i}^{i+j-1} a[k] \wedge y' = y + \sum_{k=0}^{j-1} b[k]$ $\wedge a' = a \wedge b' = b\}$
`while (j>0) { i=i+1;` `x=x+a[i]; y=y+b[j]; j=j-1;}`	$I(j \leq 0) \cup \{(s,s') \mid i \leq N \wedge i' = i + j \wedge j' = 0 \wedge$ $x' = x + \sum_{k=i+1}^{i+j} a[k] \wedge y' = y + \sum_{k=1}^{j} b[k]$ $\wedge a' = a \wedge b' = b\}$
`while (j>0) { i=i+1;` `j=j-1; x=x+a[i]; y=y+b[j];}`	$I(j \leq 0) \cup \{(s,s') \mid j > 0 \wedge i' = i + j \wedge j' = 0 \wedge$ $x' = x + \sum_{k=i+1}^{i+j} a[k] \wedge y' = y + \sum_{k=0}^{j-1} b[k]$ $\wedge a' = a \wedge b' = b\}$
`while (j>=0) { x=x+a[i];` `y=y+b[j]; i=i+1; j=j-1; }`	$I(j < 0) \cup \{(s,s') \mid j \geq 0 \wedge i' = i + j + 1 \wedge j' = -1 \wedge$ $x' = x + \sum_{k=i}^{i+j} a[k] \wedge y' = y + \sum_{k=0}^{j} b[k]$ $\wedge a' = a \wedge b' = b\}$
`while (j>=0) { j=j-1;` `x=x+a[i]; y=y+b[j]; i=i+1; }`	$I(j < 0) \cup \{(s,s') \mid j \geq 0 \wedge i' = i + j + 1 \wedge j' = -1 \wedge$ $x' = x + \sum_{k=i}^{i+j} a[k] \wedge y' = y + \sum_{k=-1}^{j-1} b[k]$ $\wedge a' = a \wedge b' = b\}$
`while (j>=0) { i=i+1;` `x=x+a[i]; y=y+b[j]; j=j-1; }`	$I(j < 0) \cup \{(s,s') \mid j \geq 0 \wedge i' = i + j + 1 \wedge j' = -1 \wedge$ $x' = x + \sum_{k=i+1}^{i+j+1} a[k] \wedge y' = y + \sum_{k=0}^{j} b[k]$ $\wedge a' = a \wedge b' = b\}$
`while (j>=0) { i=i+1;` `j=j-1; x=x+a[i]; y=y+b[j]; }`	$I(j < 0) \cup \{(s,s') \mid j \geq 0 \wedge i' = i + j + 1 \wedge j' = -1 \wedge$ $x' = x + \sum_{k=i+1}^{i+j+1} a[k] \wedge y' = y + \sum_{k=-1}^{j-1} b[k]$ $\wedge a' = a \wedge b' = b\}$

Each loop has been tested using randomly generated test data and using the associated function as an oracle; all of them have returned successful tests, as long as array indices are maintained within range.

Interest of this example: This example shows how our simple tool enables us to analyze, to a great level of precision, the impact that minor changes in the source code can have on the function of the loop. This example also illustrates that we can deal with array data structures without having to axiomatize arrays; rather, the array recognizers capture all the information we need for our purposes. Of course, these recognizers, while they are adequate for the current example, do not cover all array operations; in particular they do not cover statements that alter the array.

5.3   A Fixpoint Iteration

We consider the following loop on variables x and y of type **real**:

$$\textbf{while } (|x - y| > \epsilon)\{y := x; x := 1 + a/x; \}$$

where ϵ is a small real value and a is a variable of type **real**. We are interested to determine the function of this loop. This example is special for many reasons:

- Because it handles real numbers, this loop is prone to roundoff errors; now, roundoff errors are virtually impossible to model in the context of deriving a loop function because their behavior does not lend itself to an inductive argument.
- Even though the loop may be written to model a precise function (in the example above, the solution to an equation), we know that, due to roundoff errors, the program will produce an approximation of the target function.
- When we seek to compute the function of the loop, we do not necessarily seek to derive the exact function that the computer defines (given its specific arithmetic, its word size, its roundoff policies, etc); rather we are typically interested in an acceptable approximation of the target function.
- As a provision for roundoff errors, programs that handle real numbers specify the output only partially, producing non-deterministic results. For example, the condition of the loop is not $x = y$ but instead $|x - y| < \epsilon$, introducing a measure of non-determinacy. Our algorithm overrides this non-determinacy by selecting an arbitrary value within the interval of acceptable approximations; this is illustrated subsequently.
- This loop can be analyzed using only constructive lower bounds; in fact we would be hard pressed to derive a non-trivial invariant assertion (different from **true**) for this program (which is what creative lower bounds are useful for).

Theorem 2 provides the following lower bound:

$$T$$
$$=\qquad \{ \text{ Theorem 2 } \}$$
$$(L \circ [B] \cup I) \circ I(\neg t)$$
$$=\qquad \{ \text{ Expansion, simplification } \}$$
$$\{(s, s')|x' = 1 + \tfrac{a}{y'} \wedge |x' - y'| \leq \epsilon \wedge\} \cup I(\neg t)$$

Mathematica cannot solve the equations in x' and y' that stem from this lower bound:

$$x' = 1 + \frac{a}{y'} \wedge |x' - y'| \le \epsilon.$$

We simplify them by means of alternative interpretations of the clause $|x' - y'| \le \epsilon$. For example, if we interpret this clause as $x' = y'$, then Mathematica produces the following outcome (two solutions):

$$x' = \frac{1 - \sqrt{1 + 4a}}{2} \wedge y' = \frac{1 - \sqrt{1 + 4a}}{2}.$$

$$x' = \frac{1 + \sqrt{1 + 4a}}{2} \wedge y' = \frac{1 + \sqrt{1 + 4a}}{2}.$$

In practice, the loop converges to one of these solutions depending on the initial value of variable x. Other interpretations of the clause $|x' - y'| < \epsilon$ produce other approximations of the actual function of the loop.

Interest of this example: Though it is simple, this examples enables us to showcase many attributes of our approach. First, that we can handle loops whose function is not defined precisely, but approximately. Second, that we can handle loops that have no inductive variables. Third, that we can handle loops that have no meaningful invariant assertion. Fourth, that we can handle loops with real variables (hence are prone to roundoff errors). Fifth, that we can handle (correctly predict the behavior of) loops that have more than one possible outcome.

5.4 Non Sequential Code

We consider the following loop on integer variables x, y, z and t:

```
while (x ≠ 1)
    { if (x%4 == 0){x := x div 4; y := y + 2; t := t * 4; }
    else if (x%2 == 0){x = x div 2; y = y + 1; t = t * 2; }
    else {x := x − 1; z := z + t; }}.
```

When we transform this loop into CCA notation, we find the following code:

```
while (x ≠ 1)
    {(x mod 4 = 0) → {x := x/4, y := y + 2, t := t * 4, z := z}[]
    (x mod 4 ≠ 0 ∧ x mod 2 = 0) →
        {x := x/2, y := y + 1, t := t * 2, z := z}[]
    (x mod 2 = 1) → {x := x − 1, y := y, t := t, z := z + t}}
```

As it is written, the loop body is structured as a union, hence is a candidate for Theorem 4. This theorem calls for finding a reflexive transitive superset R of one term of the union (P) then using it to compute the product QR (where Q is the other term). Because we do not know how to compute and simplify the product of two relations, we resolve to find a common reflexive transitive superset of P and Q (whose product by itself is idempotent, since it is reflexive and transitive). But finding a common reflexive transitive superset of P and Q means having to analyze large portions of the code at the same time, which is contrary to the premise of separation of concerns. We resolve instead to find individual lower bounds for each branch of the loop body, then computing the meet of the lower bounds. To this effect, we use a new type of recognizers, which match not only the space declarations and source code, but also the condition under which each branch is applied; we call these *conditional recognizers*. Of course, all the traditional recognizers can be recast as conditional recognizers, with the default condition **true**. A sample of conditional recognizers (used in this example) is given below:

ID	State Space	Condition	Code Pattern	Lower Bound $T =$
1R5	x: int	$(x \bmod 2 = 1)$	x=x-1	$\{(s, s') \mid \lfloor \log_2(x) \rfloor = \lfloor \log_2(x') \rfloor \wedge \neg t(s')\}$
2R1:	x, y: int const a, b: int	**true**	x = x+a y = y+b	$\{(s, s') \mid ay - bx = ay' - bx' \wedge \neg t(s')\}$
2R11	x, y: int	$(x \bmod 2 = 0)$	x:=x/2, y:=y+1	$\{(s, s') \mid y + \log_2(x) = y' + \log_2(x') \wedge \neg t(s')\}$
2R12	x, y: int	$(x \bmod 2 = 0)$	x:=x/2, y:=2*y	$\{(s, s') \mid xy = x'y' \wedge \neg t(s')\}$
2R13	x, y: int	$(x \bmod 4 = 0)$	x:=x/4, y:=y+2	$\{(s, s') \mid y + \log_2(x) = y' + \log_2(x') \wedge \neg t(s')\}$
2R14	x, y: int	$(x \bmod 4 = 0)$	x:=x/4, y:=4*y	$\{(s, s') \mid xy = x'y' \wedge \neg t(s')\}$
3R2	x,y,z: int const a, b: int	$(x \bmod 2 = 0)$	x=x-a, y=y+b*z, z=z	$\{(s, s') \mid ay + bxz = ay' + bx'z' \wedge \neg t(s')\}$

Using these recognizers, we find the following lower bounds for the three branches of the loop body:

$$T_1 = \{(s, s') \mid y + \log_2(x) = y' + \log_2(x') \wedge xt = x't' \wedge \frac{t}{4^{y/2}} = \frac{t'}{4^{y'/2}}$$

$$\wedge y \bmod 2 = y' \bmod 2 \wedge z' = z\},$$

$$T_2 = \{(s, s') \mid y + \log_2(x) = y' + \log_2(x') \wedge xt = x't' \wedge \frac{t}{2^y} = \frac{t'}{2^{y'}} \wedge y \le y' \wedge z' = z\},$$

$$T_3 = \{(s, s') \mid y' = y \wedge \lfloor \log_2(x) \rfloor = \lfloor \log_2(x') \rfloor \wedge t' = t \wedge z + tx = z' + t'x' \wedge x \ge x'\}.$$

Taking the meet of these three relations (a non-trivial step, which we are trying to automate using theorem prover Otter), we find the following lower bound for the loop:

$$T = \{(s, s') | x' = 1 \wedge y + \lfloor \log_2(x) \rfloor = y' + \lfloor \log_2(x') \rfloor \wedge z + tx = z' + t'x' \wedge \frac{t}{2^y} = \frac{t'}{2^{y'}}\}.$$

When we run Mathematica to solve the equations that define T in unknowns x', y', z' and t', it returns:

$$\begin{pmatrix} x' = 1 \\ y' = y + \lfloor \log_2(x) \rfloor \\ z' = z + tx - t \times 2^{\lfloor \log_2(x) \rfloor} \\ t' = t \times 2^{\lfloor \log_2(x) \rfloor} \end{pmatrix}.$$

We have tested this loop against an oracle formed from this function definition, using more than twelve million test cases; all were successful.

Interest of this example: The interest of this example is of course that it showcases the ability of our approach to deal with loops whose loop body involves conditional statements. In this example, the generation of the individual lower bounds for the various branches of the loop's CCA notation is fairly straightforward, given that we have the right recognizers in the database. The hardest step in the derivation of this loop function was the generation of the meet of the individual lower bounds (T_1, T_2, T_3). We are exploring ways to automate this step using theorem proving technology.

6   Conclusion: Summary, Assessment, Prospects

Summary. In this paper we discuss a method for deriving the function of a loop by successive approximations in a lattice. We have presented four theorems that form the basis of our approach then we have outlined the algorithm that we are using to derive loop functions by application of the proposed theorems. Then we have illustrated the application of our algorithm on four diverse examples and have discussed for each example what attribute of our approach it showcases.

Assessment and Comparison. The derivation of loop functions using invariant relations and invariant functions is related to, but very distinct from the massive literature that exists today on generating loop invariants [2,5,4,3,1,6,7,8,9,10]. Our work is closer, in terms of its goals, to that of Dunlop and Basili [22], which seeks to derive the function of a loop by generalization from multiple special cases. It is also closer, in terms of its algebraic approach, to the work of Desharnais, Moeller and Tchier [21].

Prospects of Further research. Practical extensions of our work include, obviously, expanding its scope by finding and deploying more recognizers, to deal with more general data structures and control structures. Another focus of our future research is the resolution of the equations that stem from lower bounds; we are considering a combination of a computer algebra system (e.g. Mathematica) and a theorem prover (e.g. Otter).

The most critical theoretical extension that we must envision is to generalize the algorithm to deal with arbitrary (or more general) structures of the loop body. Another important theoretical extension consists in lifting the hypothesis of termination; this is likely to make matters much more complicated, but also to produce sounder results. One of the most crucial benefits that we are now gaining from this hypothesis is that loop functions are maximal in the lattice of refinement, hence all aproximations of the loop function are done with lower bounds. When this hypothesis is lifted, loop functions are going to lie anywhere in the lattice, and approximations may then use lower bounds as well as upper bounds; this is likely to be very interesting, though we expect it to make our job much harder.

Acknowledgement. The authors are very grateful to the anonymous reviewers for their valuable, insightful feedback, which has greatly improved the contents and presentation of this paper.

References

1. Fu, J., Bastani, F.B., Yen, I.L.: Automated discovery of loop invariants for high assurance programs synthesized using ai planning techniques. In: HASE 2008: 11th High Assurance Systems Engineering Symposium, Nanjing, China, pp. 333–342 (2008)
2. Carette, J., Janicki, R.: Computing properties of numeric iterative programs by symbolic computation. Fundamentae Informatica 80, 125–146 (2007)
3. Podelski, A., Rybalchenko, A.: Transition invariants. In: Proceedings, 19th Annual Symposium on Logic in Computer Science, pp. 132–144 (2004)
4. Hu, L., Harman, M., Hierons, R., Binkley, D.: Loop squashing transformations for amorphous slicing. In: Proceedings of 11th Working Conference on Reverse Engineering. IEEE Computer Society, Los Alamitos (2004)
5. Ernst, M.D., Perkins, J.H., Guo, P.J., McCamant, S., Pacheco, C., Tschantz, M.S., Xiao, C.: The Daikon system for dynamic detection of likely invariants. Science of Computer Programming (2006)
6. Denney, E., Fischer, B.: A generic annotation inference algorithm for the safety certification of automatically generated code. In: Proceedings of the Fifth International Conference on Generative programming and Component Engineering, Portland, Oregon (2006)
7. Carbonnell, E.R., Kapur, D.: Program verification using automatic generation of invariants. In: Liu, Z., Araki, K. (eds.) ICTAC 2004. LNCS, vol. 3407, pp. 325–340. Springer, Heidelberg (2005)
8. Colon, M.A., Sankaranarayana, S., Sipma, H.B.: Linear invariant generation using non linear constraint solving. In: Hunt Jr., W.A., Somenzi, F. (eds.) CAV 2003. LNCS, vol. 2725, pp. 420–432. Springer, Heidelberg (2003)
9. Sankaranarayana, S., Sipma, H.B., Manna, Z.: Non linear loop invariant generation using groebner bases. In: Proceedings of ACM SIGPLAN Principles of Programming Languages, POPL 2004, pp. 381–329 (2004)

10. Kovacs, L., Jebelean, T.: An algorithm for automated generation of invariants for loops with conditionals. In: Petcu, D. (ed.) Proceedings of the Computer-Aided Verification on Information Systems Workshop (CAVIS 2005), 7th International Symposium on Symbolic and Numeric Algorithms for Scientific Computing (SYNASC 2005), Department of Computer Science, West University of Timisoara, Romania, pp. 16–19 (2005)
11. Mili, A., Aharon, S., Nadkarni, C., Mraihi, O., Louhichi, A., Jilani, L.L.: Reflexive transitive invariant relations: A basis for computing loop functions. Journal of Symbolic Computation (2009)
12. Mili, A., Desharnais, J., Gagne, J.R.: Strongest invariant functions: Their use in the systematic analysis of while statements. Acta Informatica (1985)
13. Linger, R., Mills, H., Witt, B.: Structured programming. Addison-Wesley, Reading (1979)
14. Mills, H.: The new math of computer programming. Communications of the ACM 18 (1975)
15. Gries, D.: The Science of programming. Springer, Heidelberg (1981)
16. Morgan, C.: Programming from Specifications. International Series in Computer Sciences. Prentice Hall, London (1998)
17. Dijkstra, E.: A Discipline of Programming. Prentice-Hall, Englewood Cliffs (1976)
18. Manna, Z.: A Mathematical Theory of Computation. McGraw-Hill, New York (1974)
19. Boudriga, N., Elloumi, F., Mili, A.: The lattice of specifications: Applications to a specification methodology. Formal Aspects of Computing 4, 544–571 (1992)
20. Mili, A., Aharon, S., Nadkarni, C.: Mathematics for reasoning about loop functions. Technical report, New Jersey Institute of Technology (2008), http://web.njit.edu/~mili/scp.pdf
21. Desharnais, J., Mueller, B., Tchier, F.: Kleene under a modal demonic star. Journal of Logic and Algebraic Programming 66, 127–160 (2006)
22. Dunlop, D., Basili, V.R.: A heuristic for deriving loop functions. IEEE Transactions on Software Engineering 10, 275–285 (1984)

Modalities, Relations, and Learning
A Relational Interpretation of Learning Approaches

Martin Eric Müller

Dept. Computer Science, University of Applied Sciences Bonn-Rhein-Sieg
martin.mueller@h-brs.de

Abstract. While the popularity of statistical, probabilistic and exhaustive machine learning techniques still increases, relational and logic approaches are still a niche market in research. While the former approaches focus on predictive accuracy, the latter ones prove to be indispensable in knowledge discovery.

In this paper we present a relational description of machine learning problems. We demonstrate how common ensemble learning methods as used in classifier learning can be reformulated in a relational setting. It is shown that multimodal logics and relational data analysis with rough sets are closely related. Finally, we give an interpretation of logic programs as approximations of hypotheses.

It is demonstrated that at a certain level of abstraction all these methods unify into one and the same formalisation which nicely connects to multimodal operators.

1   Introduction

During the last decades, Machine Learning evolved from theories of reasoning in Artificial Intelligence to an essential component of software systems. Statistical methods outperform logic based approaches in most application domains—and with increasing computational power it became possible to *generate-and-test* classifiers. As a more sophisticated approach *ensemble learning* implements divide-and-conquer strategies on the learning problem. With the further increase of data collections (e.g. data warehouses), the problem we are facing is not concerned with *how* we can induce a classifier that supports our model assumptions on the data but rather to understand *what* kind of information there actually *is*. In Machine Learning, this approach is known as *knowledge discovery*. Its most successful approach is *inductive logic programming* (ILP). Another well-known method is rough set data analysis. It has been applied to a variety of machine learning problems and it has been studied in connection with multi-valued and multi-modal logics ([1,2,3]) and formal concept analysis, [4,5].

This motivates a deeper analysis of the relations between two different ensemble learning methods on the one hand and rough set data analysis and inductive logic programming on the other hand. This paper presents a uniform description framework borrowed from rough set data analysis, multi modal logics and relational calculus. It concludes with a list of open questions for future work.

R. Berghammer et al. (Eds.): RelMiCS/AKA 2009, LNCS 5827, pp. 260–275, 2009.

2   Machine Learning

Machine learning is concerned with the problem of inducing a general principle from a limited set of observations in order to predict properties of new, unknown cases.

2.1   Learning Hypotheses from Samples

By U, we denote the base set or *universe* that consists of all objects of a domain. Learning means that we try to acquire a certain capability, usually formulated as classification problem. More formally, learning is the process of approximating a *target function*

$$\mathbf{t} : U \to C \tag{1}$$

where C is the set of target values.[1] In order to be able to approximate $\mathbf{t}$, we need a set of support points ("*examples*"). A *sampling function*

$$S : (\mathbb{N} \times C^U) \to 2^{U \times C} \tag{2}$$

draws m objects from U and labels each with its target value. This set is called a sample $\mathbf{s} = \{\langle x_1, \mathbf{t}(x_1)\rangle, \dots, \langle x_m, \mathbf{t}(x_m)\rangle\}$. An equivalent definition is $\mathbf{s} := \mathbf{t}|_s$ where $R|_s$ is short for $R \cap s$. Complex domains build structures on U which contain more information than just a set. One way of formalising this is to assume a probability measure μ on U. The distribution μ affects S in a way that some objects are more likely to occur in a sample than others. So if $S(m, \mathbf{t})$ preserves μ, it holds that

$$\mu(\{\langle x, \mathbf{t}(x)\rangle\} \cap \mathbf{s}) = \mu(\{x\}). \tag{3}$$

For two independent sampling procedures, $S(m, \mathbf{t}) = \mathbf{s} \neq \mathbf{s}' = S(m, \mathbf{t})$.[2] The problem is that μ usually is unknown. So, especially with small size samples, we always might end up with a collection that does *not* properly represent U—but we can't tell since we do not know μ. The fundamental *inductive assumption* is that for large m, the likeliness of an event $e = \{x_0, \dots, x_{n-1}\}$ is the same as its probability:

$$\lim_{m \to |U|} \mu(\{\langle x_i, \mathbf{t}x_i\rangle : i \in \mathbf{n}\} \cap S(m, \mathbf{t})) = \mu(e) \tag{4}$$

A *learning algorithm* $\mathbf{A}$ takes a sample $\mathbf{s}$ and computes a hypothesis h such that

$$\mathbf{A}(S(m, \mathbf{t})) = \mathbf{A}(\mathbf{s}) = h \approx \mathbf{t}. \tag{5}$$

---

[1] In a binary context, $C = \{c, -c\}$ and $\mathbf{t}$ is the characteristic function $\chi(c)$ of c.

[2] This is due to the fact that S draws x's from U with *replacement*. This results in a non-deterministic behaviour which is why it is more appropriate to think of S as a procedure rather than a function.

2.2 Properties of h

Whether the demand for $h \approx \mathbf{t}$ is fulfilled or not depends on the type of similarity or quality measure we define. Let $\mathbf{s} = S(m, \mathbf{t})$ with $m \leq |U|$. If $h|_s = \mathbf{t}|_s$ for some $s \subseteq U$, we say that h *agrees with* $\mathbf{t}$ *on* s. The set

$$\mathrm{errset}_{\mathbf{t}}(h, s) = \{x \in s : h(x) \neq \mathbf{t}(x)\} \tag{6}$$

is called the *error set (of h on s w.r.t. $\mathbf{t}$)*. Based on the notion of error sets one can define a large number of error measures, the simplest being $\mathrm{error}_{\mathbf{t}}(h, s) = |\mathrm{errset}_{\mathbf{t}}(h, s)|/|s|$. The most important thing is to keep in mind that the *true* error of h is h's error with respect to μ:

$$\mathrm{error}_{\mathbf{t}}^{\mu}(h, s) = \mu(\mathrm{errset}_{\mathbf{t}}(h, s)) = \mu(\bigcup_{x \in s}\{x : h(x) \neq \mathbf{t}(x)\}) \tag{7}$$

$$= \sum_{x \in s} \mu(\{x\} \cap \mathrm{errset}_{\mathbf{t}}(h, s)) \tag{8}$$

2.3 Learning Targets and Learnability

Good learning requires an algorithm which is good at quickly picking the *best* hypothesis. However, no one ever knows, which one actually is best (both the hypothesis and the algorithm). But in an allusion to the No-Free-Lunch-Theorem [6] we can safely state that what we are looking for is an algorithm whose *average error is minimal*:

$$\mathbf{A}_{\mathrm{opt}} = \arg\min E(\mathrm{error}_{\mathbf{t}}^{\mu}(\mathbf{A}(S(m, \mathbf{t})), s)) \tag{9}$$

for arbitrary s and m and an *unknown* μ where E is the expected value. The No-Free-Lunch-Theorem also states that in general there is no such optimum. For finite domains and hypothesis spaces the case is clear; a simple enumeration guarantees an optimal solution. But in most cases, we are faced with infinite domains and hypothesis spaces. This leads to the notion of *probably approximately correct* hypotheses: A problem $\mathbf{t}$ is called *probably approximately correctly (PAC)* learnable, if there is an algorithm satisfying

$$\forall m > m_0 : \mu^m\{\mathbf{s} \in S(m, \mathbf{t}) : \mathrm{error}_{\mathbf{t}}^{\mu}(\mathbf{A}(\mathbf{s}), U) \leq \varepsilon\} \leq \delta \tag{10}$$

for $0 < \varepsilon, \delta < 1$ and for any μ and $\mathbf{t}$ where the value of m_0 only depends on ε and δ. In other words: an algorithm is PAC, if we can determine the number of minimal examples it requires to learn a hypothesis that is at most ε-bad with a probability of at least δ.

Since μ is unknown, the Free-Lunch-Theorem formulates another crucial point: The smaller $\mathrm{errset}_{\mathbf{t}}(h, s)$ (and the smaller s), the more likely it becomes that on another set s', $\mathrm{errset}_{\mathbf{t}}(h, s')$ increases. As soon as minimization of $\mathrm{error}_{\mathbf{t}}(h, s)$ implies an increase of $\mathrm{error}_{\mathbf{t}}(h, s')$, further "optimisation" is called *over-fitting*.

Accordingly, good learning means to quickly pick a hypothesis that we expect to perform with a smaller than average error on the subset of objects which we expect to occur most frequently.

2.4   Algorithms

Since most interesting learning problems are *not* PAC, we need to search for sufficiently accurate hypotheses. Hypotheses can be arranged to form a hypothesis space which can be searched more efficiently. Important ordering relations are *generality* or *entailment*, but these relations cannot always be efficiently computed. Let, for example, $\mathcal{K}$ denote a set of first order formulas (called the *background knowledge*). Usually h entails h', if $h \models h'$. But with respect to our learning target, a weaker formulation suffices: h is more general than h' with respect to $\mathbf{s}$, if from $\mathcal{K} \cup \{h\} \models \mathbf{s}$ it follows that $\mathcal{K} \cup \{h'\} \models \mathbf{s}$. To show this, we have to prove entailment—which might become very expensive. A popular workaround is to introduce an efficiently computable (syntactical) order relation $\sqsubseteq \neq \leq$. The problem is that in most cases $\sqsubseteq$ is not complete or correct (or both) w.r.t. $\leq$.[3]

```
01   PROC learn ()
02   {   h := init();
03       WHILE (not good_enough(h))
04       {   ⟨ConsiderNow, ConsiderLater⟩ := select(h);
05           h := refine(ConsiderNow) ∪ ConsiderLater;
06           h := filter(h);
07       }
08       RETURN(h);
09   }
```

Fig. 1. A biased Learning Algorithm

Fig. 1 shows an abstract pseudo–code learning algorithm. It illustrates that learning is *refinement* (l. 05) and that it is assumed that repeated refinement leads to a better result (l. 03). From this and the fact that the overall behaviour of the algorithm entirely depends on the underlined functions, it follows that nearly any implementation must be heavily biased. As a rule of thumb it holds that the more efficient the algorithm, the heavier the bias, and the higher the probability of losing effectivity.

3   Ensemble Learning

Learning complex hypotheses from very large or very small data sets may run into two different problems: First, it could be that no hypothesis of satisfactory

---

[3] For example, consider first order formulas φ and ψ and a substitution ϑ. Then, setting $\varphi \sqsubseteq \psi :\Longleftrightarrow \varphi\vartheta = \psi$ we have $\varphi \sqsubseteq \psi \Longrightarrow \varphi \models \psi \Longleftrightarrow \varphi \leq \psi$, but not vice versa.

accuracy can be learned at all. Second, if there is an accurate hypothesis, it is likely to overfit the learning set. Ensemble learning offers methods to avoid either case. One fundamental method is to split $\mathbf{s}$ into three disjoint subsets $\mathbf{s}_{train}$, $\mathbf{s}_{test}$ and $\mathbf{s}_{val}$ (where at least $\mathbf{s}_{train} \neq \emptyset$). The training set $\mathbf{s}_{train}$ is used to learn h, and $\mathbf{s}_{test}$ is used by $\mathbf{A}$ to estimate h's error on U. $\mathbf{s}_{val}$ is used to evaluate h once $\mathbf{A}$ has finished refinement. For the algorithm described in Fig. 1 this means that *selection* is implemented by considering $error_t^\mu(h, \mathbf{s}_{test})$ and the termination criterion *good_enough* by $error_t^\mu(h, \mathbf{s}_{val})$. To increase accuracy further, $\mathbf{s}$ is repeatedly divided into different subsets $\mathbf{s}_{train}$ and $\mathbf{s}_{test}$ (cross-validation).

3.1 Bagging

Bagging is a "bootstrapping and aggregation" method, [7]. The idea is as follows. Instead of learning $h = \mathbf{A}(\mathbf{s})$, we divide $\mathbf{s}$ into a set $\{\mathbf{s}_i \subseteq \mathbf{s} : i \in \mathbf{k} \wedge |\mathbf{s}_i| = M\}$ of k subsamples, each of size M.[4] We then learn $\mathbf{A}(\mathbf{s}_i) = h_i : U \to C$ and define an *aggregation function*

$$\text{agg} : C^k \to C \tag{11}$$

which takes all k classifiers' outputs and computes one single outcome h (voting or average; depending on the structure of C). An abstract Bagging algorithm is shown in Fig. 2. Bagging works quite well under the condition of *instability*. An algorithm is called *unstable*, if for two nearly equal samples the hypotheses differ significantly, [8]. As a consequence, we obtain a set of k (different, over-fit) classifiers with high predictive accuracy on a small set but a rather poor performance on

---

```
01    s := S_μ(m, t); μ_0({x}) = 1/m; k := NumberOfBags; M := BagSize;
02
03    PROC Bootstrap (s, k, M, μ_0)
04    {   FOR i ∈ k
05        {   s_i := select(s, M, μ_0);          % selects a random subset
06            h_i := A(s_i);
07        }
08        RETURN(⟨h_0, . . . , h_{k-1}⟩);
09    }
10    PROC h_agg(⃗h, x)
11    {
12        RETURN 1/ℓ(⃗h) Σ_{i∈ℓ(⃗h)} h_i(x);
13    }
```

Fig. 2. Bagging (with arithmetic average aggregation)

---

[4] Note that $\mathbf{s}$ depends on μ but when building $\mathbf{s}_i$ examples are drawn by *random (iid)* and *with replacement*. This way, the set of all $\mathbf{s}_i$ preserves μ while within each subsample this needs not to be the case.

s_{val} (or U). The higher $\mu(\{x\})$, the higher the probability that $\langle x, \mathbf{t}(x)\rangle \in \mathbf{s}_i$. As a consequence, more probable x are more likely to be classified correctly—even if $\mathbf{A}$ is just a weak learner. Roughly speaking, most h_i cover the most probable examples (correctly) and every h_i also covers an additional small set of less probable examples with an accuracy (much) better than average. This means that the aggregated result $h_{\text{agg}}(\langle \mathbf{A}(\mathbf{s}_0), \mathbf{A}(\mathbf{s}_1), \ldots, \mathbf{A}(\mathbf{s}_{k-1})\rangle)$ cannot perform worse on U than $\mathbf{A}(\mathbf{s})$ but probably even better. For stable learning algorithms, subsampling and aggregation infuse additional noise which may not only not improve the result but even lead to a worse result. A more formal description is out of scope of this paper; for a proof we refer to [7].

3.2 Boosting

Boosting takes a slightly different approach than Bagging. In order to refine h and to quickly decrease the error, it is a reasonable approach to focus on first learning those cases that cause most error. There are several methods to obtain such a behaviour; the simplest is the first version of the Ada-Boost-Algorithm which is shown in Fig. 3. Boosting basically is a three-step learning process:

```
01   μ({x}) = 1/m; k := BoostingRounds;
02
03   PROC adaBoost (μ, k)
04   {   FOR i ∈ k
05       {   h_i := A(S(m,t));
06           μ({x}) := 1/ν · μ({x})^(−α(i)·h_i(x)·t(x));
07       }
08       RETURN(⟨h_0, ..., h_{k−1}⟩);
09   }
10   PROC h_boost(x)
11   {
12       RETURN sgn (∑_{i∈k} α(i)h_i(x));
13   }
```

ν is a normalization factor to ensure μ is a probability distribution and $\alpha(t)$ is a weighting factor which determines the error gradient along which the hypothesis has to be refined, e.g.

$$\alpha(i) = \frac{\ln\left(\frac{1 - \text{error}_{\mathbf{t}}^{\mu}(h_i, \mathbf{s})}{\text{error}_{\mathbf{t}}^{\mu}(h_i, \mathbf{s})}\right)}{2}$$

Fig. 3. Boosting

1. Generate a sample $\mathbf{s} = S(m, \mathbf{t})$ w.r.t. μ and learn $h = \mathbf{A}(\mathbf{s})$ such that h is better than random.
2. Modify μ such that h becomes random by flipping a coin: On heads, S draws some x for which $h(x) = \mathbf{t}(x)$, otherwise S draws some $\bar{x}$, for which $h(\bar{x}) \neq \mathbf{t}(\bar{x})$. This results in a set s_{random}, on which h's error is 0.5. Now, learn h' on $s_{\text{random}} = \{\langle x, \mathbf{t}(x)\rangle : x \in s_{\text{random}}\}$.[5]

---

[5] Note that this sample consists of 50 per cent objects misclassified by h but assigned with their *true* target value.

3. Modify μ such that for all x drawn by S, $h(x) \neq h'(x)$. This results in a set $\bar{s}$ that consists of objects on which h and h' disagree; i.e. where the probability that h' is correct is higher than the probability that h is correct. On this set, learn $\bar{h}$ which specialises the "difference" between h and h'.

Then, given an object x, the prediction is $\text{sgn}(h(x) + h'(x) + \bar{h}(x))$. Hence, its predictive accuracy is always better than that of each single hypothesis. It can be shown that boosting allows reducing the error of any weak learner (which is only slightly better than random guesses) to arbitrarily low bounds [9,10].

The Ada-Boost algorithm (Fig. 3, [11]) performs an iterative gradient descent method to modify μ: Each time, μ is changed in a way that increases the probability of x being considered as an example in the next training cycle if it has been misclassified by the current hypothesis.

4   Relational and Logic Learning

There are many representation paradigms and corresponding algorithms for machine learning. Two of them are of special interest for us: Relational and logic learning. We assume U to be structured by a set of features F forming an *information system*. Each feature $f \in F$ maps an element $x \in U$ to a *value* $f(x) \in V_f$. Features $f \in F$ induce equivalence relations $R_f \subseteq U \times U$ where $x R_f y :\Longleftrightarrow f(x) = f(y)$ with x and y being elements of the base set U. For or very large V_f, one usually applies quantisation in order to make the quotient set U/R_f not consist of singletons only. The set of all such relations is $\mathbf{F} = \{R_f : f \in F\}$.

4.1   Sets, Classes and Rough Concepts

For any subset $\mathbf{P} \subseteq \mathbf{F}$ we call $\bigcap \mathbf{P}$ the *indiscernability relation over* $\mathbf{P}$, [12]. Two objects x and y are $\mathbf{P}$–*indiscernible on* s iff

$$\forall x, y \in s : x \bigcap \mathbf{P} y \tag{12}$$

for any $s \subseteq U$; it means that $x R y$ for all $R \in \mathbf{P}$. It holds that

$$\mathbf{P} \subseteq \mathbf{R} \Longrightarrow \bigcap \mathbf{R} \subseteq \bigcap \mathbf{P} \Longleftrightarrow [x]_{\bigcap \mathbf{R}} \subseteq [x]_{\bigcap \mathbf{P}} \tag{13}$$

where $[x]$ is the equivalence class of x. A relation $R \in \mathbf{P}$ is *dispensable on* s, iff

$$(\bigcap \mathbf{P} - \{R\})|_{s \times s} = (\bigcap \mathbf{P})|_{s \times s}. \tag{14}$$

Removing dispensable relations yields the set $\text{Red}_s(\mathbf{R})$ of *reducts* (w.r.t. s):

$$\mathbf{R} \in \text{Red}_s(\mathbf{P}) :\Longleftrightarrow \mathbf{R} \subseteq \mathbf{P} \wedge \bigcap \mathbf{R}|_s = \bigcap \mathbf{P}|_s$$
$$\wedge \forall \mathbf{Q} \subset \mathbf{R} : \bigcap \mathbf{Q}|_s \neq \bigcap \mathbf{P}|_s \tag{15}$$

The intersection of all reducts is called a *core*: $\text{Cor}_s(\mathbf{P}) = \bigcap \text{Red}_s(\mathbf{P})$. For every set s there exist a maximal subset and a minimal superset called the *lower* and *upper* $\mathbf{P}$-approximation of s:

$$\llbracket \mathbf{P} \rrbracket s :\Longleftrightarrow \{x \in U : [x]_{\bigcap \mathbf{P}} \subseteq s\} \tag{16}$$

$$\langle\!| \mathbf{P} |\!\rangle s :\Longleftrightarrow \{x \in U : [x]_{\bigcap \mathbf{P}} \cap s \neq \emptyset\}. \tag{17}$$

It holds that $\llbracket \mathbf{P} \rrbracket s \subseteq s \subseteq \langle\!| \mathbf{P} |\!\rangle s$. A set s is called $\mathbf{P}$–*definable*, iff $\langle\!| \mathbf{P} |\!\rangle s = \llbracket \mathbf{P} \rrbracket s$. Non-definable or *rough* sets are pairs of respective approximations: $s_{\mathbf{P}} := \langle \llbracket \mathbf{P} \rrbracket s, \langle\!| \mathbf{P} |\!\rangle s \rangle$. Even more interesting are propositions about the relationships between sets of relations. For any $s \subseteq U$ it holds that

$$\mathbf{P} \subseteq \mathbf{R} \Longrightarrow \llbracket \mathbf{P} \rrbracket s \subseteq \llbracket \mathbf{R} \rrbracket s \text{ and } \mathbf{P} \subseteq \mathbf{R} \Longrightarrow \langle\!| \mathbf{R} |\!\rangle s \subseteq \langle\!| \mathbf{P} |\!\rangle s.$$

It simply means that the more knowledge we add, the bigger becomes the lower approximation and the smaller the upper approximation. What we are looking for in machine learning is a *minimal* hypothesis that describes the target concept; here, a smallest set $\mathbf{P}$ such that for any relation $R \in \mathbf{F} - \mathbf{P}$ it holds that

$$\llbracket \mathbf{P} \rrbracket s = \llbracket \mathbf{P} \cup \{R\} \rrbracket s \text{ and } \langle\!| \mathbf{P} |\!\rangle c = \langle\!| \mathbf{P} \cup \{R\} |\!\rangle s. \tag{18}$$

To find a suitable $\mathbf{P}$, one would start with $\emptyset$ and add relations until the upper approximation cannot be reduced (e.g. [13]) or delete relations from $\mathbf{F}$ until the lower approximation becomes smaller (e.g. [14,15]). The characteristic function of $s_{\mathbf{P}}$ takes three values

$$\chi(s_{\mathbf{P}})(x) = \begin{cases} 1, \text{ if } x \in \llbracket \mathbf{P} \rrbracket c \\ \quad, \text{ if } x \in \neg\llbracket \mathbf{P} \rrbracket \cap \langle\!| \mathbf{P} |\!\rangle s \\ 0, \text{ if } x \in \neg\langle\!| \mathbf{P} |\!\rangle s \end{cases}$$

The quality of $\mathbf{P}$ can be measured in several ways:

$$\text{roughness}(s_{\mathbf{P}}) := \frac{|\langle\!| \mathbf{P} |\!\rangle s| - |\llbracket \mathbf{P} \rrbracket s|}{|\langle\!| \mathbf{P} |\!\rangle s|} \tag{19}$$

$$\text{errset}_t(s_{\mathbf{P}}, s') := \text{errset}_t(\llbracket \mathbf{P} \rrbracket s, s') \cup \text{errset}_t(\langle\!| \mathbf{P} |\!\rangle s, s') \tag{20}$$

where $s' \neq s$ is a test set used to estimate the error of $s_{\mathbf{P}}$ on U. This allows to reformulate the learning goal in terms of rough sets: The hypothesis $h = \mathbf{P}$ should be chosen such that its roughness and error on the validation set is minimal.

4.2 Inductive Logic Programming

In this section, we give a further interpretation of box- and diamond operators in terms of logic programs. Roughly speaking, we will define $\llbracket \mathbf{P} \rrbracket s$ as the set of all elements in s that can be derived by SLD-resolution.

Inductive Logic Programming (ILP) is an approach to machine learning with a strong focus on logic representations [16,17]. It is based on the general idea of inverting the resolution calculus [18] on Horn clauses [19]. A first definition of refinement operators by inverting resolution was given in [20], based on earlier works on generalisation operators in [21,22,23]. Simple ILP learning problems are PAC-learnable [24]; see Eq. (10). The sample consists of a set of Horn clauses with a label describing whether they are consistent with the target concept Γ or not: $\mathbf{t}(\varphi) = 1$ iff $\Gamma \cup \{\varphi\}$ is satisfiable A hypothesis for $\mathbf{t}$ is a set Ξ of Horn clauses satisfying

$$\forall \langle \varphi, \mathbf{t}(\varphi) \rangle \in \mathbf{s}_{\text{train}} : \Xi \models \varphi \text{ iff } \mathbf{t}(\varphi) = 1 \tag{21}$$

The restriction to Horn clauses allows using resolution proofs to model entailment:

$$\Xi \cup \{\neg\varphi\} \vdash_{\text{RES}} \{\} \implies \Xi \models \varphi \tag{22}$$

i.e. if $\Xi \cup \{\neg\varphi\}$ has no model, φ follows from Ξ. In order to widen the deductive closure so as to cover more positive cases, we need to *generalise* Ξ, and in order to exclude wrong conclusions, we have to *specialise* Ξ. The *least general generalisation* of two Horn clauses can be defined syntactically in terms of θ-*subsumption*:

$$\text{lgg}(\{\varphi_1, \ldots, \varphi_k\}) = \varphi \text{ iff } \exists\theta \, \forall 1 \leq i \leq k : \varphi\theta \subseteq \varphi_i \tag{23}$$

and if there are substitutions σ_i satisfying $\varphi\sigma_i \subseteq \varphi_i$, then there are substitutions ξ_i with $\sigma_i = \theta\xi_i$. Generalisation is truth preserving *with respect* to a given theory $\mathcal{K}$, which gives rise to a definition of a *relative* least generalisation:

$$\text{rlgg}(\{\varphi_1, \ldots, \varphi_k\}) = \text{lgg}(\{\varphi_0 \leftarrow \mathcal{K}, \ldots, \varphi_1 \leftarrow \mathcal{K}\})$$

Logically, it either holds that $\Xi \models \varphi$ or $\Xi \not\models \varphi$, but things are a bit different with logic programs: SLD-resolution is correct, but it is in general not complete. It turns out that this property actually defines some kind of upper and lower approximations, too: everything that cannot be proven to be either true or false belongs to the boundary region. This region is the hypothesis space that has to be searched in order to find a suitable Ξ describing Γ. The incompleteness makes resolution a refutation calculus where negation is modelled by failure. This already suggests a strong connection to intuitionistic logic. We will discuss methods for refining logic programs along the upper and lower approximations of target concepts in Sect. 5.3.

5   Relational Machine Learning

In this section, we reformulate the different techniques described in the last section in a more unified, relational way.

5.1   Relational Bagging

Bagging means to learn classifiers that specialise on subsets of the sample. In a relational setting, this means that for a set of k subsamples $\mathbf{s}_i$ of size $M \leq m$, $\mathbf{A}$

computes k hypotheses $\mathbf{P}_i$ approximating $c_i := s_i \cap c$, $s_i := \{x : \langle x, \mathbf{t}(x) \rangle \in \mathbf{s}_i\}$. Then, there exist reducts $\mathbf{P}_i \in \text{Red}_{c_i}(\mathbf{F})$ that define c_i and it holds that

$$\mathbf{P} = \bigcup_{i=1}^{k} \mathbf{P}_i \tag{24}$$

The problem here is that reducts are not unique and, even though a smaller set speeds up the process of finding reducts, it may be that some reducts are disjoint which leaves the core empty. Since the core $\text{Cor}_{c_i}(\mathbf{F})$ can be empty, we define a table $T(i,j) = |\{R \in \text{Red}_{c_j}(\mathbf{F}) : R_i \in \mathbf{P}\}|$ with $1 \leq j \leq k$ and $1 \leq i \leq |\text{Red}_{c_j}(\mathbf{F})|$. If $T(i,j) = |\text{Red}_{c_j}(\mathbf{F})|$, then R_i must be an element of $\text{Cor}_c(\mathbf{F})$. Computing this table is computationally infeasible. Accordingly, there is no canonical efficient bottom-up implementation to determine a core from reducts using a bagging-like method.

If, on the other hand, we try to construct reducts from cores (i.e. top-down), we can apply a very simple method (since cores are unique): For a set s of objects, the discernability matrix $D_{\mathbf{P}}(i,j)$ contains the names of all relations by which x_i and x_j can be discriminated: $R \in D_{\mathbf{R}}(i,j) :\Longleftrightarrow x_i R x_j$. If for some i, j it holds that $D(i,j) = \{R\}$, then R must be an element of $\text{Cor}_s(\mathbf{F})$. The runtime complexity of computing the core this way is $\mathcal{O}(nm^2)$, where the worst case is $\text{Cor}_c(\mathbf{F}) = \mathbf{F}$. Using the same method in bagging with k bags of size M, we obtain $\mathcal{O}(knM^2)$. Furthermore, the algorithm in Fig. 4 benefits from parallel computations of smaller discernability matrices.

```
01   PROC relBag (F, s, k, M)
02   {   H := F; C = {}
03       FOREACH(i ∈ k)
04       {   s_i := randomselect(M,s);
05           C_i := core(s_i, H);
06           H := H − C_i; C := H + C_i;              % + denotes concatenation
07       };
08       H := sortby(β, H);
09       WHILE (error_t(C, s) ≥ ε ∨ H = {})
10       {   R := first(H); R := tail(H);
11           IF (error_t(C, s) > error_t(C ∪ {R}, s)) THEN {C := C + {R} };
12           H := R;
13       }
14       return (C);
15   }
```

Fig. 4. Relational Bagging

```
01    PROC relBoost (H, s)
02    {    IF (good_enough(H)) THEN return(H);
03         C := sortby(β, F − H);
04         WHILE (error_t(H, s) ≥ ε)
05         {    C =: [C|R];
06              IF(error_t(H ∪ C, s) < error_t(H, s)) THEN
07              { relBoost(H ∪ C, errset_t(H ∪ C, s)); }
08              ELSE
09              { return(H ∪ relBoost(R, errset_t(H ∪ C, s))); }
10         }
11         return(⊥);
12    }
```

Fig. 5. Relational "Boosting" by Re-learning Errors Only

5.2 Relational Boosting

In a relational boosting approach we do not have a probability distribution which we could adjust in order to focus on learning error sets.

Instead of boosting the probability of those examples that are misclassified by a hypothesis h_i, we only remove the set of already correctly classified objects from the set of entities to be taken into consideration; i.e. we restrict the search for a hypothesis to the boundary region. Given a hypothesis **H**, the problem is to find a relation R which is a good candidate to rule out as many elements from the boundary region as possible by adding or removing it from **H**.

This can be done only heuristically and, as such, is a source of bias. The algorithm shown in Fig. 3 uses a function *sortby* to pick the "best" relation (determined by the heuristic function β). One possible heuristics could be information gain (see Sect. 5.3) or many other, computationally even cheaper methods (for example, choosing R whose index has a certain property). Another, though more expensive method, is to validate R against a test sample s_{test} and choose $\beta(R) = \text{error}_t(R, s_{test})^{-1}$. To learn c from **F** we chose $R \in \mathbf{F}$ with $R = \arg\max\{\beta(R) : R \in \mathbf{F}\}$, hoping that it generates a fine grained partition that has a minimal boundary region on the target concept. We then iterate this process on the boundary region only.

Note that—in contrast to standard boosting—we do not keep a sequence of hypotheses, but we iteratively build a reduct starting from the empty set. As such, it is a bottom-up learning algorithm.

5.3 Relational Logic Learning

In Sect. 4.2 we already motivated a connection between learning logic programs and refining upper and lower approximations of target concepts. If there is a set

C of propositions that we wish to describe, we want to learn a program Ξ for which $\forall \varphi \in C : \Xi \vdash \varphi$.

Let $t \subseteq U$ be our target concept—the set of all objects in U that satisfy a certain property. Then, $\chi(t) = \mathbf{t}$. At the same time, t can be described by a *predicate* $\mathbf{t}(x)$ whose interpretation is defined as $\mathrm{mng}(\mathbf{t}(x)) = \mathbf{t}(x)$. The problem is that we need to *define* $\mathbf{t}$ in terms of other predicates. To find such a definition means to find a suitable hypothesis Ξ such that (given knowledge $\mathcal{K}$),

$$\mathcal{K} \cup \Xi \models t(x) \quad \text{and} \quad \mathcal{K} \cup \Xi \not\models \neg t(x) \tag{25}$$

Horn Theories Ξ as Approximations

Motivation. Understanding Horn theories as approximations is not a new idea, [25], but there were only few contributions from the inductive logic programming community which is mostly due to the generally rather negative results, e.g. [26].

Every $R \in \mathbf{F}$ also defines a binary predicate $\mathbf{r}(x, f_R(x))$.[6] The satisfaction set of $\mathbf{r}$ is the set of all instantiations of x for which $\mathbf{r}$ holds and whose meaning equals the corresponding R-equivalence class:

$$[x]_R = \{y \in U : \mathbf{r}(y, f_R(x))\}.$$

In order to derive $\mathbf{r}(x, v)$ from a given set of clauses Ξ, that is, $\mathcal{K} \cup \Xi \vdash \mathbf{r}(x, v)$, one needs to show that there is a correct answer substitution θ such that $(\mathcal{K} \cup \Xi \cup \{\neg \mathbf{r}(x, v)\})\theta \vdash \Box$. An optimal hypothesis Ξ guarantees that

$$\forall \theta : \mathbf{r}(X, f_R(X))\theta \iff (\mathcal{K} \cup \Xi\{\neg \mathbf{r}(X, V)\})\theta \vdash \Box \tag{26}$$

Since $\vdash$ is correct but not complete, we are able to give a lower approximation of r where $[\![\Xi]\!]r$ describes a subset of the satisfaction set of r:

$$[\![\Xi]\!]r :\iff \{y : (\mathcal{K} \cup \Xi \cup \{\neg \mathbf{r}(X, V)\})\theta \vdash \Box\} \tag{27}$$

Example. Let $\mathcal{K} = \{\}, \Xi = \{(\mathsf{p}(X, 1) \leftarrow \mathsf{q}(X)), \mathsf{q}(a), \mathsf{q}(b)\}$. Furthermore, let $f_R(a) = f_R(c) = 1$ and $f_R(b) = 0$. Then, $\Xi \vdash \mathsf{p}(a, 1)$ but $\Xi \vdash \mathsf{p}(b, 1)$ and $\Xi \not\vdash \mathsf{p}(c, 1)$. The set of wrong predictions is $\mathrm{errset}_R(\Xi, \{a, b, c\}) = \{b, c\}$. This shows the two different types of errors: b is a false positive by sound derivation, and c is a false negative by negation as failure.

Binary Classification Problems. As already motivated by the example above we now describe a binary classification problem: Let $f_R = \mathbf{t} : U \rightarrow \{0, 1\}$. Then, $t = \{x \in U : f_R(x) = 1\}$ and $\mathbf{t} = \chi(t)$. It holds that

$$x \in [\![\Xi]\!]\mathbf{t} \iff \mathcal{K} \cup \Xi \vdash \mathbf{t}(x, 1) \implies \mathbf{t}(x) = 1 \tag{28}$$

$$x \notin [\![\Xi]\!]\mathbf{t} \iff \mathcal{K} \cup \Xi \not\vdash \mathbf{t}(x, 1) \tag{29}$$

---

[6] $R \in \mathbf{F}$ is an equivalence relation induced by $f_R \in F$, whereas $r \subseteq U$ denotes a concept. $\mathbf{r}$ finally is a predicate, whose definition is unknown and needs to be learned such that its satisfaction set approximates r.

and

$$x \in [\![\varXi]\!](-t) \Longleftrightarrow \mathcal{K} \cup \varXi \vdash \mathbf{t}(x,0) \Longleftrightarrow x \notin \langle\!\langle \varXi \rangle\!\rangle t \Longrightarrow \mathbf{t}(x) = 0 \qquad (30)$$
$$x \notin [\![\varXi]\!](-t) \Longleftrightarrow \mathcal{K} \cup \varXi \not\vdash \mathbf{t}(x,0) \Longleftrightarrow x \in \langle\!\langle \varXi \rangle\!\rangle t \qquad (31)$$

This shows that the definition of upper and lower approximations on sets carries over to Horn resolution and satisfaction sets defined by correct answer substitutions.

Refinement of $\varXi$. Learning means to refine $\varXi$ in order to increase its predictive accuracy with respect to $\mathbf{t}$. This means that refinement decreases the error set. The error set of a logic program is the set of wrong conclusions and the set of non-derivable propositions:

$$\begin{aligned}
\mathrm{errset}_t(\varXi, s) := &\{x \in U : \mathcal{K} \cup \varXi \vdash \mathbf{t}(x,V) \wedge V \neq \mathbf{t}(x)\} \cup \\
&\{x \in U : \mathcal{K} \cup \varXi \not\vdash \mathbf{t}(x, \mathbf{t}(x))\} \cup \\
&\{x \in U : \mathcal{K} \cup \varXi \not\vdash \neg\mathbf{t}(x, \mathbf{t}(x))\} \qquad (32)
\end{aligned}$$

In order to minimise the error one needs to generalise $\varXi$, otherwise the error set can be minimised by specialising $\varXi$—this defines the learning problem in inductive logic programming.

Generalisation Operators. The simplest method to generalise a Horn theory is to replace a set of formulas by their least general generalisation. This means that from a set of facts $\mathbf{p}(x_i, f_p(x_i))$ with $f_p(x_i) = f_p(x_j)$, we generalise $\mathbf{p}(x, f_p(x_i))$ where $x = \mathrm{lgg}(\{x_0, \ldots, x_n\})$.[7] Assuming that all x_i are atoms, generalisation by least general generalisation means that the according equivalence relation becomes to $U \times U$ or that $f_p(x) = \bot$ for all $x \in U$.

Dropping literals via subsumption (cf. Eq. (23)) is also a very common method to generalise Horn theories:

$$\mathbf{p}(X, V) \leftarrow \mathbf{q}(X, W_q), \mathbf{r}(X, W_r) \models \mathbf{p}(X, V) \leftarrow \mathbf{q}(X, W_q)$$

Because of the soundness of Horn resolution it holds that any correct answer substitution for the second clause is also a correct answer substitution for the first clause which corresponds to the intended inclusion of satisfaction sets. The last method to generalise a theory is to add the clauses that are not covered or adding generalisations thereof. For all three methods it holds that refining $\varXi$ to a more general theory $\varXi'$ means $[\![\varXi]\!]t \subseteq [\![\varXi']\!]t$.

Specialisation Operators. Accordingly, $\varXi'$ is a specialisation of $\varXi$, $\langle\!\langle \varXi' \rangle\!\rangle t \subseteq \langle\!\langle \varXi \rangle\!\rangle t$. Specialisation means to remove theorems from the deductive closure of the theory. According operators correspond to the dual operations described for generalisation: Variable instantiations limit the set of possible assignments, unification

---

[7] Note that lgg is defined for *terms*, i.e. x_i can be complex expressions with free variables.

of two (valid) predicates leads to one single but much more special predicate, and deleting entire clauses from Ξ results in a smaller deductive closure.

Now, the ILP-problem can be reformulated as follows: Given Ξ, the hypothesis we are searching for is a program Ξ' and a (possibly empty) set $\mathcal{K}$ of facts for which

$$[\![\mathcal{K} \cup \Xi]\!]t \subseteq [\![\mathcal{K} \cup \Xi' \cup \Xi]\!]t \subseteq \langle\!|\mathcal{K} \cup \Xi' \cup \Xi|\!\rangle t \subseteq \langle\!|\mathcal{K} \cup \Xi|\!\rangle t \qquad (33)$$

Reducts of Logic Programs. As in Sec. 5.3, one can encode each feature f into a predicate symbol $\mathbf{f}$. Then a Horn theory $\mathcal{K} \cup \Xi$ models an information system, if:

$$\forall x \in U \; \forall f \in F : \mathcal{K} \cup \Xi \vdash \mathbf{f}(x, y) \iff f(x) = y$$

A *Horn reduct* Ξ' of Ξ is a set of clauses where for each clause $\varphi \in \Xi'$ there is a clause $\psi \in \Xi$ and a substitution θ such that $\varphi \subseteq \psi\theta$ and Ξ and Ξ' induce the same theory. Translating the original definition of a reduct (see Eq. (15)), $\Xi' \in \mathrm{Red}_s(\Xi)$, if $\Xi' \in \mathrm{Red}_s(\Xi)$ and the satisfaction set of Ξ' equals the satisfaction set of Ξ on s.

As an example, let us consider the case of *literal dropping* and its relationship to building reducts. Let there be two clauses,

$$\mathbf{t}(x, 1) \leftarrow \mathbf{p}_1(x, f_1(x)) \wedge \cdots \wedge \mathbf{p}_k(x, f_k(x)) \wedge \mathbf{p}_{k+1}(x, f_{k+1}(x)) \text{ and}$$
$$\mathbf{t}(x, 1) \leftarrow \mathbf{p}_1(x, f_1(x)) \wedge \cdots \wedge \mathbf{p}_k(x, f_k(x)).$$

Obviously, the former implies the latter. If the satisfaction sets of both are the same, then the second clause is a *reduct* of the first one by dropping one literal, or, equivalently, by dropping one feature or equivalence relation. If

$$\mathbf{t}(x, 1) \leftarrow \mathbf{p}_1(x, a) \wedge \mathbf{p}_2(x, b)$$
$$\mathbf{t}(x, 1) \leftarrow \mathbf{p}_1(x, a) \wedge \mathbf{p}_2(x, c)$$

one can induce $\mathbf{t}(x, 1) \leftarrow \mathbf{p}_1(x, a) \wedge \mathbf{p}_2(x, y)$ or even $\mathbf{t}(x, 1) \leftarrow \mathbf{p}_1(x, a)$. Now that we can translate ILP into a relational learning framework, we can estimate the quality of each of these hypotheses by the error measures as defined in Sect. 4.1.

Efficient Refinement. Refinement approaches usually are very expensive. The best hypothesis we are looking for is hidden somewhere in the structure defined by $[\![\cdot]\!]$ and $\langle\!|\cdot|\!\rangle$. Therefore, it is infeasible to search through the entire space of hypotheses. Instead, there are several measures to guide this search. One of the most popular measures is *information gain* as used in decision tree induction [27] and also in the induction of logic programs [28]. There exist many other measures (for example, minimum description length was used in Progol [29]) but all of them always infuse a bias into the learning process. Ensemble learning methods allow us to focus on small parts of the learning problem only—hoping that solving all the subproblems and aggregating the resulting hypotheses does not only perform better but also is quicker. All of these search heuristics can be applied in any of the approaches described here, since we have established a connection between the ordering relations on hypotheses in the different paradigms.

6 Conclusion

This paper does not contribute to the knowledge of the scientific community in a sense that we added analytic knowledge. It is rather the other way around: We showed that at a certain level of abstraction, all those *different* approaches in machine learning can be reformulated and unified in a relational framework. Besides the contribution of a descriptive formalism, a second focus of our current work is in implementing an efficient system based on the ideas presented in this paper. Presently, we develop an RSDA toolkit which makes use of the improved efficiency in relational database systems [13]. There remains a lot of work to be done: Boosting and Bagging algorithms are based on error measures which in turn are multi-valued. This means that we need to extend our formalism to probably approximately correct hypotheses. One perspective is then to extend ILP by several different logic reasoning approaches [2,5].

Similarly, ensemble learning methods have been applied to increase a weak learner's predictive accuracy. Our reformulations delivered heuristically guided algorithms for extraction reducts from information systems that do not perform better, but quicker. Here, a deeper analysis of runtime complexity is required.

Throughout the paper, we motivated a strong connection between multi-modal logics, relational learning and their algebraic counterparts. It remains to analyse whether inductive logic programming as theory refinement can be expressed algebraically (Heyting, Lindenbaum-Tarski).

Acknowledgements. The author wishes to thank Peter Höfner, Bernhard Möller and all anonymous reviewers for many helpful comments and discussions.

References

1. Orlowska, E.: Reasoning with incomplete information: Rough set based information logics. In: Proceedings of the SOFTEKS Workshop on Incompleteness and Uncertainty in Information Systems, pp. 16–33 (1993)
2. Yao, Y.Y.: On generalizing rough set theory. In: Wang, G., Liu, Q., Yao, Y., Skowron, A. (eds.) RSFDGrC 2003. LNCS (LNAI), vol. 2639, p. 579. Springer, Heidelberg (2003)
3. Düntsch, I.: A logic for rough sets. Theoretical Computer Science 179, 427–436 (1997)
4. Xu, F., Yao, Y., Miao, D.: Rough set approximations in formal concept analysis and knowledge spaces. In: An, A., Matwin, S., Raś, Z.W., Ślęzak, D. (eds.) Foundations of Intelligent Systems. LNCS (LNAI), vol. 4994, pp. 319–328. Springer, Heidelberg (2008)
5. Düntsch, I., Gediga, G., Orłowska, E.: Relational attribute systems II: Reasoning with relations in information structures. In: Peters, J.F., Skowron, A., Marek, V.W., Orłowska, E., Słowiński, R., Ziarko, W.P. (eds.) Transactions on Rough Sets VII. LNCS, vol. 4400, pp. 16–35. Springer, Heidelberg (2007)
6. Wolpert, M.: No free lunch theorems for optimization. IEEE Transactions on Evolutionary Computation 1, 67–82 (1997)
7. Breiman, L.: Bagging predictors. Technical Report 421, University of California, Berkeley (1994)

8. Breiman, L.: Heuristics of instability and stabilization in model selection. The Annals of Statistics 24 (1996)
9. Schapire, R.E.: The strength of weak learnability. Machine Learning 5, 197–227 (1990)
10. Schapire, R.E.: The boosting approach to machine learning: An overview. In: MSRI Workshop on Nonlinear Estimation and Classification (2002)
11. Freund, Y., Schapire, R.E.: Experiments with a new boosting algorithm. In: Proc. 19th Intl. Conf. Machine Learning (1996)
12. Pawlak, Z.: On rough sets. Bulletin of the EATCS 24, 94–184 (1984)
13. Han, X., Lin, T.Y., Han, J.: A new rough sets model based on database systems. Fundamenta Informaticae 59, 135–152 (2003)
14. Wróblewski, J.: Finding minimal reducts using genetic algorithms. In: Proc. of the Second Annual Joint Conference on Information Sciences (1995)
15. Øhrn, A.: Discernibility and Rough Sets in Medicine: Tools and Applications. PhD thesis, Norwegian University of Science and Technology, Department of Computer and Information Science (1999)
16. Lavrač, N., Džeroski, S.: Inductive Logic Programming: Techniques and Applications. Ellis-Horwood (1994)
17. Raedt, L.D.: Logical and Relational Learning. In: Cognitive Technologies. Springer, Heidelberg (2008)
18. Robinson, J.: A machine-oriented logic based on the resolution principle. J. ACM 12, 23–41 (1965)
19. Warren, D.H.D.: An abstract prolog instruction set. Technical Note 309, SRI International, Menlo Park, CA (1983)
20. Muggleton, S., Buntine, W.: Machine invention of first-order predicates by inverting resolution. In: Proceedings of the 5th International Conference on Machine Learning, pp. 339–352. Kaufmann, San Francisco (1988)
21. Plotkin, G.: A further note on inductive generalization. In: Machine Intelligence, vol. 6. Edinburgh University Press, Edinburgh (1971)
22. Plotkin, G.: A note on inductive generalisation. In: Meltzer, B., Michie, D. (eds.) Machine Intelligence, vol. 5, pp. 153–163. Edinburgh University Press, Edinburgh (1969)
23. Shapiro, E.: Inductive inference of theories from facts. In: Lassez, J.L., Plotkin, G. (eds.) Computational logic: essays in honor of Alan Robinson. The MIT Press, Cambridge (1991)
24. Džeroski, S., Muggleton, S., Russell, S.: PAC-learnability of determinate logic programs. In: Proceedings of the 5th ACM Workshop on Computational Learning Theory, pp. 128–135. ACM Press, New York (1992)
25. Kautz, H., Kearns, M., Selman, B.: Horn approximations of empirical data. Artificial Intelligence 74 (1995)
26. Nock, R., Jappy, P.: Function-free Horn clauses are hard to approximate. In: Page, D. (ed.) ILP 1998. LNCS (LNAI), vol. 1446. Springer, Heidelberg (1998)
27. Quinlan, J.: Induction of decision trees. Machine Learning 1, 81–106 (1986)
28. Quinlan, J., Cameron, R.: Induction of logic programs: FOIL and related systems. New Generation Computing 13, 287–312 (1995)
29. Muggleton, S.H.: Inverse entailment and Progol. New Generation Computing 13, 245–286 (1995)

The Cube of Kleene Algebras and the Triangular Prism of Multirelations

Koki Nishizawa[1], Norihiro Tsumagari[2], and Hitoshi Furusawa[3]

[1] Department of Information Systems, Faculty of Environmental and Information Studies, Tottori University of Environmental Studies
koki@kankyo-u.ac.jp
[2] Graduate School of Science and Engineering, Kagoshima University
k1852610@kadai.jp
[3] Department of Mathematics and Computer Science, Kagoshima University
furusawa@sci.kagoshima-u.ac.jp

Abstract. We refine and extend the known results that the set of ordinary binary relations forms a Kleene algebra, the set of up-closed multirelations forms a lazy Kleene algebra, the set of up-closed finite multirelations forms a monodic tree Kleene algebra, and the set of total up-closed finite multirelations forms a probabilistic Kleene algebra. For the refinement, we introduce a notion of type of multirelations. For each of eight classes of relaxation of Kleene algebra, we give a sufficient condition on type T so that the set of up-closed multirelations of T belongs to the class. Some of the conditions are not only sufficient, but also necessary.

1 Introduction

A notion of Kleene algebras is introduced by Kozen [1] in order to handle regular languages algebraically. Recently, some relaxations of Kleene algebras have been introduced in order to treat various systems algebraically. For example, *lazy Kleene algebras* are introduced by Möller [2] to treat both finite and infinite streams. *Monodic tree Kleene algebras* are introduced by Takai and Furusawa [3] to develop Kleene-like algebras for some class of tree languages. *Probabilistic Kleene algebras* are introduced by McIver and Weber [4] to analyze probabilistic distributive systems without numerical calculations. The notion of lazy Kleene algebra is a generalization of monodic tree Kleene algebra, the notion of monodic tree Kleene algebra is a relaxation of probabilistic Kleene algebra, and the notion of probabilistic Kleene algebra is a relaxation of Kleene algebra.

On the other hand, *multirelations* are studied as a semantic domain of programs. Up-closed multirelations provide models of game logic introduced by Parikh [5,6]. Operations of game logic have been studied from an algebraic point of view by Goranko [7] and Venema [8]. They have given a complete axiomatization of iteration-free game logic.

We study the relationship between the two different research topics. It is known that the set of ordinary binary relations on a set forms a Kleene algebra. However, it does not seem that there are enough results about what

R. Berghammer et al. (Eds.): RelMiCS/AKA 2009, LNCS 5827, pp. 276–290, 2009.

class of multirelations forms what relaxation of Kleene algebras. In recent papers [9,10,11,12], the authors show that the set of up-closed multirelations forms a lazy Kleene algebra, that the set of finitary up-closed multirelations forms a monodic tree Kleene algebra, and that the set of total finitary up-closed multirelations forms a probabilistic Kleene algebra. This paper extends these results as follows.

First, we define a cube consisting of eight classes of lazy Kleene algebras, by introducing three axioms (the 0-axiom, the +-axiom, and the D-axiom) on a lazy Kleene algebra. A lazy Kleene algebra satisfies all of the three axioms if and only if it is a Kleene algebra. We define a cube consisting of eight classes of complete IL-semirings, by introducing three conditions (preservation of the right 0, the right +, and all right directed joins) on a complete IL-semiring. A complete IL-semiring satisfies all of the three conditions if and only if it is a complete I-semiring (or quantale). And we obtain a mapping from the second cube to the first cube, by proving that a complete IL-semiring forms a lazy Kleene algebra and that preservation of the right 0, the right +, and all right directed joins on a complete IL-semiring imply the 0-axiom, the +-axiom, and the D-axiom on a lazy Kleene algebra, respectively. This is a new explanation of the fact that a complete I-semiring forms a Kleene algebra.

Second, we focus on a notion of multirelations. While a multirelation over a set A is defined to be a subset of $A \times \wp(A)$, this paper extends this notion. We call a subfunctor T of the covariant powerset functor $\wp \colon \mathbf{Set} \to \mathbf{Set}$ a *type* and call a subset of $A \times T(A)$ a *multirelation of type T over A*. And we give a sufficient condition on T such that the set of up-closed multirelations of T forms a complete IL-semiring. We call a type satisfying this condition a *closed type*. We also define a cube consisting of eight classes of closed types, by introducing three conditions (total, affine and finite) on a closed type. The cube is actually a triangular prism, since affineness implies finiteness. We show that a closed type T is total, affine, and finite *if and only if* the set of up-closed multirelations of type T over an arbitrary set A forms a complete IL-semiring preserving the right 0, the right +, and all right directed joins, respectively.

Combining the above results, we show which type of up-closed multirelations forms a lazy Kleene algebra satisfying which axiom. The result includes the results for ordinary binary relations, up-closed multirelations, finitary up-closed multirelations, and total finitary up-closed multirelations.

This paper is organized as follows. Section 2 shows that the set of up-closed multirelations forms a complete IL-semiring. In Section 3, we show that the set of up-closed multirelations forms a lazy Kleene algebra. Section 4 defines the cube consisting of eight classes of lazy Kleene algebras. In Section 5, we define the cube consisting of eight classes of complete IL-semirings and define a mapping from it to the cube of lazy Kleene algebras. Section 6 defines a notion of types, a triangular prism consisting of six classes of types, and a mapping from it to the cube of complete IL-semirings. Section 7 summarizes this work and future work.

2 Multirelational Model of Complete IL-Semiring

In this section, we show that the set of up-closed multirelations forms a complete IL-semiring.

Complete IL-semirings are relaxations of complete I-semirings (or quantales).

Definition 1 (IL-semiring). *An* idempotent left semiring (IL-semiring) *[2] is a tuple* $(K, +, 0, \cdot, 1)$ *with the following properties:*

1. $(K, +, 0)$ *is an idempotent commutative monoid, or a partially ordered set* $(K, \leq)$ *with the binary join (least upper bound)* $+$ *and the least element* 0.
2. $(K, \cdot, 1)$ *is a monoid.*
3. $a \leq a'$ *and* $b \leq b'$ *imply* $a \cdot b \leq a' \cdot b'$.
4. $(a + b) \cdot c = a \cdot c + b \cdot c$ *and* $0 \cdot a = 0$.

Definition 2 (Complete IL-semiring). *A* complete IL-semiring *is a tuple* $(K, +, 0, \cdot, 1, \bigvee)$ *with the following properties:*

1. $(K, +, 0, \cdot, 1)$ *is an IL-semiring.*
2. $(K, \leq)$ *has the join* $\bigvee S$ *for each subset* S *of* K.
3. $(\bigvee S) \cdot a = \bigvee \{x \cdot a | x \in S\}$.

A complete IL-semiring also has the meet (greatest lower bound) for each subset. We write $\bigwedge S$ for the meet of subset S.

Example 1. For a set A, a tuple $(K, +, 0, \cdot, 1, \bigvee)$ forms a complete IL-semiring where

- K is the set of all ordinary binary relations over A,
- $R + Q$ is the binary union of R and Q,
- 0 is the empty relation,
- $R \cdot Q$ is the composition of R and Q,
- 1 is the identity (diagonal) relation on A, and
- $\bigvee$ is the union operator.

The leading example of complete IL-semirings in this paper is formed by up-closed multirelations. Multirelations are relaxations of ordinary binary relations.

Definition 3 (Multirelation). *A multirelation over a set* A *is a subset of* $A \times \wp(A)$ *where* $\wp(A)$ *is the power set of* A.

Definition 4 (Up-closed multirelation). *A multirelation* R *over* A *is called up-closed if* $(a, X) \in R$ *and* $X \subseteq Y$ *imply* $(a, Y) \in R$ *for each* $a \in A$, $X, Y \in \wp(A)$.

Example 2. The empty set and $A \times \wp(A)$ are up-closed multirelations over A. The set $\{(a, X) | a \in X, X \subseteq A\}$ is also an up-closed multirelation. The set $\{(a, \{b\}) | a, b \in A\}$ is a multirelation, but not always up-closed.

The following proposition is proved in paper [9].

Proposition 1. *For a set A, a tuple $\mathbf{UMR}(A) = (K, +, 0, \cdot, 1, \bigvee)$ forms a complete IL-semiring where*

- K *is the set of all up-closed multirelations over A,*
- $R + Q$ *is the binary union of R and Q,*
- 0 *is the empty set,*
- $(a, X) \in R \cdot Q \iff \exists Y.(a, Y) \in R$ *and* $\forall y \in Y.(y, X) \in Q$,
- $1 = \{(a, X) | a \in X, X \subseteq A\}$, *and*
- $\bigvee$ *is the union operator.*

Up-closed multirelations can not be composed in the same way as ordinary binary relations. The above operation $R \cdot Q$ is called the *composition* of up-closed multirelations R, Q.

3 Multirelational Model of Lazy Kleene Algebra

In this section, we show that the set of up-closed multirelations forms a lazy Kleene algebra. It is proved as a corollary of the theorem that a complete IL-semiring forms a lazy Kleene algebra.

Definition 5 (Lazy Kleene algebra). *A lazy Kleene algebra [2] is a tuple $(K, +, 0, \cdot, 1, {}^*)$ with the following properties:*

1. $(K, +, 0, \cdot, 1)$ *is an IL-semiring.*
2. $1 + a \cdot a^* \leq a^*$.
3. $b + a \cdot c \leq c$ *implies* $a^* \cdot b \leq c$.

Every complete IL-semiring $(K, +, 0, \cdot, 1, \bigvee)$ satisfies $a \cdot b \leq c \iff a \leq c/b$ where $c/b = \bigvee\{x \in K \mid x \cdot b \leq c\}$ (left residual). Note that $(c/b) \cdot b \leq c$ holds, since $(c/b) \cdot b \leq c \iff c/b \leq c/b$.

Theorem 1. *Every complete IL-semiring forms a lazy Kleene algebra. Moreover, every homomorphism between complete IL-semirings is also a homomorphism between the induced lazy Kleene algebras.*

Proof. Consider a complete IL-semiring $(K, +, 0, \cdot, 1, \bigvee)$. For each $a \in K$, the function $f(x) = 1 + a \cdot x$ is monotone, since $+$ and $\cdot$ are monotone. By Tarski's fixed point theorem, we have

- $1 + a \cdot a^* \leq a^*$ and
- $1 + a \cdot b \leq b$ implies $a^* \leq b$

where $a^* = \bigwedge\{x | 1 + a \cdot x \leq x\}$. Therefore, the second property of Definition 5 is satisfied. The third property is satisfied, since

$$
\begin{aligned}
a^* \cdot b \leq c &\iff a^* \leq c/b \\
&\Longleftarrow 1 + a \cdot (c/b) \leq c/b \\
&\iff (1 + a \cdot (c/b)) \cdot b \leq c \\
&\iff 1 \cdot b + a \cdot (c/b) \cdot b \leq c \\
&\Longleftarrow b + a \cdot c \leq c.
\end{aligned}
$$

Therefore, it is proved that $(K, +, 0, \cdot, 1, {}^*)$ forms a lazy Kleene algebra.

Next, we prove the property about homomorphisms. Let $(K, +, 0, \cdot, 1, \bigvee)$ and $(L, +, 0, \cdot, 1, \bigvee)$ be complete IL-semirings. Let $(K, +, 0, \cdot, 1, {}^*)$ and $(L, +, 0, \cdot, 1, {}^*)$ be the induced lazy Kleene algebras. Let f be a function $f\colon K \to L$ preserving $+$, 0, $\cdot$, 1, and $\bigvee$. The following $g\colon L \to K$ is called the right adjoint to f and it satisfies $f(x) \leq y \iff x \leq g(y)$.

$$g(y) = \bigvee\{x \in K \mid f(x) \leq y\}$$

Now, f preserves * as follows.

$$
\begin{aligned}
(f(a))^* \leq f(a^*) &\Longleftarrow 1 + f(a) \cdot f(a^*) \leq f(a^*) \\
&\Longleftrightarrow f(1 + a \cdot a^*) \leq f(a^*) \\
&\Longleftarrow 1 + a \cdot a^* \leq a^*
\end{aligned}
$$

$$
\begin{aligned}
f(a^*) \leq (f(a))^* &\Longleftrightarrow a^* \leq g((f(a))^*) \\
&\Longleftarrow 1 + a \cdot g((f(a))^*) \leq g((f(a))^*) \\
&\Longleftrightarrow f(1 + a \cdot g((f(a))^*)) \leq (f(a))^* \\
&\Longleftrightarrow 1 + f(a) \cdot f(g((f(a))^*)) \leq (f(a))^* \\
&\Longleftarrow 1 + f(a) \cdot (f(a))^* \leq (f(a))^* \qquad \square
\end{aligned}
$$

Corollary 1. *The set of up-closed multirelations over a fixed set forms a lazy Kleene algebra.*

4   Cube of Kleene Algebras

In this section, we define a cube consisting of eight classes of lazy Kleene algebras, by defining three independent axioms. A lazy Kleene algebra satisfying all of the three axioms is a Kleene algebra. Therefore, the cube consists of eight classes between lazy Kleene algebras and Kleene algebras.

Definition 6 (Cube of lazy Kleene algebra). *A tuple $(K, +, 0, \cdot, 1, {}^*)$ is called a lazy Kleene algebra satisfying*

- *the 0-axiom if $a \cdot 0 = 0$ for each $a \in K$,*
- *the +-axiom if $a \cdot (b + c) = a \cdot b + a \cdot c$ for each $a, b, c \in K$, and*
- *the D-axiom if $a \cdot (b + 1) \leq a$ implies $a \cdot b^* \leq a$ for each $a, b \in K$,*

respectively.

The reason why we call the third axiom the D-axiom is that this axiom has a relationship with directed sets (explained in the next section).

We write **LKA** for the category whose objects are lazy Kleene algebras and whose arrows are homomorphisms between them. We write $\mathbf{LKA_0}$ for the full subcategory of **LKA** whose objects are lazy Kleene algebras satisfying the 0-axiom. Similarly, we define $\mathbf{LKA_{0,+,D}}$, $\mathbf{LKA_{0,+}}$, and so on. The eight categories and forgetful functors between them form the cube of Fig. 1.

Objects of $\mathbf{LKA_D}$, $\mathbf{LKA_{0,D}}$, and $\mathbf{LKA_{0,+,D}}$ are known as monodic tree Kleene algebras, probabilistic Kleene algebras, and Kleene algebras, respectively.

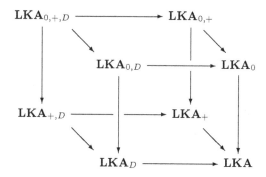

Fig. 1. The cube of lazy Kleene algebras

Proposition 2. *A lazy Kleene algebra satisfies the D-axiom if and only if it is a monodic tree Kleene algebra [3]. A lazy Kleene algebra satisfies the 0-axiom and the D-axiom if and only if it is a probabilistic Kleene algebra [4]. A lazy Kleene algebra satisfies the 0-axiom, the +-axiom, and the D-axiom if and only if it is a Kleene algebra [1].*

5   Cube of Complete IL-Semirings

In this section, we define a cube consisting of eight classes of complete IL-semirings, by introducing three independent axioms on a complete IL-semiring. We also obtain a mapping from it to the cube of the previous section, by using Theorem 1 and proving that the three conditions on a complete IL-semiring imply the three axioms on a lazy Kleene algebra, respectively.

Definition 7 (Directed set). *A subset S of a lattice is called* directed *if each finite subset of S has an upper bound in S.*

A directed set always has an element, since a directed set must have an upper bound of the empty subset.

Definition 8 (Cube of complete IL-semiring). *A tuple $(K, +, 0, \cdot, 1, \bigvee)$ is called a complete IL-semiring preserving*

- *the* right 0 *if $a \cdot 0 = 0$ for each $a \in K$,*
- *the* right + *if $a \cdot (b + c) = a \cdot b + a \cdot c$ for each $a, b, c \in K$, and*
- *all* right directed joins *if $a \cdot \bigvee S = \bigvee \{a \cdot x | x \in S\}$ for each $a \in K$ and each directed $S \subseteq K$,*

respectively.

We write **CILS** for the category whose objects are complete IL-semirings and whose arrows are homomorphisms between them. We write **CILS**$_D$ for the full subcategory of **CILS** whose objects are complete IL-semirings preserving all right directed joins. Similarly, we define **CILS**$_{0,+,D}$, **CILS**$_{0,+}$, and so on. The eight categories and forgetful functors between them form the cube of Fig. 2.

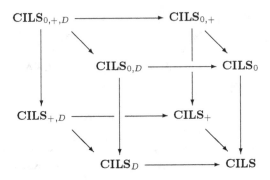

Fig. 2. The cube of complete IL-semirings

Proposition 3. *A tuple* $(K, +, 0, \cdot, 1, \bigvee)$ *is a complete I-semiring [2] (or, quantale) if and only if it is a complete IL-semiring preserving the right 0, the right* $+$, *and all right directed joins.*

Proof. A complete I-semiring is defined to be a complete IL-semiring satisfying $a \cdot (\bigvee S) = \bigvee \{a \cdot x \mid x \in S\}$. Trivially, a complete I-semiring is a complete IL-semiring preserving the right 0, the right $+$, and all right directed joins. Conversely, let K be a complete IL-semiring preserving the right 0, the right $+$, and all right directed joins. For an arbitrary subset S of K, $\bigvee S = \bigvee \{\bigvee X \mid X \subseteq S, X \text{ is finite}\}$ and the set $\{\bigvee X \mid X \subseteq S, X \text{ is finite}\}$ is directed. Therefore,

$$
\begin{aligned}
a \cdot (\bigvee S) &= a \cdot (\bigvee \{\bigvee X \mid X \subseteq S, X \text{ is finite}\}) \\
&= \bigvee \{a \cdot (\bigvee X) \mid X \subseteq S, X \text{ is finite}\} \\
&= \bigvee \{\bigvee \{a \cdot x \mid x \in X\} \mid X \subseteq S, X \text{ is finite}\} \\
&= \bigvee \{a \cdot x \mid x \in S\}.
\end{aligned}
$$
□

Theorem 2. *Every complete IL-semiring* C *forms a lazy Kleene algebra* L. *Moreover, the following hold.*

1. *L satisfies the 0-axiom if and only if* C *preserves the right 0.*
2. *L satisfies the +-axiom if and only if* C *preserves the right* $+$.
3. *L satisfies the D-axiom if* C *preserves all right directed joins.*

Proof. Similarly to the proof of Theorem 1, we construct L from C. By the construction, the case 1 and the case 2 trivially hold. We show the case 3. Assume that C preserves all right directed joins. Each function $f_b(x) = 1 + b \cdot x$ preserves the join of an arbitrary directed subset. Therefore, by the fixed point theorem, the least fixed point b^* of f_b is equal to $\bigvee \{f_b^n(0) \mid n \in \mathbf{Nat}\}$. Assume $a \cdot (b + 1) \leq a$. We show that $a \cdot f_b^n(0) \leq a$ holds for each $n \in \mathbf{Nat}$ by induction on n.

$(n = 0)$ $a \cdot f_b^0(0) = a \cdot 0 \leq a \cdot 1 = a$.
$(n = 1)$ $a \cdot f_b^1(0) = a \cdot (1 + b \cdot 0) \leq a \cdot (1 + b \cdot 1) \leq a$.
$(n \geq 2)$ Note that $1 \leq 1 + b \cdot f_b^{n-2}(0) = f_b^{n-1}(0)$. Assume $a \cdot f_b^{n-1}(0) \leq a$. Then, we have

$$a \cdot f_b^n(0) = a \cdot (1 + b \cdot f_b^{n-1}(0))$$
$$\leq a \cdot (f_b^{n-1}(0) + b \cdot f_b^{n-1}(0))$$
$$\leq a \cdot (1 + b) \cdot f_b^{n-1}(0)$$
$$\leq a \cdot f_b^{n-1}(0)$$
$$\leq a.$$

Therefore, we have $\bigvee\{a \cdot f_b^n(0) | n \in \mathbf{Nat}\} \leq a$. Since the set $\{f_b^n(0) | n \in \mathbf{Nat}\}$ is directed and C preserves all right directed joins, we have

$$a \cdot b^* = a \cdot \bigvee\{f_b^n(0) | n \in \mathbf{Nat}\} = \bigvee\{a \cdot f_b^n(0) | n \in \mathbf{Nat}\} \leq a. \qquad \square$$

In the above theorem, 1 and 2 give necessary and sufficient conditions respectively, but 3 does not. In fact, all authors do not know whether if L satisfies the D-axiom then C preserves all right directed joins.

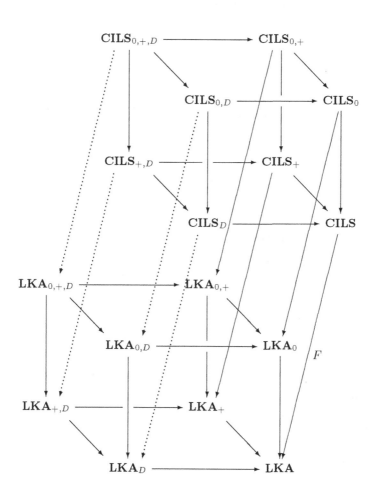

Fig. 3. The maps from the cube of **CILS** to the cube of **LKA**

This theorem can be represented by Fig. 3. The functor F from **CILS** to **LKA** is given by Theorem 1. The other seven functors from the cube of complete IL-semirings to the cube of lazy Kleene algebras are given by Theorem 2. By 1 of Theorem 2, the square consisting of **CILS**, **CILS$_0$**, **LKA**, and **LKA$_0$** is not only a commutative square, but also a pullback square in **Cat**, that is, an object C of **CILS** belongs to **CILS$_0$** if and only if $F(C)$ belongs to **LKA$_0$**. Similarly, every square in Fig. 3 is a pullback square, except for squares consisting of three solid arrows and one dotted arrow.

6   Triangular Prism of Multirelations

In recent papers [9,10,11,12], the authors show that the set of up-closed multire-lations forms a lazy Kleene algebra, that the set of finitary up-closed multire-lations forms a monodic tree Kleene algebra, and that the set of total finitary up-closed multirelations forms a probabilistic Kleene algebra. To extend these results, in this section, we define a triangular prism consisting of six classes of multirelations and obtain a mapping from it to the cube of lazy Kleene algebras.

First, we extend the notion of multirelations.

Definition 9 (Typed multirelation). *A type T of multirelation is a subfunctor of the powerset functor $\wp \colon$ **Set** $\to$ **Set** (i.e., $T(A) \subseteq \wp(A)$ for each set A). A multirelation of type T over A is a subset of $A \times T(A)$. A multirelation R of type T over A is called* up-closed *if $(a, X) \in R$ and $X \subseteq Y$ imply $(a, Y) \in R$ for each $a \in A$, $X, Y \in T(A)$.*

We give a sufficient condition on a type T such that the set of up-closed mul-tirelations of type T over an arbitrary set A forms a complete IL-semiring. We call a type satisfying this condition *closed*.

Definition 10 (Closed type). *A type T is called* closed *if for each set A,*

1. *$\forall a \in A.\{a\} \in T(A)$, and*
2. *if a family $\{X_i\}_{i \in I}$ of subsets of A satisfies $I \in T(A)$ and $\forall i \in I.X_i \in T(A)$, then $\bigcup_{i \in I} X_i \in T(A)$.*

Example 3. Every submonad of the powerset monad forms a closed type. In fact, all closed types mentioned in this section are submonads of the powerset monad.

Proposition 4. *For an arbitrary set A, a tuple T-**UMR**$(A) = (K, +, 0, \cdot, 1, \bigvee)$ forms a complete IL-semiring where*

- K *is the set of all up-closed multirelations of type T over A,*
- $R + Q$ *is the binary union of R and Q,*
- 0 *is the empty set,*
- $(a, X) \in R \cdot Q \iff \exists Y \in T(A).(a, Y) \in R$ and $\forall y \in Y.(y, X) \in Q$,
- $1 = \{(a, X) | a \in X, X \in T(A)\}$, and
- $\bigvee$ *is the union operator,*

if and only if T is a closed type or the constant functor to the empty set.

Proof

($\Longleftarrow$) If T is the constant functor to the empty set, then T-**UMR**(A) is the trivial complete IL-semiring.

On the other hand, let T be a closed type. We show $R \subseteq 1 \cdot R$. If $T(A) = \emptyset$, then $R = 0 \subseteq 1 \cdot R$. Assume $T(A) \neq \emptyset$ and $(a, X) \in R$. By the first condition of Definition 10, we have $\{a\} \in T(A)$. Therefore, $(a, X) \in 1 \cdot R$. Therefore, $R \subseteq 1 \cdot R$.

Next, we show $R \cdot (Q \cdot P) \subseteq (R \cdot Q) \cdot P$. Assume $(a, X) \in R \cdot (Q \cdot P)$. Then, there exists $Y \in T(A)$ such that $(a, Y) \in R$ and $\forall y \in Y.\exists Z_y \in T(A).(y, Z_y) \in Q$ and $\forall z \in Z_y.(z, X) \in P$. By the second condition of Definition 10, we have $\bigcup_{y \in Y} Z_y \in T(A)$. It satisfies $(a, \bigcup_{y \in Y} Z_y) \in R \cdot Q$, since $(a, Y) \in R$ and $\forall y \in Y.(y, \bigcup_{y \in Y} Z_y) \in Q$. Since $\forall z \in \bigcup_{y \in Y} Z_y.(z, X) \in P$, we have $(a, X) \in (R \cdot Q) \cdot P$. Therefore, $R \cdot (Q \cdot P) \subseteq (R \cdot Q) \cdot P$.

The other conditions for complete IL-semirings are easy to prove.

($\Longrightarrow$) Assume that T is neither a closed type nor the constant functor to the empty set. There exists a set A which does not satisfy 1 of Definition 10 or which does not satisfy 2 of Definition 10. We show that T-**UMR**(A) does not form a complete IL-semiring.

- Assume that 1 of Definition 10 does not hold. We can take $a \in A$ satisfying $\{a\} \notin T(A)$. Since T is a functor on **Set**, if $T(A)$ is empty, then $T(X)$ is empty for each set X, that is, T is the constant functor to the empty set. Therefore, $T(A)$ is not empty. We can take $X \in T(A)$. Let $R = \{(a, Y) | Y \in T(A)\}$. We have $(a, X) \in R$ but $(a, X) \notin 1 \cdot R$. Therefore, $R \not\subseteq 1 \cdot R$.

- Assume that 2 of Definition 10 does not hold. We can take $\{X_i\}_{i \in I}$ satisfying $I \in T(A)$, $\forall i \in I.X_i \in T(A)$, and $\bigcup_{i \in I} X_i \notin T(A)$. If $A = \emptyset$ then every $T(A) \subseteq \wp(A)$ (i.e., $T(A) = \emptyset$ or $T(A) = \{\emptyset\}$) satisfies the second condition of Definition 10. Therefore, A is not empty. We can take $a \in A$. Let $R = \{(a, X) | X \in T(A), I \subseteq X\}$, $Q = \{(i, X) | i \in I, X \in T(A), X_i \subseteq X\}$, and $P = \{(x, X) | x \in \bigcup_{i \in I} X_i, X \in T(A)\}$. Then, we have $(a, I) \in R \cdot (Q \cdot P)$, since $(a, I) \in R$, $\forall i \in I.(i, X_i) \in Q$, and $\forall i \in I.\forall x \in X_i.(x, I) \in P$. But we have $(a, I) \notin (R \cdot Q) \cdot P$, since $(a, I) \in (R \cdot Q) \cdot P$ implies $\bigcup_{i \in I} X_i \in T(A)$. Therefore, $R \cdot (Q \cdot P) \not\subseteq (R \cdot Q) \cdot P$. $\qquad\Box$

By the above proposition, T is not always closed, even if T-**UMR**(A) forms a complete IL-semiring for an arbitrary set A. However, this is not a problem, since the trivial complete IL-semiring is not such an important counterexample.

We write $|X|$ for the number of elements of X.

Definition 11 (Cube of type of multirelation). *A closed type T of multirelations is called*

- *total if for an arbitrary set A, $A \neq \emptyset$ implies $\emptyset \notin T(A)$,*
- *affine if for an arbitrary set A, $\forall X \in T(A).|X| \leq 1$, and*
- *finite if for an arbitrary set A, $\forall X \in T(A).X$ is finite,*

respectively.

We write **UMR** for the category whose objects are complete IL-semirings T-**UMR**(A) for some closed type T and some set A and whose arrows are homomorphisms between them. We write **UMR**$_t$ for the full subcategory of **UMR** whose object is a complete IL-semiring T-**UMR**(A) for some total closed type T and some set A. Similarly, we write **UMR**$_a$ for the case of affine closed types and **UMR**$_f$ for the case of finite closed types. The eight categories and forgetful functors between them form the cube of Fig. 4. Note that **UMR**$_{a,f}$ = **UMR**$_a$ and **UMR**$_{t,a,f}$ = **UMR**$_{t,a}$ since affineness implies finiteness. Therefore, this cube is actually a triangular prism.

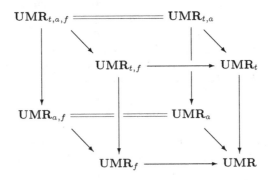

Fig. 4. The cube of complete IL-semirings of multirelations

Example 4. $T(A) = \wp(A)$ is a closed type. In this case, T-**UMR**(A) is equal to the complete IL-semiring **UMR**(A) consisting of all up-closed multirelations on A (defined in Proposition 1).

Example 5. $T(A) = \{\{a\}|a \in A\}$ is a closed, total, and affine (and finite) type. In this case, T-**UMR**(A) is isomorphic to the complete IL-semiring consisting of all ordinary binary relations on A (defined in Example 1).

We obtain the correspondence between the cube of Fig. 4 and the cube of complete IL-semirings.

Theorem 3. *Let T be a closed type. T-**UMR**(A) forms a complete IL-semiring for each A. Moreover, the following hold.*

1. *T-**UMR**(A) preserves the right 0 for each A if and only if T is total,*
2. *T-**UMR**(A) preserves the right $+$ for each A if and only if T is affine, and*
3. *T-**UMR**(A) preserves all right directed joins for each A if and only if T is finite.*

Proof

1. ($\Longleftarrow$) Assume that T is total. If $A = \emptyset$, then $R \cdot 0 = 0 \cdot 0 = 0$. If $A \neq \emptyset$, then $R \cdot 0 = \{(a, X) \mid X \in T(A), (a, \emptyset) \in R\} = 0$.
   ($\Longrightarrow$) Conversely, assume that T is not total. Let R be $A \times T(A)$. There exists a set A satisfying $A \neq \emptyset$ and $\emptyset \in T(A)$. There exists $a \in A$ such that $(a, \emptyset) \in R$. Therefore, $R \cdot 0 = \{(a, X) \mid X \in T(A), (a, \emptyset) \in R\} \neq 0$.

2. ($\Longleftarrow$) Assume that T is affine. $R \cdot Q + R \cdot P \subseteq R \cdot (Q + P)$ holds trivially. We show $R \cdot (Q + P) \subseteq R \cdot Q + R \cdot P$. Let (a, X) be an element of $R \cdot (Q + P)$. We can take $Y \in T(A)$ such that $(a, Y) \in R$ and $\forall y \in Y.(y, X) \in Q + P$. If $|Y| = 0$, then $(a, \emptyset) \in R$. Therefore, $(a, X) \in R \cdot Q \subseteq R \cdot Q + R \cdot P$. If $|Y| = 1$, then $Y = \{y\}$ and $(y, X) \in Q + P$. Moreover, if $(y, X) \in Q$, then $(a, X) \in R \cdot Q \subseteq R \cdot Q + R \cdot P$. If $(y, X) \in P$, then $(a, X) \in R \cdot P \subseteq R \cdot Q + R \cdot P$.
($\Longrightarrow$) Conversely, assume that T is not affine. Take a set A and $X \in T(A)$ satisfying $2 \le |X|$, and take $a \in X$. Let $R = \{(a, Y)|X \subseteq Y, Y \in T(A)\}$, $Q = \{(a, Y)|a \in Y, Y \in T(A)\}$, and $P = \bigcup_{y \in X \setminus \{a\}} \{(y, Y)|y \in Y, Y \in T(A)\}$. Then, $(a, X) \in R \cdot (Q + P)$ but $(a, X) \notin R \cdot Q + R \cdot P$. Therefore, $R \cdot (Q + P) \not\subseteq R \cdot Q + R \cdot P$.

3. ($\Longleftarrow$) Assume that T is finite. Let D be a directed subset of T-**UMR**(A). For each $R \in T$-**UMR**(A), $\bigvee \{R \cdot Q \,|\, Q \in D\} \subseteq R \cdot \bigvee D$ holds trivially. We show $R \cdot \bigvee D \subseteq \bigvee \{R \cdot Q \,|\, Q \in D\}$. Let (a, X) be an element of $R \cdot \bigvee D$. We can take $Y \in T(A)$ such that $(a, Y) \in R$ and $\forall y \in Y.\exists Q_y \in D.(y, X) \in Q_y$. Since Y is finite and D is directed, there exists $P \in D$ such that $\forall y \in Y.Q_y \subseteq P$. Therefore, $(a, X) \in R \cdot P \subseteq \bigvee \{R \cdot Q \,|\, Q \in D\}$.
($\Longrightarrow$) Conversely, assume that T is not finite. There exist a set A and an infinite set X satisfying $X \in T(A)$. Let $R_x = \{(x, Y)|X \subseteq Y, Y \in T(A)\}$ for each $x \in X$. Let $D = \{\bigcup_{x \in I} R_x \,|\, I \subseteq X, I \text{ is finite}\}$. Then, D is directed. Take $a \in X$. We have $(a, X) \in R_a \cdot \bigvee D$ but $(a, X) \notin \bigvee \{R_a \cdot Q|Q \in D\}$. Therefore, $R_a \cdot \bigvee D \not\subseteq \bigvee \{R_a \cdot Q|Q \in D\}$.   □

This theorem can be represented by Fig. 5. The functor G from **UMR** to **CILS** is given by Proposition 4. The other seven functors are given by Theorem 3. Every square in Fig. 5 is a pullback square.

As a corollary of Theorem 2 and Theorem 3, we get the mapping from the cube of complete IL-semirings consisting of up-closed typed multirelations to the cube of lazy Kleene algebras. The case 1 and the case 2 of this corollary give necessary and sufficient conditions, since the case 1 and the case 2 of Theorem 2 do so.

Corollary 2. *Let T be a closed type. T-**UMR**(A) forms a lazy Kleene algebra for each A. Moreover, the following hold.*

1. *T-**UMR**(A) satisfies the 0-axiom for each A if and only if T is total,*
2. *T-**UMR**(A) satisfies the +-axiom for each A if and only if T is affine, and*
3. *T-**UMR**(A) satisfies the D-axiom for each A if T is finite.*

This corollary includes many results about multirelational models of lazy Kleene algebras.

Example 6. $T(A) = \wp(A)$ is a closed type. In this case, T-**UMR**(A) is a lazy Kleene algebra. Therefore, the set of up-closed multirelations over A forms a lazy Kleene algebra.

Example 7. $T(A) = \{X \subseteq A \,|\, X \text{ is finite}\}$ is a closed type. Since this type T is finite, T-**UMR**(A) is a monodic tree Kleene algebra.

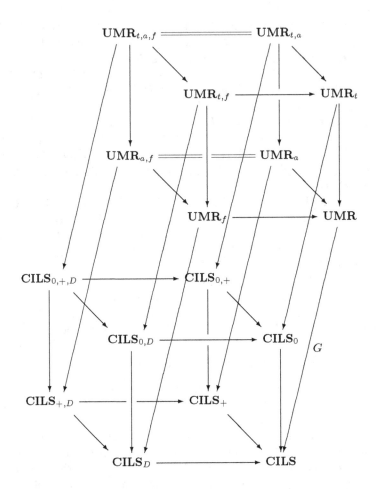

Fig. 5. The maps from the cube of **UMR** to the cube of **CILS**

Example 8. $T(A) = \{X \subseteq A \mid X$ is finite and non-empty$\}$ is a closed type. Since this type T is total and finite, T-**UMR**(A) is a probabilistic Kleene algebra.

Example 9. $T(A) = \{\{a\} \mid a \in A\}$ is a closed type. Since this type T is total, affine, and finite, T-**UMR**(A) is a Kleene algebra, Therefore, the set of ordinary binary relations over A forms a Kleene algebra.

Example 10. In paper [9], a multirelation $R \subseteq A \times \wp(A)$ is called *finitary up-closed* if it satisfies the following.

1. $\forall(a, X) \in R.\forall Y \in \wp(A).X \subseteq Y$ implies $(a, Y) \in R$.
2. $\forall(a, X) \in R.\exists Y \in \wp(A).$ such that Y is a finite subset of X and $(a, Y) \in R$.

We compare this notion with the notion of up-closed multirelations of type $T(A) = \{X \subseteq A \mid X$ is finite$\}$. A finitary up-closed multirelation is not always an up-closed multirelation of this type T over A, since the former may include (a, X) for an infinite X, but the latter must not. Conversely, an up-closed multirelation of type T over A is not always a finitary up-closed multirelation. However, the set of finitary up-closed multirelations forms a complete IL-semiring and it is isomorphic to T-**UMR**(A) as a complete IL-semiring [11]. The isomorphism maps a finitary up-closed multirelation R to $\{(a, X) \in R \mid X \text{ is finite}\}$ and an up-closed multirelation R of type T over A to $\{(a, X) \in A \times \wp(A) \mid \exists Y \subseteq X.(a, Y) \in R\}$. Therefore, the set of finitary up-closed multirelations also forms a lazy Kleene algebra satisfying the D-axiom.

Example 11. $T(A) = \{X \subseteq A \mid X \text{ is non-empty}\}$ is a closed type. Since this type T is total, T-**UMR**(A) is a lazy Kleene algebra satisfying the 0-axiom.

Example 12. $T(A) = \{\emptyset\} \cup \{\{a\} \mid a \in A\}$ is a closed type. Since this type T is affine, T-**UMR**(A) is a lazy Kleene algebra satisfying the +-axiom.

7   Conclusion

We studied the relationship between relaxations of Kleene algebras and classes of multirelations.

We extended the notion of multirelations by introducing types of multirelations. For each of eight classes of relaxation of Kleene algebra, we gave a sufficient condition on type T such that the set of up-closed multirelations of type T belongs to the class. In particular, the affineness condition and the totality condition of a type are not only sufficient, but also necessary.

This paper includes the result that the set of ordinary binary relations forms a Kleene algebra, the set of up-closed multirelations forms a lazy Kleene algebra, the set of up-closed finite multirelations forms a monodic tree Kleene algebra, and the set of total up-closed finite multirelations forms a probabilistic Kleene algebra. The cube consisting of eight conditions of type of multirelation is actually a triangular prism. It is strange but interesting.

The paper [2] extends the notion of lazy Kleene algebras to treat both finite and infinite streams, by adding the notion of meets and greatest fixed points. We are also going to extend our cube by adding conditions about meets and greatest fixed points.

We showed that if a complete IL-semiring preserves all right directed joins, then it forms a lazy Kleene algebra satisfying the D-axiom. However, the converse direction has not been proved yet. It is future work.

Acknowledgments

This work was supported by Grant-in-Aid for Young Scientists(B)(20700004). Some mistakes in the previous version of this paper were found by discussions with Shin-ya Katsumata. The authors thank to Ichiro Hasuo for valuable comments about submonads of the covariant powerset monad.

References

1. Kozen, D.: A Completeness Theorem for Kleene Algebras and the Algebra of Regular Events. Inf. Comput. 110(2), 366–390 (1994)
2. Möller, B.: Lazy Kleene Algebra. In: Kozen, D., Shankland, C. (eds.) MPC 2004. LNCS, vol. 3125, pp. 252–273. Springer, Heidelberg (2004)
3. Takai, T., Furusawa, H.: Monodic Tree Kleene Algebra. In: Schmidt, R.A. (ed.) RelMiCS/AKA 2006. LNCS, vol. 4136, pp. 402–416. Springer, Heidelberg (2006)
4. McIver, A., Weber, T.: Towards Automated Proof Support for Probabilistic Distributed Systems. In: Sutcliffe, G., Voronkov, A. (eds.) LPAR 2005. LNCS (LNAI), vol. 3835, pp. 534–548. Springer, Heidelberg (2005)
5. Parikh, R.: The Logic of Games. Annals of Discrete Mathematics 24, 111–140 (1985)
6. Pauly, M., Parikh, R.: Game Logic - An Overview. Studia Logica 75(2), 165–182 (2003)
7. Goranko, V.: The Basic Algebra of Game Equivalences. Studia Logica 75(2), 221–238 (2003)
8. Venema, Y.: Representation of Game Algebras. Studia Logica 75(2), 239–256 (2003)
9. Furusawa, H., Tsumagari, N., Nishizawa, K.: A Non-probabilistic Relational Model of Probabilistic Kleene Algebras. In: Berghammer, R., Möller, B., Struth, G. (eds.) RelMiCS/AKA 2008. LNCS, vol. 4988, pp. 110–122. Springer, Heidelberg (2008)
10. Tsumagari, N., Nishizawa, K., Furusawa, H.: Multirelational Model of Lazy Kleene Algebra. In: Berghammer, R., Möller, B., Struth, G. (eds.) Relations and Kleene Algebra in Computer Science, PhD Programme at RelMiCS10/AKA5 2008-04, Institut für Informatik, Universität Augsburg, Germany (April 2008)
11. Tsumagari, N., Nishizawa, K., Furusawa, H.: Reflexive Transitive Closure of Binary Multirelations. In: Proc. of 25th Conference of Japan Society for Software Science and Technology, Tokyo, Japan (September 2008) (in Japanese)
12. Furusawa, H., Nishizawa, K., Tsumagari, N.: Multirelational Models of Lazy, Monodic Tree, and Probabilistic Kleene Algebras. Bulletin of Informatics and Cybernetics (to appear)

Discrete Duality for Relation Algebras and Cylindric Algebras

Ewa Orłowska[1,*] and Ingrid Rewitzky[2,**]

[1] National Institute of Telecommunications, Warsaw
[2] Department of Mathematical Sciences, University of Stellenbosch, South Africa

Abstract. Following the representation theorems for relation algebras and cylindric algebras presented in [5] and [7] we develop discrete duality for relation algebras and relation frames, and for cylindric algebras and cylindric frames.

1 Introduction

Duality theory emerged from the work by Marshall Stone [29] on Boolean algebras and distributive lattices in the 1930s. Jónsson and Tarski [10] extended the Stone's results to Boolean algebras with operators which are also known as polymodal algebras with possibility operators. Later in the early 1970s Larisa Maksimova [16,17] and Hilary Priestley [27] developed the analogous results for Heyting algebras, topological Boolean algebras, and distributive lattices. Since then establishing a duality has become an important methodological problem both in algebra and in logic. All the above mentioned classical duality results are developed using topological spaces as dual spaces of algebras.

Discrete duality is a duality where a class of abstract relational systems is a dual counterpart to a class of algebras. These relational systems are referred to as frames following the terminology of non-classical logics. A topology is not involved in the construction of these frames and hence they may be thought of as having a discrete topology. Establishing discrete duality involves the following steps. Given a class Alg of algebras (resp. a class Frm of frames) we define a class Frm of frames (resp. a class Alg of algebras). Next, for an algebra $L \in Alg$ we define its canonical frame $\mathcal{X}(L)$ and for each frame $X \in Frm$ we define its complex algebra $\mathcal{C}(X)$. Then we prove that $\mathcal{X}(L) \in Frm$ and $\mathcal{C}(X) \in Alg$. A duality between Alg and Frm holds provided that the following facts are proved:

(D1) Every algebra $L \in Alg$ is embeddable into the complex algebra of its canonical frame, i.e., $\mathcal{C}(\mathcal{X}(L))$.

(D2) Every frame $X \in Frm$ is embeddable in the canonical frame of its complex algebra, i.e., $\mathcal{X}(\mathcal{C}(X))$.

---

* Partial support from the Polish Ministry of Science and Higher Education grant N206 399134 is gratefully acknowledged.
** Support from the Poland/SA Bilateral Collaboration Programme and the National Research Foundation of South Africa is gratefully acknowledged.

R. Berghammer et al. (Eds.): RelMiCS/AKA 2009, LNCS 5827, pp. 291–305, 2009.

Our terminology follows that used in modal logic and it is well suited to the study of discrete duality in case of signature extensions of not-necessarily Boolean algebras, where the complex algebras need not be the powerset algebras of all the subsets of the universes of the frames. As indicated in [19], the complex algebra of the canonical frame of a Boolean algebra with operators in the sense of [10] is the canonical extension (or a perfect extension or a canonical embedding algebra). Also, in that case canonical frames are sometimes referred to as ultrafilter extensions.

A distinguishing feature of this framework for establishing a discrete duality is that the algebraic and the logical notions involved in the proofs are defined in an autonomous way, we do not mix the algebraic and logical methodologies. The separation of logical and algebraic constructs enables us to view classes of algebras and frames as two types of semantic structures of a formal language. As a consequence we easily obtain what we call duality via truth [22]. Given a formal language Lan, a class of frames Frm which determines a relational semantics for Lan and a class Alg of algebras which determines its algebraic semantics, a duality via truth theorem says that these two kinds of semantics are equivalent in the following sense:

(DvT)   A formula $\phi \in Lan$ is true in every algebra of Alg iff it is true in every frame of Frm.

In this paper we develop discrete duality for relation algebras and cylindric algebras of finite dimension. The basis of our work are the developments in [8,10,11,12,13]. The notions of relation frame, cylindric frame, and complex algebras of such frames presented in the present paper are derived from the developments there. In the representation theorems for relation algebras and cylindric algebras presented in [8] the representation algebras are the complex algebras of the atom structures of the given algebras. In [13] a representation theorem for relation algebras and cylindric algebras is announced without a proof showing that the representation algebra is the complex algebra of the canonical frame of the given algebra, as it is usual in the discrete duality framework. In this paper we prove both the representation theorems for relation algebras and cylindric algebras and the representation theorems for relation frames and cylindric frames following the methodology of discrete duality. Holding of the theorem of the form (D1) is guaranteed by the Sahlqvist theorem [28] adapted to Boolean algebras with operators in the sense of [10] in [3].

From the foundational perspectives, discrete duality and duality via truth contribute to a formal explanation of the interaction between logics and algebras. On the application side, discrete duality results have a number of applications in computer science in the development of relational dual tableaux deduction systems. Given a theory presented as a class of algebras, having a discrete duality for that class we may present the theory as a logic with a Kripke-style relational semantics. Once a relational semantics is provided the methods of construction of dual tableaux can be applied. Relational dual tableaux have been constructed for a great variety of theories, ranging from well established non-classical logics such as intuitionistic, modal, relevant, and multiple-valued logics, to important applied theories such as, among others, temporal, in particular interval temporal logics, various logics of programs, fuzzy logics, logics of rough sets, theories of spatial reasoning including region connection calculus, theories of order of magnitude reasoning, and formal concept analysis. Relational dual tableaux are powerful tools which perform not only verification of validity (i.e., verification of truth of the statements in all the models of a theory) but often they can also be used for proving entailment (i.e., verification that truth of a finite number of statements implies truth of some other statement), model checking (i.e., verification of truth of a statement in a particular fixed model), and satisfaction (i.e., verification that a statement is satisfied by some fixed objects of a model). An exhaustive presentation of these applications can be found in [18].

2   Relation Algebra

Many properties of relations discussed already in [2] and [26] are captured in terms of an abstract algebraic structure defined as a Boolean algebra with operators as follows (see e.g., [30,15]).

Definition 1. A relation algebra $(L, \vee, \wedge, -, 0, 1, ;, \smile, 1')$ is such that, for any $a, b, c \in L$,

(RA1) $(L, \vee, \wedge, -, 0, 1)$ is a Boolean algebra
(RA2) $(L, ;, 1')$ is a monoid
(RA3) $(a \vee b); c = a; c \vee b; c$
(RA4) $a^{\smile\smile} = a$
(RA5) $(a \vee b)^{\smile} = a^{\smile} \vee b^{\smile}$
(RA6) $(a; b)^{\smile} = b^{\smile}; a^{\smile}$
(RA7) $a^{\smile}; (-(a; b)) \leq -b.$

The proofs for the following properties of relations can be found in [1].

Lemma 2. For any $a, b, c, d \in L$,

(a) $c; (a \vee b) = c; a \vee c; b.$
(b) $1'; a = a = a; 1'.$
(c) $(a \wedge b)^{\smile} = a^{\smile} \wedge b^{\smile}.$
(d) $a \leq b$ iff $a^{\smile} \leq b^{\smile}.$
(e) If $a \leq b$ and $c \leq d$ then $a; c \leq b; d.$
(f) (De Morgan Theorem K, see [2])
$$(a; b) \wedge c = 0 \quad \text{iff} \quad (a^{\smile}; c) \wedge b = 0 \quad \text{iff} \quad (c; b^{\smile}) \wedge a = 0.$$

For the representation results we need the following definition and variation of the Prime Filter Theorem for Boolean algebras, as proved by Urquhart [31].
For any subsets F and G of L, $F; G = \{c \in L \mid \exists a \in F, \exists b \in G, \ a; b \leq c\}.$

Lemma 3. *Let* $(L, \vee, \wedge, -, 0, 1, ;, \breve{}, 1')$ *be a relation algebra.*

(a) *If F and G are filters of L then so is $F; G$.*
(b) *Let F and G be filters and let P be a prime filter of L. If $F; G \subseteq P$ then there exist prime filters F' and G' such that $F \subseteq F'$ and $G \subseteq G'$ and $F'; G' \subseteq P$.*

Following [8,10,11,14] the notion of relation frame is defined as follows. However, as mentioned in the introduction, here the frames are abstract relational structures whose universes are not necessarily sets of atoms of algebras.

Definition 4. A relation frame (X, C, f, I) is a non-empty set X endowed with a ternary relation $C \subseteq X^3$, a unary operation $f : X \to X$, and a designated set $I \subseteq X$ such that

(RF1) $f(f(x)) = x$
(RF2) $C(x, y, z) \Rightarrow C(f(x), z, y)$
(RF3) $C(x, y, z) \Rightarrow C(z, f(y), x)$
(RF4) $x = y$ iff $\exists z \in I,\ C(x, z, y)$
(RF5) $C(x, y, z) \wedge C(z, v, w) \Rightarrow \exists u, C(x, u, w) \wedge C(y, v, u)$
(RF6) $C(x, y, z) \wedge C(v, z, w) \Rightarrow \exists u, C(u, y, w) \wedge C(v, x, u).$

Lemma 5. *For any $x, y, z \in X$,*

(a) $C(x, y, z)$ iff $C(f(x), z, y)$.
(b) $C(x, y, z)$ iff $C(z, f(y), x))$.

Proof. For (a) note that, by (RF1) and (RF2), for any $x, y, z \in X$,

$$C(x, y, z) \Rightarrow C(f(x), z, y) \Rightarrow C(f(f(x)), y, z) \Leftrightarrow C(x, y, z).$$

Similarly, (b) follows using (RF1) and (RF3).   □

Given a relation frame (X, C, f, I), its complex algebra is

$$(2^X, \cup, \cap, -, \emptyset, X, ;^c, \breve{}^c, 1'^c),$$

where $(2^X, \cup, \cap, -, \emptyset, X)$ is the powerset Boolean algebra of X and, for any $A, B \subseteq X$,

$$A ;^c B = \{z \in X \mid \exists x \in A, \exists y \in B,\ C(x, y, z)\}$$
$$A^{\breve{}c} = \{f(x) \mid x \in A\}$$
$$1'^c = I.$$

Theorem 6. *The complex algebra of a relation frame is a relation algebra.*

Proof. By definition the power set algebra of the non-empty set X is a Boolean algebra.

For (RA2), we use (RF5) $(A ;^c B_1) ;^c B_2 \subseteq A ;^c (B_1 ;^c B_2)$ as follows. Assume $z \in (A ;^c B_1) ;^c B_2$. Then, for some $w \in A ;^c B_1$ and some $v \in B_2$, $C(w, v, z)$. So, expanding further, for some $x \in A$, some $y \in B_1$ and some $v \in B_2$, $C(x, y, w)$

and $C(w, v, z)$. Thus, by (RF5), for some $u \in L$, $C(x, u, z)$ and $C(y, v, u)$. Hence, for some $x \in A$ and some $u \in L$, $C(x, u, z)$ and $u \in B_1;^c B_2$. Therefore, $z \in A;^c (B_1;^c B_2)$, as required. Similarly, using (RF6), $A;^c (B_1;^c B_2) \subseteq (A;^c B_1);^c B_2$. Using (RF4), we have

$$z \in A;^c 1'^c \; \Leftrightarrow \; \exists x \in A, \exists y \in I, \; C(x, y, z) \; \Leftrightarrow \; \exists x \in A, \; x = z \; \Leftrightarrow \; z \in A.$$

Similarly, using (RF1) to (RF4), we have

$$\begin{aligned} z \in 1'^c;^c A \; &\Leftrightarrow \; \exists x \in I, \exists y \in A, \; C(x, y, z) \\ &\Leftrightarrow \; \exists x \in I, \exists y \in A, \; C(z, f(y), x) \\ &\Leftrightarrow \; \exists x \in I, \exists y \in A, \; C(f(z), x, f(y)) \\ &\Leftrightarrow \; \exists y \in A, \; f(y) = f(z) \\ &\Leftrightarrow \; \exists y \in A, \; f(f(y)) = f(f(z)) \\ &\Leftrightarrow \; \exists y \in A, \; y = z \\ &\Leftrightarrow \; z \in A. \end{aligned}$$

For (RA3), by definition of union, it follows that $;^c$ distributes over $\cup$.
   For (RA4), using (RF1) we have

$$\begin{aligned} z \in (A^{\smile c})^{\smile c} \; &\Leftrightarrow \; \exists y \in A^{\smile c}, \; z = f(y) \\ &\Leftrightarrow \; \exists x \in A, \; z = f(f(x)) \\ &\Leftrightarrow \; \exists x \in A, \; z = x \\ &\Leftrightarrow \; z \in A. \end{aligned}$$

(RA5) holds since, for any $z \in X$,

$$\exists x \in A \cup B, \; z = f(x) \quad \Leftrightarrow \quad \exists x \in A, \; z = f(x) \text{ or } \exists x \in B, \; z = f(x).$$

For (RA6), we have that for any $z \in X$,

$$\begin{aligned} z \in (A;^c B)^{\smile c} \quad &\Leftrightarrow \quad \exists u \in A;^c B, \; z = f(u) \\ &\Leftrightarrow \quad \exists u, \exists x \in A, \exists y \in B, \; C(x, y, u) \wedge z = f(u) \\ &\Leftrightarrow \quad \exists x \in A, \exists y \in B, \; C(x, y, f(z)) \end{aligned}$$

since $z = f(u)$ iff $u = f(z)$. Also,

$$\begin{aligned} z \in B^{\smile c};^c A^{\smile c} \quad &\Leftrightarrow \quad \exists w \in A^{\smile c}, \exists v \in B^{\smile c}, \; C(v, w, z) \\ &\Leftrightarrow \quad \exists x \in A, \exists y \in B, \; C(f(y), f(x), z) \end{aligned}$$

Now, by Lemma 5(a) and (b) and (RF1),

$$C(x, y, f(z)) \; \Leftrightarrow \; C(f(x), f(z), y) \; \Leftrightarrow \; C(y, z, f(x)) \; \Leftrightarrow \; C(f(y), f(x), z).$$

Hence, the result follows.

For (RA7), assume $z \in (A^\smile{}^c) ;^c (-(A;^c B))$. Then, there is some $x \in A$ and there is some $v \in X$ such that

$$(\forall u, \forall y, \ u \in A \wedge C(u, y, v) \Rightarrow y \notin B) \wedge C(f(x), v, z).$$

Suppose $z \in B$. Take u to be x and y to be z. Then, since $x \in A$, $C(x, z, v)$ does not hold. Thus, by Lemma 5(a), not $C(f(x), v, z)$ which gives the required contradiction.   □

Given a relation algebra $(L, \vee, \wedge, -, 0, 1, ; , \smile, 1')$, its canonical frame

$$(\mathcal{X}(L), C^c, f^c, I^c)$$

is the set $\mathcal{X}(L)$ of all prime filters of the Boolean algebra $(L, \vee, \wedge, -, 0, 1)$ endowed with a ternary relation C^c, a unary operation f^c and a designated set I^c where, for any $F, G, H \in \mathcal{X}(L)$,

$$C^c(F, G, H) \text{ iff } F; G \subseteq H \quad \text{where } F; G = \{c \in L \mid \exists a \in F, \exists b \in G, \ a; b \le c\}$$
$$f^c(F) = \{a^\smile \mid a \in F\}$$
$$I^c = \{F \mid 1' \in F\}.$$

Note that, since $(L, \vee, \wedge, -, 0, 1)$ is a Boolean algebra, f^c is well-defined, that is, $f^c(F) \in \mathcal{X}(L)$ for $F \in \mathcal{X}(L)$.

Theorem 7. *The canonical frame of a relation algebra is a relation frame.*

Proof. For (RF1), by definition and using (RA4),

$$f^c(f^c(F)) = \{a^\smile \mid a \in f^c(F)\} = \{a^\smile \mid \exists b \in F, a = b^\smile\} = \{b^{\smile\smile} \mid b \in F\},$$

hence $f^c(f^c(F)) = F$.

For (RF2), assume $C^c(F, G, H)$, that is, $F; G \subseteq H$. Take any $z \in L$ such that $z \in f^c(F); H$ and $z \notin G$. Then, for some $a \in F$ and some $b \in H$, $a^\smile; b \le z$ and $z \notin G$, and hence $a^\smile; b \notin G$. Now $a \in F$ and $-(a^\smile; b) \in G$ imply $a; (-(a^\smile; b)) \in F; G \subseteq H$. Thus, by (RA7), $-b \in H$, which gives the required contradiction.

For (RF3), assume $C^c(F, G, H)$, that is, $F; G \subseteq H$. Take any $z \in L$ such that $z \in H; f^c(G)$ and $z \notin F$. Then, for some $a \in G$ and some $b \in H$, $b; a^\smile \le z$ and $z \notin F$, and hence $b; a^\smile \notin F$. Now $-(b; a^\smile); a \in F; G \subseteq H$. Now $b; a^\smile = (b; a^\smile)^{\smile\smile} = (a; b^\smile)^\smile$. By (RA7), $a^\smile; (-(a; b^\smile)) \le -b^\smile$ and, by Lemma 2(d), $(a^\smile; (-(a; b^\smile)))^\smile \le (-b^\smile)^\smile$, that is, $-(a; b^\smile)^\smile; a \le -b$. Since $-(a; b^\smile)^\smile; a \in H$, $-b \in H$ which gives the contradiction.

For (RF4), assume $F = G$. Let $\uparrow 1'$ be the principal filter generated by $1'$. Then $F; \uparrow 1' \subseteq G$. Hence, by Lemma 3(b), there is some $H \in \mathcal{X}(L)$ such that $1' \in \uparrow 1' \subseteq H$ and $F; H \subseteq G$. On the other hand, assume that for some $H \in \mathcal{X}(L)$, $1' \in H$ and $F; H \subseteq G$. Take any $a \in F$. Then $a = a; 1' \in F; H \subseteq G$. So $F \subseteq G$. Now take any $a \notin F$. Then, since $(L, \vee, \wedge, -, 0, 1)$ is a Boolean algebra, $-a \in F$ and hence $-a = -a; 1' \in F; H \subseteq G$. Thus $a \notin G$ and hence $G \subseteq F$.

For (RF5), assume $C^c(F, G, H)$ and $C^c(H, K, M)$, that is, $F; G \subseteq H$ and $H; K \subseteq M$. Then $F; (G; K) \subseteq M$ where, by Lemma 3(a), $G; K$ is a filter.

Therefore, by Lemma 3(b), there is a prime filter $U \in \mathcal{X}(W)$ such that $G; K \subseteq U$ and $F; U \subseteq M$. Hence $C^c(G, K, U)$ and $C^c(F, U, M)$, as required.

The proof of (RF6) is similar to that for (RF5).   □

We establish the representation theorem for relation frames using the mapping $k : X \to \mathcal{X}(2^X)$, defined, for any $x \in X$, by

$$k(x) = \{A \in 2^X \mid x \in A\}.$$

This mapping is an embedding [10] of X into $\mathcal{X}(2^X)$. All that remains is to show that k preserves the relation C, function f and special set I on X.

Theorem 8. *For any* $x, y, z \in X$,

(a) $C(x, y, z)$ *iff* $C^c(k(x), k(y), k(z))$
(b) $k(f(x)) = f^c(k(x))$
(c) $k(I) = I^c$

Proof

(a) For any $x, y, z \in X$,

$$C^c(k(x), k(y), k(z))$$
$$\Leftrightarrow \quad k(x);^c k(y) \subseteq k(z)$$
$$\Leftrightarrow \quad \forall A \in 2^X, \ (\exists B_1 \in k(x), \exists B_2 \in k(y), \ B_1;^c B_2 \subseteq A) \Rightarrow A \in k(z)$$
$$\Leftrightarrow \quad \forall A \in 2^X, (\exists B_1, B_2 \in 2^X, \ x \in B_1 \ \wedge \ y \in B_2 \ \wedge$$
$$\{t \in X \mid \exists p \in B_1, \exists q \in B_2, \ C(p, q, t)\} \subseteq A)$$
$$\Rightarrow \quad z \in A.$$

Take any $x, y, z \in X$ such that $C(x, y, z)$, and take any $A \in 2^X$ such that for some $B_1, B_2 \in 2^X$, $x \in B_1$, $y \in B_2$, $B_1;^c B_2 \subseteq A$. Now, since $x \in B_1$, $y \in B_2$ and $C(x, y, z)$, $z \in B_1;^c B_2$ and hence $z \in A$, as required.

On the other hand, take any $x, y, z \in X$ such that $C^c(k(x), k(y), k(z))$. Consider $B_1 = \{x\}$ and $B_2 = \{y\}$. Then $B_1;^c B_2 = \{t \in X \mid C(x, y, t)\}$. Take $A = \{t \in X \mid C(x, y, t)\}$. Then, since $C^c(k(x), k(y), k(z))$, $z \in A$. Hence $C(x, y, z)$.

(b) For any $B \subseteq X$,

$$B \in f^c(k(x)) \quad \Leftrightarrow \quad \exists A, A \in k(x) \ \wedge \ B = A^{\smile c}$$
$$\Leftrightarrow \quad \exists A, x \in A \ \wedge \ B = \{f(x) \mid x \in A\}$$
$$\Leftrightarrow \quad f(x) \in B$$
$$\Leftrightarrow \quad B \in k(f(x)).$$

(c) For any $\mathsf{F} \in \mathcal{X}(2^X)$,

$$\mathsf{F} \in I^c \quad \Leftrightarrow \quad I \in \mathsf{F} \ \wedge \ \exists x \in X, \ \mathsf{F} = k(x)$$
$$\Leftrightarrow \quad \exists x \in X, \ \mathsf{F} = k(x) \ \wedge \ x \in I$$
$$\Leftrightarrow \quad \exists x \in I, \ \mathsf{F} = k(x)$$
$$\Leftrightarrow \quad \mathsf{F} \in k(I).$$ □

We establish the representation theorem for relation algebras using the Stone mapping $h : L \to 2^{(\mathcal{X}(L))}$, defined, for any $a \in L$, by

$$h(a) = \{F \in \mathcal{X}(L) \mid a \in F\}.$$

This is an embedding [10] between the Boolean algebras underlying the relation algebras. To show that h is an embedding of the relation algebras, it suffices to show that h preserves $;,\smile$ and $1'$, that is,

Theorem 9. *For any $a, b \in L$,*

(a) $h(a; b) = h(a);^c h(b)$
(b) $h(a^\smile) = h(a)^{\smile c}$
(c) $h(1') = 1'^c$.

Proof

(a) We need to show that, for any $H \in \mathcal{X}(L)$ and any $a, b \in L$,

$$a; b \in H \quad \text{iff} \quad \exists F, G \in \mathcal{X}(L), a \in F, b \in G \text{ and } F; G \subseteq H.$$

Assume for some $F, G \in \mathcal{X}(L), a \in F, b \in G$ and $F; G \subseteq H$. Take $c = a; b$. Then $c \in F; G$ and hence $c = a; b \in H$. On the other hand, assume $a; b \in H$. Then $\uparrow a; \uparrow b \subseteq H$ where $\uparrow a$ and $\uparrow b$ are principal filters generated by a and b respectively. So, by Lemma 3, there is a prime filter F such that $a \in \uparrow a \subseteq F$ and $F; \uparrow b \subseteq H$. Hence, by another application of Lemma 3, there is a prime filter G such that $b \in \uparrow b \subseteq G$ and $F; G \subseteq H$.

(b) For any $H \in \mathcal{X}(L)$ and any $a \in L$,

$$H \in h(a^\smile) \Leftrightarrow a^\smile \in H \Leftrightarrow a = a^{\smile\smile} \in f^c(H) \Leftrightarrow H = f^c(f^c(H)) \in h(a)^{\smile c}.$$

(c) For any $H \in \mathcal{X}(L)$, $H \in h(1') \Leftrightarrow 1' \in H \Leftrightarrow H \in I^c \Leftrightarrow H \in 1'^c$. □

3 Cylindric Algebras of Finite Dimension

The notion of cylindric algebra, invented by Alfred Tarski and presented in [8], plays a role comparable for first-order logic as Boolean algebras play for propositional logic. In this section we consider cylindric algebras of dimension n where $3 \le n < \omega$. They enable us to study relations of arbitrary finite rank $n > 2$ in an analogous way as relation algebras for binary relations. These are Boolean algebras with additional (cylindrification) operators that model quantification.

Definition 10. A cylindric algebra of dimension n where $3 \le n < \omega$,

$$(L, \vee, \wedge, -, 0, 1, \{d_{ij} \in L \mid i, j \le n\}, \{c_i : L \to L \mid i \le n\})$$

is such that, for any $a, b \in L$,

(CA1) $(L, \vee, \wedge, -, 0, 1)$ is a Boolean algebra.
(CA2) $c_i(0) = 0$
(CA3) $a \leq c_i(a)$
(CA4) $c_i(a \wedge c_i(b)) = c_i(a) \wedge c_i(b)$
(CA5) $c_i(c_j(a)) = c_j(c_i(a))$
(CA6) $d_{ii} = 1$
(CA7) $k \neq i, j$ implies $d_{ij} = c_k(d_{ik} \wedge d_{kj})$
(CA8) $i \neq j$ implies $c_i(d_{ij} \wedge a) \wedge c_i(d_{ij} \wedge -a) = 0$.

The proofs of the following properties of the cylindrification can be found in [8].

Lemma 11. *For any $a, b \in L$ and for every $i \leq n$,*

(a) $c_i(c_i(a)) = c_i(a)$
(b) $a \leq b$ *implies* $c_i(a) \leq c_i(b)$
(c) $c_i(a \vee b) = c_i(a) \vee c_i(b)$
(d) $c_i(1) = 1$
(e) $c_i(a) \wedge b = 0$ *iff* $a \wedge c_i(b) = 0$
(f) $c_i(-c_i(a)) = -c_i(a)$.

Following [8] and [11] the notion of cylindric frame is defined as follows.

Definition 12. A cylindric frame is

$$(X, \{D_{ij} \subseteq X \mid i, j \leq n\}, \{E_i \subseteq X^2 \mid i \leq n\})$$

where X is a non-empty set and

(CF1) E_i $(i \leq n)$ is an equivalence relation on X
(CF2) $E_i; E_j = E_j; E_i$
(CF3) $D_{ii} = X$
(CF4) $D_{ij} = \{y \in X \mid \forall k \neq i, j, \exists x \in D_{ik} \cap D_{kj}, \, xE_k y\}$
(CF5) $i \neq j, x, y \in D_{ij}$ and $xE_i y$ imply $x = y$.

Given a cylindric frame $(X, \{D_{ij} \subseteq X \mid i, j \leq n\}, \{E_i \subseteq X^2 \mid i \leq n\})$, its complex algebra is

$$(2^X, \cup, \cap, -, \emptyset, X, \{d_{ij}^c \in 2^X \mid i, j \leq n\}, \{c_i^c : 2^X \rightarrow 2^X \mid i \leq n\})$$

where $(2^X, \cup, \cap, -, \emptyset, X)$ is the powerset Boolean algebra of X, $d_{ij}^c = D_{ij}$ and, for any $A \subseteq X$,

$$c_i^c(A) = \{y \in X \mid \exists x \in A, xE_i y\}.$$

Theorem 13. *The complex algebra of a cylindric frame is a cylindric algebra.*

Proof. By definition the power set algebra of the non-empty set X is a Boolean algebra.

For (CA2), $c_i^c(\emptyset) = \{y \in X \mid \exists x, \, x \in \emptyset \wedge xE_i y\} = \emptyset$.

For (CA3), note that for any $y \in A$, $yE_i y$ since E_i is reflexive. Hence $y \in c_i^c(A)$.

For (CA4), by transitivity of E_i ($i \leq n$), for any $A, B \subseteq X$ and any $y \in X$,

$$
\begin{aligned}
y \in c_i^c(A \cap c_i^c(B)) \quad &\Leftrightarrow \quad \exists x \in A \cap c_i^c(B), \; xE_iy \\
&\Leftrightarrow \quad \exists x, \; x \in A \wedge \exists z, \; z \in B \wedge zE_ix \wedge xE_iy \\
&\Rightarrow \quad (\exists x, \; x \in A \wedge xE_iy) \wedge (\exists z, \; z \in B \wedge zE_iy) \\
&\Leftrightarrow \quad y \in c_i^c(A) \text{ and } y \in c_i^c(B).
\end{aligned}
$$

On the other hand, assume that for some $x \in A$ xE_iy and for some $z \in B$ zE_iy. Since E_i is symmetric and transitive, xE_iz. Suppose $y \notin c_i^c(A \cap c_i^c(B))$. Then, for all $t \in X$, $t \in A \cap c_i^c(B)$ implies tE_iy does not hold, that is,

$$
\forall t, \; t \in A \wedge \exists p, \; p \in B \wedge pE_it \Rightarrow \text{ not } tE_iy.
$$

Taking t to be x and p to be z we get that xE_iz does not hold, which gives the required contradiction.

For (CA5), using (CF2) we have that for any $A \subseteq X$ and any $y \in X$,

$$
\begin{aligned}
y \in c_i^c(c_j^c(A)) \quad &\Leftrightarrow \quad \exists x, \; x \in c_j^c(A) \wedge xE_iy \\
&\Leftrightarrow \quad \exists x, z, \; z \in A \wedge zE_jx \wedge xE_iy \\
&\Leftrightarrow \quad \exists z, \; z \in A \wedge zE_j; E_iy \\
&\Leftrightarrow \quad \exists z, \; z \in A \wedge zE_i; E_jy \\
&\Leftrightarrow \quad \exists z, x, \; z \in A \wedge zE_ix \wedge xE_jy \\
&\Leftrightarrow \quad \exists x, \; x \in c_i^c(A) \wedge xE_jy \\
&\Leftrightarrow \quad y \in c_j^c(c_i^c(A)).
\end{aligned}
$$

For (CA6), by (CF3), $d_{ii}^c = D_{ii} = X$.

For (CA7), assume $k \neq i, j$. Then, by (CF4), for any $y \in X$,

$$
y \in d_{ij}^c = D_{ij} \Leftrightarrow \exists x \in D_{ik} \cap D_{kj}, \; xE_ky \Leftrightarrow y \in c_k^c(D_{ik} \cap D_{kj}).
$$

For (CA8), assume $i \neq j$. Take any $y \in c_i^c(d_{ij}^c \cap A)$. Then, for some $x \in d_{ij}^c \cap A$, xE_iy. Suppose $y \in c_i^c(d_{ij}^c \cap -A)$. Then, for some $z \in d_{ij}^c \cap -A$, zE_iy. Thus $x, z \in d_{ij}^c$ and by symmetry and transitivity of E_i, xE_iz. So, by (CF5), $z = x$. Then $z \in A$ and $x \notin A$, which give the required contradiction. $\square$

Given a cylindric algebra $(L, \vee, \wedge, -, 0, 1, \{d_{ij} \in L \mid i, j \leq n\}, \{c_i : L \to L \mid i \leq n\})$, its canonical frame is

$$
(\mathcal{X}(L), \{D_{ij}^c \subseteq \mathcal{X}(L) \mid i, j \leq n\}, \{E_i^c \subseteq \mathcal{X}(L)^2 \mid i \leq n\}),
$$

where $\mathcal{X}(L)$ is the set of all prime filters of the Boolean algebra $(L, \vee, \wedge, -, 0, 1)$, $D_{ij}^c = \{F \in \mathcal{X}(L) \mid d_{ij} \in F\}$ and, for any $F, G \in \mathcal{X}(L)$,

$$
FE_i^cG \quad \text{iff} \quad c_i(F) \subseteq G, \quad \text{where } c_i(F) = \{c_i(a) \mid a \in F\}.
$$

Theorem 14. *The canonical frame of a cylindric algebra is a cylindric frame.*

Proof. For (CF1), for any $F \in \mathcal{X}(L)$ and $b \in L$,

$$b \in c_i(F) \Leftrightarrow \exists a, a \in F \wedge b = c_i(a) \Rightarrow \exists a, c_i(a) \in F \wedge b = c_i(a) \Rightarrow b \in F.$$

Assume E_i^c is not symmetric. Then, for some $F, G \in \mathcal{X}(L)$, FE_i^cG and not GE_i^cF, that is, $c_i(F) \subseteq G$ and $c_i(G) \cap -F \neq \emptyset$. Then, for some $a \in X$, $a \in -F$ and $a \in c_i(G)$. Hence, $-a \in F$ and, for some $b \in G$, $a = c_i(b)$. Now, by Lemma 11(f), $-c_i(b) = c_i(-c_i(b)) = c_i(-a)$. Also $c_i(-a) \in c_i(F)$. Thus $-c_i(b) \in c_i(F) \subseteq G$, and hence $c_i(b) \notin G$. Also, since $b \in G$ we have by (CA3) that $c_i(b) \in G$ which gives the required contradiction. Therefore, E_i^c is symmetric.

For transitivity, assume that $c_i(F) \subseteq G$ and $c_i(G) \subseteq H$. Take any $b \in L$ such that $b \in F$. Then $c_i(b) \in G$ and hence $c_i(c_i(b)) \in H$. Now, by Lemma 11(a), $c_i(b) \in H$, as required.

For (CF2), by (CA5) we have that, for any $F, G \in \mathcal{X}(L)$,

$$
\begin{aligned}
F(E_i; E_j)G &\Leftrightarrow \exists H, \; FE_iH \wedge HE_jG \\
&\Leftrightarrow \exists H, \; c_i(F) \subseteq H \wedge c_j(H) \subseteq G \\
&\Leftrightarrow c_j(c_i(F)) \subseteq G \\
&\Leftrightarrow c_i(c_j(F)) \subseteq G \\
&\Leftrightarrow \exists H, \; c_j(F) \subseteq H \wedge c_i(H) \subseteq G \\
&\Leftrightarrow \exists H, \; FE_jH \wedge HE_iG \\
&\Leftrightarrow F(E_j; E_i)G.
\end{aligned}
$$

For (CF3), using (CA6) it follows that $D_{ii}^c = \{F \in \mathcal{X}(L) \mid 1 = d_{ii} \in F\} = \mathcal{X}(L)$.

For (CF4), we need to show, for any $G \in \mathcal{X}(L)$, that

$$G \in D_{ij}^c \quad \text{iff} \quad \forall k \neq i, j, \; \exists F \in \mathcal{X}(L), \; d_{ik} \wedge d_{kj} \in F \wedge FE_k^cG.$$

Assume $G \in D_{ij}^c$ and $k \neq i, j$. Then, by definition of D_{ij}^c and (CA7), $c_k(d_{ik} \wedge d_{kj}) \in G$. Then $d_{ik} \wedge d_{kj} \in c_k^{-1}(G)$ and, since $c_k^{-1}(G)$ is an upset, $\uparrow (d_{ik} \wedge d_{kj}) \subseteq c_k^{-1}(G)$. Also $-c_k^{-1}(G)$ is an ideal disjoint from the principal filter $\uparrow (d_{ik} \wedge d_{kj})$. Therefore, by the Prime Filter Theorem, there is a prime filter $F \in \mathcal{X}(L)$ such that $\uparrow (d_{ik} \wedge d_{kj}) \subseteq F$ and $F \subseteq c_k^{-1}(G)$. Hence, $d_{ik} \wedge d_{kj} \in F$ and $c_k(F) \subseteq c_k(c_k^{-1}(G)) \subseteq G$. On the other hand, take any $G \in \mathcal{X}(L)$ such that, for $k \neq i, j$, there is some $F \in \mathcal{X}(L)$ such that $d_{ik} \wedge d_{kj} \in F$ and $c_k(F) \subseteq G$. Hence, $c_k(d_{ik} \wedge d_{kj}) \in c_k(F)$ and $c_k(F) \subseteq G$. Thus, by (CA7), $d_{ij} \in G$, that is, $G \in D_{ij}^c$.

For (CF5), assume $i \neq j$, $F, G \in D_{ij}^c$ and $c_i(F) \subseteq G$. Take any $a \in G$. Then $a \wedge d_{ij} \in G$ so, by (CA3), $c_i(a \wedge d_{ij}) \in G$. Hence, by (CA8), $-c_i(-a \wedge d_{ij}) \in G$ and so $c_i(-a \wedge d_{ij}) \notin G$. Thus $-a \wedge d_{ij} \notin F$, that is, $-a \notin F$ (since $d_{ij} \in F$). Therefore, $a \in F$. Thus $G \subseteq F$. On the other hand, take any $a \in F$. Then $c_i(a \wedge d_{ij}) \in c_i(F) \subseteq G$, so $c_i(a \wedge d_{ij}) \in G$. Hence, by (CA8), $c_i(-a \wedge d_{ij}) \notin G$ and hence, by (CA3), $-a \wedge d_{ij} \notin G$, that is, $-a \notin G$. Therefore, $a \in G$. Thus $F \subseteq G$. $\square$

We establish the representation theorem for cylindric frames using the mapping $k : X \to \mathcal{X}(2^X)$, defined, for any $x \in X$, by

$$k(x) = \{A \in 2^X \mid x \in A\}.$$

It is an embedding of X into $\mathcal{X}(2^X)$. All that remains is to show that:

Theorem 15
(a) $k(D_{ij}) = D_{ij}^c$
(b) *For any $x, y \in X$, $xE_i y$ iff $k(x)E_i^c k(y)$.*

Proof

(a) Take any $\mathsf{F} \in \mathcal{X}(2^X)$. Then

$$
\begin{aligned}
\mathsf{F} \in D_{ij}^c &\Leftrightarrow \exists x \in X,\ \mathsf{F} = k(x) \wedge d_{ij}^c \in k(x) \\
&\Leftrightarrow \exists x \in X,\ \mathsf{F} = k(x) \wedge x \in d_{ij}^c = D_{ij} \\
&\Leftrightarrow \exists x \in X,\ \mathsf{F} = k(x) \wedge k(x) \in k(D_{ij}) \\
&\Leftrightarrow \mathsf{F} \in k(D_{ij}).
\end{aligned}
$$

(b) Note that for any $x, y \in X$,

$$
\begin{aligned}
k(x)E_i^c k(y) &\Leftrightarrow c_i^c(k(x)) \subseteq k(y) \\
&\Leftrightarrow \{c_i^c(A) \mid x \in A\} \subseteq k(y) \\
&\Leftrightarrow \forall A \subseteq X,\ x \in A \Rightarrow c_i^c(A) \in k(y) \\
&\Leftrightarrow \forall A \subseteq X,\ x \in A \Rightarrow y \in c_i^c(A) \\
&\Leftrightarrow \forall A \subseteq X,\ x \in A \Rightarrow \exists u \in A, uE_i y.
\end{aligned}
$$

Assume $xE_i y$. Take any $A \subseteq X$ such that $x \in A$. Then, for some $u \in A$, $uE_i y$. In particular, take u to be x, $uE_i y$. On the other hand, assume $xE_i y$ does not hold. Let $A = \{x\} \subseteq X$. Then $x \in A$ and, for every $u \in A$, $uE_i y$ does not hold. Hence $k(x)E_i^c k(y)$ does not hold.   □

We establish the representation theorem for cylindric algebras using the mapping $h : L \to 2^{\mathcal{X}(L)}$, defined, for any $a \in L$, by

$$
h(a) = \{F \in \mathcal{X}(L) \mid a \in F\}.
$$

It is an embedding between the underlying Boolean algebras of the relation algebras. It remains to show that:

Theorem 16
(a) $h(d_{ij}) = d_{ij}^c$
(b) *For any $a \in L$, $h(c_i(a)) = c_i^c(h(a))$.*

Proof

(a) By definition, for any $F \in \mathcal{X}(L)$,

$$
F \in h(d_{ij}) \quad \Leftrightarrow \quad d_{ij} \in F \quad \Leftrightarrow \quad F \in d_{ij}^c.
$$

(b) We need to show, for any $G \in \mathcal{X}(L)$ and any $a \in L$,

$$c_i(a) \in G \quad \Leftrightarrow \quad \exists F \in h(a),\ F E_i^c G \quad \Leftrightarrow \quad \exists F \in \mathcal{X}(L),\ a \in F \wedge c_i(F) \subseteq G.$$

The right-to-left direction is trivial. For the left-to-right direction, assume that $c_i(a) \in G$. We need to show that there is some $F \in \mathcal{X}(L)$ such that $a \in F$ and $c_i(F) \subseteq G$. Then $a \in c_i^{-1}(G)$. Now $c_i^{-1}(G)$ is an upset containing the principle filter $\uparrow a$ and $-(c_i^{-1}(G))$ is an ideal. By the Prime Filter Theorem, there is a prime filter $F \in \mathcal{X}(L)$ such that $\uparrow a \subseteq F$ and $F \subseteq c_i^{-1}(G)$. Hence, for some $F \in \mathcal{X}(L)$, $a \in F$ and $c_i(F) \subseteq c_i(c_i^{-1}(G)) \subseteq G$. $\qquad \square$

4 Conclusion

In this paper we established discrete dualities between relation algebras and relation frames and between cylindric algebras and cylindric frames. This work is part of a broader project aimed at providing systematically discrete dualities for lattice structures with operators. The results of this paper are meaningful not only from a theoretical perspective, but also for the various theories in computer science based on relation algebras or cylindric algebras, in particular for automated theorem proving in such theories, as mentioned in Section 1.

Discrete dualites for Boolean algebras with operators in the sense of [10] and also for Boolean algebras with some other kinds of operators can be found in [19,23,24]. Discrete dualites for distributive lattices with operators are studied in [25] and [4,24] Discrete representation theorems of the form (D1) for not necessarily distributive lattices can be found in [5,6,7,20,21,32]. Some correspondence theory results studied from the perspective of discrete duality can be found in [9]. Our next goal is to develop discrete duality for Kleene algebras and for rough relation algebras, that is, the relation algebras based on double regular Stone algebras.

Acknowledgement. The authors gratefully acknowledge the helpful comments of the anonymous referees.

References

1. Chin, L., Tarski, A.: Distributive and modular laws in the arithmetic of relation algebras. University of California Publications (1951)
2. De Morgan, A.: On the syllogism: IV, and on the logic of relations. Transactions of the Cambridge Philosophical Society 10, 331–358 (1864)
3. De Rijke, M., Venema, Y.: Salqvists theorem for Boolean algebras with operators with applications to cylindric algebras. Studia Logica 54, 61–78 (1995)
4. Düntsch, I., Orłowska, E.: A discrete duality between the apartness algebras and apartness frames. Journal of Applied Non-classical Logics 18(2-3), 209–223 (2008)
5. Düntsch, I., Orłowska, E., Radzikowska, A.: Lattice-based relation algebras II. In: de Swart, H., Orłowska, E., Schmidt, G., Roubens, M. (eds.) TARSKI 2006. LNCS (LNAI), vol. 4342, pp. 267–289. Springer, Heidelberg (2006)

6. Dzik, W., Orłowska, E., van Alten, C.: Relational representation theorems for general lattices with negations. In: Schmidt, R.A. (ed.) RelMiCS/AKA 2006. LNCS, vol. 4136, pp. 162–176. Springer, Heidelberg (2006)

7. Düntsch, I., Orłowska, E., Radzikowska, A., Vakarelov, D.: Relational representation theorems for some lattice-based structures. Journal of Relational Methods in Computer Science 1, 132–160 (2005)

8. Henkin, L., Monk, J.D., Tarski, A.: Cylindric Algebras. Part I, Part II. North Holland, Amsterdam (1971/1985)

9. Järvinen, J., Orłowska, E.: Relational correspondences for lattices with operators. In: MacCaull, W., Winter, M., Düntsch, I. (eds.) RelMiCS 2005. LNCS, vol. 3929, pp. 134–146. Springer, Heidelberg (2006)

10. Jónsson, B., Tarski, A.: Boolean algebras with operators. Part I: American Journal of Mathematics 73, 891–939 (1951), Part II: ibidem 74, 127–162 (1952)

11. Maddux, R.: Some varieties containing relation algebras. Transactions of the American Mathematical Society 272, 501–526 (1982)

12. Maddux, R.: Finite Integral Relation Algebras. Lecture Notes in Mathematics, vol. 1149, pp. 175–197 (1985)

13. Maddux, R.: Introductory course on relation algebras, finite-dimensional cylindric algebras, and their interconnections. In: Andreka, H., Monk, J.D., Nemeti, I. (eds.) Algebraic Logic (Proc. Conf. Budapest 1988). Colloq. Math. Soc. J. Bolyai, vol. 54, pp. 361–392. North-Holland, Amsterdam (1991)

14. Maddux, R.: Relation algebras. In: Abramsky, S., Artemov, S., Gabbay, D.M., et al. (eds.). Studies in Logic and the Foundations of Mathematics, vol. 150. Elsevier, Amsterdam (1996)

15. Maddux, R.: Relation algebras. In: Brink, C., Kahland, W., Schmidt, G. (eds.) Relational Methods in Computer Sciences. Advances in Computer Science. Springer, New York (1997)

16. Maksimova, L.L.: Pretabular superintuitionistic logics. Algebra and Logic 11(5), 558–570 (1972)

17. Maksimova, L.L.: Pretabular extensions of the Lewis' logic S4. Algebra and Logic 14(1), 28–55 (1975)

18. Orłowska, E., Golinska-Pilarek, J.: Dual Tableaux: Foundations, Methodolody, Case Studies, Draft of the book (2009)

19. Orłowska, E., Rewitzky, I., Düntsch, I.: Relational semantics through duality. In: MacCaull, W., Winter, M., Düntsch, I. (eds.) RelMiCS 2005. LNCS, vol. 3929, pp. 17–32. Springer, Heidelberg (2006)

20. Orłowska, E., Radzikowska, A.: Relational representability for algebras of substructural logics. In: MacCaull, W., Winter, M., Düntsch, I. (eds.) RelMiCS 2005. LNCS, vol. 3929, pp. 212–224. Springer, Heidelberg (2006)

21. Orłowska, E., Radzikowska, A.: Representation theorems for some fuzzy logics based on residuated non-distributive lattices. Fuzzy Sets and Systems 159, 1247–1259 (2008)

22. Orłowska, E., Rewitzky, I.: Duality via Truth: Semantic frameworks for lattice-based logics. Logic Journal of the IGPL 13(4), 467–490 (2005)

23. Orłowska, E., Rewitzky, I.: Context algebras, context frames and their discrete duality. In: Peters, J.F., Skowron, A., Rybiński, H. (eds.) Transactions on Rough Sets IX. LNCS, vol. 5390, pp. 212–229. Springer, Heidelberg (2008)

24. Orłowska, E., Rewitzky, I.: Discrete duality and its applications to reasoning with incomplete information. In: Kryszkiewicz, M., Peters, J.F., Rybiński, H., Skowron, A. (eds.) RSEISP 2007. LNCS (LNAI), vol. 4585, pp. 51–56. Springer, Heidelberg (2007)

25. Orłowska, E., Rewitzky, I.: Algebras for Galois-style connections and their discrete duality (submitted, 2008)

26. Peirce, C.S.: Note B: the logic of relatives. In: Peirce, C.S. (ed.) Studies in Logic by Members of the Johns Hopkins University, pp. 187–203. Little, Brown, and Co., Boston (1883)

27. Priestley, H.A.: Representation of distributive lattices by means of ordered Stone spaces. Bulletin of the London Mathematical Society 2, 186–190 (1970)

28. Sahlqvist, H.: Completeness and correspondence in the first and second order semantics for modal logics. In: Kanger, S. (ed.) 3rd Skandinavian Logic Symposium, Uppsala, Sweden, 1973, pp. 110–143. North-Holland, Amsterdam (1975)

29. Stone, M.H.: The theory of representations for Boolean algebras. Transactions of the American Mathematical Society 40, 37–111 (1936)

30. Tarski, A.: On the calculus of relations. Journal of Symbolic Logic 6, 73–89 (1941)

31. Urquhart, A.: Duality for algebras of relevant logics. Studia Logica 56, 263–276 (1996)

32. Vakarelov, D., Orłowska, E.: Lattice-based modal algebras and modal logics. In: Hajek, P., Valds-Villanueva, L.M., Westerstahl, D. (eds.) Logic, Methodology and Philosophy of Science. Proceedings of the 12th International Congress, pp. 147–170. Kings College London Publications (2005)

Contact Relations with Applications

Gunther Schmidt[1] and Rudolf Berghammer[2]

[1] Fakultät für Informatik, Universität der Bundeswehr München
85577 Neubiberg, Germany
gunther.schmidt@unibw.de
[2] Institut für Informatik, Christian-Albrechts-Universität Kiel
Olshausenstraße 40, 24098 Kiel, Germany
rub@informatik.uni-kiel.de

Abstract. Using relation algebra, we generalize Aumann's notion of a contact relation and that of a closure operation from powersets to general membership relations and their induced partial orders. We also investigate the relationship between contacts and closures in this general setting and use contacts to establish a one-to-one correspondence between the column space and the row space of a relation.

1 Introduction

Forming closures of subsets of a set X is a very basic technique in various disciplines. Typically this is combined with some predicate that holds for X and is ∩-hereditary, like "being transitive" or "being convex". Such predicates lead to closure systems, i.e., subsets $\mathfrak{C}$ of the powerset 2^X of X that contain X and any intersection of subsets collected in $\mathfrak{C}$. It is well known that there is a one-to-one correspondence between the set of closure systems of 2^X and the set of extensive, monotone, and idempotent functions on 2^X (the closure operations on 2^X).

According to G. Aumann, [1], closures always come with a relation, namely a contact. When introducing this concept, one intention was to formalize the essential properties of a contact between objects and sets of objects, mainly to obtain for beginners a more suggestive access to topology than "traditional" axiom systems provide. In the introduction of his paper, Aumann also mentions sociological applications as motivation, but in fact all examples of [1] are from mathematics. A main result of [1] is that, like closure systems and closure operations, also closure operations and contact relations are cryptomorphic mathematical structures in the sense of [6].

In this paper, we generalize Aumann's concept of a contact between sets and their powersets to contacts given by an (almost) arbitrary relation M, that may be interpreted as "individual is a member of a group of individuals". Such an approach allows to treat also examples from sociology, political science and so forth. As we will show, each group membership relation M induces a partial order Ω_M on the groups of individuals. With respect to Ω_M, we consider a notion of closure operation that directly arises out of the original one by replacing set inclusion by Ω_M. In this very general setting, we investigate contacts, their

R. Berghammer et al. (Eds.): RelMiCS/AKA 2009, LNCS 5827, pp. 306–321, 2009.
© Springer-Verlag Berlin Heidelberg 2009

properties, and a construction similar to the lower/upper-derivative construction of formal concept analysis. The latter leads to a fixed point description of the set of contacts. Guided by Aumann's main result, we also study the relationship between general M-contacts and Ω_M-closures. Finally, we use contacts to establish a one-to-one correspondence between the column space and the row space of a relation (or a Boolean matrix).

To carry out our investigations, we use abstract relation algebra in the sense of [13,12]. This allows very concise and precise specifications and algebraic proofs that drastically reduce the danger of making mistakes. To give an example, when constructing closures from contacts, a subtle definedness condition plays a decisive role that easily can be overlooked when using the customary approach with closures being functions. Relation-algebraic specifications also allow to use tool support. For obtaining the results of this paper, the use of the RELVIEW tool (see [3]) for computing contacts and closures, testing properties, experimenting with concepts etc. was very helpful.

2   Relation-Algebraic Preliminaries

We denote the set (or type) of relations with domain X and range Y by $[X \leftrightarrow Y]$ and write $R : X \leftrightarrow Y$ instead of $R \in [X \leftrightarrow Y]$. If the sets X and Y are finite, we may consider R as a Boolean matrix. Since this interpretation is well suited for many purposes, we will often use matrix notation and terminology in this paper. In particular, we talk about rows, columns and entries of relations, and write $R_{x,y}$ instead of $\langle x, y \rangle \in R$ or $x\,R\,y$.

We assume the reader to be familiar with the basic operations on relations, viz. R^{T} (transposition), $\overline{R}$ (complement), $R \cup S$ (join), $R \cap S$ (meet), $R;S$ (composition), the predicate indicating $R \subseteq S$ (inclusion), and the special relations O (empty relation), L (universal relation) and I (identity relation). Each type $[X \leftrightarrow Y]$ with the operations $^{-}$, $\cup$, $\cap$, the ordering $\subseteq$ and the constants O and L forms a complete Boolean lattice. Further well-known rules are, e.g., $R^{\mathsf{T}^{\mathsf{T}}} = R$, $\overline{R^{\mathsf{T}}} = \overline{R}^{\mathsf{T}}$ and that $R \subseteq S$ implies $R^{\mathsf{T}} \subseteq S^{\mathsf{T}}$. The theoretical framework for these rules and many others to hold is that of an (axiomatic) relation algebra. The axioms of a relation algebra are those of a complete Boolean lattice for the Boolean part, the associativity and neutrality of identity relations for composition, the equivalence of $Q;R \subseteq S$, $Q^{\mathsf{T}};\overline{S} \subseteq \overline{R}$, and $\overline{S};R^{\mathsf{T}} \subseteq \overline{Q}$ (Schröder rule), and that $R \neq \mathsf{O}$ implies $\mathsf{L};R;\mathsf{L} = \mathsf{L}$ (Tarski rule).

Furthermore, we assume the reader to be familiar with relation-algebraic specifications of the most fundamental properties of relations, like univalence $R^{\mathsf{T}};R \subseteq \mathsf{I}$, totality $R;\mathsf{L} = \mathsf{L}$, transitivity $R;R \subseteq R$, and the symmetric quotient construction $\mathrm{syq}(R, S) := \overline{R^{\mathsf{T}};\overline{S}} \cap \overline{\overline{R}^{\mathsf{T}};S}$ together with its main properties like the following ones.

$$\mathrm{syq}(R, S) = \mathrm{syq}(\overline{R}, \overline{S}) \qquad [\mathrm{syq}(R, S)]^{\mathsf{T}} = \mathrm{syq}(S, R) \qquad (1)$$

$$R;\mathrm{syq}(R, R) = R \qquad \mathrm{syq}(Q, R);\mathrm{syq}(R, S)] \subseteq \mathrm{syq}(Q, S) \qquad (2)$$

Otherwise, he may consult e.g., [12], Sections 3.1, 4.2, and 4.4.

The set-theoretic symbol $\in$ gives rise to powerset relations $\varepsilon : X \leftrightarrow 2^X$ that relate $x \in X$ and $Y \in 2^X$ iff $x \in Y$. In [4,5] it is shown that for ε the formulae of (3) hold and these even characterize the powerset relation ε up to isomorphism.

$$\mathsf{syq}(\varepsilon, \varepsilon) = \mathsf{I} \qquad \forall R : \mathsf{L}; \mathsf{syq}(\varepsilon, R) = \mathsf{L} \tag{3}$$

Based on (3), a lot of further set-theoretic constructions can be formalized in terms of relation algebra. In this paper, we need the following.

$$\imath := \mathsf{syq}(\mathsf{I}, \varepsilon) : X \leftrightarrow 2^X \qquad \varOmega := \overline{\varepsilon^{\mathsf{T}}; \overline{\varepsilon}} : 2^X \leftrightarrow 2^X \tag{4}$$

The relation $\imath$ is called singleton-set former, since it associates $x \in X$ with $Y \in 2^X$ iff $Y = \{x\}$. The relation $\varOmega$ specifies the inclusion order on sets. Based on (3) and (4), the following properties are shown in [4]:

Lemma 2.1. *If* $\varepsilon : X \leftrightarrow 2^X$ *is a powerset relation, then* $\imath : X \leftrightarrow 2^X$ *is an injective mapping*[1], $\varOmega : 2^X \leftrightarrow 2^X$ *is a partial order, and* $\imath; \varOmega = \varepsilon = \varepsilon; \varOmega$. □

The construction used in the definition of $\varOmega$ can be generalized to arbitrary relations $R : X \leftrightarrow Y$. Then $\varOmega_R := \overline{R^{\mathsf{T}}; \overline{R}} : Y \leftrightarrow Y$ is reflexive and transitive due to the Schröder rule; it shows the "column-is-contained-preorder". In case of $\mathsf{syq}(R, R) = \mathsf{I}$, i.e., without multiple columns, it is even antisymmetric and, thus, a partial order. Besides these partial order properties, we will apply the following fact.

Lemma 2.2. *For all relations* $R : X \leftrightarrow Y$ *we have* $R; \varOmega_R = R$.

Proof. The inclusion $R \subseteq R; \varOmega_R$ follows from the reflexivity of $\varOmega_R$, and with the help of the Schröder rule $R; \varOmega_R \subseteq R$ is shown by

$$R^{\mathsf{T}}; \overline{R} \subseteq R^{\mathsf{T}}; \overline{R} \iff R; \overline{R^{\mathsf{T}}; \overline{R}} \subseteq R. \qquad \square$$

As a last construction, we need the canonical epimorphism $\eta_E : X \leftrightarrow X/E$ induced by an equivalence relation $E : X \leftrightarrow X$. It relates each element $x \in X$ to the equivalence class $c \in X/E$ it belongs to. The following properties are immediate consequences of this component-wise specification; it can even be shown that they characterize canonical epimorphisms up to isomorphism.

$$\eta_E; \eta_E^{\mathsf{T}} = E \qquad \eta_E^{\mathsf{T}}; \eta_E = \mathsf{I} \tag{5}$$

In Sections 3 and 5 we will apply canonical epimorphisms induced by the two equivalence relations $\varPsi_R := \mathsf{syq}(R, R)$ and $\varPhi_R := \mathsf{syq}(R^{\mathsf{T}}, R^{\mathsf{T}})$, respectively. In this context, the following additional property will be used.

Lemma 2.3. *For all* $R : X \leftrightarrow Y$, *the canonical epimorphism* $\eta_{\varPsi_R} : Y \leftrightarrow Y/\varPsi_R$ *induced by* $\varPsi_R$ *fulfils* $\overline{R; \eta_{\varPsi_R}} = \overline{R}; \eta_{\varPsi_R}$.

---

[1] ...in the relational sense of Def. 4.2.1 of [12].

Proof. We abbreviate η_{Ψ_R} by η. Then, inclusion "$\subseteq$" follows from

$$\eta^\mathsf{T} \text{ total} \implies \overline{\eta^\mathsf{T}; R^\mathsf{T}} \subseteq \eta^\mathsf{T}; \overline{R^\mathsf{T}} \iff \overline{R; \eta} \subseteq \overline{R}; \eta$$

using Prop. 4.2.4.i of [12], and inclusion "$\supseteq$" from

$$R \subseteq R \iff R; \mathrm{syq}(R, R) \subseteq R \iff R; \eta; \eta^\mathsf{T} \subseteq R \iff \overline{R}; \eta \subseteq \overline{R; \eta}$$

using the first rule of (2), the first axiom of (5), and the Schröder rule.   □

3   Contact Relations

If we formulate Aumann's original definition of a contact relation given in [1] in our notation, then a relation $A : X \leftrightarrow 2^X$ is an *(Aumann) contact relation* if the following conditions hold.

(A_1) $\forall x : A_{x,\{x\}}$
(A_2) $\forall x, Y, Z : A_{x,Y} \wedge Y \subseteq Z \rightarrow A_{x,Z}$
(A_3) $\forall x, Y, Z : A_{x,Y} \wedge (\forall y : y \in Y \rightarrow A_{y,Z}) \rightarrow A_{x,Z}$

Our aim is to investigate contact relations by relation-algebraic means and supporting tools (like the manipulation system RELVIEW [3]), thereby generalizing Aumann's original approach by replacing the powerset by a set G (of groups of individuals, political parties, alliances, organizations, ...) and the set-theoretic membership relation $\varepsilon : X \leftrightarrow 2^X$ by a generalized membership relation $M : X \leftrightarrow G$ with regard to G. The latter point not only allows to treat mathematical examples for contact relationships as [1] does, but also examples from sociology, political science and so forth. In the following theorem, we present relation-algebraic versions of the above axioms. The proof of their correspondence consists of step-wise transformations of (A_1) to (A_3) into point-free versions using well-known correspondences between logical and relation-algebraic constructions. Doing so, (A_1) leads to a singleton-former $\imath$ and (A_2) to an inclusion order Ω as specified in (4).

Theorem 3.1. *A relation $A : X \leftrightarrow 2^X$ is an Aumann contact relation iff $\imath \subseteq A$, $A; \Omega \subseteq A$, and $A; \overline{\varepsilon^\mathsf{T}; \overline{A}} \subseteq A$.*

Proof. We only show the equivalence of (A_3) and $A; \overline{\varepsilon^\mathsf{T}; \overline{A}} \subseteq A$; the other equivalences are calculated in quite a similar way.

$$\begin{aligned}
& \forall x, Y, Z : A_{x,Y} \wedge (\forall y : y \in Y \rightarrow A_{y,Z}) \rightarrow A_{x,Z} \\
&\iff \forall x, Y, Z : A_{x,Y} \wedge \neg(\exists y : y \in Y \wedge \overline{A}_{y,Z}) \rightarrow A_{x,Z} \\
&\iff \forall x, Y, Z : A_{x,Y} \wedge \overline{\varepsilon^\mathsf{T}; \overline{A}}_{Y,Z} \rightarrow A_{x,Z} \\
&\iff \forall x, Z : (\exists Y : A_{x,Y} \wedge \overline{\varepsilon^\mathsf{T}; \overline{A}}_{Y,Z}) \rightarrow A_{x,Z} \\
&\iff \forall x, Z : (A; \overline{\varepsilon^\mathsf{T}; \overline{A}})_{x,Z} \rightarrow A_{x,Z} \\
&\iff A; \overline{\varepsilon^\mathsf{T}; \overline{A}} \subseteq A
\end{aligned}$$

□

The relation-algebraic characterization of contacts just developed does not yet allow the generalization intended. We still have to remove the singleton-former, since such a construct need not exist in the general case of membership we want to deal with. The next theorem shows how this is possible.

Theorem 3.2. *A relation $A : X \leftrightarrow 2^X$ is an Aumann contact relation iff $\varepsilon \subseteq A$ and $A^\mathsf{T}; \overline{A} \subseteq \varepsilon^\mathsf{T}; \overline{A}$.*

Proof. We show that the relation-algebraic specification of an Aumann contact relation of Theorem 3.1 is equivalent to $\varepsilon \subseteq A$ and $A^\mathsf{T}; \overline{A} \subseteq \varepsilon^\mathsf{T}; \overline{A}$. Starting with "$\Longrightarrow$", we use Lemma 2.1 to show $\varepsilon \subseteq A$ by

$$\imath \subseteq A \implies \imath; \Omega \subseteq A; \Omega \iff \varepsilon \subseteq A; \Omega \implies \varepsilon \subseteq A.$$

Because of the Schröder rule, $A^\mathsf{T}; \overline{A} \subseteq \varepsilon^\mathsf{T}; \overline{A}$ is equivalent with $A; \overline{\varepsilon^\mathsf{T}; \overline{A}} \subseteq A$. In the case "$\Longleftarrow$", property $\imath \subseteq A$ follows from $\imath \subseteq \varepsilon$ and $\varepsilon \subseteq A$. Using the Schröder rule, we obtain $A; \Omega \subseteq A$ from

$$A^\mathsf{T}; \overline{A} \subseteq \varepsilon^\mathsf{T}; \overline{A} \subseteq \varepsilon^\mathsf{T}; \overline{\varepsilon} = \overline{\Omega}.$$

For the last property, cf. the proof of "$\Longrightarrow$".   □

Hence, we have that membership implies contact and for all $Y, Z \in 2^X$ from the existence of an element that is in contact with Y but not in contact with Z it follows that even a member of Y is not in contact with Z. In the literature such relations are also known as dependence or entailment relations and in particular considered in combination with so-called exchange properties. See [7,6] for example. And here is our generalization of Aumann's concept of a contact.

Definition 3.1. *A relation $K : X \leftrightarrow G$ is called an* (Aumann) contact relation *with respect to the relation $M : X \leftrightarrow G$ — in short: an M-contact — if the following properties hold:*

$$(\text{K}_1) \;\; M \subseteq K \qquad\qquad (\text{K}_2) \;\; K^\mathsf{T}; \overline{K} \subseteq M^\mathsf{T}; \overline{K}$$

Axiom (K_2) is called the infectivity of a contact. We have chosen this form since it proved to be particularly suitable for relation-algebraic reasoning. For concrete sociological or similar applications, frequently the equivalent version $K; \overline{M^\mathsf{T}; \overline{K}} \subseteq K$ is more appropriate. E.g., in the case of persons and syndicates it says that if a person x is in contact to a syndicate Y_1 all of whose members are in contact to a syndicate Y_2, then also x is in contact to Y_2.

In real life, contacts are frequently established by common interests. As an example, we consider a protesters scene of non-governmental organizations. There exist persons willing to protest against several topics $t \in T$. Then typically a person $x \in X$ will get in touch with an activist group $g \in G$ iff for all topics he is in opposition to, there is at least one supporter for it in the group g. If we formalize the situation in predicate logic and afterwards translate this version into a relation-algebraic expression, we arrive at $\text{mi}_J(\text{ma}_J(M))_{x,g}$, where $M : X \leftrightarrow G$

denotes activist group membership, the complement of the relation $J : X \leftrightarrow T$ specifies the relationship "is in opposition to", and the functions mi_J and ma_J are defined as follows:

$$\mathrm{mi}_J(R) = \overline{\overline{J}; R} \qquad \mathrm{ma}_J(S) = \overline{\overline{J}^\mathsf{T}; S} \qquad (6)$$

If J is a partial order, then mi_J and ma_J column-wise compute lower bounds and upper bounds, respectively; in the general case, they column-wise compute lower derivatives and upper derivatives, respectively, in the sense of formal concept analysis (see [9]). The next theorem shows that the above construction based on interest-relations J always leads to M-contacts.

Theorem 3.3. *For all relations $M : X \leftrightarrow G$ and $J : X \leftrightarrow T$, we obtain an M-contact K if we define $K := \mathrm{mi}_J(\mathrm{ma}_J(M))$.*

Proof. Property (K_1) follows from

$$\overline{J}^\mathsf{T}; M \subseteq \overline{J}^\mathsf{T}; M \iff \overline{J}; \overline{\overline{J}^\mathsf{T}; M} \subseteq \overline{M} \qquad \text{Schröder rule}$$
$$\iff M \subseteq \overline{\overline{J}; \overline{\overline{J}^\mathsf{T}; M}}$$
$$\iff M \subseteq \mathrm{mi}_J(\mathrm{ma}_J(M)) \qquad \text{by (6)}$$
$$\iff M \subseteq K,$$

and property (K_2) from

$$\overline{M^\mathsf{T}; \overline{J}; \overline{J}^\mathsf{T}} \subseteq \overline{M^\mathsf{T}; \overline{J}; \overline{J}^\mathsf{T}}$$
$$\iff \overline{M^\mathsf{T}; \overline{J}; \overline{J}^\mathsf{T}} \subseteq \left(\overline{\overline{J}; \overline{\overline{J}^\mathsf{T}; M}}\right)^\mathsf{T}$$
$$\iff \overline{\overline{J}; \overline{\overline{J}^\mathsf{T}; M}}^\mathsf{T}; \overline{J} \subseteq M^\mathsf{T}; \overline{J} \qquad \text{Schröder rule}$$
$$\iff \left[\mathrm{mi}_J(\mathrm{ma}_J(M))\right]^\mathsf{T}; \overline{J} \subseteq M^\mathsf{T}; \overline{J} \qquad \text{by (6)}$$
$$\iff K^\mathsf{T}; \overline{J} \subseteq M^\mathsf{T}; \overline{J}$$
$$\implies K^\mathsf{T}; \overline{J}; \overline{\overline{J}^\mathsf{T}; M} \subseteq M^\mathsf{T}; \overline{J}; \overline{\overline{J}^\mathsf{T}; M}$$
$$\iff K^\mathsf{T}; \overline{\mathrm{mi}_J(\mathrm{ma}_J(M))} \subseteq M^\mathsf{T}; \overline{\mathrm{mi}_J(\mathrm{ma}_J(M))} \qquad \text{by (6)}$$
$$\iff K^\mathsf{T}; \overline{K} \subseteq M^\mathsf{T}; \overline{K}. \qquad \qquad \square$$

Next, we give a concrete application of the construction of Theorem 3.3. We assume four persons, denoted by the natural numbers 1 to 4, three groups g_1, g_2 and g_3, and six topics A, B, C, D, E and F. If group membership is described by the left-most of the following three RELVIEW-matrices and the persons' interests by the RELVIEW-matrix in the middle, then these relations lead to the contact specified by the right RELVIEW-matrix.

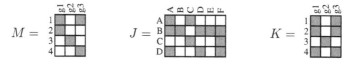

In these pictures, a black square means 1 and a white square means 0 so that, e.g., the first person is a member of g_1 and g_3. By definition, $M \subseteq K$. In addition, $(4, g_1) \in K$, because wherever *all* persons of the group g_1 are jointly J-interested in a couple of topics (here $\{1, 2\} \times \{A\}$), then also person 4 is J-interested in these topics. Also $(2, g_3) \in K$: the rectangle $\{1, 4\} \times \{A\}$ indicates that all members of the group are jointly J-interested in topic set $\{A\}$ and so is person 2.

We even can prove completeness of the construction of Theorem 3.3, i.e., that every M-contact K can be represented as an expression $\mathrm{mi}_J(\mathrm{ma}_J(M))$. As the next theorem shows, we only have to take the groups as topics and K itself as interest relation J.

Theorem 3.4. *For all relations $M : X \leftrightarrow G$ and all M-contacts $K : X \leftrightarrow G$ the equation $K = \mathrm{mi}_K(\mathrm{ma}_K(M))$ holds.*

Proof. "$\subseteq$": This inclusion is equivalent to property (K_2), since

$$
\begin{aligned}
K \subseteq \mathrm{mi}_K(\mathrm{ma}_K(M)) &\iff K \subseteq \overline{\overline{K}; \overline{K}^{\mathsf{T}}; M} \qquad \text{by (6)} \\
&\iff \overline{K}; \overline{K}^{\mathsf{T}}; M \subseteq \overline{K} \\
&\iff \overline{K}^{\mathsf{T}}; K \subseteq \overline{K}^{\mathsf{T}}; M \qquad \text{Schröder rule} \\
&\iff K^{\mathsf{T}}; \overline{K} \subseteq M^{\mathsf{T}}; \overline{K}.
\end{aligned}
$$

"$\supseteq$": Starting with (K_1), we get the result by

$$
\begin{aligned}
M \subseteq K &\iff \overline{K}; \mathsf{I} \subseteq \overline{M} \\
&\iff \overline{K}^{\mathsf{T}}; M \subseteq \overline{\mathsf{I}} \qquad \text{Schröder rule} \\
&\iff \mathsf{I} \subseteq \overline{\overline{K}^{\mathsf{T}}; M} \\
&\implies \overline{K} \subseteq \overline{K}; \overline{\overline{K}^{\mathsf{T}}; M} \\
&\iff \overline{\overline{K}; \overline{K}^{\mathsf{T}}; M} \subseteq K \\
&\iff \mathrm{mi}_K(\mathrm{ma}_K(M)) \subseteq K \qquad \text{by (6).} \qquad \square
\end{aligned}
$$

From the Theorems 3.3 and 3.4, we immediately obtain a fixed point characterization of the set of M-contacts.

Theorem 3.5. *Assume a generalized membership relation $M : X \leftrightarrow G$ to be given and consider all relations $R : X \leftrightarrow T$ for some set T. Then the function*

$$
\tau_M : [X \leftrightarrow T] \to [X \leftrightarrow G] \qquad \tau_M(R) = \mathrm{mi}_R(\mathrm{ma}_R(M)),
$$

will always produce an M-contact. The set $\mathfrak{K}_M$ of all M-contacts equals the set of fixed points of τ_M in case $T = G$. $\square$

Using relational fixed point enumeration techniques (cf. [2]), this property can be used to compute for small relations M all M-contacts by a tool like RELVIEW.

Since the underlying relation M is contained in each M-contact K, normally in K a lot of columns coincide. The column equivalence relation $\Psi_K = \mathrm{syq}(K, K)$

relates two groups iff the corresponding columns of K are equal. Hence, we can remove duplicates of columns of K by multiplying it with the canonical epimorphism η_{Ψ_K} induced by Ψ_K from the right. In the next theorem we prove that in the construction $\mathrm{mi}_K(\mathrm{ma}_K(M))$, instead of K also its revised form can be used.

Theorem 3.6. *For all relations* $M : X \leftrightarrow G$ *and all* M-*contacts* $K : X \leftrightarrow G$ *we have that* $K = \mathrm{mi}_{K;\eta_{\Psi_K}}(\mathrm{ma}_{K;\eta_{\Psi_K}}(M))$.

Proof. In the following calculation we abbreviate η_{Ψ_K} by η.

$$
\begin{aligned}
\mathrm{mi}_{K;\eta}(\mathrm{ma}_{K;\eta}(M)) &= \overline{\overline{K;\eta};\ \overline{\overline{K;\eta}^{\mathsf{T}};M}} & \text{by (6)} \\
&= \overline{\overline{K};\eta;\ (\overline{K};\eta)^{\mathsf{T}};M} & \text{Lemma 2.3} \\
&= \overline{\overline{K};\eta;\ \eta^{\mathsf{T}};\overline{K}^{\mathsf{T}};M} & \\
&= \overline{\overline{K};\ \eta;\eta^{\mathsf{T}};\overline{K}^{\mathsf{T}};M} & \text{[12] Prop. 4.2.4.ii} \\
&= \overline{\overline{K};\ \mathrm{syq}(K,K);\overline{K}^{\mathsf{T}};M} & \text{by (5)} \\
&= \overline{\overline{K};\ \left[\overline{K};\mathrm{syq}(\overline{K},\overline{K})\right]^{\mathsf{T}};M} & \text{by (1)} \\
&= \overline{\overline{K};\ \overline{K}^{\mathsf{T}};M} & \text{by (2)} \\
&= K & \text{Theorem 3.4} \qquad \square
\end{aligned}
$$

4   Contacts and Closures

Closure operations appear in many fields in computer science and mathematics. Usually, they are defined as extensive, monotone, and idempotent functions on powersets, i.e., functions $h : 2^X \to 2^X$ such that the following conditions hold.

$$
\begin{aligned}
&(\mathrm{H}_1) \quad \forall Y : Y \subseteq h(Y) \\
&(\mathrm{H}_2) \quad \forall Y, Z : Y \subseteq Z \to h(Y) \subseteq h(Z) \\
&(\mathrm{H}_3) \quad \forall Y : h(h(Y)) \subseteq h(Y)
\end{aligned}
$$

As in the case of Aumann contact relations, we start our investigations with a relation-algebraic characterization of closure operations. In the next theorem, the relation Ω denotes set inclusion on the powerset 2^X as specified in (4).

Theorem 4.1. *A mapping* $H : 2^X \leftrightarrow 2^X$ *is a closure operation iff* $H \subseteq \Omega$, $\Omega \subseteq H;\Omega;H^{\mathsf{T}}$, *and* $H;H \subseteq H$.

Proof. As in the case of Theorem 3.1, we only treat one case, viz. the equivalence of (H_3) and $H;H \subseteq H$. To enhance readability, in the following calculations, we apply the common notation of function application also for H.

$$
\begin{aligned}
&\forall Y : H(H(Y)) \subseteq H(Y) \\
\Longleftrightarrow\ &\forall Y, Z, U : H(Y) = U \land H(U) = Z \to (\exists W : H(Y) = W \supseteq Z) \\
\Longleftrightarrow\ &\forall Y, Z, U : H_{Y,U} \land H_{U,Z} \to (\exists W : H_{Y,W} \land \Omega_{Z,W}) \\
\Longleftrightarrow\ &\forall Y, Z : (\exists U : H_{Y,U} \land H_{U,Z}) \to (\exists W : H_{Y,W} \land \Omega^{\mathsf{T}}_{W,Z}) \\
\Longleftrightarrow\ &\forall Y, Z : (H;H)_{Y,Z} \to (H;\Omega^{\mathsf{T}})_{Y,Z} \\
\Longleftrightarrow\ &H;H \subseteq H;\Omega^{\mathsf{T}}
\end{aligned}
$$

(H_1) equals $H \subseteq \Omega$, so that with antisymmetry and univalency of H we get

$$H; H \subseteq H; \Omega \cap H; \Omega^\mathsf{T} = H; (\Omega \cap \Omega^\mathsf{T}) \subseteq H; \mathsf{I} = H. \qquad \square$$

A simple relation-algebraic reasoning shows that $H; H \subseteq H$ in fact is equivalent to the equation $H; H = H$ when $H \subseteq \Omega \subseteq H; \Omega; H^\mathsf{T}$. This corresponds to the well-known property that in (H_3), due to (H_1) and (H_2), even equality holds.

Because of Theorem 4.1, we are able to generalize the concept of a closure operation from powerset lattices to arbitrary partial order relations within the language of relation algebra as follows.

Definition 4.1. *Given a partial order* $P : X \leftrightarrow X$, *a mapping* $H : X \leftrightarrow X$ *is called a* closure operation *with respect to* P — *in short: a* P-closure — *if the following conditions hold:*

$$(\mathrm{C}_1) \ \ H \subseteq P \qquad (\mathrm{C}_2) \ \ P \subseteq H; P; H^\mathsf{T} \qquad (\mathrm{C}_3) \ \ H; H \subseteq H$$

In [1] it is shown that there is a one-to-one correspondence between the set of all Aumann contact relations between X and 2^X and the set of all closure operations on 2^X. Without proof and reference to its origin, this correspondence is also mentioned in [6]. In the remainder of this section, we investigate the relationship between contact relations and closure operations in our general setting, i.e., in conjunction with M-contacts and Ω_M-closures, and using relation-algebraic means. As the only basic prerequisite on the relation $M : X \leftrightarrow G$ we assume $\mathrm{syq}(M, M) = \mathsf{I}$, i.e., pairwise different columns, to ensure that Ω_M is a partial order (see Section 2). (Even this is not a really essential requirement.)

How to obtain M-contacts from Ω_M-closures is shown in the following theorem. In words, the theorem states that $x \in X$ is in contact with $g \in G$ iff x is a member of the closure of g.

Theorem 4.2. *For all relations* $M : X \leftrightarrow G$ *such that* $\mathrm{syq}(M, M) = \mathsf{I}$ *and all* Ω_M-*closures* $H : G \leftrightarrow G$, *the relation* $K := M; H^\mathsf{T} : X \leftrightarrow G$ *is an* M-*contact.*

Proof. For proving (K_1), we use (C_1) and Prop. 4.2.3 of [12] in

$$M; \Omega_M \subseteq M \implies M; H \subseteq M \iff M \subseteq M; H^\mathsf{T} \iff M \subseteq K.$$

Now, Lemma 2.2 yields the result. The verification of property (K_2) bases on the following calculation.

$$
\begin{aligned}
K; \overline{M^\mathsf{T}}; \overline{K} &= M; H^\mathsf{T}; \overline{M^\mathsf{T}}; \overline{M; H^\mathsf{T}} \\
&= M; H^\mathsf{T}; \overline{M^\mathsf{T}}; \overline{M} ; H^\mathsf{T} && \text{[12] Prop. 4.2.4.iii} \\
&= M; H^\mathsf{T}; \Omega_M; H^\mathsf{T} \\
&\subseteq M; \Omega_M; H^\mathsf{T}; H^\mathsf{T} && \text{by (C}_2\text{) (cf. [12] p. 143)} \\
&\subseteq M; \Omega_M; H^\mathsf{T} && \text{by (C}_3\text{)} \\
&= M; H^\mathsf{T} && \text{Lemma 2.2} \\
&= K
\end{aligned}
$$

An application of the Schröder rule to this inclusion completes the proof. $\qquad \square$

To obtain a closure operation h from a contact relation A, in [1] the closure $h(Y)$ of a set Y is defined as the set of elements being in contact with Y. Relation-algebraically, this leads to the expression $\mathrm{syq}(A, \varepsilon)$ for the closure operation. Contrary to the transition from closure operations to contact relations, which also works in our general setting, the transition from M-contacts K to Ω_M-closures is problematic. The reason is that $\mathrm{syq}(K, M)$ may be non-total. But if $\mathrm{syq}(K, M)$ is total, it is indeed an Ω_M-closure as the following theorem shows.

Theorem 4.3. *For all relations $M : X \leftrightarrow G$ such that $\mathrm{syq}(M, M) = \mathsf{I}$ and all M-contacts $K : X \leftrightarrow G$, the relation $H := \mathrm{syq}(K, M) : G \leftrightarrow G$ is an Ω_M-closure provided it is total.*

Proof. Since totality of H has been assumed as a prerequisite, we show univalence to establish H as a mapping:

$$
\begin{aligned}
H^{\mathsf{T}}; H &= [\mathrm{syq}(K, M)]^{\mathsf{T}}; \mathrm{syq}(K, M) \\
&= \mathrm{syq}(M, K); \mathrm{syq}(K, M) && \text{by (1)} \\
&\subseteq \mathrm{syq}(M, M) && \text{by (2)} \\
&= \mathsf{I}.
\end{aligned}
$$

Property (C_1) follows from (K_1), since

$$
H = \mathrm{syq}(K, M) \subseteq \overline{K^{\mathsf{T}}; \overline{M}} \subseteq \overline{M^{\mathsf{T}}; \overline{M}} = \Omega_M.
$$

In the proof of (C_2) we use that totality of $\mathrm{syq}(K, M)$ implies surjectivity of $\mathrm{syq}(M, K) = \mathrm{syq}(\overline{M}, \overline{K})$ (cf. Prop. 4.4.1.i,ii of [12]). We start with

$$
\begin{aligned}
H; \overline{\Omega_M}; H^{\mathsf{T}} &= [\mathrm{syq}(M, K)]^{\mathsf{T}}; M^{\mathsf{T}}; \overline{M}; \mathrm{syq}(M, K) && \text{by (1)} \\
&= [M; \mathrm{syq}(M, K)]^{\mathsf{T}}; \overline{M}; \mathrm{syq}(\overline{M}, \overline{K}) && \text{by (1)} \\
&= K^{\mathsf{T}}; \overline{K} && \text{[12] Prop. 4.4.2.ii} \\
&\subseteq M^{\mathsf{T}}; \overline{K} && \text{by } (K_2) \\
&\subseteq M^{\mathsf{T}}; \overline{M} && \text{by } (K_1).
\end{aligned}
$$

Using that H is a mapping, we get from this $\overline{H; \Omega_M; H^{\mathsf{T}}} \subseteq \overline{\Omega_M}$, i.e., the desired inclusion $\Omega_M \subseteq H; \Omega_M; H^{\mathsf{T}}$. Also the first two calculations of the subsequent proof of property (C_3) use the surjectivity of $\mathrm{syq}(M, K) = \mathrm{syq}(\overline{M}, \overline{K})$. From (1) and Prop. 4.4.2.ii of [12] and (K_1) we get

$$
K^{\mathsf{T}}; \overline{M}; \mathrm{syq}(M, K) = K^{\mathsf{T}}; \overline{M}; \mathrm{syq}(\overline{M}, \overline{K}) = K^{\mathsf{T}}; \overline{K} \subseteq K^{\mathsf{T}}; \overline{M}
$$

and Prop. 4.4.2.ii of [12] and (K_2) yield

$$
\overline{K}^{\mathsf{T}}; M; \mathrm{syq}(M, K) = \overline{K}^{\mathsf{T}}; K = (K^{\mathsf{T}}; \overline{K})^{\mathsf{T}} \subseteq (M^{\mathsf{T}}; \overline{K})^{\mathsf{T}} = \overline{K}^{\mathsf{T}}; M.
$$

Putting these inclusions together, we obtain

$$
(K^{\mathsf{T}}; \overline{M} \cup \overline{K}^{\mathsf{T}}; M); \mathrm{syq}(M, K) \subseteq K^{\mathsf{T}}; \overline{M} \cup \overline{K}^{\mathsf{T}}; M
$$

that, due to the definition of $\text{syq}(K, M)$ and (1), holds iff

$$\overline{\text{syq}(K, M)} \, ; [\text{syq}(K, M)]^\mathsf{T} \subseteq \overline{\text{syq}(K, M)} \, .$$

An application of the Schröder rule to this result followed by the definition of H, finally, shows $H; H \subseteq H$.   □

Combining the last two theorems, we obtain for our general setting an injective embedding of the Ω_M-closures into the M-contacts.

Corollary 4.1. *Assume a relation $M : X \leftrightarrow G$ such that $\text{syq}(M, M) = \mathsf{I}$ and let $\mathfrak{K}_M$ and $\mathfrak{H}_{\Omega_M}$ denote the set of M-contacts and Ω_M-closures, respectively. Then the function $\text{con}_M : \mathfrak{H}_{\Omega_M} \to \mathfrak{K}_M$, where $\text{con}_M(H) = M; H^\mathsf{T}$, is injective.*

Proof. First we show that $\text{syq}(\text{con}_M(H), M)$ is total for all $H \in \mathfrak{H}_{\Omega_M}$.

$$\begin{aligned}
\text{syq}(\text{con}_M(H), M); \mathsf{L} &= \text{syq}(M; H^\mathsf{T}, M); \mathsf{L} && \text{definition of } \text{con}_M(H) \\
&= H; \text{syq}(M, M); \mathsf{L} && \text{[12] Prop. 4.4.1.vi} \\
&= H; \mathsf{L} && \text{since } \text{syq}(M, M) = \mathsf{I} \\
&= \mathsf{L} && H \text{ total}
\end{aligned}$$

Hence, $\text{syq}(\text{con}_M(H), M)$ is an Ω_M-closure due to Theorems 4.2 and 4.3. The above calculation, furthermore, shows that the function

$$\text{clo}_M : \text{con}_M(\mathfrak{H}_{\Omega_M}) \to \mathfrak{H}_{\Omega_M} \qquad \text{clo}_M(K) = \text{syq}(K, M)$$

fulfils $\text{clo}_M(\text{con}_M(H)) = H$ for all $H \in \mathfrak{H}_{\Omega_M}$, and we are done.   □

Specifying the point-wise ordering of mappings relation-algebraically, we obtain for $H_1, H_2 \in \mathfrak{H}_{\Omega_M}$ that $H_1 \leq H_2$ iff $H_1 \subseteq H_2; \Omega_M{}^\mathsf{T}$. In respect thereof, the following theorem shows that the function con_M is even an order embedding from the ordered set $(\mathfrak{H}_{\Omega_M}, \leq)$ into the ordered set $(\mathfrak{K}_M, \subseteq)$.

Theorem 4.4. *Under the assumptions of Corollary 4.1 we have $H_1 \subseteq H_2; \Omega_M{}^\mathsf{T}$ iff $M; H_1{}^\mathsf{T} \subseteq M; H_2{}^\mathsf{T}$.*

Proof. In the following calculation we combine the fact that H_1 and H_2 are mappings with Prop. 4.2.4.iii of [12].

$$\begin{aligned}
H_1 \subseteq H_2; \Omega_M{}^\mathsf{T} &\iff H_1 \subseteq H_2; \overline{M^\mathsf{T}; \overline{M}}^\mathsf{T} \\
&\iff H_1 \subseteq H_2; \overline{\overline{M}^\mathsf{T}; M} \\
&\iff H_1 \subseteq \overline{H_2; M^\mathsf{T}}; M && \text{Prop. 4.2.4.ii of [12]} \\
&\iff \overline{H_2; M^\mathsf{T}}; M \subseteq \overline{H_1} \\
&\iff H_1; M^\mathsf{T} \subseteq H_2; M^\mathsf{T} && \text{Schröder rule} \qquad □
\end{aligned}$$

A little reflection shows that $(\mathfrak{K}_M, \subseteq)$ is a complete lattice. For the ordered set $(\mathfrak{H}_{\Omega_M}, \leq)$ this is not true in general. It is, however, true if the underlying set G on which the closure operations work is finite [10]. In general, we are not able to establish a one-to-one correspondence between contact relations and closure operations in our general setting without further assumptions on the underlying relation $M : X \leftrightarrow G$. For instance, for the example of Section 3, RELVIEW computed for the membership relation M and M-contact K given there the following matrices for Ω_M and $\mathrm{syq}(K, M)$.

$$\Omega_M = \begin{array}{c} \\ \text{g1} \\ \text{g2} \\ \text{g3} \end{array} \begin{array}{ccc} \text{go1} & \text{go2} & \text{go3} \\ \end{array} \qquad \mathrm{syq}(K, M) = \begin{array}{c} \\ \text{g1} \\ \text{g2} \\ \text{g3} \end{array} \begin{array}{ccc} \text{go1} & \text{go2} & \text{go3} \\ \end{array}$$

The relation Ω may be described as being the column-is-contained-preorder for M, while $\mathrm{syq}(K, M)$ compares columns of K and M for being identical. Furthermore, the tool ascertained that there exist exactly 128 relations containing M and exactly 66 of them are M-contacts. Since Ω_M is the identity relation, however, there exists only one Ω_M-closure, viz. Ω_M.

In matrix terminology, totality of $\mathrm{syq}(K, M)$ means that each column of $K : X \leftrightarrow G$ appears also as a column of M. Hence, this property should hold for G being a powerset 2^X and M being the powerset relation $\varepsilon : X \leftrightarrow 2^X$. And, in fact, totality of $\mathrm{syq}(K, \varepsilon)$ can be shown so that, together with the already obtained results, we are able to give not only a completely relation-algebraic proof of the above mentioned result of Aumann but also to show that the sets are isomorphic complete lattices.

Corollary 4.2. *For all powerset relations $\varepsilon : X \leftrightarrow 2^X$, the ordered sets $(\mathfrak{K}_\varepsilon, \subseteq)$ and $(\mathfrak{H}_\Omega, \leq)$ are isomorphic via the function $\mathrm{con}_\varepsilon : \mathfrak{H}_\Omega \to \mathfrak{K}_\varepsilon$ of Corollary 4.1 and its inverse function $\mathrm{clo}_\varepsilon : \mathfrak{K}_\varepsilon \to \mathfrak{H}_\Omega$, where $\mathrm{clo}_\varepsilon(K) = \mathrm{syq}(K, \varepsilon)$.*

Proof. For each $K \in \mathfrak{K}_\varepsilon$, (1) and the second axiom of (3) imply

$$\mathrm{syq}(K, \varepsilon); \mathsf{L} = \left(\mathsf{L}; \mathrm{syq}(K, \varepsilon)^\mathsf{T}\right)^\mathsf{T} = \left(\mathsf{L}; \mathrm{syq}(\varepsilon, K)\right)^\mathsf{T} = \mathsf{L}.$$

Because of Theorem 4.3, therefore, $\mathrm{clo}_\varepsilon(K)$ is defined for all $K \in \mathfrak{K}_\varepsilon$. From the proof of Corollary 4.1 we know already that

$$\mathrm{clo}_\varepsilon(\mathrm{con}_\varepsilon(H)) = H$$

holds for all $H \in \mathfrak{H}_\Omega$. Furthermore, we obtain for all $K \in \mathfrak{K}_\varepsilon$ the equation

$$\mathrm{con}_\varepsilon(\mathrm{clo}_\varepsilon(K)) = \varepsilon; \mathrm{syq}(K, \varepsilon)^\mathsf{T} = \varepsilon; \mathrm{syq}(\varepsilon, K) = K$$

using the second axiom of (3) in combination with Prop. 4.4.2.ii of [12]. These two properties show that the functions are bijective and mutually inverses. That the two mappings are order isomorphisms follows from Theorem 4.4.   □

One might conjecture that in the case $\mathrm{syq}(M, M) = \mathsf{I}$ from an isomorphism between the sets $\mathfrak{K}_M$ and $\mathfrak{H}_{\Omega_M}$ also the second axiom of (3) follows, i.e., M is essentially a powerset relation. Unfortunately, this speculation is false, as the simple example with a single group, i.e., $G := \mathbf{1}$, and M as $\mathsf{L} : X \leftrightarrow \mathbf{1}$ shows.

5   Linking Column and Row Types of a Relation

Considering a relation $M : X \leftrightarrow Y$ as a Boolean matrix, rows and columns may be joined or intersected in much the same way as one may form sums of rows or columns of real-valued matrices. In comparison with the vector space spanned by the real-valued rows, on will then obtain unions of rows, or intersections, respectively. Unions of rows of M may, of course, also be considered as complements of intersections of complemented rows. For the following, we decide to treat mainly intersections. Although this looks more complicated introducing complements, it gives better guidance along residuation.

By the following four RELVIEW-pictures we want to decribe the situation. We consider a 4×4 Boolean matrix M. The 4×7 matrix right besides M shows all possible intersections of sets of columns of M, each of the seven results represented by a column of the matrix. Note, that the universal vector is obtained by intersecting the empty set of columns. In the same way the 7×4 matrix below M enumerates all intersections of sets of rows of M. Again we have seven different results, now represented by the matrix's rows. Finally, the 7×7 matrix β bijectively links the column intersections and the row intersections of M.

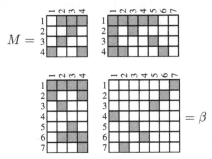

It is evident that several combinations of rows may produce the same union. When considering ε_X^T, multiplied from the left, where $\varepsilon_X : X \leftrightarrow 2^X$ is the powerset relation of X, one will probably obtain many identical unions of rows. In order to eliminate multiply occurring unions, one may, of course, wish to identify them. A little reflection shows in an analogous way that all intersections of rows of M are given by the rows of $R := \overline{\varepsilon_X^\mathsf{T}; \overline{M}} : 2^X \leftrightarrow X$. The elimination of multiple rows of R is obtained via $\eta_\Xi^\mathsf{T}; R : 2^X/\Xi \leftrightarrow Y$, where $\eta_\Xi : 2^X \leftrightarrow 2^X/\Xi$ is the canonical epimorphism induced by the row equivalence relation $\Xi := \mathrm{syq}(R^\mathsf{T}, R^\mathsf{T}) : 2^X \leftrightarrow 2^X$. Equivalence classes of rows so obtained will be called row types.

We will use contacts for linking the row types of a relation with its column types. The corresponding reflection, namely, shows that all intersections of columns of M are given by the columns of $C := \overline{\overline{M}; \varepsilon_Y} : X \leftrightarrow 2^Y$, where $\varepsilon_Y : Y \leftrightarrow 2^Y$ is the powerset relation of Y, so that we proceed with

Definition 5.1. *Given* $M : X \leftrightarrow Y$, $C := \overline{\overline{M}; \varepsilon_Y}$, *and* $R := \overline{\varepsilon_X^\mathsf{T}; \overline{M}}$, *we define the* column intersection types relation *as* $C; \eta_\Psi : X \leftrightarrow 2^Y/\Psi$ *and the* row intersection types relation *as* $\eta_\Xi^\mathsf{T}; R : 2^X/\Xi \leftrightarrow Y$.

To visualize the constructions, we consider again the 4×4 matrix M of the above example. The following RELVIEW-matrices represent the membership relation ε_Y and the relation C, respectively.

If we transform C into the column intersection types relation $C; \eta_\Psi$ by the elimination of all multiple occurrences of columns, we exactly obtain the result already shown above.

It is a remarkable fact that there exists a close connection between the row and the column types relation. By the following bijection, one may feel reminded that for a real-valued matrix the row rank equals the column rank. Some ideas from the approach stem from real valued matrices as presented e.g., in [11]. For the proof we need that symmetric quotients are difunctional in the sense that

$$\mathrm{syq}(P,Q); \left[\mathrm{syq}(P,Q)\right]^{\mathsf{T}}; \mathrm{syq}(P,Q) \subseteq \mathrm{syq}(P,Q), \tag{7}$$

which immediately follows from (1), (2) and Prop. 4.4.1.iv of [12].

Theorem 5.1. *Given a relation $M : X \leftrightarrow Y$ together with the derived relations $C := \overline{M}; \varepsilon_Y$, $R := \overline{\varepsilon_X^{\mathsf{T}}; M}$, $\Psi := \mathrm{syq}(C, C)$, and $\Xi := \mathrm{syq}(R^{\mathsf{T}}, R^{\mathsf{T}})$, there exists a bijective mapping (in the relational sense) of type $[2^X/\Xi \leftrightarrow 2^Y/\Psi]$.*

Proof. The idea is to compare the contact relation $\mathrm{mi}_M(\mathrm{ma}_M(\varepsilon_X)) = \mathrm{mi}_M(R^{\mathsf{T}})$ and the lower derivative $\mathrm{mi}_M(\varepsilon_Y) = C$ via a symmetric quotient construction; so we define (equality of the two versions is easy to prove by expansion):

$$A := \mathrm{syq}(\mathrm{mi}_M(R^{\mathsf{T}}), C) = \mathrm{syq}(R^{\mathsf{T}}, \mathrm{ma}_M(C)) : 2^X \leftrightarrow 2^Y$$

The relation A is total and surjective. For totality, we calculate

$$\begin{aligned}
A &= \mathrm{syq}(\mathrm{mi}_M(R^{\mathsf{T}}), C) \\
&= \mathrm{syq}(\,\overline{\overline{M}\,;\,\overline{\overline{M}}^{\mathsf{T}}; \varepsilon_X}\,,\,\overline{M}; \varepsilon_Y\,) \\
&= \mathrm{syq}(\overline{M}\,;\,\overline{\overline{M}}^{\mathsf{T}}; \varepsilon_X\,,\,\overline{M}; \varepsilon_Y) && \text{by (1)} \\
&\supseteq \mathrm{syq}(\,\overline{\overline{M}}^{\mathsf{T}}; \varepsilon_X\,,\,\varepsilon_Y) && \text{[12] Prop. 4.4.1.v}
\end{aligned}$$

and apply then that $\mathrm{syq}(\,\overline{\overline{M}}^{\mathsf{T}}; \varepsilon_X\,,\,\varepsilon_Y)$ is total by (3) and (1) . To prove surjectivity, we reason in the same way, but use the other variant of A.

Next, we have a look at the row equivalence relation $\Xi' := \mathrm{syq}(A^{\mathsf{T}}, A^{\mathsf{T}})$ and the column equivalence relation $\Psi' := \mathrm{syq}(A, A)$. It so happens that $\Xi = \Xi'$ and $\Psi = \Psi'$ via a general cancelling rule for symmetric quotients that follows from the laws of [12], Section 4.4. E.g., the second equality is shown by

$$\begin{aligned}
\Psi' &= \mathrm{syq}(A, A) \\
&= \mathrm{syq}(\mathrm{syq}(\mathrm{mi}_M(R^{\mathsf{T}}), C), \mathrm{syq}(\mathrm{mi}_M(R^{\mathsf{T}}), C)) \\
&= \mathrm{syq}(C, C) && \text{cancelling} \\
&= \Psi.
\end{aligned}$$

Based on $A : 2^X \leftrightarrow 2^Y$ and the canonical epimorphisms $\eta_\Xi : 2^X \leftrightarrow 2^X/\Xi$ and $\eta_\Psi : 2^Y \leftrightarrow 2^Y/\Psi$, now we define the following relation by simple composition:

$$\beta := \eta_\Xi{}^\mathsf{T} ; A ; \eta_\Psi : 2^X/\Xi \leftrightarrow 2^Y/\Psi$$

This is a matching, defined as a relation that is at the same time univalent and injective. Using the Schröder rule, for the proof of univalency we start with

$$A^\mathsf{T} ; A \subseteq \overline{\overline{A}^\mathsf{T} ; A} \Longleftrightarrow A ; \overline{A}^\mathsf{T} ; A \subseteq \overline{A} \Longleftrightarrow A ; A^\mathsf{T} ; \overline{A} \subseteq \overline{A} \Longleftrightarrow A ; A^\mathsf{T} ; A \subseteq A.$$

This yields $A^\mathsf{T} ; A \subseteq \overline{\overline{A}^\mathsf{T} ; A}$ and, by transposition, also $A^\mathsf{T} ; A \subseteq \overline{A^\mathsf{T} ; \overline{A}}$, since symmetric quotients are difunctional due to (7). So, we have $A^\mathsf{T} ; A \subseteq \mathrm{syq}(A, A)$. If we combine this with $\Xi = \Xi' = \mathrm{syq}(A^\mathsf{T}, A^\mathsf{T})$ and Prop 4.4.1.iii of [12], we get

$$A^\mathsf{T} ; \Xi ; A = A^\mathsf{T} ; \mathrm{syq}(A^\mathsf{T}, A^\mathsf{T}) ; A = A^\mathsf{T} ; A \subseteq \mathrm{syq}(A, A) = \Psi' = \Psi.$$

Now, the univalency of the relation β can be shown as follows:

$$
\begin{aligned}
\beta^\mathsf{T} ; \beta &= \left[\eta_\Xi{}^\mathsf{T} ; A ; \eta_\Psi\right]^\mathsf{T} ; \eta_\Xi{}^\mathsf{T} ; A ; \eta_\Psi \\
&= \eta_\Psi{}^\mathsf{T} ; A^\mathsf{T} ; \eta_\Xi ; \eta_\Xi{}^\mathsf{T} ; A ; \eta_\Psi \\
&= \eta_\Psi{}^\mathsf{T} ; A^\mathsf{T} ; \Xi ; A ; \eta_\Psi && \text{by (5)} \\
&\subseteq \eta_\Psi{}^\mathsf{T} ; \Psi ; \eta_\Psi && \text{see above} \\
&= \eta_\Psi{}^\mathsf{T} ; \eta_\Psi ; \eta_\Psi{}^\mathsf{T} ; \eta_\Psi && \text{by (5)} \\
&= \mathsf{I} && \text{by (5)}
\end{aligned}
$$

Transpositions of difunctional relations obviously are also difunctional. This implies $A ; A^\mathsf{T} \subseteq \mathrm{syq}(A^\mathsf{T}, A^\mathsf{T})$ and from this fact we obtain, analogously to the above calculations, first $A ; \Psi ; A^\mathsf{T} \subseteq \Xi$ and then injectivity $\beta ; \beta^\mathsf{T} \subseteq \mathsf{I}$.

Since canonical epimorphisms and their transpositions are total and surjective and these properties pass on to compositions, by construction β is also total and surjective, i.e., the bijective mapping we have searched for. □

Let, for $M : X \leftrightarrow Y$ and $y \in Y$, by $M^{(y)} : Y \leftrightarrow \mathbf{1}$ the y-column of M be denoted. Then $M_c^\cap := \{\bigcap_{y \in I} M^{(y)} \mid I \in 2^Y\}$ is the set of all intersections of sets of columns of M and $M_r^\cap := (M^\mathsf{T})_c^\cap$ that of all intersections of sets of rows. It is easy to show that $\bigcap_{y \in I} M^{(y)} \mapsto [I]$ is a bijective function from $M_c^\cap$ to $2^Y/\Psi$ in the usual mathematical sense and, hence, $|M_c^\cap| = |2^Y/\Psi|$ and $|M_r^\cap| = |2^X/\Xi|$. Now, from the above theorem we get $|M_c^\cap| = |M_r^\cap|$, as already demonstrated by means of the introductionary example of this section.

Note that all constructions of Theorem 5.1 and its proof are relation-algebraic expressions, that is, algorithmic. As a consequence, they immediately can be translated into RELVIEW code, such that the tool can be used to compute for a given relation its column intersection types relation as well as its row intersection types relation and also the mapping that bijectively links the rows of the latter with the columns of the first one. Similar to Definition 5.1 also column union types relations and row union types relations can be introduced and then an analogon of Theorem 5.1 holds for these constructions.

6   Conclusion

At the end of Section 4, we have remarked that a one-to-one correspondence between M-contacts and Ω_M-closures also may exist for M not being (isomorphic to) a set-theoretic membership relation. Presently, we are looking for simple conditions on M which ensure that the sets $\mathfrak{K}_M$ and $\mathfrak{H}_{\Omega_M}$ are isomorphic. In this context, it is also interesting to study whether these conditions imply that Ω_M belongs to a specific class of orders. In respect thereof, a first result is that each relation M that, using matrix terminology, is obtained from a powerset relation ϵ by adding additional rows consisting of 1's only has as many M-contacts as Ω_M-closures and in this case Ω_M is isomorphic to Ω.

Besides Aumann contacts, another concept of contacts is discussed in the literature, mainly for reasoning about spatial regions. In most cases (see e.g.,[8]), the underlying structure is a Boolean lattice, i.e., essentially a powerset ordered by set inclusion. This fact leads in a natural way to the task of detecting the interdependencies between the two concepts (if such are) and whether it is also possible and reasonable to generalize the latter one similar to our generalization of Aumann contacts to M-contacts, a work that is planned for the future. Another future work is the relation-algebraic treatment of other closure objects, like implicational structures, join-congruences, Moore families and so on.

References

1. Aumann, G.: Kontaktrelationen. Bayerische Akademie der Wissenschaften. Mathematisch-Naturwissenschaftliche Klasse Sitzungsberichte 1970, 67–77 (1970)
2. Berghammer, R.: Relation-algebraic computation of fixed points with applications. Journal of Logic and Algebraic Programming 66, 112–126 (2006)
3. Berghammer, R., Neumann, F.: RELVIEW – An OBDD-Based Computer Algebra System for Relations. In: Ganzha, V.G., Mayr, E.W., Vorozhtsov, E.V. (eds.) CASC 2005. LNCS, vol. 3718, pp. 40–51. Springer, Heidelberg (2005)
4. Berghammer, R., Schmidt, G., Zierer, H.: Symmetric quotients. Report TUM-I8620, Institut für Informatik, Technische Universität München (1986)
5. Berghammer, R., Schmidt, G., Zierer, H.: Symmetric quotients and domain constructions. Information Processing Letters 33, 163–168 (1989/1990)
6. Caspard, N., Monjardet, B.: The lattices of closure systems, closure operators, and implicational systems on a finite set: a survey. Discrete Applied Mathematics 127, 241–269 (2003)
7. Doignon, J.P., Falmagne, J.C.: Knowledge spaces. Springer, Heidelberg (1999)
8. Düntsch, I., Winter, M.: A representation theorem for Boolean contact algebras. Theoretical Computer Science B 347, 498–512 (2005)
9. Ganter, B., Wille, R.: Formal concept analysis. Mathematical foundations. Springer, Heidelberg (1998)
10. Private communication with P. Jipsen (2008)
11. Kim, K.H.: Boolean matrix theory and applications. Marcel Dekker, New York (1982)
12. Schmidt, G., Ströhlein, T.: Relationen und Graphen. Springer, Heidelberg (1989). Available also in English: Relations and graphs. Discrete Mathematics for Computer Scientists. EATCS Monographs on Theoretical Computer Science (1993)
13. Tarski, A.: On the calculus of relations. Journal of Symbolic Logic 6, 73–89 (1941)

A While Program Normal Form Theorem in Total Correctness

Kim Solin

Uppsala Universitet, Uppsala, Sweden

Abstract. A classical while-program normal-form theorem is derived in demonic refinement algebra. In contrast to Kozen's partial-correctness proof of the theorem in Kleene algebra with tests, the derivation in demonic refinement algebra provides a proof that the theorem holds in total correctness.

1 A Rather Brief Introduction

A classical folk theorem says that any while program can be simulated by a while program consisting of at most one loop, provided extra Boolean variables are allowed. The normal-form theorem for while loops was first published by Böhm and Jacopini [2], but according to Harel [6] this theorem was known to Kleene before that. Kozen [8] – based on a proof of Mirkowska [9] – showed how this theorem can be perspicuously proved in Kleene algebra with tests by elegant calculational derivations, and was the first to prove the normal form theorem without introducing an explicit assignment mechanism. However, Kleene algebra with tests only provides partial-correctness proofs. In this paper, we show how to obtain a *total-correctness* normal-form theorem. Our novel total-correctness proof is based on that of Kozen, but differs in that *refinement algebra* is used in the proof. Refinement algebras are abstract algebras intended for reasoning about program refinement in a *total-correctness setting*; the creator of the first such algebra is Joakim von Wright [10,11].

We proceed as follows. First, we present demonic refinement algebra and its use for reasoning about programs. Then we consider commutativity conditions and a preservation technique, upon which we conclude by proving the normal-form theorem.

2 Demonic Refinement Algebra

The demonic refinement algebra of von Wright [10] is axiomatised over four operators and two constants. The first two operators, denoted ; and $\sqcap$, respectively, are binary infix and the last two, denoted $*$ and $^\omega$ are unary postfix. The constants are denoted 1 and $\top$.

The intended intuition behind the operators and the constants is as follows. First of all, the carrier set of the algebra is to be seen as consisting of *programs*

R. Berghammer et al. (Eds.): RelMiCS/AKA 2009, LNCS 5827, pp. 322–336, 2009.
© Springer-Verlag Berlin Heidelberg 2009

possibly containing demonic nondeterminism. This means that the operators are operators *on* programs and the constants are special programs. The *demonic-choice* operator $\sqcap$ applied to two programs, $x \sqcap y$, is to be seen as a choice between x and y made by a demon. That the choice is made by a demon means that we have no influence over it and that it can be done in the, for us, most undesirable way: striving to abortion. We will extensively use this way of looking at demonic choice in the sequel. The operator ; is *sequential composition*. It denotes sequential composition of programs: if x and y are programs, then $x; y$ denotes a program where, first, x is executed and, then, y is executed. The *weak-iteration* operator $*$ is to be seen as an iteration of any length that does terminate, whereas the *strong-iteration* operator ω is to be seen as an iteration that either terminates or goes on infinitely – which means abortion. The special program denoted by the constant $\top$ is the fictitious program magic that can establish any postcondition, and the special program denoted by 1 is skip, the immediately terminating program. Our terminology will be in the vein of Back and von Wright [1] and we will, for example, talk about execution of magic, although it is fictitious and cannot be implemented.[1]

We can now formulate the basic refinement algebra, which we call *demonic refinement algebra*.

Definition 1. (von Wright 2002) A *demonic refinement algebra* (dRA) is a structure over the signature

$$(\sqcap, ; , {}^{*}, {}^{\omega}, \top, 1)$$

satisfying the following axioms and rules ($\sqcap$ has weakest precedence, followed by ;, and then $*$ and ω, which have equal precedence – we omit ; so that $x; y$ is written xy when no confusion can arise):

$$x \sqcap (y \sqcap z) = (x \sqcap y) \sqcap z, \tag{1}$$
$$x \sqcap y = y \sqcap x, \tag{2}$$
$$x \sqcap \top = x, \tag{3}$$
$$x \sqcap x = x, \tag{4}$$
$$x(yz) = (xy)z, \tag{5}$$
$$1x = x = x1, \tag{6}$$
$$\top x = \top, \tag{7}$$
$$x(y \sqcap z) = xy \sqcap xz, \tag{8}$$
$$(x \sqcap y)z = xz \sqcap yz, \tag{9}$$

---

[1] Some would object that talking about execution of magic is nonsense, since it is nonsensical to talk about execution of a nonimplementable program: a program that cannot be implemented is no program at all and can certainly not be executed. Some would not. (The idea of using magic was introduced, independently of each other, by C.C. Morgan, J.M. Morris and G. Nelson in the 1980s).

$$x^* = 1 \sqcap xx^*, \tag{10}$$

$$x \sqsubseteq yx \sqcap z \Rightarrow x \sqsubseteq y^*z, \tag{11}$$

$$x \sqsubseteq xy \sqcap z \Rightarrow x \sqsubseteq zy^*, \tag{12}$$

$$x^\omega = 1 \sqcap xx^\omega, \tag{13}$$

$$yx \sqcap z \sqsubseteq x \Rightarrow y^\omega z \sqsubseteq x \text{ and} \tag{14}$$

$$x^\omega = x^* \sqcap x^\omega \top, \tag{15}$$

where the order $\sqsubseteq$ is defined by $x \sqsubseteq y \Leftrightarrow_{df} x \sqcap y = x$. $\lhd$

The *refinement ordering* on the algebra defined above by

$$x \sqsubseteq y \Leftrightarrow_{df} x \sqcap y = x$$

is to be read "y establishes anything that x does and possibly more" (intuitively, if x is refined by y, then a demon would always choose x since y can do anything that x does and possibly more; by choosing x the demon has a better chance of winning).

It can be shown that all the operators are isotone with respect to the refinement ordering $\sqsubseteq$ and that $\sqsubseteq$ is a partial order. The reduct structure over the signature $(\sqcap, ; , \top, 1)$ is an idempotent semiring, and the reduct structure over the signature $(\sqcap)$ is a bounded greatest-lower-bound semilattice, with $\top$ as the greatest element. In comparison to Kleene algebra [7], the axiom preventing reasoning about total correctness $(x\top = \top)$ has been removed and strong iteration has been added.

We define a syntactic constant $\bot$ with the intuition that it stands for an always nonterminating program, an abort statement [10]:

$$\bot =_{df} 1^\omega.$$

We thus equate abortion and (idle) nontermination. The syntactic constant $\bot$ is a least element and a left annihilator, as the following proposition states.

Proposition 1. *(von Wright 2002) Let x be an element in the carrier set of a* dRA. *Then*

$$\bot \sqsubseteq x \text{ and} \tag{16}$$

$$\bot x = \bot \tag{17}$$

hold.

Axioms will usually be referred to by number, but for convenience the properties will sometimes only be referred to by their canonical name, such as *associativity*, *commutativity*, *idempotence*, *skip* or *annihilation*. Axioms (8) and (9) will be referred to as *distributivity* and axioms (10) and (13) will be referred to as *unfolding*; axioms (11–12) and (14) as *induction*; and axiom (15) as *isolation*.

Let us look at the program-theoretic intuition behind some of the axioms. The third axiom says that a demon choosing between a miracle and a program

x will always choose x. This is because the demon always wants to establish abortion, and if magic is executed then this is not possible. The seventh axiom says that after magic has been executed nothing affects the program any more. The tenth axiom says that a finite iteration can be seen as an unfolding of the iterated statement: x is repeated any finite number of times until, finally, 1 is chosen, the program skips and, so, the iteration ends. Axiom (11) says that if $x \sqsubseteq yx \sqcap z$, then x can be refined by a succession of ys, ending with z:

$$x \sqsubseteq yx \sqsubseteq yyx \sqsubseteq yyyx \sqsubseteq \cdots \sqsubseteq yyy \cdots z.$$

That is, y can be repeated any finite number of times and then followed by z, in other words y^*z. Axiom (12) is analogous. Axiom (14) says the same thing as axiom (11), but now the iteration might possibly not terminate. The reason we do not have a strong-iteration axiom analogous to (12) is related to our ability to express magic: Take for example $x = y = 1$ and $z = \top$. Then the left-hand side of a strong-iteration induction rule analogous to (12) would hold, whereas the the right-hand side would not. Axiom (15) separates an iteration into its finite and infinite parts: the finite part is given by weak iteration and the infinite part is given by the $x^\omega \top$. The intuition behind $x^\omega \top$ denoting the infinite part is that unless the iteration goes on forever, and thus aborts, a demon would not choose that alternative, since this would result in a miracle. The remaining axioms can easily be given similar interpretations.

So spoke the intuition. But one could argue that this concordance of the axioms with the every-day understanding of program-theoretic constructs does not actually justify anything. It is yet to be shown how this algebra mathematically relates to how one traditionally has understood programs formally. One, but not necessarily the only, such relation is given by the following fact: the set of conjunctive predicate transformers over a fixed state space [5] equipped with standard operators (matching the above interpreation in a predicate-transformer setting) forms a demonic refinement algebra [10,11].

The *leapfrog* and *decomposition* properties

$$x(yx)^\omega = (xy)^\omega x \text{ and} \tag{18}$$
$$(x \sqcap y)^\omega = x^\omega (yx^\omega)^\omega, \tag{19}$$

respectively, have been proved by von Wright [10] and will be used later on.

An element g of the carrier set that has a complement $\bar{g}$ satisfying

$$g\bar{g} = \bar{g}g = \top \quad \text{and} \quad g \sqcap \bar{g} = 1 \tag{20}$$

is called a *guard*. We collectively refer to the guards as *the set of guards* and we will use the symbols f, g and h for denoting guards (if needed, indexed with natural numbers). Intuitively, guards are statements that check whether a predicate holds and, if so, skip, otherwise do magic. The first guard axiom says that either a predicate or its negation holds, so a sequential composition of a guard and its complement is always a miracle. The second guard axiom says that a demon will always be able to skip when choosing between a guard or the guard's complement.

As the following proposition states, the set of guards forms a Boolean algebra.

Proposition 2. *(von Wright 2002) Let G be the set of guards of a* dRA. *Then*

$$(G, \sqcap, ;, ^-, 1, \top)$$

is a Boolean algebra, where $\sqcap$ *is meet,* ; *is join,* $^-$ *is complement, 1 is the bottom element, and* $\top$ *is the top element.*

Every guard g is defined to have a corresponding *assertion*

$$g^\circ = \bar{g} \bot \sqcap 1. \tag{21}$$

This means that $^\circ$ is a mapping from guards to a subset of the carrier set, the *set of assertions*. Assertions are similar to guards, but abort if the predicate does not hold. If the predicate does not hold, then a demon would choose the left-hand side of the demonic choice and the negated guard would skip and the whole program abort (which is what a demon wants). If, on the other hand, the predicate holds, then a demon would choose the right-hand side, since otherwise the negated guard would do magic and the demon could then no longer establish abortion. Note that $\bar{g}^\circ$ means that the assertion operator $^\circ$ is applied to the guard $\bar{g}$ (and not to be read the other way around, that the complement operator is applied to the assertion g°). Assertions will not be used in this paper, but are mentioned to make the presentation of demonic refinement algebra complete. The abstract-algebraic guards and assertions have canonical interpretations as predicate transformers (see [10]).

The assertion-skip-guard property (the *asg property*) is an especially important property, which states that

$$g^\circ \sqsubseteq 1 \sqsubseteq h \tag{22}$$

holds for any guards g and h. The property is immediate from the definition of assertion and the fact that the guards form a Boolean algebra. We will also make use of the fact that

$$\bar{g}(gx)^\omega = \bar{g} \tag{23}$$

holds (by unfolding of strong iteration (13), the definition of guards and the fact that $\top$ is the top element (3)).

3   Conditionals and Loops

To show how the algebra relates to traditional program constructs we here show how to do encodings of conditionals and while loops. This means that we can express the classical while language: sequential composition, choice and iteration.

Conditionals are one of the basic building blocks of any programming language. A conditional checks whether a predicate holds, and depending on this it chooses between two actions. Traditionally, it is written

if g then x else y fi,

and is to be understood so that if g holds, then x is executed, otherwise y is executed. In the algebra we can encode this as $gx \sqcap \bar{g}y$ with the rationale that the demon always chooses the statement for which the guard holds. If the guard does not hold, then the guard performs a miracle. If the demon, then, would choose to execute that statement the demon could never establish abortion. Note that the falsum predicate is thus represented by a guard that is always miraculous, that is, by $\top$.

Another central construct in any programming language is the while loop, or loop for short. A loop

while g do x od

iterates a program statement x any number of times as long as the predicate g holds. If the predicate always holds, then the loop will iterate infinitely – a usually undesirable scenario. In dRA we have two possibilities of modeling a loop: one using weak iteration, $(gx)^* \bar{g}$, and another using strong iteration, $(gx)^\omega \bar{g}$. If weak iteration is used to model the loop we are assuming that the iteration terminates. All that is proved about a loop encoded using weak iteration thus assumes that the iteration is terminating (nevertheless, the loop might still be nonterminating (aborting) if the iterated program statement is aborting). If, on the other hand, strong iteration is used, we do not need to assume that the loop is terminating – indeed, everything we prove about a loop using the strong iteration operator holds when the loop is terminating *as well as* when it is nonterminating.

4   Commutativity Conditions

In order to state the theorem, a commutativity condition of the form

$$\begin{cases} gxg = gx \\ \bar{g}x\bar{g} = \bar{g}x \end{cases} \text{, or equivalently, of the form } \begin{cases} xg \sqsubseteq gx \\ x\bar{g} \sqsubseteq \bar{g}x \end{cases} \text{,}$$

must be made on the programs involved. Intuitively, the condition above says that "if the program x terminates, it preserves g." If the program aborts, anything can happen. We will say that x *preserves* g if x and g meet the above condition. The two equivalent conditions above correspond to Kozen's commutativity conditions, but since we want to prove total correctness, it makes sense not to assume termination of the programs. This means that, unlike Kozen, we cannot make assumptions of the form $gx \sqsubseteq xg$, since this would imply that x must terminate (cf. the total-correctness condition in [10,11]; in Kleene algebra, the characterisations of total and weak correctness coincide). Note that the first part (the first line) of the condition does not imply the second and vice versa (a concrete counterexample can be constructed).

To illustrate this technique, Kozen [8] uses

if g then $x; y_1$ else $x; y_2$ fi

as an example of a conditional that is to be simplified. We now reuse this example and resettle the properties in total correctness. First assume that g is preserved by x, that is assume that

$$\begin{cases} gxg = gx \\ \bar{g}x\bar{g} = \bar{g}x \end{cases}$$

holds. Then the program can be rewritten into a more separated form as

x; if g then y_1 else y_2 fi ,

which can be formulated and proved in dRA by

$$x(gy_1 \sqcap \bar{g}y_2)$$
$$= \{\text{distributivity}\}$$
$$xgy_1 \sqcap x\bar{g}y_2$$
$$= \{\text{axiom (6), definition of guards (20)}\}$$
$$(g \sqcap \bar{g})xgy_1 \sqcap (g \sqcap \bar{g})x\bar{g}y_2$$
$$= \{\text{distributivity (9)}\}$$
$$gxgy_1 \sqcap \bar{g}xgy_1 \sqcap gx\bar{g}y_2 \sqcap \bar{g}x\bar{g}y_2$$
$$= \{\text{preservation assumption}\}$$
$$gxy_1 \sqcap \bar{g}x\bar{g}gy_1 \sqcap gxg\bar{g}y_2 \sqcap \bar{g}xy_2$$
$$= \{\text{definition of guards, axiom (7)}\}$$
$$gxy_1 \sqcap \bar{g}x\top \sqcap gx\top \sqcap \bar{g}xy_2$$
$$= \{\text{axiom (2), distributivity (8)}\}$$
$$gx(y_1 \sqcap \top) \sqcap \bar{g}x(y_2 \sqcap \top)$$
$$= \{\text{axiom (3)}\}$$
$$gxy_1 \sqcap \bar{g}xy_2.$$

Although similar, the proof is different from Kozen's, since we work in total correctness and thus in demonic refinement algebra and with the above preservation conditions.

It is easy to show that if e is a well-formed expression consisting of elements from the carrier set and the operators ; and $\sqcap$ and all the carrier-set elements preserve a guard g, then the whole expression e preserves g. This, in turn, means that x^ω preserves g (this is the invariant rule for strong iteration in weak correctness). We will sometimes refer to this fact as "move guard."

Moreover, assuming that x preserves g, it can be shown – by induction, the assumption, distributivity, and unfolding – that

$$(gx)^\omega g \sqsubseteq gx^\omega \tag{24}$$

holds, a fact which we will employ later on.

5   Kozen's Preservation Technique

Suppose we would like to preserve the value of g across the program x, but we cannot assume that x preserves g. To do this we need to first introduce a new guard h and assume that x preserves h. Then we can set h to g by a special program $z; (g \leftrightarrow h)$, where $g \leftrightarrow h =_{df} hg \sqcap \bar{h}\bar{g}$, and this program can then be injected into an appropriate place. This corresponds, on an abstract level, to adding extra Boolean variables. As Kozen [8] notes, the intuition is that z assigns the value of g to some new Boolean variable that is tested by h. The guard $g \leftrightarrow h$ says that g and h have the same Boolean value just after execution of z [8]. This technique was used by Kozen [8] in his seminal paper on Kleene algebra with tests.

Consider again the example with the conditional from the previous section, but assume now that g is not preserved by x. Then we can use Kozen's technique to first "store" the value of g in a new guard h which is preserved by x, and then prove that

$$z; (g \leftrightarrow h); \text{if } g \text{ then } x; y_1 \text{ else } x; y_2 \text{ fi}$$

is equivalent to

$$z; (g \leftrightarrow h); x; \text{if } h \text{ then } y_1 \text{ else } y_2 \text{ fi} .$$

This is done as follows. Assume

$$\begin{cases} hxh = hx \\ \bar{h}x\bar{h} = \bar{h}x, \end{cases}$$

and derive

$(g \leftrightarrow h)(gxy_1 \sqcap \bar{g}xy_2)$
$= \{\text{definition}\}$
$(gh \sqcap \bar{g}\bar{h})(gxy_1 \sqcap \bar{g}xy_2)$
$= \{\text{distributivity}\}$
$ghgxy_1 \sqcap \bar{g}\bar{h}gxy_1 \sqcap ghg\bar{g}xy_2 \sqcap \bar{g}\bar{h}\bar{g}xy_2$
$= \{\text{guards form a Boolean algebra}\}$
$ghgxy_1 \sqcap \bar{g}g\bar{h}xy_1 \sqcap g\bar{g}hxy_2 \sqcap \bar{g}\bar{h}\bar{g}xy_2$
$= \{\text{definition of guards (20), axiom (7)}\}$
$ghgxy_1 \sqcap \top \sqcap \top \sqcap \bar{g}\bar{h}\bar{g}xy_2$
$= \{\text{axiom (3)}\}$
$ghgxy_1 \sqcap \bar{g}\bar{h}\bar{g}xy_2$
$= \{\text{guards form a Boolean algebra}\}$
$ghxy_1 \sqcap \bar{g}\bar{h}xy_2$
$= \{\text{axiom (3)}\}$
$ghx(y_1 \sqcap \top) \sqcap \bar{g}\bar{h}x(y_2 \sqcap \top)$
$= \{\text{preservation assumption, guards form a BA, distributivity}\}$
$ghxhy_1 \sqcap ghx\bar{h}y_2 \sqcap \bar{g}\bar{h}xhy_1 \sqcap \bar{g}\bar{h}x\bar{h}y_2$
$= \{\text{distributivity}\}$
$(gh \sqcap \bar{g}\bar{h})x(hy_1 \sqcap \bar{h}y_2)$
$= \{\text{definition}\}$
$(g \leftrightarrow h)x(hy_1 \sqcap \bar{h}y_2).$

By isotony, we have then proved the claim.

6   The Normal Form Theorem

We will say that a while program is in *normal form* if it is of the form

x; while g do y od,

where x and y do not contain while loops. Using Kleene algebra with tests
Kozen [8] proved that every while program can be written in normal form. Kozen
thus proved the theorem in partial correctness, but here we use refinement algebra to obtain a theorem in total correctness.

Theorem 1. *Every (possibly nonterminating) while program, appropriately aug-
mented with subprograms of the form $z; (g \leftrightarrow h)$ and when reasoning under
preservation assumptions of the form*

$$\begin{cases} gxg = gx \\ \bar{g}x\bar{g} = \bar{g}x \end{cases},$$

is equivalent to a while program in normal form.

Proof. The theorem is proved by induction on the structure of while programs.
Following Kozen [8], who in turn follows Mirkowska [9], we give a method for
moving an inner while loop to the outside for every program construct. That
we are working in demonic refinement algebra and encode the loop with strong
iteration in order to obtain total correctness means that several of the individual
steps must be done quite differently from Kozen's.

Step 1: Conditional. Consider the program

if g then x_1 while f_1 do y_1 od
        else x_2 while f_2 do y_2 od fi .

To show how to move the while loops outside, we first introduce a new test h and
program z that sets h to g. We also assume that h is preserved by the programs
x_1, x_2, y_1 and y_2. Having taken on these assumptions, we prove that

$$z; (g \leftrightarrow h); \text{if } g \text{ then } x_1 \text{ while } f_1 \text{ do } y_1 \text{ od} \qquad (25)$$
$$\text{else } x_2 \text{ while } f_2 \text{ do } y_2 \text{ od fi}$$

and

$$z; (g \leftrightarrow h); \text{if } h \text{ then } x_1 \text{ else } x_2 \text{ fi };$$
$$\text{while } (hf_1 \sqcap \bar{h}f_2) \text{ do} \qquad (26)$$
$$\text{if } h \text{ then } y_1 \text{ else } y_2 \text{ fi}$$
$$\text{od}$$

are equivalent. To prove that (25) and (26) are equivalent, the beginning z can
be removed and the remaining parts be shown equivalent – this follows from
isotony. Encoding into demonic refinement algebra and then using distributivity,

the guard definition, axiom (7) and Boolean algebra for simplifying, the first expression (25) takes the form

$$ghx_1(f_1y_1)^\omega \bar{f}_1 \;\sqcap\; \bar{g}hx_2(f_2y_2)^\omega \bar{f}_2. \tag{27}$$

Similarly, the second expression (26) becomes

$$(ghx_1 \sqcap \bar{g}hx_2)(hf_1y_1 \sqcap \bar{h}f_2y_2)^\omega (h\bar{f}_1 \sqcap \bar{h}\bar{f}_2). \tag{28}$$

To see that this is indeed so, use the basic equality

$$hf_1 \sqcap \bar{h}f_2 = (\bar{h} \sqcap f_1)(h \sqcap f_2),$$

de Morgan rules and double negation on the subexpression $\overline{hf_1 \sqcap \bar{h}f_2}$, and then distributivity on the remaining subexpression; this is exactly like in Kozen's paper [8]. The second expression (28) is in fact equivalent to

$$\begin{aligned}
&ghx_1(hf_1y_1 \sqcap \bar{h}f_2y_2)^\omega h\bar{f}_1\\
\sqcap\;\\
&ghx_1(hf_1y_1 \sqcap \bar{h}f_2y_2)^\omega \bar{h}\bar{f}_2\\
\sqcap\;\\
&\bar{g}hx_2(hf_1y_1 \sqcap \bar{h}f_2y_2)^\omega h\bar{f}_1\\
\sqcap\;\\
&\bar{g}hx_2(hf_1y_1 \sqcap \bar{h}f_2y_2)^\omega \bar{h}\bar{f}_2
\end{aligned} \tag{29}$$

by distributivity (cf. again Kozen [8]). We now show how the equality of (27) and (29) can be derived.

For the reverse refinement, $\sqsupseteq$, it suffices to derive

$$\begin{aligned}
&ghx_1(hf_1y_1 \sqcap \bar{h}f_2y_2)^\omega h\bar{f}_1\\
&\sqsubseteq \{\text{isotony}\}\\
&ghx_1(hf_1y_1)^\omega h\bar{f}_1\\
&\sqsubseteq \{\text{property (24)}\}\\
&ghx_1 h(hf_1y_1)^\omega \bar{f}_1\\
&= \{\text{preservation assumption}\}\\
&ghx_1(hf_1y_1)^\omega \bar{f}_1,
\end{aligned}$$

that is, to derive the left-hand side (with respect to $\sqcap$) of (27) from the first part of (29). The right-hand side of (27) follows symmetrically from the fourth part of (29). By this and isotony we have shown the reverse refinement.

For the refinement, $\sqsubseteq$, we derive

$$\begin{aligned}
&ghx_1(f_1y_1)^\omega \bar{f}_1\\
&\sqsubseteq \{\text{asg property (22)}\}\\
&ghx_1(hf_1y_1)^\omega h\bar{f}_1\\
&= \{\text{guards form a Boolean algebra, axiom (3)}\}\\
&ghx_1(hhf_1y_1 \sqcap \top)^\omega h\bar{f}_1\\
&= \{\text{guards form a Boolean algebra, axiom (7), distributivity}\}\\
&ghx_1(h(hf_1y_1 \sqcap \bar{h}f_2y_2))^\omega h\bar{f}_1\\
&\sqsubseteq \{\text{property (24)}\}
\end{aligned}$$

$$ghx_1h(hf_1y_1 \sqcap \bar{h}f_2y_2)^\omega \bar{f}_1$$
$= \{\text{preservation assumption}\}$
$$ghx_1(hf_1y_1 \sqcap \bar{h}f_2y_2)^\omega \bar{f}_1$$
$\sqsubseteq \{\text{asg property (22)}\}$
$$ghx_1(hf_1y_1 \sqcap \bar{h}f_2y_2)^\omega h\bar{f}_1.$$

We also derive

$$ghx_1(f_1y_1)^\omega \bar{f}_1$$
$\sqsubseteq \{\text{asg property (22), isotony}\}$
$$ghx_1(hf_1y_1)^\omega \top$$
$= \{\text{definition of guards (20), axiom (7)}\}$
$$ghx_1(hf_1y_1)^\omega h\bar{h}\bar{f}_2$$
$\sqsubseteq \{\text{preservation assumption, move guard}\}$
$$ghx_1h(hf_1y_1)^\omega \bar{h}\bar{f}_2$$
$\sqsubseteq \{\text{assumption A, see the appendix}\}$
$$ghx_1h(hf_1y_1 \sqcap \bar{h}f_2y_2)^\omega \bar{h}\bar{f}_2$$
$= \{\text{preservation assumption}\}$
$$ghx_1(hf_1y_1 \sqcap \bar{h}f_2y_2)^\omega \bar{h}\bar{f}_2.$$

By this, we have established that

$$ghx_1(f_1y_1)^\omega \bar{f}_1 \sqsubseteq ghx_1(hf_1y_1 \sqcap \bar{h}f_2y_2)^\omega h\bar{f}_1 \sqcap ghx_1(hf_1y_1 \sqcap \bar{h}f_2y_2)^\omega \bar{h}\bar{f}_2,$$

and symmetric reasoning can be used to show that the right-hand side of (27) derives the two remaining parts of (29).

Step 2: Nested loops. In this step, we can follow Kozen [8], since no commutativity conditions are needed – nevertheless, strong iteration is different from weak iteration and this must be taken into consideration.

We first show that

while f do x; while g do y od od   (30)

is equal to

if f
then x; while $f \sqcap g$ do if g then y else x fi od   (31)
else skip
fi

and then the normal form follows from applying the rule in Step 1. It is easy to show that the skip clause in the conditional is equivalent to

skip; while $\top$ do skip od,

so that the second form exactly matches that of Step 1.

Program (30) takes the form

$$(fx(gy)^\omega \bar{g})^\omega \bar{f}$$

in refinement algebra, and (31) becomes

$$fx(gy \sqcap \bar{g}fx)^\omega \bar{f}\bar{g} \sqcap \bar{f},$$

after simplification with distributivity and Boolean algebra. We now calculate

$$(fx(gy)^\omega \bar{g})^\omega \bar{f} = fx(gy \sqcap \bar{g}fx)^\omega \bar{f}\bar{g} \sqcap \bar{f}$$
$$\Leftarrow \{\text{unfolding (13), isotony}\}$$
$$(gy)^\omega \bar{g}(fx(gy)^\omega \bar{g})^\omega \bar{f} = (gy \sqcap \bar{g}fx)^\omega \bar{f}\bar{g}$$
$$\Leftrightarrow \{\text{leapfrog (18)}\}$$
$$(gy)^\omega (\bar{g}fx(gy)^\omega)^\omega \bar{g}\bar{f} = (gy \sqcap \bar{g}fx)^\omega \bar{f}\bar{g}$$
$$\Leftrightarrow \{\text{guards form a Boolean algebra}\}$$
$$(gy)^\omega (\bar{g}fx(gy)^\omega)^\omega \bar{f}\bar{g} = (gy \sqcap \bar{g}fx)^\omega \bar{f}\bar{g}$$
$$\Leftrightarrow \{\text{decomposition}\}$$
$$\text{true}$$

and have so proved what we wanted to establish.

Step 3: Eliminating postcomputations. In this step of the proof we show that a computation that is to be executed after a while loop can be included in the while loop. More precisely, we first show that

$$\textsf{while } g \textsf{ do } x \textsf{ od; } y \tag{32}$$

is equal to

$$\begin{aligned}
&\textsf{if } \bar{g} \textsf{ then } y; \\
&\quad \textsf{else while } g \textsf{ do } x \\
&\qquad\qquad\qquad \textsf{if } \bar{g} \textsf{ then } y \textsf{ else skip fi} \\
&\qquad\qquad \textsf{od}
\end{aligned} \tag{33}$$

under the assumption that y preserves g. (The case where g is not preserved by y is dealt with later.)

If x and y are while free, then the else clause of (33) is in normal form and the whole program can thus be transformed in to normal form by Step 1.

Expression (32) can be encoded into refinement algebra as $(gx)^\omega \bar{g}y$, which after unfolding the omega once by axiom (13) and then distributing becomes $gx(gx)^\omega \bar{g}y \sqcap \bar{g}y$. The second expression becomes $\bar{g}y \sqcap g(gx(\bar{g}y \sqcap g))^\omega \bar{g}$ after applying axiom (6). We now show that

$$(gx)^\omega \bar{g}y \sqsubseteq \bar{g}y \sqcap g(gx(\bar{g}y \sqcap g))^\omega \bar{g}$$

and

$$\bar{g}y \sqcap g(gx(\bar{g}y \sqcap g))^\omega \bar{g} \sqsubseteq gx(gx)^\omega \bar{g}y \sqcap \bar{g}y$$

and have thereby shown what we want.

For the first claim, it suffices to calculate

$$(gx)^\omega \bar{g}y \sqsubseteq \bar{g}y \sqcap g(gx(\bar{g}y \sqcap g))^\omega \bar{g}$$
$\Leftrightarrow$ {distributivity}
$$(gx)^\omega \bar{g}y \sqsubseteq \bar{g}y \sqcap g(gx\bar{g}y \sqcap gxg)^\omega \bar{g}$$
$\Leftarrow$ {induction (14)}
$$gx(\bar{g}y \sqcap g(gx\bar{g}y \sqcap gxg)^\omega \bar{g}) \sqcap \bar{g}y \sqsubseteq \bar{g}y \sqcap g(gx\bar{g}y \sqcap gxg)^\omega \bar{g}$$
$\Leftarrow$ {isotony}
$$gx(\bar{g}y \sqcap g(gx\bar{g}y \sqcap gxg)^\omega \bar{g}) \sqsubseteq g(gx\bar{g}y \sqcap gxg)^\omega \bar{g}$$
$\Leftrightarrow$ {short hand: $K =_{df} (gx\bar{g}y \sqcap gxg)^\omega \bar{g}$, unfolding (13), distributivity, guards form a Boolean algebra}
$$gx(\bar{g}y \sqcap gK) \sqsubseteq gx\bar{g}yK \sqcap gxgK \sqcap g\bar{g}$$
$\Leftrightarrow$ {guards form a Boolean algebra, axiom (3)}
$$gx(\bar{g}y \sqcap gK) \sqsubseteq gx\bar{g}yK \sqcap gxgK$$
$\Leftarrow$ {distributivity, isotony}
$$gx\bar{g}y \sqsubseteq gx\bar{g}yK$$
$\Leftrightarrow$ {preservation assumption}
$$gx\bar{g}y \sqsubseteq gx\bar{g}y\bar{g}K$$
$\Leftrightarrow$ {distributivity, property (23)}
$$gx\bar{g}y \sqsubseteq gx\bar{g}y\bar{g}$$
$\Leftrightarrow$ {preservation assumption}
$$gx\bar{g}y \sqsubseteq gx\bar{g}y$$
$\Leftrightarrow$ {reflexivity of refinement}
true.

For the second claim, we have that

$$g(gx(\bar{g}y \sqcap g))^\omega \bar{g}$$
$=$ {unfolding (13), distributivity, guards form a BA, axiom (3)}
$$gx(\bar{g}y \sqcap g)(gx(\bar{g}y \sqcap g))^\omega \bar{g}$$
$=$ {leapfrog (18)}
$$gx((\bar{g}y \sqcap g)gx)^\omega (\bar{g}y \sqcap g)\bar{g}$$
$\sqsubseteq$ {isotony, guards form a Boolean algebra}
$$gx(gx)^\omega \bar{g}y\bar{g}$$
$=$ {preservation assumption}
$$gx(gx)^\omega \bar{g}y,$$

which by isotony and idempotency of demonic choice establishes what we wanted to prove (this direction is similar to that of Kozen [8]).

If y does not preserve g, then we can introduce a new test f that is preserved by y and a program z that sets f to g. We can then insert the program $z; (f \leftrightarrow g)$ before the loop and into the loop body. This means that the programs

$$z; (f \leftrightarrow g); \text{ while } g \text{ do } x; z; (f \leftrightarrow g) \text{ od}; y$$

and

$$z; (f \leftrightarrow g); \text{ while } f \text{ do } x; z; (f \leftrightarrow g) \text{ od}; y$$

are equivalent, and we can replace the former with the latter – for which the commutativity assumption that y preserves f holds. The proof that these two programs are indeed equivalent is left to the reader (for the weak iteration version, see Kozen [8]).

Step 4: Composition. The last step we need to consider before finishing our proof is that of composing two programs in normal form. In this step, we can almost exactly follow Kozen's proof. What we want to do is, then, to transform the program

$$x_1; \text{while } g_1 \text{ do } y_1 \text{od}; x_2; \text{while } g_2 \text{ do } y_2 \text{ od} \tag{34}$$

into a program in normal form. First, by Step 3, we can move x_2 into the first while loop. The program x_1 can, as Kozen notes, be ignored as it can be included in the precomputation of the resulting normal-form program. It thus suffices to show that

$$\text{while } g \text{ do } x \text{ od}; \text{while } h \text{ do } y \text{ od} \tag{35}$$

can be turned into normal form.

We can assume that g commutes with y without loss of generality, just like in Step 3. This means that g commutes with the second while loop, since

$$yg \sqsubseteq gy$$
$\Rightarrow \{\text{isotony}\}$
$$hyg \sqsubseteq hgy$$
$\Leftrightarrow \{\text{guards form a Boolean algebra}\}$
$$hyg \sqsubseteq ghy$$
$\Rightarrow \{\text{strong iteration preserves outer commutativity, see von Wright [10]}\}$
$$(hy)^\omega g \sqsubseteq g(hy)^\omega$$
$\Rightarrow \{\text{isotony}\}$
$$(hy)^\omega g\bar{h} \sqsubseteq g(hy)^\omega\bar{h}$$
$\Leftrightarrow \{\text{guards form a Boolean algebra}\}$
$$(hy)^\omega \bar{h}g \sqsubseteq g(hy)^\omega\bar{h}$$

holds. This, in turn, means that we can use Step 3 to turn the program into

$$\begin{aligned}
&\text{if } \bar{g} \text{ then while } h \text{ do } y \text{ od} \\
&\quad\quad \text{else while } g \text{ do } x \\
&\quad\quad\quad\quad \text{if } \bar{g} \text{ then while } h \text{ do } y \text{ od else skip fi} \\
&\quad\quad \text{od.}
\end{aligned} \tag{36}$$

We can now apply Step 1 to the inner conditional, which yields two nested while loops. These nested while loops can be transformed with Step 2. Finally, we can apply Step 1, which gives us a program in normal form.

The transformations of the steps above yield a systematic method for transforming any program into normal form by inductively moving while loops outwards, starting from the innermost loop.   □

7   The Proof Is Concluded and So Is the Paper

The abstract-algebraic method both initiated and revived by the work of Kozen, Cohen, von Wright, the Desharnais-Möller-Struth trio and others should be interesting for several different communities. As we hope to have shown in this paper, abstract algebra can be useful for proving properties of programs, and much of the work done in related frameworks can be reused, or only slightly modified, to yield interesting results. The abstract-algebraic method is thus a method worth learning if one is looking for an efficient and practical reasoning tool.

Acknowledgements. Thanks to L.A. Meinicke, R.J.R. Back, Jules Desharnais, E.C.R. Hehner and Bernhard Möller for valuable discussions, and to three anonymous reviewers for helpful suggestions.

References

1. Back, R.J.R., von Wright, J.: Refinement calculus: A systematic introduction. Springer, Heidelberg (1998)
2. Böhm, C., Jacopini, G.: Flow diagrams, Turing machines and languages with only two formation rules. Commun. ACM 9(5), 366–371 (1966)
3. Cohen, E.: Separation and reduction. In: Backhouse, R., Oliveira, J.N. (eds.) MPC 2000. LNCS, vol. 1837, pp. 45–59. Springer, Heidelberg (2000)
4. Desharnais, J., Möller, B., Struth, G.: Kleene algebra with domain. ACM Trans. Comput. Log. 7(4), 798–833 (2006)
5. Dijkstra, E.W.: A discipline of programming. Prentice-Hall International, Englewood Cliffs (1976)
6. Harel, D.: On folk theorems. Commun. ACM 23(7), 379–389 (1980)
7. Kozen, D.: A completeness theorem for Kleene algebras and the algebra of regular events. Inf. Comput. 110(2), 366–390 (1994)
8. Kozen, D.: Kleene algebra with tests. ACM Trans. Prog. Lang. Syst. 19(3), 427–443 (1997)
9. Mirkowska, G.: Algorithmic logic and its applications. PhD thesis. Univ. of Warsaw, Warsaw, Poland (1972) (in Polish)
10. von Wright, J.: From Kleene algebra to refinement algebra. In: Boiten, E.A., Möller, B. (eds.) MPC 2002. LNCS, vol. 2386, pp. 233–262. Springer, Heidelberg (2002)
11. von Wright, J.: Towards a refinement algebra. Science of Computer Programming 51, 23–45 (2004)

A   Proof of Assumption A

Assumption A follows from the general fact that for any guard g and any elements x and y in the carrier set such that x preserves g

$$g(gx)^\omega \sqsubseteq g(gx \sqcap \bar{g}y)^\omega$$
$\Leftarrow$ {guards form a Boolean algebra, isotony, induction (14)}
$$gxg(gx \sqcap \bar{g}y)^\omega \sqcap 1 \sqsubseteq g(gx \sqcap \bar{g}y)^\omega$$
$\Leftrightarrow$ {unfold (13), distributivity, guards form a BA, axiom (7)}
$$gxg(gx \sqcap \bar{g}y)^\omega \sqcap 1 \sqsubseteq (gx \sqcap \top)(gx \sqcap \bar{g}y)^\omega \sqcap g$$
$\Leftrightarrow$ {axiom (3), preservation assumption, asg property (22), isotony}
true.

Complements in Distributive Allegories

Michael Winter*

Department of Computer Science,
Brock University,
St. Catharines, Ontario, Canada, L2S 3A1
mwinter@brocku.ca

Abstract. It is known in topos theory that the axiom of choice implies that the topos is Boolean. In this paper we want to prove and generalize this result in the context of allegories. In particular, we will show that partial identities do have complements in distributive allegories with relational sums and total splittings assuming the axiom of choice. Furthermore, we will discuss possible modifications of the assumptions used in that theorem.

1 Introduction

The calculus of relations, and its categorical versions in particular, are often used to model programming languages, classical and non-classical logics and different methods of data mining (see for example [1,2,10,12]). In many applications the notion of a complement is essential to specify and/or solve a given problem. Even though a lot of properties can be formalized using pseudocomplements or residuals a classical negation is sometimes needed.

It is well-known in topos theory [3,5,8] that the axiom of choice implies that the underlying logic is classical, i.e. that the topos is Boolean. In particular, it is clear from the proof (see [3]) that this result already holds in any pretopos. This fact immediately implies the same result for distributive allegories with total splittings and relational sums that are also tabular. Notice that it is well-known that a tabular allegory is also representable [4] but not necessarily vice versa. The current paper was motivated by this situation. We will generalize the previous result to arbitrary, not necessarily tabular, allegories. The motivation for this is threefold. First of all, relation algebras are closely related to models of the 3-variable fragment of various kinds of logics [11]. In distributive allegories, besides being categories instead of algebras, the complement operation has been removed. Therefore, they can be used to model the 3-variable fragment of the corresponding restricted logic which provides an immediate motivation to study the axiom of choice. Notice that allegories with relational products, or pretabular allegories, are already sufficient to model logics with 4 or more variables. We will provide a simple example of an allegory that has relational products and

* The author gratefully acknowledges support from the Natural Sciences and Engineering Research Council of Canada.

R. Berghammer et al. (Eds.): RelMiCS/AKA 2009, LNCS 5827, pp. 337–350, 2009.

is representable but is not tabular. Even though the original theorem cannot be applied, it follows from the theory developed in this paper that the allegory in question is indeed Boolean. Secondly, in certain applications of relational methods the axiom of choice plays an important role. For example, a decomposition of a graph using cardinal maximal dicliques uses obviously cardinality properties of relations while simultaneoulsy staying in a 3-variable environment. Certain properties of the underlying cardinal arithmetic depend on the axiom of choice [13]. Last but not least, it has been shown that some non-representable allegories have interesting properties such as the so-called unsharpness of relational products [9]. Those properties, and, hence, non-representable allegories, might be of particular interest for certain applications such as models of concurrency and quantum computing.

In addition, we will discuss weakenings of the assumptions used in the main theorem. In particular, we will be interested whether it is possible to replace the existence of total splittings and/or relational sums by stronger versions of the axiom of choice. Notice that those considerations are trivial for tabular allegories, i.e. a tabular allegory satisfying ACC, introduced in Section 4, already has all total splittings. However, already for pre-tabular allegories, which are still representable, this is not the case.

The paper is organized as follows. In Section 2 we introduce some basic notions needed throughout the paper. The main Section 3 focuses on the relational (and generalized) version of the topos theoretic result mentioned above. Finally, in Section 4 we discuss possible modifications of the assumptions used in the main theorem.

2   Categories of Relations

Throughout this paper we assume that the reader is familiar with the basic notions from category and lattice theory. For notions not defined here we refer to [4,6].

Given a category $\mathcal{C}$ we denote its collection of objects by $\mathrm{Obj}_\mathcal{C}$ and its collection of morphisms by $\mathrm{Mor}_\mathcal{C}$. To indicate that a morphism f has source A and target B we usually write $f : A \to B$. The collection of all morphisms between A and B is denoted by $\mathcal{C}[A, B]$. We use ; for composition of morphisms, which has to be read from left to right, i.e. $f; g$ means first f then g. The identity morphism on the object A is written as $\mathbb{I}_A$.

Definition 1. *An allegory $\mathcal{R}$ is a category satisfying the following:*

1. *For all objects A and B the class $\mathcal{R}[A, B]$ is a lower semi-lattice. Meet and the induced ordering are denoted by $\sqcap, \sqsubseteq$, respectively. The elements in $\mathcal{R}[A, B]$ are called relations.*
2. *There is a monotone operation $\smile$ (called the converse operation) such that for all relations $Q, R : A \to B$ and $S : B \to C$ the following holds*

$$(Q; S)^\smile = S^\smile; Q^\smile \quad and \quad (Q^\smile)^\smile = Q.$$

3. *For all relations* $Q : A \to B$, $R, S : B \to C$ *we have* $Q; (R \sqcap S) \sqsubseteq Q; R \sqcap Q; S$.
4. *For all relations* $Q : A \to B, R : B \to C$ *and* $S : A \to C$ *the following modular law holds* $Q; R \sqcap S \sqsubseteq Q; (R \sqcap Q^\smile; S)$.

A relation $R : A \to B$ is called univalent (or a partial function) iff $R^\smile; R \sqsubseteq \mathbb{I}_B$ and total iff $\mathbb{I}_A \sqsubseteq R; R^\smile$. Functions are total and univalent relations. In the remainder of the paper we will denote univalent relations usually by lowercase letters. R is called injective iff $R^\smile$ is univalent and surjective iff $R^\smile$ is total.

In the following lemma we have summarized several basic properties of relations used in this paper. A proof can be found in [4,10,12].

Lemma 1. *Let* $\mathcal{R}$ *be an allegory. Then we have:*

1. $Q; R \sqcap S \sqsubseteq (Q \sqcap S; R^\smile); (R \sqcap Q^\smile; S)$ *for all relations* $Q : A \to B, R : B \to C$ *and* $S : A \to C$ *(Dedekind formula),*
2. $R = (\mathbb{I}_A \sqcap R; R^\smile); R$ *for all* $R : A \to B$,
3. *If* $Q : A \to B$ *is univalent, then* $Q; (R \sqcap S) = Q; R \sqcap Q; S$ *for all relations* $R, S : B \to C$,
4. *If* $R : B \to C$ *is univalent, then* $Q; R \sqcap S = (Q \sqcap S; R^\smile); R$ *for all relations* $Q : A \to B$ *and* $S : A \to C$.

Another important class of relations is given by partial identities, i.e. relations $i : A \to A$ such that $i \sqsubseteq \mathbb{I}_A$. Partial identities is one possible way of abstractly describing subsets of a given object.

Lemma 2. *Let* $\mathcal{R}$ *be an allegory, and* $i, j : A \to A$ *be partial identities. Then we have:*

1. i *is symmetric and idempotent, i.e.* $i^\smile = i$ *and* $i; i = i$,
2. $i; j = i \sqcap j$.

A proof may be found in [4,12].

A subset M of a set N may also be described by the canonical injection $f : M \to N$. Furthermore, the set of equivalence classes of an equivalence relation is fully determined by the function mapping each element to its equivalence class. Combining both concepts we aim at the notion of a splitting.

Definition 2. *Let* $Q : A \to A$ *be a symmetric idempotent relation, i.e.* $Q^\smile = Q$ *and* $Q; Q = Q$. *An object* B *together with a relation* $R : B \to A$ *is called a splitting of* Q *(or* R *splits* Q*) iff* $R; R^\smile = \mathbb{I}_B$ *and* $R^\smile; R = Q$.

A splitting is unique up to isomorphism. If Q is a partial identity, the object B of the splitting corresponds to the subset given by Q. Analogously, if Q is an equivalence relation, i.e. $\mathbb{I}_A \sqsubseteq Q, Q; Q \sqsubseteq Q$ and $Q^\smile = Q$, B corresponds to the set of equivalence classes. Throughout this paper we are mainly interested in splitting equivalence relations. Therefore, we say an allegory has total splittings if every equivalence relation Q, i.e. $\mathbb{I}_A \sqsubseteq Q$ and $Q; Q \sqsubseteq Q$, splits.

The next lemma collects some properties of symmetric and idempotent relations needed in this paper.

Lemma 3. *Let $\mathcal{R}$ be an allegory, $X : B \to B$ be a symmetric and idempotent relation and $R : A \to B$ with $R; X = X$. Then we have:*

1. *If $S \sqsubseteq R$, then $S \sqsubseteq S; X$,*
2. *If $Q \sqsubseteq X$, then $R; Q^{\smile} \sqsubseteq R$.*

Proof

1. follows immediately from

$$
\begin{aligned}
S &= S \sqcap R \\
&= S \sqcap R; X \\
&\sqsubseteq (S; X^{\smile} \sqcap R); X \\
&\sqsubseteq S; X^{\smile}; X \\
&= S; X.
\end{aligned}
$$

2. We get $R; Q^{\smile} \sqsubseteq R; X^{\smile} = R; X = R$. $\square$

The notion of a distributive allegory normally does not include the existence of a greatest element for every collection of relations [4,7,12]. For convenience we will include this element in the current paper.

Definition 3. *An allegory $\mathcal{R}$ is called distributive if each class of morphisms $\mathcal{R}[A, B]$ is a distributive lattice with least element $\bot\!\!\!\bot_{AB}$ and greatest element $\top\!\!\!\top_{AB}$. The union operation is denoted by $\sqcup$.*

Two functions $f : C \to A$ and $g : C \to B$ with common source are said to tabulate a relation $R : A \to B$ iff $R = f^{\smile}; g$ and $f; f^{\smile} \sqcap g; g^{\smile} = \mathbb{I}_C$. If for all relations of an allegory $\mathcal{R}$ there is tabulation, then $\mathcal{R}$ is called tabular. Notice that a function $f : A \to B$ and its converse $f^{\smile} : B \to A$ always have a tabulation. The tabulation is given by $(\mathbb{I}_A, f)$ and $(f, \mathbb{I}_B)$, respectively. A tabulation of the greatest element $\top\!\!\!\top_{AB}$ is called a relational product of A and B. In this case the the object of the tabulation is usually denoted by $A \times B$ and the two projection functions by $\pi : A \times B \to A$ and $\rho : A \times B \to B$. If $\mathcal{R}$ has all relational products but not necessarily tabulations for every relation, $\mathcal{R}$ is called pre-tabular.

The dual concept of a relational product is a relational sum.

Definition 4. *Let $\mathcal{R}$ be a distributive allegory and A and B be objects of $\mathcal{R}$. Then an object $A+B$ together with two relations $\iota : A \to A+B$ and $\kappa : B \to A+B$ is called a relational sum if it satisfies the following:*

$$
\iota; \iota^{\smile} = \mathbb{I}_A, \qquad \kappa; \kappa^{\smile} = \mathbb{I}_B, \qquad \iota; \kappa^{\smile} = \bot\!\!\!\bot_{AB}, \qquad \iota^{\smile}; \iota \sqcup \kappa^{\smile}; \kappa = \mathbb{I}_{A+B}.
$$

The next lemma is of particular interest if R and S are the injections ι and κ of a relational sum.

Lemma 4. *Let $\mathcal{R}$ be a distributive allegory, $R : A \to C$ and $S : B \to C$ with $R; S^{\smile} = \bot\!\!\!\bot_{AB}$. Then we have for all $Q_1 : D \to A, Q_2 : D \to B$ and $T_1 : A \to E, T_2 : B \to E$:*

1. $Q_1; R \sqcap Q_2; S = \mathbb{1}_{DC}$,
2. $(Q_1; R \sqcup Q_2; S); (R^{\smile}; T_1 \sqcup S^{\smile}; T_2) = Q_1; T_1 \sqcup Q_2; T_2$.

Proof

1. follows immediately from $Q_1; R \sqcap Q_2; S \sqsubseteq (Q_1; R; S^{\smile} \sqcap Q_2); S = \mathbb{1}_{CD}$.
2. was already shown several times, e.g. [10].   □

Throughout this paper we are interested in the implication of the axiom of choice in the context of allegories. One of the possible version of this axiom is as follows:

(AC) For all total relations $R : A \rightarrow B$ there is a function $f : A \rightarrow B$ with $f \sqsubseteq R$.

In the case that the subcategory of functions constitutes a topos the property above is equivalent to the topos theoretic version of this axiom (see Lemma 4.5.6 in [8]).

3   Complements in Allegories

In this section we want to prove the main theorems about the existence of complements of certain relations of a distributive allegory in the presence of the axiom of choice. As usual, we call $X : A \rightarrow B$ a complement of $Y : A \rightarrow B$ with respect to a downward closed subclass of $\mathcal{R}[A, B]$ with greatest element $Z : A \rightarrow B$ if $X \sqcap Y = \mathbb{1}_{AB}$ and $X \sqcup Y = Z$. In particular, we will be interested in the full class $\mathcal{R}[A, B]$ with its greatest element $\mathbb{T}_{AB}$ and in the class of partial identities on an object A with its greatest element $\mathbb{I}_A$. Notice that complements will be unique since all allegories considered are assumed to be distributive.

The following theorem, and its proof, was motivated by a similar property in topos theory. As a consequence, the proof follows closely the proof given in [3].

Theorem 1. *Let $\mathcal{R}$ be a distributive allegory with total splittings and relational sums. If $\mathcal{R}$ satisfies (AC), then every partial identity has a complement.*

Before we prove this theorem we want to illustrate the construction in the proof by an example. In this example we will work with concrete relations between finite sets which we are going to represent by Boolean matrices. Let the following partial identity

$$j = \begin{pmatrix} 1 & 0 & 0 \\ 0 & 1 & 0 \\ 0 & 0 & 0 \end{pmatrix}$$

on the set $M = \{a, b, c\}$ be given, i.e. $j = \{(a, a), (b, b)\}$. We now construct an equivalence relation Ξ on the relational sum of M by $\Xi := \mathbb{I}_{D+D} \sqcup \iota^{\smile}; j; \kappa \sqcup \kappa^{\smile}; j; \iota$, its splitting R and choose a function f in R using the axiom of choice. In our example we obtain

$$\Xi = \begin{pmatrix} 1&0&0&1&0&0 \\ 0&1&0&0&1&0 \\ 0&0&1&0&0&0 \\ 1&0&0&1&0&0 \\ 0&1&0&0&1&0 \\ 0&0&0&0&0&1 \end{pmatrix}, \quad R = \begin{pmatrix} 1&0&0&1&0&0 \\ 0&1&0&0&1&0 \\ 0&0&1&0&0&0 \\ 0&0&0&0&0&1 \end{pmatrix}, \quad f = \begin{pmatrix} 1&0&0&0&0&0 \\ 0&0&0&0&1&0 \\ 0&0&1&0&0&0 \\ 0&0&0&0&0&1 \end{pmatrix}.$$

Next we define the four relations

$$x_0 := \kappa; R^\smile; f; \iota^\smile, \qquad\qquad x_1 := \iota; R^\smile; f; \kappa^\smile,$$
$$x_2 := \iota; R^\smile; f; \iota^\smile, \qquad\qquad x_3 := \kappa; R^\smile; f; \kappa^\smile.$$

In the example we obtain

$$x_0 = \begin{pmatrix} 1 & 0 & 0 \\ 0 & 0 & 0 \\ 0 & 0 & 0 \end{pmatrix}, \quad x_1 = \begin{pmatrix} 0 & 0 & 0 \\ 0 & 1 & 0 \\ 0 & 0 & 0 \end{pmatrix}, \quad x_2 = \begin{pmatrix} 1 & 0 & 0 \\ 0 & 0 & 0 \\ 0 & 0 & 1 \end{pmatrix}, \quad x_3 = \begin{pmatrix} 0 & 0 & 0 \\ 0 & 1 & 0 \\ 0 & 0 & 1 \end{pmatrix}.$$

From those relations we get $j = x_0 \sqcup x_1$ and that its complement is given by $x_2 \sqcap x_3$. We now formally prove the theorem.

Proof of Theorem 1. Let $j \sqsubseteq \mathbb{I}_D$ be a partial identity, and define $\varXi := \mathbb{I}_{D+D} \sqcup \iota^\smile; j; \kappa \sqcup \kappa^\smile; j; \iota$ where $\iota, \kappa : D \to D+D$ are the injections of D into the relational sum $D + D$. Then $\varXi$ is an equivalence relation, since $\varXi$ is obviously reflexive and symmetric and we have

$\varXi; \varXi$

$= (\mathbb{I}_{D+D} \sqcup \iota^\smile; j; \kappa \sqcup \kappa^\smile; j; \iota); (\mathbb{I}_{D+D} \sqcup \iota^\smile; j; \kappa \sqcup \kappa^\smile; j; \iota)$

$= \varXi \sqcup \iota^\smile; j; \kappa; \iota^\smile; j; \kappa \sqcup \iota^\smile; j; \kappa; \kappa^\smile; j; \iota \sqcup \kappa^\smile; j; \iota; \iota^\smile; j; \kappa \sqcup \iota^\smile; j; \kappa; \iota^\smile; j; \kappa$

$= \varXi \sqcup \iota^\smile; j; j; \iota \sqcup \kappa^\smile; j; j; \kappa$

$\sqsubseteq \varXi \sqcup \iota^\smile; \iota \sqcup \kappa^\smile; \kappa$

$= \varXi.$

Furthermore, we obtain

$(*_1) \qquad \iota; \varXi; \iota^\smile = \iota; (\mathbb{I}_{D+D} \sqcup \iota^\smile; j; \kappa \sqcup \kappa^\smile; j; \iota); \iota^\smile$

$\qquad\qquad\qquad = \iota; \iota^\smile \sqcup \iota; \iota^\smile; j; \kappa; \iota^\smile \sqcup \iota; \kappa^\smile; j; \iota; \iota^\smile$

$\qquad\qquad\qquad = \mathbb{I}_D,$

$(*_2) \qquad \kappa; \varXi; \iota^\smile = \kappa; (\mathbb{I}_{D+D} \sqcup \iota^\smile; j; \kappa \sqcup \kappa^\smile; j; \iota); \iota^\smile$

$\qquad\qquad\qquad = \kappa; \iota^\smile \sqcup \kappa; \iota^\smile; j; \kappa; \iota^\smile \sqcup \kappa; \kappa^\smile; j; \iota; \iota^\smile$

$\qquad\qquad\qquad = j,$

$(*_3) \qquad \kappa; \varXi; \kappa = \mathbb{I}_D,$

$(*_4) \qquad \iota; \varXi; \kappa^\smile = j,$

where $(*_3)$ and $(*_4)$ are shown similar to $(*_1)$ and $(*_2)$, respectively. Now, let $R : A \to D+D$ be a splitting of $\varXi$. Since R is total, (AC) implies that there is a function $f : A \to D+D$ with $f \sqsubseteq R$. From $f; f^\smile \sqsubseteq R; R^\smile = \mathbb{I}_A$ we conclude that f is also injective. The situation so far is illustrated in the following diagram:

Define $x_i : D \rightarrow D$ for $i = 1, 2, 3, 4$ by

$$x_0 := \kappa; R^{\smile}; f; \iota^{\smile}, \qquad x_1 := \iota; R^{\smile}; f; \kappa^{\smile},$$
$$x_2 := \iota; R^{\smile}; f; \iota^{\smile}, \qquad x_3 := \kappa; R^{\smile}; f; \kappa^{\smile}.$$

First, we want to show that $x_0 \sqcup x_1 = j$. We obtain

$$
\begin{aligned}
x_0 &= \kappa; R^{\smile}; f; \iota^{\smile} \\
&\sqsubseteq \kappa; R^{\smile}; R; \iota^{\smile} \\
&= \kappa; \varXi; \iota^{\smile} \\
&= j \qquad\qquad\qquad (*_2)
\end{aligned}
$$

and $x_1 \sqsubseteq j$, analogously. This implies $x_0 \sqcup x_1 \sqsubseteq j$. Now, consider the following computation

$$
\begin{aligned}
j; \iota; R^{\smile} &= j; \iota; R^{\smile}; R; R^{\smile} && R \text{ splitting} \\
&= j; \iota; \varXi; R^{\smile} && R \text{ splitting} \\
&= j; \iota; \varXi; (\iota^{\smile}; \iota \sqcup \kappa^{\smile}; \kappa); R^{\smile} \\
&= j; (\iota; \varXi; \iota^{\smile}; \iota \sqcup \iota; \varXi; \kappa^{\smile}; \kappa); R^{\smile} \\
&= j; (\iota \sqcup j; \kappa); R^{\smile} && (*_{1,4}) \\
&= (j; \iota \sqcup j; j; \kappa); R^{\smile} \\
&= (j; \iota \sqcup j; \kappa); R^{\smile} && \text{Lemma 2(1)} \\
&= j; (\iota \sqcup \kappa); R^{\smile}.
\end{aligned}
$$

Analogously, we get $j; \kappa; R^{\smile} = j; (\iota \sqcup \kappa); R^{\smile}$, and, hence, $j; \iota; R^{\smile} = j; \kappa; R^{\smile}$. Finally, the other inclusion follows from

$$
\begin{aligned}
&j \sqcap (x_0 \sqcup x_1) \\
&= j; (x_0 \sqcup x_1); j && \text{Lemma 2(2)} \\
&= j; (x_0; x_0^{\smile} \sqcup x_1; x_1^{\smile}); j && \text{Lemma 2(1)} \\
&= j; (\kappa; R^{\smile}; f; \iota^{\smile}; \iota; f^{\smile}; R; \kappa^{\smile} \sqcup \iota; R^{\smile}; f; \kappa^{\smile}; \kappa; f^{\smile}; R; \iota^{\smile}); j \\
&= j; \kappa; R^{\smile}; f; \iota^{\smile}; \iota; f^{\smile}; R; \kappa^{\smile}; j \sqcup j; \iota; R^{\smile}; f; \kappa^{\smile}; \kappa; f^{\smile}; R; \iota^{\smile}; j \\
&= j; \iota; R^{\smile}; f; \iota^{\smile}; \iota; f^{\smile}; R; \iota^{\smile}; j \sqcup j; \iota; R^{\smile}; f; \kappa^{\smile}; \kappa; f^{\smile}; R; \iota^{\smile}; j && \text{see above} \\
&= j; \iota; R^{\smile}; f; (\iota^{\smile}; \iota \sqcup \kappa^{\smile}; \kappa); f^{\smile}; R; \iota^{\smile}; j \\
&= j; \iota; R^{\smile}; f; f^{\smile}; R; \iota^{\smile}; j \\
&= j; \iota; R^{\smile}; R; \iota^{\smile}; j && f \text{ injective} \\
&= j; \iota; \varXi; \iota^{\smile}; j && R \text{ splitting} \\
&= j. && (*_1)
\end{aligned}
$$

Now, define $-j := x_2 \sqcap x_3$. From

$$x_2 = \iota; R^{\smile}; f; \iota^{\smile}$$
$$\sqsubseteq \iota; R^{\smile}; R; \iota^{\smile}$$
$$= \iota; \Xi; \iota^{\smile}$$
$$= \mathbb{I}_D \qquad\qquad (*_1)$$

and a similar computation for x_3 we conclude $-j \sqsubseteq \mathbb{I}_D$. Furthermore, we obtain

$$x_0 \sqcap -j \sqsubseteq x_0 \sqcap x_3$$
$$= \kappa; R^{\smile}; f; \iota^{\smile} \sqcap \kappa; R^{\smile}; f; \kappa^{\smile}$$
$$= \kappa; R^{\smile}; f; (\iota^{\smile} \sqcap \kappa^{\smile}) \qquad f, \kappa \text{ univalent, } R \text{ injective}$$
$$= \bot\!\!\!\bot_{DD} \qquad\qquad \text{Lemma 4(1)}$$

and $x_1 \sqcap -j = \bot\!\!\!\bot_{DD}$, analogously. We conclude $j \sqcap -j = \bot\!\!\!\bot_{DD}$. It remains to show that $j \sqcup -j = \mathbb{I}_D$. Therefore, consider the computation

$$j \sqcup x_2 \sqsupseteq x_1 \sqcup x_2$$
$$= x_1; x_1^{\smile} \sqcup x_2; x_2^{\smile} \qquad\qquad \text{Lemma 2(1)}$$
$$= \iota; R^{\smile}; f; \kappa^{\smile}; \kappa; f^{\smile}; R; \iota^{\smile} \sqcup \iota; R^{\smile}; f; \iota^{\smile}; \iota; f^{\smile}; R; \iota^{\smile}$$
$$= \iota; R^{\smile}; f; (\kappa^{\smile}; \kappa \sqcup \iota^{\smile}; \iota); f^{\smile}; R; \iota^{\smile}$$
$$= \iota; R^{\smile}; f; f^{\smile}; R; \iota^{\smile}$$
$$= \iota; R^{\smile}; R; \iota^{\smile} \qquad\qquad f \text{ injective}$$
$$= \iota; \Xi; \iota^{\smile} \qquad\qquad R \text{ splitting}$$
$$= \mathbb{I}_D \qquad\qquad (*_1),$$

which immediately implies $j \sqcup x_2 = \mathbb{I}_D$ since j and x_2 are partial identities. Analogously, we obtain $j \sqcup x_3 = \mathbb{I}_D$. Consequently, we get

$$j \sqcup -j = (j \sqcup x_2) \sqcap (j \sqcup x_3) = \mathbb{I}_D \sqcup \mathbb{I}_D = \mathbb{I}_D.$$

This completes the proof. □

In the presence of relational products every relation $R : A \to B$ corresponds to a partial identity on the product $A \times B$ of the source and target of R. Notice that it is equivalent for an allegory to be tabular or to have (all) splittings and relational products. Even though we now require tabulations, the following result needs neither power objects nor a unit.

Theorem 2. *Let $\mathcal{R}$ be a distributive allegory with total splittings, relational sums and products. If $\mathcal{R}$ satisfies (AC), then every relation has a complement.*

Proof. Let $R : A \to B$ be given. Then $j := \mathbb{I}_{A \times B} \sqcap \pi; R; \rho^{\smile} : A \times B \to A \times B$ is a partial identity. By Theorem 1 the partial identity j has a complement $-j$.

Define $-R := \pi^\smile; -j; \rho$ and show that $-R$ is the complement of R:

$$R \sqcap -R = R \sqcap \pi^\smile; -j; \rho$$
$$\sqsubseteq \pi^\smile; (\pi; R; \rho^\smile \sqcap -j); \rho$$
$$= \pi^\smile; (\mathbb{I}_{A \times B} \sqcap \pi; R; \rho^\smile \sqcap -j); \rho \qquad -j \text{ partial identity}$$
$$= \pi^\smile; (j \sqcap -j); \rho$$
$$= \pi^\smile; \amalg_{A \times B\, A \times B}; \rho \qquad -j \text{ complement of } j$$
$$= \amalg_{AB},$$

$$R \sqcup -R = R \sqcup \pi^\smile; -j; \rho$$
$$\sqsupseteq \pi^\smile; \pi; R; \rho^\smile; \rho \sqcup \pi^\smile; -j; \rho \qquad \pi, \rho \text{ univalent}$$
$$= \pi^\smile; (\pi; R; \rho^\smile \sqcup -j); \rho$$
$$\sqsupseteq \pi^\smile; ((\mathbb{I}_{A \times B} \sqcap \pi; R; \rho^\smile) \sqcup -j); \rho$$
$$= \pi^\smile; (j \sqcup -j); \rho$$
$$= \pi^\smile; \rho \qquad -j \text{ complement of } j$$
$$= \mathbb{T}_{AB}. \qquad \pi, \rho \text{ tabulate } \mathbb{T}$$

This completes the proof.   □

4   Alternative Versions of the Axiom of Choice

In this section we want to investigate whether the assumptions of Theorem 1 can be weakened. In particular, we are interested whether it is possible to replace the existence of total splittings and/or relational sums by other properties.

Every allegory $\mathcal{R}$ can be fully embedded into an allegory $\mathcal{R}_{\mathrm{Eq}}$ that has total splittings [4]. The allegory $\mathcal{R}_{\mathrm{Eq}}$ and the full embedding $E : \mathcal{R} \to \mathcal{R}_{\mathrm{Eq}}$ are given by:

1. The objects of $\mathcal{R}_{\mathrm{Eq}}$ are the equivalence relations from $\mathcal{R}$.
2. A relation in $\mathcal{R}_{\mathrm{Eq}}$ with source $X : A \to A$ and target $Y : B \to B$ is a relation $R : A \to B$ from $\mathcal{R}$ with $X; R; Y = R$.
3. The full embedding E is defined by $E(A) = \mathbb{I}_A$ and $E(R) = R$.

Notice that the identity relation $\mathbb{I}_X$ of an object X in $\mathcal{R}_{\mathrm{Eq}}$ is the relation X itself.

Unfortunately, $\mathcal{R}_{\mathrm{Eq}}$ may not satisfy (AC) even if $\mathcal{R}$ does, as the following example shows.

Example 1. Let $M = \{a, b\}$ be a set with two elements. Then M together with the three relations $\amalg, \mathbb{I}, \mathbb{T}$ is a distributive allegory $\mathcal{R}$ with one object M. This allegory satisfies (AC), since $\mathbb{T}$ is the only total relation that is not univalent but includes the function $\mathbb{I}$. On the other hand, $\mathbb{T}$ does not split.

The allegory $\mathcal{R}_{\mathrm{Eq}}$ does not satisfy (AC). For example, the relation $\mathbb{T} : \mathbb{T} \to \mathbb{I}$ is total but does not include a function.

Fig. 1. The allegories $\mathcal{R}$ and $\mathcal{R}_{\mathrm{Eq}}$

There is a simple reason for the fact that (AC) fails in $\mathcal{R}_{\mathrm{Eq}}$. In $\mathcal{R}$ there is a function, namely $\mathbb{I}$, that is included in the equivalence relation $\mathbb{T}$. But this function does not map the two elements a, b of the equivalence class induced by $\mathbb{T}$ to the same element. A function with such a behavior is not in $\mathcal{R}$. Consequently, $\mathbb{T}$ as a relation between one equivalence class (induced by $\mathbb{T}$) and two equivalence classes (induced by $\mathbb{I}$) does not contain a function in $\mathcal{R}_{\mathrm{Eq}}$.

As shown by the previous example we need a stronger version of the axiom of choice in order to guarantee that (AC) holds in $\mathcal{R}_{\mathrm{Eq}}$. We call this axiom *Axiom of consistent choice* (ACC). It is given by (AC) plus the following:

– For every equivalence relation $X : A \to A$ there is a function $f : A \to A$ with $f \sqsubseteq X$ and $X; f \sqsubseteq f$.

The only extra condition is given by $X; f \sqsubseteq f$ which requires that f maps every element of the same equivalence class to a single element. The next lemma provides some basic properties of the relation f in a slightly more general context.

Lemma 5. *Let $\mathcal{R}$ be an allegory, $X : A \to A$ be a symmetric and idempotent relation and $f : A \to A$ univalent with $f \sqsubseteq X, \mathbb{I}_A \sqcap X = \mathbb{I}_A \sqcap f; f^{\smile}$ and $X; f \sqsubseteq f$. Then we have*

1. $X; f = f$,
2. $X = f; f^{\smile}$.

Proof

1. Using Lemma 1(2) this follows immediately from

$$f = (\mathbb{I}_A \sqcap f; f^{\smile}); f = (\mathbb{I}_A \sqcap X); f \sqsubseteq X; f.$$

2. Consider the following computation

$$\begin{aligned} X &= X; (\mathbb{I}_A \sqcap X) && \text{Lemma 1(2)} \\ &= X; (\mathbb{I}_A \sqcap f; f^{\smile}) \\ &\sqsubseteq X; f; f^{\smile} \\ &= f; f^{\smile}. && \text{by 1.} \end{aligned}$$

The converse inclusion follows immediately from $f \sqsubseteq X$ and the fact that X is symmetric and idempotent. □

Before we prove the main property about (ACC) we want to study the relationship of this axiom to other properties of the allegory.

Lemma 6. *Let $\mathcal{R}$ be a tabular allegory satisying (ACC). Then $\mathcal{R}$ has all total splittings.*

Proof. Suppose $X : A \to A$ is an equivalence relation. By (ACC) there is a function $f : A \to A$ with $f \sqsubseteq X$ and $X; f \sqsubseteq f$. Let $g : B \to A$ and $h : B \to A$ be a tabulation of $f^\smile; f$, i.e. $g^\smile; h = f^\smile; f$ and $g; g^\smile \sqcap h; h^\smile = \mathbb{I}_B$. Then we have

$$h \sqsubseteq g; g^\smile; h \qquad\qquad g \text{ total}$$
$$= g; f^\smile; f$$
$$\sqsubseteq g \qquad\qquad f \text{ univalent}$$

and $g \sqsubseteq h$ analogously. This implies $f^\smile; f = g^\smile; h = g^\smile; g$ and $\mathbb{I}_B = g; g^\smile \sqcap h; h^\smile = g; g^\smile$. Define $R := g; f^\smile$ and compute

$$R; R^\smile = g; f^\smile; f; g^\smile$$
$$= g; g^\smile; g; g^\smile$$
$$= \mathbb{I}_B, R^\smile; R \qquad\qquad = f; g^\smile; g; f^\smile$$
$$= f; f^\smile; f; f^\smile$$
$$= X; X \qquad\qquad\qquad \text{Lemma 5(2)}$$
$$= X,$$

i.e. R splits X. $\qquad\qquad\qquad\qquad\qquad\qquad\qquad\qquad\qquad\qquad\qquad\qquad \square$

Notice that the same result can be proven if every map in $\mathcal{R}$ factors as a surjection followed by an injection. However, already for pre-tabular allegories the previous result is not longer true as the following examples shows.

Example 2. Again, consider the set $M = \{a, b\}$. Let $\mathcal{R}$ be the full allegory of binary relations whose objects are n-ary cartesian products of M, i.e. an object A of $\mathcal{R}$ satisfies $A = M^i$ for an $i \geq 1$. Since all morphism sets are finite this allegory satisfies (AC) and also (ACC). For example, $\mathbb{T}_M = M \times M : M \to M$ is an equivalence relation on M. The constant valued function f, mapping every element to a, is a function required by (ACC). Furthermore, this allegory is pre-tabular. It is easy to verify that M^{i+j} constitues a relational product of the objects M^i and M^j. However, $\mathcal{R}$ does not have all total splitting since $\mathbb{T}_M$ does not split.

We are now ready to prove the main property about (ACC).

Lemma 7. *Let $\mathcal{R}$ be an allegory satisfying (ACC). Then $\mathcal{R}_{\mathrm{Eq}}$ satisfies (AC).*

Proof. Let $R : X \to Y$ be a total relation in $\mathcal{R}_{\mathrm{Eq}}$, i.e. $X : A \to A$ and $Y : B \to B$ are equivalence relations from $\mathcal{R}$ with $X; R; Y = R$ and we have $\mathbb{I}_X \sqsubseteq R; R^\smile$. Since $\mathbb{I}_X = X \sqsupseteq \mathbb{I}_A$ the latter shows that R is also total in $\mathcal{R}$. Therefore, (AC) implies that there is a function $h : A \to B$ with $h \sqsubseteq R$. Furthermore, (ACC) implies that there is a function $f : A \to A$ with $f \sqsubseteq X$ and $X; f \sqsubseteq X$. Define

$g := f; h; Y$. We want to show that g is the required function for R in $\mathcal{R}_{\mathrm{Eq}}$. First, we have

$$
\begin{aligned}
X; g; Y = X; f; h; Y; Y \\
= X; f; h; Y \qquad\qquad && Y \text{ idempotent} \\
= f; h; Y \qquad\qquad && \text{by (ACC)} \\
= g
\end{aligned}
$$

so that g is indeed a relation in $\mathcal{R}_{\mathrm{Eq}}$ with source X and target Y. Furthermore, we have $g = f; h; Y \sqsubseteq X; R; Y = R$. That g is total follows from

$$
\begin{aligned}
\mathbb{I}_X = X \qquad\qquad && \text{identities in } \mathcal{R}_{\mathrm{SId}} \\
= f; f^\smile \qquad\qquad && \text{Lemma 5(2)} \\
\sqsubseteq f; h; h^\smile; f^\smile \qquad\qquad && h \text{ total} \\
\sqsubseteq f; h; Y; h^\smile; f^\smile \qquad\qquad && Y \text{ reflexive} \\
= f; h; Y; Y^\smile; h^\smile; f^\smile \qquad\qquad && Y \text{ symmetric and idempotent} \\
= g; g^\smile.
\end{aligned}
$$

Finally, the following computation

$$
\begin{aligned}
g^\smile; g = Y; h^\smile; f^\smile; f; h; Y \qquad\qquad && Y \text{ symmetric} \\
\sqsubseteq Y; Y \qquad\qquad && f, h \text{ univalent} \\
= Y \qquad\qquad && Y \text{ idempotent}
\end{aligned}
$$

shows that g is univalent since Y is the identity on the object Y in $\mathcal{R}_{\mathrm{Eq}}$. $\qquad\square$

Corollary 1. *Let $\mathcal{R}$ be a distributive allegory with relational sums satisfying (ACC). Then every partial identity has a complement. Furthermore, if $\mathcal{R}$ has relational products, then every relation has a complement.*

Proof. This follows immediately from the fact that $\mathcal{R}_{\mathrm{Eq}}$ has relational sums [4], Lemma 7, Theorem 1 and 2 and that $E : \mathcal{R} \to \mathcal{R}_{\mathrm{Eq}}$ is a full embedding. $\qquad\square$

Now, we want to focus on the second assumption in Theorem 1 – the existence of relational sums. Every distributive allegory $\mathcal{R}$ can be fully embedded into an allegory $\mathcal{R}^+$ that has relational sums [4]. The allegory $\mathcal{R}^+$ and the full embedding $E : \mathcal{R} \to \mathcal{R}^+$ are given by:

1. The objects of $\mathcal{R}^+$ are finite lists of objects from $\mathcal{R}$.
2. A relation in $\mathcal{R}^+$ with source $[A_1, \ldots, A_m]$ and target $[B_1, \ldots, B_n]$ is a $m \times n$-matrix R such that $R_{i,j}$ is a relation from $\mathcal{R}$ with source A_i and target B_j.
3. The full embedding E is defined by $E(A) = [A]$ and $E(R) = (R)$, i.e. the singleton matrix containing R.

Unfortunately, $\mathcal{R}^+$ may not satisfy (AC) even if $\mathcal{R}$ does, as the following example shows.

Example 3. Let $M = \{a, b, c\}$ be a set with three elements. Consider the following five partial identities on M:

$$\bot = \emptyset, x = \{(a, a), (b, b)\}, y = \{(b, b), (c, c)\}, z = \{(b, b)\}, \mathbb{I} = \{(a, a), (b, b), (c, c)\}$$

or written as Boolean matrices

$$\begin{pmatrix} 000 \\ 000 \\ 000 \end{pmatrix}, \begin{pmatrix} 100 \\ 010 \\ 000 \end{pmatrix}, \begin{pmatrix} 000 \\ 010 \\ 001 \end{pmatrix}, \begin{pmatrix} 000 \\ 010 \\ 000 \end{pmatrix}, \begin{pmatrix} 100 \\ 010 \\ 001 \end{pmatrix}.$$
$$\quad\;\, \bot \qquad\quad x \qquad\quad y \qquad\quad z \qquad\quad \mathbb{I}$$

The relations above are closed under all relational operations, and, hence, form a distributive allegory $\mathcal{R}$ with one object M. Notice that since all relations are partial identities we have $V^{\smile} = V$ and $U; V = U \sqcap V$ for all U, V in $\{\bot, x, y, z, \mathbb{I}\}$.

Now, let $R : [M] \rightarrow [M, M]$ be the relation in $\mathcal{R}^+$ defined by $R = (x\,y)$. From

$$R; R^{\smile} = (x\,y) ; \begin{pmatrix} x \\ y \end{pmatrix}$$
$$= (x; x \sqcup y; y)$$
$$= (\mathbb{I}),$$
$$R^{\smile}; R = \begin{pmatrix} x \\ y \end{pmatrix} ; (x\,y)$$
$$= \begin{pmatrix} x; x & x; y \\ y; x & y; y \end{pmatrix}$$
$$= \begin{pmatrix} x & z \\ z & y \end{pmatrix}$$
$$\not\sqsubseteq \begin{pmatrix} \mathbb{I} & \bot \\ \bot & \mathbb{I} \end{pmatrix}$$

we conclude that R is total but not univalent. The only two relations of $\mathcal{R}$ included in x are $\bot$ and c. The same applies to y. Therefore, the greatest relation $S : [M] \rightarrow [M, M]$ included in R is $S = (z\,z)$. But

$$S; S^{\smile} = (z\,z) ; \begin{pmatrix} z \\ z \end{pmatrix}$$
$$= (z; z \sqcup z; z)$$
$$= (z)$$

shows that S is not total. Therefore, $\mathcal{R}^+$ does not satisfy (AC).

Unlike the previous example it does not seem possible to strengthen (AC) slightly in order to obtain the desired result. In the current example (AC) fails in $\mathcal{R}^+$ because the pair (b, b) is in both relations x and y, but there is no relation corresponding to x (or y) without that pair. In terms of the operations of an allegory, we would need some version of a relative complement u of z in x (or y). More precisely, assume that $T = (u\,y)$ with a partial identity u is the function included in R as required by (AC). Then

$$T;T^\smile = (uy)\,;\begin{pmatrix}u\\y\end{pmatrix}$$
$$= (u;u\sqcup y;y)$$
$$= (u\sqcup y)\,,$$
$$T^\smile;T = \begin{pmatrix}u\\y\end{pmatrix};(uy)$$
$$= \begin{pmatrix}u;u\ u;y\\y;u\ y;y\end{pmatrix}$$
$$= \begin{pmatrix}u & u\sqcap y\\u\sqcap y & y\end{pmatrix}$$

implies that $u\sqcup y = \mathbb{I}$ and $u\sqcap y = \perp\!\!\!\perp$, i.e. u is the complement of y.

5 Conclusion

In this paper we have shown that the axiom of choice implies that partial identities in distributive allegories with relational sums and total splittings have complements. Furthermore, we have shown that the assumption of the existence of total splittings can be dropped by requiring a stronger version of the axiom of choice, namely the axiom of consistent choice. Unfortunately, a similar result does not seem possible for the existence of relational sums. Our example has shown that sums are essential, i.e. that generating sums and preserving the axiom of choice basically requires complements.

References

1. Bird, R., de Moor, O.: Algebra of Programming. Prentice-Hall, Englewood Cliffs (1997)
2. Brink, C., Kahl, W., Schmidt, G. (eds.): Relational Methods in Computer Science. Advances in Computer Science. Springer, Vienna (1997)
3. Diaconescu, R.: Axiom of Choice and Complementation. Proc. AMS 51, 176–178 (1975)
4. Freyd, P., Scedrov, A.: Categories, Allegories. North-Holland, Amsterdam (1990)
5. Goldblatt, R.: Topoi: The Categorical Analysis of Logic. Studies in Logic and the Foundation of Mathematics, vol. 98. Elsevier, Amsterdam (1984)
6. Grätzer, G.: General lattice theory, 2nd edn. Birkhäuser, Basel (2003)
7. Johnstone, P.: Sketches of an Elephant: A Topos Theory Compendium. Oxford Logic Guides 43, vol. 1. Oxford University Press, Oxford (2002)
8. Johnstone, P.: Sketches of an Elephant: A Topos Theory Compendium. Oxford Logic Guides 44, vol. 2. Oxford University Press, Oxford (2002)
9. Maddux, R.: On the derivation of identities involving projection functions. In: Csirmaz, Gabbay, de Rijke (eds.) Logic Colloquium 1992, pp. 145–173. Center for the Study of Languages and Information Publications, Stanford (1995)
10. Schmidt, G., Ströhlein, T.: Relationen und Graphen. Springer, Heidelberg (1989); English version: Relations and Graphs. Discrete Mathematics for Computer Scientists. EATCS Monographs on Theoret. Comput. Sci. Springer, Heidelberg (1993)
11. Tarski, A., Givant, S.: A Formalization of Set Theory without Variables. Colloquium Publications, vol. 41. AMS (1987)
12. Winter, M.: Goguen Categories. A Categorical Approach to L-Fuzzy Relations. Trends in Logic, vol. 25. Springer, Heidelberg (2007)
13. Kawahara, Y., Winter, M.: Cardinal Addition in Distributive Allegories. Submitted to RelMiCS 11/AKA 6

On the Skeleton of Stonian p-Ortholattices

Michael Winter[1,*], Torsten Hahmann[2,*], and Michael Gruninger[3,*]

[1] Department of Computer Science,
Brock University,
St. Catharines, ON, Canada
mwinter@brocku.ca
[2] Department of Computer Science,
University of Toronto,
torsten@cs.toronto.edu
[3] Department of Mechanical and Industrial Engineering,
Department of Computer Science,
University of Toronto
gruninger@mie.utoronto.ca

Abstract. Boolean Contact Algebras (BCA) establish the algebraic counterpart of the mereotopology induced by the Region Connection Calculus (RCC). Similarly, Stonian p-ortholattices serve as a lattice theoretic version of the ontology RT^- of Asher and Vieu. In this paper we study the relationship between BCAs and Stonian p-ortholattices. We show that the skeleton of every Stonian p-ortholattice is a BCA, and, conversely, that every BCA is isomorphic to the skeleton of a Stonian p-ortholattice. Furthermore, we prove the equivalence between algebraic conditions on Stonian p-ortholattices and the axioms C5, C6, and C7 for BCAs.

1 Introduction

Region-based theories of space play a crucial role in qualitative spatial reasoning (QSR) within Artificial Intelligence (cf. [4]). Mereotopology – consisting of some topological notion of *contact* and a mereological notion of *parthood* – is the common core to most region-based theories of space. Instead of points as in classical point-set topology, mereotopology uses regions as primitives and focuses on the qualitative relations between different regions, such as contact, overlap, external contact, and parthood. In allowing to define part-whole relations such as self-connectedness of regions, the combination of topology with mereology is more expressive than either theory by itself.

As long as AI has been interested in mereotopology, different first-order mereotopological theories have been proposed. Most prominent amongst them is the Region-Connection Calculus (RCC) [3], which originated from Clarke's theory [2]. Another theory of the same origin, the RT_0 by Asher and Vieu [1], has

---

* The authors gratefully acknowledge support from the Natural Sciences and Engineering Research Council of Canada.

R. Berghammer et al. (Eds.): RelMiCS/AKA 2009, LNCS 5827, pp. 351–365, 2009.

received less attention although the theory is fairly similar to the RCC. The two theories differ mainly in their intended topological interpretations: RCC models include only regular closed sets while RT models allow any kind of regular sets (closed, open, clopen, or neither). A very fruitful way of understanding these theories of qualitative space is by looking at their algebraic counterparts. For the RCC, it was shown that the models can be defined in terms of Boolean Contact Algebras (BCA) for which topological representations have been given for various subsets of the original RCC axioms [6,7,11,12]. Recently, an algebraic representation of the theory RT^- as Stonian p-ortholattices [10] has been proved which allows to compare the models of RCC and RT^- in a purely algebraic way. A comparative study of the topological models of the two theories is beyond the scope of this paper, it will be part of our future work.

In this work, we exhibit the relationship between BCAs and Stonian p-ortholattices by using the skeleton of Stonian p-ortholattices as bridging structure. We show that the skeleton $S(L)$ of an arbitrary Stonian p-ortholattice L is a BCA when defining the contact relation of the BCA in terms of the lattice L. In addition we prove the equivalence between algebraic conditions on Stonian p-ortholattices and the axioms C5, C6, and C7 for BCAs. On the reverse, we prove that every BCA can be embedded in a Stonian p-ortholattice. This theoretical work provides semantic mappings between the two theories; it specifies which class of models of the RT^- can be mapped to which class of BCAs and vice versa.

The paper is structured as following. Section 2 introduces Stonian p-ortholattices and their algebraic properties. We define standard topological models and the notion of a skeleton for Stonian p-ortholattices. Afterwards, we briefly review BCAs and their topological representation. The following two sections contain the main results of this paper. In Section 4 we establish that the skeleton of a Stonian p-ortholattice is a BCA when choosing the contact relation accordingly, and in Section 5 we construct a Stonian p-ortholattice from any BCA by using the Boolean algebra of a BCA as the skeleton of the Stonian p-ortholattice. However, examples demonstrate that there is no unique embedding of BCAs into Stonian p-ortholattices.

These constructive embedding theorems verify in an algebraic way that the models of the theory RT^- are indeed more general than BCAs. Most significantly for QSR, the results imply that every model of RT^- that is connected, *-normal, and has a dense skeleton is in fact a model of the RCC. However, arbitrary Stonian p-ortholattices L of RT^- models do not adhere to the extensionality, interpolation, and connection axioms. Their skeletons $S(L)$ are arbitrary BCAs as axiomatized by C0-C4.

2   Stonian p-Ortholattices

In [1] Asher and Vieu introduced the mereotopology RT_0. This theory was intended to cover exactly those regions that have full interior and smooth boundaries. Even though this theory does not include all possible sets of a topological space, the notion of interior and closure are available. In [10] it was shown that the models of Asher and Vieu's theory RT^- are equivalent to Stonian (or Stonean) p-ortholattices. This observation now allows an algebraic treatment of that theory. First, recall pseudocomplemented and orthocomplemented lattices.

Definition 1. *A pseudocomplemented lattice (or p-algebra) is an algebraic structure $\langle L, +, \cdot, {}^*, 0, 1 \rangle$ of type $\langle 2, 2, 1, 0, 0 \rangle$ such that*

    P0. $\langle L, +, \cdot, 0, 1 \rangle$ *is a bounded lattice,*
    P1. a^* *is the pseudocomplement of a, i.e. $a \cdot x = 0 \iff x \leq a^*$.*

Definition 2. *An ortholattice (or orthocomplemented lattice) is an algebraic structure $\langle L, +, \cdot, {}^\perp, 0, 1 \rangle$ of type $\langle 2, 2, 1, 0, 0 \rangle$ such that*

    O0. $\langle L, +, \cdot, 0, 1 \rangle$ *is a bounded lattice,*
    O1. $a^\perp$ *is an orthocomplement of a, i.e. for all $a, b \in L$ we have*
        (a) $a^{\perp\perp} = a$,
        (b) $a \cdot a^\perp = 0$,
        (c) $a \leq b$ implies $b^\perp \leq a^\perp$.

A lattice that is both pseudocomplemented and orthocomplemented is called a p-ortholattice. A Stonian p-ortholattice additionally satisfies the Stone identity (PO.2). Notice that p-ortholattices are not necessarily distributive. In fact any distributive Stonian p-ortholattice is a Boolean algebra (cf. [10]).

Definition 3. *A Stonian p-ortholattice is a structure $\langle L, +, \cdot, {}^*, {}^\perp, 0, 1 \rangle$ of type $\langle 2, 2, 1, 1, 0, 0 \rangle$ such that*

    PO0. $\langle L, +, \cdot, {}^*, 0, 1 \rangle$ *is a pseudocomplemented lattice,*
    PO1. $\langle L, +, \cdot, {}^\perp, 0, 1 \rangle$ *is an ortholattice,*
    PO2. $(a \cdot b)^* = a^* + b^*$ *holds for all $a, b \in L$.*

P-ortholattices are always quasicomplemented (also known as 'dually pseudocomplemented') and thus double p-algebras. In a Stonian p-ortholattice one may define the quasicomplement a^+ of a, i.e. the smallest element b such that $a + b = 1$, as $a^+ = a^{\perp * \perp}$.

    The following basic properties of Stonian p-ortholattices were shown in [10].

Lemma 1. *Let $\langle L, +, \cdot, {}^*, {}^\perp, 0, 1 \rangle$ be a Stonian p-ortholattice. Then:*

1. $0^+ = 0^\perp = 0^* = 1$ *and* $1^+ = 1^\perp = 1^* = 0$,
2. $a \cdot a^+ = a \cdot a^\perp = a \cdot a^* = 0$ *and* $a + a^+ = a + a^\perp = a + a^* = 1$,
3. $a^+ \leq a^\perp \leq a^*$ *and* $a^{++} \leq a \leq a^{**}$,
4. $a^{+++} = a^+$ *and* $a^{***} = a^*$,
5. $a \leq b$ *implies* $b^* \leq a^*$, $b^+ \leq a^+$, *and* $b^\perp \leq a^\perp$,
6. $(a+b)^* = a^* \cdot b^*$, $(a \cdot b)^+ = a^+ + b^+$, $(a+b)^\perp = a^\perp \cdot b^\perp$ *and* $(a \cdot b)^\perp = a^\perp + b^\perp$,
7. $a^{*\perp} = a^{\perp +} = a^{*+} = a^{++}$ *and* $a^{+\perp} = a^{\perp *} = a^{+*} = a^{**}$.

Throughout this paper we will use the properties above without mentioning.

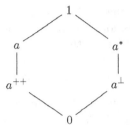

Fig. 1. The non-modular, non-distributive Stonian p-ortholattice C_6

2.1   Topological Models

Topological models of the theory of Stonian p-ortholattices are given by those sets that have full interior and smooth boundaries, i.e. are based on $\mathrm{RT}(X) = \{a \subseteq X \mid \mathrm{int}(a) = \mathrm{int}(\mathrm{cl}(a)) \wedge \mathrm{cl}(a) = \mathrm{cl}(\mathrm{int}(a))\}$ where int and cl are the interior and closure operation of the topological space $\langle X, \tau \rangle$. On those elements we define the following operations. The notations $x \cap^* y$ and $x \cup^* y$ are maintained from [1].

$$x \cap^* y = x \cap y \cap \mathrm{cl}(\mathrm{int}(x \cap y)),$$
$$x \cup^* y = x \cup y \cup \mathrm{int}(\mathrm{cl}(x \cup y)),$$
$$x^* = \mathrm{cl}(X \setminus x),$$
$$x^\perp = X \setminus x.$$

The next lemma provides some basic properties of those operations.

Lemma 2. Let $\langle X, \tau \rangle$ be a topological space. Then we have:

1. $\mathrm{cl}(x \cup^* y) = \mathrm{cl}(x) \cup \mathrm{cl}(y)$,
2. $\mathrm{int}(x \cap^* y) = \mathrm{int}(x) \cap \mathrm{int}(y)$,
3. $X \setminus (x \cap^* y) = (X \setminus x) \cup^* (X \setminus y)$,
4. $X \setminus (x \cup^* y) = (X \setminus x) \cap^* (X \setminus y)$.

Proof

1. Consider the following computation

$$\mathrm{cl}(x \cup^* y) = \mathrm{cl}(x \cup y \cup \mathrm{int}(\mathrm{cl}(x \cup y)))$$
$$= \mathrm{cl}(x \cup y) \cup \mathrm{cl}(\mathrm{int}(\mathrm{cl}(x \cup y)))$$
$$= \mathrm{cl}(x \cup y) \qquad\qquad \mathrm{cl}(\mathrm{int}(\mathrm{cl}(z))) \subseteq \mathrm{cl}(z)$$
$$= \mathrm{cl}(x) \cup \mathrm{cl}(y).$$

2. is shown analogously.

3. This property is shown by

$$X \setminus (x \cap^* y) = X \setminus (x \cap y \cap \mathrm{cl}(\mathrm{int}(x \cap y)))$$
$$= (X \setminus x) \cup (X \setminus y) \cup (X \setminus \mathrm{cl}(\mathrm{int}(x \cap y)))$$
$$= (X \setminus x) \cup (X \setminus y) \cup \mathrm{int}(\mathrm{cl}(X \setminus (x \cap y)))$$
$$= (X \setminus x) \cup (X \setminus y) \cup \mathrm{int}(\mathrm{cl}((X \setminus x) \cup (X \setminus y)))$$
$$= (X \setminus x) \cup^* (X \setminus y).$$

4. is shown analogously.   □

The next theorem verifies that the class of all structures $\mathrm{RT}(X)$ can be seen as the class of standard topological models of this kind of mereotopology.

Theorem 1. *Let* $\langle X, \tau \rangle$ *be a topological space. Then* $\langle \mathrm{RT}(X), \cup^*, \cap^*, {}^*, {}^\perp, \emptyset, X \rangle$ *is a Stonian p-ortholattice.*

Proof. First, we have to show that $\mathrm{RT}(X)$ is closed under all operations. Consider the following computations

$$\begin{aligned} \mathrm{cl}(x \cup^* y) &= \mathrm{cl}(x) \cup \mathrm{cl}(y) & \text{Lemma 2(1)} \\ &= \mathrm{cl}(\mathrm{int}(x)) \cup \mathrm{cl}(\mathrm{int}(y)) & x, y \in \mathrm{RT}(X) \\ &= \mathrm{cl}(\mathrm{int}(x) \cup \mathrm{int}(y)) \\ &\subseteq \mathrm{cl}(\mathrm{int}(x \cup y)) \\ &\subseteq \mathrm{cl}(\mathrm{int}(x \cup^* y)), \end{aligned}$$

and

$$\begin{aligned} &\mathrm{int}(\mathrm{cl}(x \cup^* y)) \\ &= \mathrm{int}(\mathrm{cl}(x) \cup \mathrm{cl}(y)) & \text{Lemma 2(1)} \\ &= \mathrm{int}(\mathrm{int}(\mathrm{cl}(x) \cup \mathrm{cl}(y))) \\ &= \mathrm{int}(\mathrm{int}(\mathrm{cl}(x)) \cup \mathrm{int}(\mathrm{cl}(y)) \cup \mathrm{int}(\mathrm{cl}(x) \cup \mathrm{cl}(y))) & \mathrm{int}(z_1) \cup \mathrm{int}(z_2) \\ & & \subseteq \mathrm{int}(z_1 \cup z_2) \\ &= \mathrm{int}(\mathrm{int}(x) \cup \mathrm{int}(y) \cup \mathrm{int}(\mathrm{cl}(x) \cup \mathrm{cl}(y))) & x, y \in \mathrm{RT}(X) \\ &= \mathrm{int}(\mathrm{int}(x) \cup \mathrm{int}(y) \cup \mathrm{int}(\mathrm{cl}(x \cup y))) \\ &\subseteq \mathrm{int}(x \cup y \cup \mathrm{int}(\mathrm{cl}(x \cup y))) \\ &= \mathrm{int}(x \cup^* y). \end{aligned}$$

In both cases the converse inclusion is trivial. The properties $\mathrm{int}(x \cup^* y) = \mathrm{int}(\mathrm{cl}(x \cup^* y))$ and $\mathrm{int}(x \cap^* y) = \mathrm{int}(\mathrm{cl}(x \cap^* y))$ are shown analogously.

$$\begin{aligned} \mathrm{cl}(x^\perp) &= \mathrm{cl}(X \setminus x) \\ &= X \setminus \mathrm{int}(x) \\ &= X \setminus \mathrm{int}(\mathrm{cl}(x)) & x \in \mathrm{RT}(X) \\ &= \mathrm{cl}(\mathrm{int}(X \setminus x)) \\ &= \mathrm{cl}(\mathrm{int}(x^\perp)), \end{aligned}$$

$$
\begin{aligned}
\mathrm{int}(x^{\perp}) &= \mathrm{int}(X \setminus x) \\
&= X \setminus \mathrm{cl}(x) \\
&= X \setminus \mathrm{cl}(\mathrm{int}(x)) \qquad\qquad x \in \mathrm{RT}(X) \\
&= \mathrm{int}(\mathrm{cl}(X \setminus x)) \\
&= \mathrm{int}(\mathrm{cl}(x^{\perp})), \\
\mathrm{cl}(x^{*}) &= \mathrm{cl}(\mathrm{cl}(X \setminus x)) \\
&= \mathrm{cl}(X \setminus x) \\
&= \mathrm{cl}(x^{\perp}) \\
&= \mathrm{cl}(\mathrm{int}(x^{\perp})) \qquad\qquad \text{see above} \\
&= \mathrm{cl}(\mathrm{int}(X \setminus x)) \\
&\subseteq \mathrm{cl}(\mathrm{int}(\mathrm{cl}(X \setminus x))) \\
&= \mathrm{cl}(\mathrm{int}(x^{*})), \\
\mathrm{int}(x^{*}) &= \mathrm{int}(\mathrm{cl}(X \setminus x)) \\
&= \mathrm{int}(\mathrm{cl}(\mathrm{cl}(X \setminus x))) \\
&= \mathrm{int}(\mathrm{cl}(x^{*})).
\end{aligned}
$$

Now, assume $x, y, z \in \mathrm{RT}(X)$ with $z \subseteq x$ and $z \subseteq y$. Then $z \subseteq x \cap y$, and we have $z = z \cap \mathrm{cl}(z) = z \cap \mathrm{cl}(\mathrm{int}(z)) \subseteq x \cap y \cap \mathrm{cl}(\mathrm{int}(x \cap y)) = x \cap^{*} y$. This verifies that $x \cap^{*} y$ is the greatest lower bound of x and y in $\mathrm{RT}(X)$. It is shown analogously that $x \cup^{*} y$ is the least upper bound of x and y in $\mathrm{RT}(X)$.

It is easy to verify that $x^{\perp}$ is an orthocomplement of x. In order to prove that x^{*} is a pseudocomplement consider the following computation

$$
\begin{aligned}
x \cap^{*} x^{*} &= x \cap \mathrm{cl}(X \setminus x) \cap \mathrm{cl}(\mathrm{int}(x \cap \mathrm{cl}(X \setminus x))) \\
&\subseteq \mathrm{cl}(\mathrm{int}(x \cap \mathrm{cl}(X \setminus x))) \\
&= \mathrm{cl}(\mathrm{int}(x) \cap \mathrm{int}(\mathrm{cl}(X \setminus x))) \\
&= \mathrm{cl}(\mathrm{int}(x) \cap \mathrm{int}(X \setminus x)) \qquad\qquad X \setminus x \in \mathrm{RT}(X) \\
&= \mathrm{cl}(\mathrm{int}(x \cap (X \setminus x))) \\
&= \mathrm{cl}(\mathrm{int}(\emptyset)) \\
&= \emptyset.
\end{aligned}
$$

In order to verify that x^{*} is the pseudocomplement of X it remains to show that x^{*} is the largest element z with $x \cap^{*} z = \emptyset$. Therefore, assume $z \in \mathrm{RT}(X)$ with $x \cap^{*} z = \emptyset$. Then we have $\mathrm{int}(x) \cap \mathrm{int}(z) = \mathrm{int}(x \cap^{*} z) = \mathrm{int}(\emptyset) = \emptyset$ using Lemma 2(2). We conclude $\mathrm{int}(z) \subseteq X \setminus \mathrm{int}(x) = \mathrm{cl}(X \setminus x) = x^{*}$. This immediately implies $z \subseteq \mathrm{cl}(z) = \mathrm{cl}(\mathrm{int}(z)) \subseteq \mathrm{cl}(x^{*}) = x^{*}$.

The following computation verifies the Stone property

$$(x \cap^* y)^* = \mathrm{cl}(X \setminus (x \cap^* y))$$
$$= \mathrm{cl}((X \setminus x) \cup^* (X \setminus y)) \qquad \text{Lemma 2(3)}$$
$$= \mathrm{cl}(X \setminus x) \cup \mathrm{cl}(X \setminus y) \qquad \text{Lemma 2(1)}$$
$$= x^* \cup y^*$$
$$= x^* \cup y^* \cup \mathrm{int}(x^* \cup y^*)$$
$$= x^* \cup y^* \cup \mathrm{int}(\mathrm{cl}(X \setminus x) \cup \mathrm{cl}(X \setminus y))$$
$$= x^* \cup y^* \cup \mathrm{int}(\mathrm{cl}(\mathrm{cl}(X \setminus x) \cup \mathrm{cl}(X \setminus y)))$$
$$= x^* \cup y^* \cup \mathrm{int}(\mathrm{cl}(x^* \cup y^*))$$
$$= x^* \cup^* y^*.$$

This completes the proof. □

2.2 Skeleton

Skeletons (also called centers) have been first defined by Glivenko in 1929 for Brouwerian lattices [9], showing that pseudocomplementation is a closure mapping. Frink [8] generalized this result by showing that the skeleton of a pseudo-complemented meet-semilattices is always a Boolean algebra.

Definition 4. *Let* $\langle L, \cdot,^* , 0 \rangle$ *be a pseudocomplemented semilattice. Let* $S(L) = \{a^* | a \in L\}$ *be the skeleton of* L, *maintaining the order relation of* L *and with meet* $a \wedge b = a \cdot b$ *and union* $a \vee b = (a^* \cdot b^*)^*$.

Theorem 2 (Glivenko-Frink Theorem). *[8] Let* L *be a pseudocomplemented semilattice. Then* $S(L)$ *is a Boolean algebra. The (unique) complement of an element* $a \in S(L)$ *is its pseudocomplement* $a^* \in L$.

Since Stonian p-ortholattices form a subclass of the class of pseudocomplemented meet-semilattices, the previous theorem immediately implies the following corollary (cf. [10]). Notice that here we have a stronger notion: the skeleton is not just a Boolean algebra, but a Boolean subalgebra.

Corollary 1. *If* $\langle L, +, \cdot,^* ,^\perp, 0, 1 \rangle$ *is a Stonian p-ortholattice, then* $S(L)$ *is a Boolean subalgebra of* L.

2.3 Additional Properties of Stonian p-Ortholattices

Motivated by the topological interpretation of the operations (cf. [10]), we call an element $a \in S(L)$, i.e. an element with $a^{**} = a$, closed. Dually, we call a open if $a^{++} = a$, and clopen if it is open and closed.

  L is called connected iff $0, 1$ are the only clopen elements of L.

  A topological space is called normal if any two disjoint closed sets can be separated by disjoint open sets. Following this definition we call a Stonian p-ortholattice L *-normal if for all $a, b \in L$ with $a^{**} \leq b^+$ there is an element

$c \in L$ with $a^{**} \le c^{++}$ and $b^{**} \le c^{+}$. Notice that in this case $c^{+} = c^{\perp\perp+} = c^{\perp++}$. Then $c \cdot c^{\perp} = 0$ implies $c^{++} \cdot c^{\perp++} = 0$ and hence $c^{++} \cdot c^{+} = 0$, ensuring that the open sets c^{+} and c^{++} are disjoint.

A bounded sublattice L' of L is called (downwards) dense in L if for every $0 \ne a \in L$ there is a $0 \ne b \in L'$ with $b \le a$.

In Section 4 we are going to show that denseness, $*$-normality and connectedness correspond to well-known additional properties of Boolean contact algebras. But beforehand, we review Boolean contact algebras and their embedding into the Boolean algebra of regular closed sets of a topological space.

3   Boolean Contact Algebras

Boolean contact algebras were introduced as the algebraic counterpart of mereotopologies induced by the *Region Connection Calculus* RCC [3]. Therefore, they are intended to cover closed sets with full interior and smooth boundaries, i.e. regular closed sets.

Definition 5. *A binary relation C on a Boolean algebra $\langle B, +, \cdot, ^{*}, 0, 1 \rangle$ is called a* contact relation *if it satisfies:*

C0. $(\forall a)0(-C)a$;
C1. $(\forall a)[a \ne 0 \Rightarrow aCa]$;
C2. $(\forall a)(\forall b)[aCb \Rightarrow bCa]$;
C3. $(\forall a)(\forall b)(\forall c)[(aCb \land b \le c) \Rightarrow aCc]$;
C4. $(\forall a)(\forall b)(\forall c)[aC(b+c) \Rightarrow (aCb \lor aCc)]$.

The pair $\langle B, C \rangle$ is called a Boolean Contact Algebra *(BCA).*

Additionally, the following properties are of importance:

C5. $(\forall a)(\forall b)[(\forall c)(aCc \Rightarrow bCc) \Leftrightarrow a = b]$. (The extensionality axiom).
C6. $(\forall a)(\forall b)[(\forall c)(aCc \lor bCc^{*}) \Rightarrow aCb]$ (The interpolation axiom).
C7. $(\forall a)[(a \ne 0 \land a \ne 1) \Rightarrow aCa^{*}]$ (The connection axiom).

As shown in [12], in the presence of the other axioms we can replace C5 by

C5'. $(\forall a \ne 1)(\exists b \ne 0)[a(-C)b]$.

As already mentioned above, the standard models of Boolean contact algebras are given by the regular closed sets of a topological space together with the following operations:

$$x + y := x \cup y,$$
$$x \cdot y := \mathrm{cl}(\mathrm{int}(x \cap y)),$$
$$x^{*} = \mathrm{cl}(X \setminus x).$$

The contact relation is given by the standard Whiteheadean contact relation xCy iff $x \cap y \ne \emptyset$.

Since their introduction several representation theorems for BCA's were proven. The most general version is the following:

Theorem 3 (Representation Theorem [5]). *For each Boolean contact alge-
bra* $\langle B, C\rangle$ *there exists an embedding* $h : B \rightarrow RC(X)$ *into the Boolean algebra
of regular closed sets of a topological space* $\langle X, \tau\rangle$ *with* aCb *iff* $h(a) \cap h(b) \neq \emptyset$.
h *is an isomorphism if* B *is complete.*

Notice that the original theorem lists further properties of the topological space
which are not important for the current work.

4 The Skeleton as a BCA

As already mentioned in Section 2.2 the skeleton of a Stonian p-ortholattice is a
Boolean algebra. In this section we verify that it is in fact a BCA with a contact
relation induced by the outer lattice.

Theorem 4. *Let* $\langle L, +, \cdot, ^*, ^\perp, 0, 1\rangle$ *be a Stonian p-ortholattice, then* $S(L)$ *together with*

$$aCb \iff a \not\leq b^\perp$$

is a Boolean contact algebra.

Proof

C0. Assume $0Ca$ for an $a \in S(L)$. Then $0 \not\leq a^\perp$, a contradiction.
C1. From $a \leq a^\perp$ we conclude $a = 0$, and, hence, C1.
C2. aCb implies $a \not\leq b^\perp$, which is equivalent to $b \not\leq a^\perp$. The latter shows bCa.
C3. Let aCb and $b \leq c$. This implies $a \not\leq b^\perp$ and $c^\perp \leq b^\perp$. Together we conclude
   $a \not\leq c^\perp$, and , hence, aCc.
C4. Assume $aC(b+c)$. Then we have $a \not\leq (b+c)^\perp = b^\perp \cdot c^\perp$. This implies $a \not\leq b^\perp$
   or $a \not\leq c^\perp$, and, hence, aCb or aCc.   □

Notice that the definition of C in the theorem above uses an element $b^\perp$ that is
not necessarily in the skeleton, i.e. the definition of C is external to the Boolean
algebra $S(L)$. By definition of the skeleton, all elements in $S(L)$ are regular
closed.

Lemma 3. *Let* $\langle L, +, \cdot, ^*, ^\perp, 0, 1\rangle$ *be a Stonian p-ortholattice and* $\langle S(L), C\rangle$ *its
skeleton BCA. Then we have:*

1. $S(L)$ *is dense in* L *iff* C *satisfies* C5.
2. L *is* $*$-*normal iff* C *satisfies* C6.
3. L *is connected iff* C *satisfies* C7.

Proof

1. Assume $S(L)$ is dense in L. We want to show that C5' holds. Therefore, let
   $1 \neq x \in S(L)$. Then $x^\perp \neq 0$ which implies that there is an element $0 \neq y \in
   S(L)$ with $y \leq x^\perp$, i.e. $x(-C)y$. Conversely, assume C5', and let $0 \neq y \in L$.
   If $y^{++} = 0$ we conclude $y \leq y^{**} = (y^{\perp\perp})^{**} = y^{\perp+**} = y^{++**} = 0^{**} = 0$, a
   contradiction. This implies $y^* = y^{\perp**} = y^{++\perp} \neq 1$ and $y^* \in S(L)$. By C5'
   there is an element $0 \neq x \in S(L)$ with $y^*(-C)x$, i.e. $y^* \leq x^\perp$. The latter
   implies $x \leq y^{*\perp} = y^{++} \leq y$.

2. Assume L is $*$-normal, and let $x, y \in S(L)$ with $x(-C)y$. Then we have $x^{**} = x \leq y^{\perp} = y^{**\perp} = y^{+}$. We obtain an element $c \in L$ with $x^{**} \leq c^{++}$ and $y^{**} \leq c^{+}$. The elements x, y and c^{*} are closed and we get

$$x = x^{**} \leq c^{++} = c^{*\perp} \quad \text{and} \quad y = y^{**} \leq c^{+} = c^{+++} = c^{**\perp},$$

which implies $x(-C)c^{*}$ and $y(-C)c^{**}$ and thus C6 holds. Conversely, let $a^{**} \not\leq c^{++} = c^{*\perp}$ or $b^{**} \not\leq c^{+} = c^{**\perp}$ for all $c \in L$. Then $a^{**}Cc^{*}$ and $b^{**}Cc^{**}$ for all $c \in L$. Since $\{c^{*} \mid c \in L\} = S(L)$ we conclude by C6 that $a^{**}Cb^{**}$, and, hence, $a^{**} \not\leq b^{**\perp} = b^{+}$.

3. Assume C does not satisfy C7. Then there is a closed element $x \neq 0, 1$ with $x(-C)x^{*}$, i.e. $x \leq x^{*\perp} = x^{++}$. The latter shows that x is also open, and, hence, L is not connected.

Conversely, assume L is not connected. Then there is a clopen element $x \neq 0, 1$. We conclude $x = x^{++} = x^{*\perp}$ which implies $x(-C)x^{*}$.   □

We want to illustrate the previous lemma by some examples.

Example 1. Consider the Stonian p-ortholattices C_{18} and C_{14} from Figure 3 and 4. The pairs $(1, 0)$, (d, c^{++}), (d^{++}, c), (a, f^{++}), (a^{++}, f), (e, b^{++}), (e^{++}, b) define the orthocomplements of each other. The pseudocomplements are given by $1^{*} = 1$, $0^{*} = 0$ and for all other elements $\{x, x^{++}\}^{*} = x$. The closed elements are $0, a, b, c, d, e, f, 1$ and the open elements are $0, a^{++}, b^{++}, c^{++}, d^{++}, e^{++}, f^{++}, 1$. Consequently, the only clopen elements are 0 and 1. On the other hand every element of the skeleton is in contact to its complement (within the skeleton). For example, we have $a^{*} = f$ and $f^{\perp} = a^{++}$. Since a is not open, i.e. $a \neq a^{++}$, we obtain $a \not\leq a^{++} = f^{\perp} = a^{*\perp}$.

$S(L)$ is not dense in either of those Stonian p-ortholattices. For example, $a^{++} \neq 0$ but there is no closed element between a^{++} and 0. As a consequence the skeleton is not extensional. We have $f \neq 1$ and no non-zero closed element is smaller than $f^{\perp} = a^{++}$.

Finally, both lattices are not $*$-normal. For example, the closed elements a and c satisfy $a = a^{**} \leq d^{++} = c^{+}$. The open elements above a are $d^{++}, 1$ but none of $d^{+} = c^{++}, 1^{+} = 0$ is above c. Consequently, the skeleton does not satisfy C6. Indeed, $a \leq d^{++} = c^{\perp}$, and, hence, $a(-C)c$, but $C(a) = S(L) \setminus \{0, c\}$ and $C(b) = S(L) \setminus \{0, a\}$ so that we have aCb or cCb^{*} for all $b \in S(L)$.

Example 2. For the second example consider the structure $RT(\mathbb{R})$ of the real line with the usual topology. Notice that, for example, the set of all rationals r between 0 and 1 is not in $RT(\mathbb{R})$ since $cl(r) = [0, 1]$ and $cl(int(r)) = cl(\emptyset) = \emptyset$. The skeleton of this Stonian p-ortholattice is the Boolean algebra of all regular closed sets. Since $RT(\mathbb{R})$ has just two clopen elements, namely $\emptyset$ and $\mathbb{R}$, its skeleton satisfies C7. Furthermore, for every element $x \in RT(\mathbb{R})$ there is a non-empty regular closed set included in $int(x)$. Therefore, the skeleton is extensional. Finally, the space is normal so that any pair of disjoint regular closed sets can be separated by disjoint open sets, i.e. $RT(\mathbb{R})$ is $*$-normal, and, hence, its skeleton satisfies C6.

5   Embedding a BCA into a Stonian p-Ortholattice

In this section we focus on the converse process. We verify that every BCA is isomorphic to the skeleton of some Stonian p-ortholattice. The following theorem shows a way how to construct the Stonian p-ortholattice.

Theorem 5. *Let $\langle B, C \rangle$ be an arbitrary BCA. Then there is a Stonian p-ortho-lattice $\langle L, +, \cdot, ^*, ^\perp, 0, 1 \rangle$ so that the skeleton $S(L)$ is isomorphic to $\langle B, C \rangle$.*

Proof. Let $\langle X, \tau \rangle$ be the topological space induced by Theorem 3 and let $h : B \to RC(X)$. Define $L = \{x \in RT(X) \mid \exists b \in B : \mathrm{cl}(x) = h(b)\}$. Notice that if $x \in L$, i.e. $\mathrm{cl}(x) = h(b)$ for some $b \in B$, then we have

$$\begin{aligned}
h(b^*) &= \mathrm{cl}(X \setminus h(b)) & & h \text{ homomorphism}\\
&= \mathrm{cl}(X \setminus \mathrm{cl}(x))\\
&= X \setminus \mathrm{int}(\mathrm{cl}(x))\\
&= X \setminus \mathrm{int}(x) & & x \in RT(X)\\
&= \mathrm{cl}(X \setminus x).
\end{aligned}$$

We have to show that the skeleton $S(L)$ is exactly the image of h, that L is closed with respect to all operations of $RT(X)$, and that aCb iff $h(a) \not\subseteq X \setminus h(b)$ for all $a, b \in B$.

Obviously every $h(a)$ is closed, i.e. $h(a) \in S(L)$. Conversely, suppose x is closed. Then $x = \mathrm{cl}(x) = h(b)$ for some $b \in B$, i.e. x is in the image of h.

Now, suppose there are elements $b_1, b_2 \in B$ with $\mathrm{cl}(x) = h(b_1)$ and $\mathrm{cl}(y) = h(b_2)$. and consider the following computations:

$$\begin{aligned}
\mathrm{cl}(x \cap^* y) &= \mathrm{cl}(\mathrm{int}(x \cap^* y)) & & x \cap^* y \in RT(X)\\
&= \mathrm{cl}(\mathrm{int}(x) \cap \mathrm{int}(y)) & & \text{Lemma } 2(2)\\
&= \mathrm{cl}((X \setminus \mathrm{cl}(X \setminus x)) \cap (X \setminus \mathrm{cl}(X \setminus y)))\\
&= \mathrm{cl}((X \setminus h(b_1^*)) \cap (X \setminus h(b_2^*))) & & \text{see above}\\
&= \mathrm{cl}(X \setminus (h(b_1^*) \cup h(b_2^*)))\\
&= \mathrm{cl}(X \setminus h(b_1^* + b_2^*)) & & h \text{ homomorphism}\\
&= h((b_1^* + b_2^*)^*) & & h \text{ homomorphism}\\
&= h(b_1 \cdot b_2),\\
\mathrm{cl}(x \cup^* y) &= \mathrm{cl}(x) \cup \mathrm{cl}(y) & & \text{Lemma } 2(1)\\
&= h(b_1) \cup h(b_2)\\
&= h(b_1 + b_2), & & h \text{ homomorphism}\\
\mathrm{cl}(x^*) &= \mathrm{cl}(\mathrm{cl}(X \setminus x))\\
&= \mathrm{cl}(X \setminus x)\\
&= h(b_1^*), & & \text{see above}\\
\mathrm{cl}(x^\perp) &= \mathrm{cl}(X \setminus x)\\
&= h(b_1^*). & & \text{see above}
\end{aligned}$$

Finally, using Theorem 3 we immediately conclude aCb iff $h(a) \cap h(b) \neq \emptyset$ iff $h(a) \not\subseteq X \setminus h(b)$. $\qquad\square$

The Stonian p-ortholattice from the previous theorem is not necessarily the only lattice that has $\langle B, C \rangle$ as its skeleton BCA.

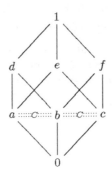

Fig. 2. A Boolean Contact Algebra

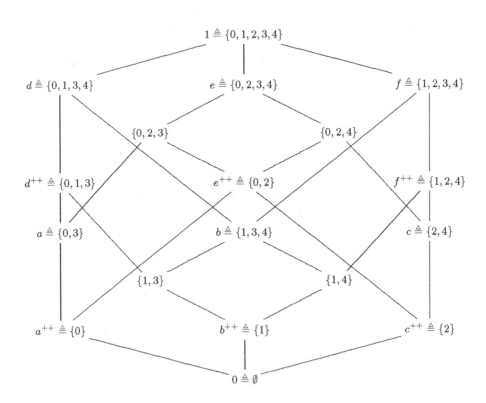

Fig. 3. The Stonian p-Ortholattice C_{18}

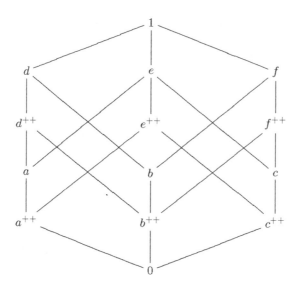

Fig. 4. The Stonian p-Ortholattice C_{14}

Example 3. Consider the BCA from Figure 2. The diagram just shows external connection between atoms ($\cdots c \cdots$ edges). The actual contact relation C on this Boolean algebra is given as the smallest relation that contains those edges, overlap and is upwards closed, i.e. closed with respect to C3. Notice that this BCA satisfies C7 but neither C5 nor C6.

The topological space that is constructed in the proof of Theorem 3 (see [5]) for this example is based on a set isomorphic to $X = \{0, 1, 2, 3, 4\}$ with open sets

$$\tau = \{\emptyset, \{0\}, \{1\}, \{2\}, \{0, 1\}, \{0, 2\}, \{1, 2\}, \{0, 1, 2\}, \{0, 1, 3\},$$
$$\{1, 2, 4\}, \{0, 1, 2, 3\}, \{0, 1, 2, 4\}, \{0, 1, 2, 3, 4\}\}.$$

From those open sets $\{0, 1\}, \{1, 2\}, \{0, 1, 2\}, \{0, 1, 2, 4\}$ and $\{0, 1, 2, 3\}$ are not regular open, e.g. we have $\text{int}(\text{cl}(\{1, 2\})) = \text{int}(\{1, 2, 3, 4\}) = \{1, 2, 4\}$. We obtain the Stonian p-ortholattice from Figure 3. Notice that this is the lattice C_{18}, one of the four structures characterizing models of RT [10].

On the other hand, the Stonian p-ortholattice C_{14} from Figure 4 has the same skeleton as C_{18}. A careful investigation also shows that the contact relations induced on the skeleton is the same for both lattices.

6   Conclusion and Future Work

In this paper we have established the relationship between BCAs and Stonian p-ortholattices. Due to the equivalence of those theories to subtheories of RCC and RT_0 we obtain similar results for those mereotopologies. Our theoretical work directly implies that every connected, $*$-normal model of RT^- with a dense skeleton is a model of the full RCC. On the other extreme, any model of the RT^- is a model of the RCC without the axioms C5, C6, and C7. Conversely, every model of the RCC is a model of RT^-: The BCA corresponding to an RCC model is isomorphic to the skeleton $S(L)$ of some Stonian p-ortholattice L by Theorem 5. However, the skeleton $S(L)$ itself is a Stonian p-ortholattice, since the Boolean algebras are a subclass of the Stonian p-ortholattices. Consequently, every RCC model is a RT^- model as well. With little effort we can show the relation to models of full RT_0: if the RCC model contains some minimal set of regular open sets, it can always be extended to a model of the full theory RT_0.

More generally speaking, by using previously published algebraic representations of the theories RCC and RT_0 and clarifying the relationship between their algebraic representations, this work contributes to the understanding of the relationship between different logical theories of mereotopology. Establishing a formal relationship between models of subtheories of RCC and RT_0 would have been extremely difficult without the lattice-theoretic account of their models. This emphasizes the benefit of algebraic representations of logical theories, in particular of mereotopological theories. Ultimately, we want to gain a deeper understanding of the relationship between the major theories of mereotopology. Part of our future work will focus on algebraic representations of other mereotopologies. In the long-term, this will allow to obtain similar relationships between the various mereotopological theories. By doing so, we hope to foster a deeper understanding of the different mereotopologies, their models, and the relationships amongst them.

As a separate issue, even though we verified that the structure RT(X) for a topological space X is indeed a Stonian p-ortholattice, a topological representation theorem has not yet been established. Future work will concentrate on this aspect as well. In particular, it is of interest whether a representation theorem can be developed that corresponds on the skeleton to the known results for BCAs.

Acknowledgement

We thank the anonymous reviewers for their suggestions to improve the paper.

References

1. Asher, N., Vieu, L.: Toward a geometry of common sense: a semantics and a complete axiomatization for mereotopology. In: Proc. of the 14th Int. Joint Conf. on Artificial Intelligence (IJCAI 1995), pp. 846–852. Morgan Kaufmann, San Francisco (1995)
2. Clarke, B.: A calculus of individuals based on 'Connection'. Notre Dame Journal of Formal Logic 22(3), 204–218 (1981)
3. Cohn, A.G., Bennett, B., Gooday, J.M., Gotts, N.M.: RCC: a calculus for region based qualitative spatial reasoning. GeoInformatica 1, 275–316 (1997)
4. Cohn, A.G., Renz, J.: Qualitative spatial representation and reasoning. In: Handbook of Knowledge Representation. Elsevier, Amsterdam (2008)

5. Dimov, G., Vakarelov, D.: Contact Algebras and Region-based Theory of Space: A Proximity Approach. Fundamenta Informaticae 74(2-3), 209–282 (2006)
6. Düntsch, I., Winter, M.: A representation theorem for boolean contact algebras. Theoretical Computer Science 347, 498–512 (2005)
7. Düntsch, I., Winter, M.: Weak contact structures. In: MacCaull, W., Winter, M., Düntsch, I. (eds.) RelMiCS 2005. LNCS, vol. 3929, pp. 73–82. Springer, Heidelberg (2006)
8. Frink, O.: Pseudo-Complements in Semi-Lattices. Duke Mathematical Journal 29(4), 505–514 (1962)
9. Glivenko, V.: Sur quelque points de la logique de M. Brouwer. Bulletin de la Classe des Sciences, 5e série, Académie royale de Belgique 15, 183–188 (1929)
10. Hahmann, T., Winter, M., Gruninger, M.: Stonian p-ortholattices: A new approach to the mereotopology RT_0. Artificial Intelligence (May 2009) (accepted)
11. Stell, J.G.: Boolean connection algebras: a new approach to the Region-Connection Calculus. Artificial Intelligence 122, 111–136 (2000)
12. Vakarelov, D., Dimov, G., Düntsch, I., Bennett, B.: A proximity approach to some region-based theories of space. Journal of Applied Non-classical Logics 12(3-4), 527–559 (2002)

Author Index